Dynamic Brain Imaging

METHODS IN MOLECULAR BIOLOGY™

John M. Walker, SERIES EDITOR

489. **Dynamic Brain Imaging:** Methods and Protocols, edited by Fahmeed Hyder, 2008
484. **Functional Proteomics:** Methods and Protocols, edited by Julie D. Thompson, Christine Schaeffer-Reiss, and Marius Ueffing, 2008
483. **Recombinant Proteins From Plants:** Methods and Protocols, edited by Loïc Faye and Veronique Gomord, 2008
482. **Stem Cells in Regenerative Medicine:** Methods and Protocols, edited by Julie Audet and William L. Stanford, 2008
481. **Hepatocyte Transplantation:** Methods and Protocols, edited by Anil Dhawan and Robin D. Hughes, 2008
480. **Macromolecular Drug Delivery:** Methods and Protocols, edited by Mattias Belting, 2008
479. **Plant Signal Transduction:** Methods and Protocols, edited by Thomas Pfannschmidt, 2008
478. **Transgenic Wheat, Barley and Oats: Production and Characterization Protocols**, edited by Huw D. Jones and Peter R. Shewry, 2008
477. **Advanced Protocols in Oxidative Stress I**, edited by Donald Armstrong, 2008
476. **Redox-Mediated Signal Transduction:** Methods and Protocols, edited by John T. Hancock, 2008
475. **Cell Fusion:** Overviews and Methods, edited by Elizabeth H. Chen, 2008
474. **Nanostructure Design:** Methods and Protocols, edited by Ehud Gazit and Ruth Nussinov, 2008
473. **Clinical Epidemiology:** Practice and Methods, edited by Patrick Parfrey and Brendon Barrett, 2008
472. **Cancer Epidemiology, Volume 2:** Modifiable Factors, edited by Mukesh Verma, 2008
471. **Cancer Epidemiology, Volume 1:** Host Susceptibility Factors, edited by Mukesh Verma, 2008
470. **Host-Pathogen Interactions:** Methods and Protocols, edited by Steffen Rupp and Kai Sohn, 2008
469. **Wnt Signaling, Volume 2:** Pathway Models, edited by Elizabeth Vincan, 2008
468. **Wnt Signaling, Volume 1:** Pathway Methods and Mammalian Models, edited by Elizabeth Vincan, 2008
467. **Angiogenesis Protocols:** Second Edition, edited by Stewart Martin and Cliff Murray, 2008
466. **Kidney Research:** Experimental Protocols, edited by Tim D. Hewitson and Gavin J. Becker, 2008.
465. **Mycobacteria**, *Second Edition*, edited by Tanya Parish and Amanda Claire Brown, 2008
464. **The Nucleus, Volume 2:** Physical Properties and Imaging Methods, edited by Ronald Hancock, 2008
463. **The Nucleus, Volume 1:** Nuclei and Subnuclear Components, edited by Ronald Hancock, 2008
462. **Lipid Signaling Protocols**, edited by Banafshe Larijani, Rudiger Woscholski, and Colin A. Rosser, 2008
461. **Molecular Embryology:** Methods and Protocols, Second Edition, edited by Paul Sharpe and Ivor Mason, 2008
460. **Essential Concepts in Toxicogenomics**, edited by Donna L. Mendrick and William B. Mattes, 2008
459. **Prion Protein Protocols**, edited by Andrew F. Hill, 2008
458. **Artificial Neural Networks:** Methods and Applications, edited by David S. Livingstone, 2008
457. **Membrane Trafficking**, edited by Ales Vancura, 2008
456. **Adipose Tissue Protocols**, Second Edition, edited by Kaiping Yang, 2008
455. **Osteoporosis**, edited by Jennifer J. Westendorf, 2008
454. **SARS- and Other Coronaviruses:** Laboratory Protocols, edited by Dave Cavanagh, 2008
453. **Bioinformatics, Volume 2:** Structure, Function, and Applications, edited by Jonathan M. Keith, 2008
452. **Bioinformatics, Volume 1:** Data, Sequence Analysis, and Evolution, edited by Jonathan M. Keith, 2008
451. **Plant Virology Protocols:** From Viral Sequence to Protein Function, edited by Gary Foster, Elisabeth Johansen, Yiguo Hong, and Peter Nagy, 2008
450. **Germline Stem Cells**, edited by Steven X. Hou and Shree Ram Singh, 2008
449. **Mesenchymal Stem Cells:** Methods and Protocols, edited by Darwin J. Prockop, Douglas G. Phinney, and Bruce A. Brunnell, 2008
448. **Pharmacogenomics in Drug Discovery and Development**, edited by Qing Yan, 2008.
447. **Alcohol:** Methods and Protocols, edited by Laura E. Nagy, 2008
446. **Post-translational Modifications of Proteins:** Tools for Functional Proteomics, *Second Edition*, edited by Christoph Kannicht, 2008.
445. **Autophagosome and Phagosome**, edited by Vojo Deretic, 2008
444. **Prenatal Diagnosis**, edited by Sinhue Hahn and Laird G. Jackson, 2008.
443. **Molecular Modeling of Proteins**, edited by Andreas Kukol, 2008.
442. **RNAi:** Design and Application, edited by Sailen Barik, 2008.
441. **Tissue Proteomics:** Pathways, Biomarkers, and Drug Discovery, edited by Brian Liu, 2008
440. **Exocytosis and Endocytosis**, edited by Andrei I. Ivanov, 2008
439. **Genomics Protocols**, Second Edition, edited by Mike Starkey and Ramnanth Elaswarapu, 2008
438. **Neural Stem Cells:** Methods and Protocols, Second Edition, edited by Leslie P. Weiner, 2008
437. **Drug Delivery Systems**, edited by Kewal K. Jain, 2008
436. **Avian Influenza Virus**, edited by Erica Spackman, 2008
435. **Chromosomal Mutagenesis**, edited by Greg Davis and Kevin J. Kayser, 2008
434. **Gene Therapy Protocols: Volume 2:** Design and Characterization of Gene Transfer Vectors, edited by Joseph M. Le Doux, 2008
433. **Gene Therapy Protocols: Volume 1:** Production and In Vivo Applications of Gene Transfer Vectors, edited by Joseph M. Le Doux, 2008
432. **Organelle Proteomics**, edited by Delphine Pflieger and Jean Rossier, 2008
431. **Bacterial Pathogenesis:** Methods and Protocols, edited by Frank DeLeo and Michael Otto, 2008
430. **Hematopoietic Stem Cell Protocols**, edited by Kevin D. Bunting, 2008
429. **Molecular Beacons:** Signalling Nucleic Acid Probes, Methods and Protocols, edited by Andreas Marx and Oliver Seitz, 2008
428. **Clinical Proteomics:** Methods and Protocols, edited by Antonia Vlahou, 2008
427. **Plant Embryogenesis**, edited by Maria Fernanda Suarez and Peter Bozhkov, 2008
426. **Structural Proteomics:** High-Throughput Methods, edited by Bostjan Kobe, Mitchell Guss, and Thomas Huber, 2008
425. **2D PAGE:** Sample Preparation and Fractionation, Volume 2, edited by Anton Posch, 2008

METHODS IN MOLECULAR BIOLOGY™

Dynamic Brain Imaging

Multi-Modal Methods and In Vivo Applications

Edited By

Fahmeed Hyder, PhD

Yale University, New Haven, CT, USA

Editor
Fahmeed Hyder
Yale University
New Haven, CT, USA
fahmeed.hyder@yale.edu

Series Editor
John M. Walker
University of Hertfordshire
Hatfield Herts
UK

ISBN: 978-1-934115-74-9
e-ISBN: 978-1-59745-543-5
ISSN: 1064-3745
e-ISSN: 1940-6029
DOI 10.1007/978-1-59745-543-5

Library of Congress Control Number: 2008937515

Printed on acid-free paper

9 8 7 6 5 4 3 2 1

springer.com

for Anita, LeiLa, Arman Hyder

Preface

If *a* picture is worth a thousand words, then *dynamic* images of brain activity certainly warrant many, many more. This book will help users learn to decipher the dynamic imaging data that will be critical to our future understanding of complex brain functions.

In recent years, there have been unprecedented methodological advancements in the imaging of brain activity. These techniques allow the measurement of everything from neural activity (e.g., membrane potential, ion flux, neurotransmitter flux) to energy metabolism (e.g., glucose consumption, oxygen consumption, creatine kinase flux) and functional hyperemia (e.g., blood flow, volume, oxygenation). This book deals with a variety of magnetic resonance, electrophysiology, and optical methods that are often used to measure some of these dynamic processes.

All chapters were written by leading experts, spanning three continents, specializing in state-of-the-art methods. Briefly, the book has five sections. In the introductory section, there are two chapters; the first one contains a brief preamble to dynamic brain imaging and the other presents a novel, analytical approach to processing of dynamically acquired data. The second section has four chapters and delves into a wide range of optical imaging methods. I am privileged to include a chapter from Lawrence B. Cohen, considered by many to be *the* authority on optical imaging and spectroscopy, both in vitro and in vivo [Cohen LB (1973) *Physiol Rev.* 53:373–418]. The next two sections have three chapters each, whereas the last section has four chapters. The third section discusses many of the most frequently used electrophysiological methods. The fourth section presents diverse uses and applications of functional magnetic resonance imaging (fMRI). This part contains a chapter from Seiji Ogawa, *the* discoverer of the technique [Ogawa S et al. (1990) *Proc Natl Acad Sci USA*. 87:9868–9872]. The final section brings together some up-and-coming alternate magnetic resonance methods with the potential to strengthen non-invasive neuroimaging.

In the very first chapter, I give a brief introduction to dynamic imaging of brain activity in vivo as measured by various magnetic resonance, electrophysiological, and optical methods. I describe the different tissue-specific components that are measured by the multi-modal methods and compare the spatiotemporal resolution ranges to be expected from these techniques, as well as advantages and disadvantages of each method. In the next chapter, Andras Eke and his colleagues tackle analysis of the complexity in dynamic signals. They discuss fractal characterization of multi-dimensional hemodynamic-based signals (e.g., as measured by optical imaging and fMRI). They argue that the seemingly disorderly spontaneous fluctuations in these signals (from both animals and humans) actually are organized according to a self-similar (i.e., fractal correlation) pattern which, in turn, can be used to differentiate between different tissue types or disease states.

The next four chapters are on optical imaging. In the third chapter, Lawrence B. Cohen and his colleagues illustrate some state-of-the-art examples of dynamic imaging of olfactory bulb activity with voltage– and calcium-sensitive dyes. They describe the utilities of both single– and two-photon imaging techniques and discuss the types of noise that may confound detection of these optical signals. In the fourth chapter, Serge Charpak and his colleagues discuss the use of two-photon laser scanning microscopy to measure blood flow. They combine these optical recordings with localized intra– and extracellular electrical recordings to examine the coupling between vascular and electrical signals at the level of individual functional units. The fifth chapter by Maiken Nedergaard and her colleagues deals with neurogliovascular coupling. They use novel optical imaging results to demonstrate the involvement of astrocytic calcium signaling and discuss activation-flow coupling in the cerebral cortex. To achieve this, they combined optical recordings from glia with electrical and blood flow measurements in a tour de force of multimodal dynamic brain imaging. The sixth chapter is from Manabu Tanifuji and colleagues and it focuses on the light scattering component of optical intrinsic signals from tissues. Optical intrinsic signal imaging has been widely used for visualizing cortical functional structures. They demonstrate that the optical properties unique to light scattering enable us to visualize spatial patterns of retinal activity noninvasively and resolve functional structures in depth.

The next three chapters focus on different electrical techniques. In the seventh chapter, Mark Laubach and his colleague deal with measurements of large populations of cortical neurons and their relationship with observed behavior. They describe the power of using novel analytical tools (e.g., principal component analysis, statistical pattern recognition) to quantify relationships among neuronal ensembles during behavior. The eighth chapter is by Andreas A. Ioannides. He introduces magnetoencephalography (MEG) as an unsurpassed technique with high temporal resolution for mapping biomagnetic (i.e., electrical) brain activity in a noninvasive manner. He discusses the signal source of MEG and demonstrates numerous examples of using this method to understand dynamic brain function in humans. The ninth chapter by Hal Blumenfeld and his colleague demonstrates imaging of generalized spike-wave seizures in both animals and humans. They demonstrate the power of combining conventional electrical methods (e.g., electroencephalogram and extracellular recordings) with fMRI to provide better understanding of specific brain regions involved in generating spike-wave seizures.

The following three chapters concentrate on fMRI. The tenth chapter is from my group. Where we describe different rodent sensory paradigms that have been studied with fMRI and neurophysiologic recordings. We demonstrate similarities and differences across different sensory modalities and discuss advantages these models may have for studying neuroscientific questions. The eleventh chapter is by Seiji Ogawa and his colleague. Together, they demonstrate how conventional fMRI can be used to elucidate fast and dynamic interactions between different brain regions, both in animals and humans. They propose the use of paired stimulation paradigms to overcome the temporal limitation of fMRI and demonstrate examples where multi-site interactions can be selectively probed. In the twelfth chapter, Bharat B. Biswal and his colleague wrestle with the resting state low frequency fluctuations in fMRI and blood flow signals which are used to reveal correlations across different brain regions.

They demonstrate inter-hemispheric synchrony in anesthetized rodents — similar to that observed in the resting, awake human brain — the power of which is affected by exsanguination.

The last three chapters are on emerging magnetic resonance techniques that have great potential for future neuroscience studies. In the thirteenth chapter, Afonso C. Silva and his colleague describe the magnetic resonance imaging (MRI) measurement of blood flow in the brain using magnetically labeled water. They describe novel manipulations of the conventional MRI used to measure steady-state blood flow in order to probe dynamic changes in perfusion during function. The fourteenth chapter is by Allen W. Song and his colleagues. They describe an MRI method that detects the Lorentz effect induced by neuroelectric activity. They use an array of in vitro models to reveal that this effect can be titrated by synchronization of the MRI pulse sequence with the stimulation paradigm. An example from the human arm is used to demonstrate its potential use in brain imaging in the future. In the last and fifteenth chapter, Wei Chen and his colleagues deal with ^{31}P and ^{17}O magnetic resonance spectroscopy (MRS) methods that probe cerebral energetics in both animals and humans. They demonstrate that the significant gains in MRS detection sensitivity at high magnetic fields leads to improvement in spectral and spatial resolutions, thus leading to novel insights about the dynamics of cerebral energetics.

This prologue, I hope, demonstrates the appealing potential of combining some of these different dynamic imaging techniques to answer important neurobiological questions. The book was initially planned to be useful for biological scientists new to dynamic brain imaging. The end result, however, is likely to be attractive to professionals and students at all levels because it contains up-to-date material with balanced emphasis on methods and applications to neuroscience. I sincerely hope that this book finds a convenient location in your bookshelf so that you may refer to it, again and again.

A book is a result of patience and persistence and no one showed more than John M. Walker, the Series Editor. I thank Susan Froetschel and Peter Herman who helped with proof reading and indexing, respectively. I owe him much. I am also indebted to each and every one of the contributors for committing to the idea and vision of this book. Without their generosity, this book would obviously not be possible. Inputs from the named reviewers, plus a few who chose to remain anonymous, made editing much more tolerable. I thank my colleagues and friends for their help and support over the past few months. Last, but certainly not least, this book is an outcome of familial sacrifices. My entire family, near and far, allowed me to take time off when it seemed limited. I thank them for their infinite understanding.

Dewan Syed Fahmeed Hyder

Contents

Contributors

CHRISTOPHER J. BAILEY • *Center of Functionally Integrative Neuroscience, Aarhus University, Aarhus, Denmark* and *Department of Diagnostic Radiology, Yale University, School of Medicine, New Haven, CT*

BRADLEY J. BAKER • *Department of Cellular and Molecular Physiology, Yale University, School of Medicine, New Haven, CT*

BHARAT B. BISWAL • *Department of Radiology, UMDNJ-New Jersey Medical School, Newark, NJ*

CHUN X. BLEAU • *Department of Cellular and Molecular Physiology, Yale University School of Medicine, New Haven, CT; RedShirtImaging, LLC, Decatur, GA*

HAL BLUMENFELD • *Departments of Neurology, Neurobiology, Neurosurgery and Program in Quantitative Neuroscience with Magnetic Resonance, Yale University, School of Medicine, New Haven, CT*

EMMANUELLE CHAIGNEAU • *Laboratory of Neurophysiology; Université Paris Descartes, INSERM U603, Paris, France*

SERGE CHARPAK • *Laboratory of Neurophysiology; Université Paris Descartes, INSERM U603, Paris, France*

WEI CHEN • *Center for Magnetic Resonance Research, Department of Radiology, University of Minnesota, School of Medicine, Minneapolis, MN*

LAWRENCE B. COHEN • *Department of Cellular and Molecular Physiology, Yale University, School of Medicine, New Haven, CT; RedShirtImaging, LLC, Decatur, GA*

FEI DU • *Center for Magnetic Resonance Research, Department of Radiology, University of Minnesota, School of Medicine, Minneapolis, MN*

ANDRAS EKE • *Institute of Human Physiology and Clinical Experimental Research, Semmelweis University, Budapest, Hungary*

OLGA GARASCHUK • *Institute of Neuroscience, Technical University Munich, Munich, Germany*

PETER HERMAN • *Department of Diagnostic Radiology, Program in Quantitative Neuroscience with Magnetic Resonance, Magnetic Resonance Research Center, Yale University, School of Medicine, New Haven, CT; Institute of Human Physiology and Clinical Experimental Research, Semmelweis University, Budapest, Hungary*

RYOTA HOMMA • *Department of Cellular and Molecular Physiology, Yale University, School of Medicine, New Haven, CT*

FAHMEED HYDER • *Departments of Diagnostic Radiology and Biomedical Engineering, Program in Quantitative Neuroscience with Magnetic Resonance, Magnetic Resonance Research Center, Yale University, New Haven, CT*

ANDREAS A. IOANNIDES • *Laboratory for Human Brain Dynamics, RIKEN Brain Science Institute, Saitama, Japan Laboratory for Human Brain Dynamics, AAI Scientific and Cultural Services Ltd, Nicosia, Cyprus*

LEI JIN • *Department of Cellular and Molecular Physiology, Yale University, School of Medicine, New Haven, CT*
SRIDHAR S. KANNURPATTI • *Department of Radiology, UMDNJ-New Jersey Medical School, Newark, NJ*
LASZLO KOCSIS • *Institute of Human Physiology and Clinical Experimental Research, Semmelweis University, Budapest, Hungary*
ARTHUR KONNERTH • *Institute of Neuroscience, Technical University Munich, Munich, Germany*
MARK LAUBACH • *The John B. Pierce Laboratory, New Haven, CT; Department of Neurobiology, Yale University, School of Medicine, New Haven, CT*
JÉRÔME LECOQ • *Laboratory of Neurophysiology; Université Paris Descartes, INSERM U603, Paris, France*
HAO LEI • *State Key Laboratory of Magnetic Resonance and Molecular and Atomic Physics, Wuhan Institute of Physics and Mathematics, The Chinese Academy of Sciences, Wuhan, China*
JOSHUA E. MOTELOW • *Department of Neurology, Yale University, School of Medicine, New Haven, CT*
NANDAKUMAR S. NARAYANAN • *The John B. Pierce Laboratory, New Haven, CT; Interdepartmental Neuroscience Program, Yale University, School of Medicine, New Haven, CT*
MAIKEN NEDERGAARD • *Center for Aging and Developmental Biology, Department of Neurosurgery, University of Rochester Medical Center, Rochester, NY*
SEIJI OGAWA • *Ogawa Laboratories for Brain Function Research, Hamano Life Science Research Foundation, Tokyo, Japan*
FERNANDO F. PAIVA • *Cerebral Microcirculation Unit, Laboratory of Functional and Molecular Imaging, National Institute of Neurological Disorders and Stroke, Bethesda, MD*
HONGYAN QIAO • *Center for Magnetic Resonance Research, Department of Radiology, University of Minnesota, School of Medicine, Minneapolis, MN*
UMA MAHESWARI RAJAGOPALAN • *Laboratory for Integrative Neural Systems, RIKEN Brain Science Institute, Saitama, Japan*
BASAVARAJU G. SANGANAHALLI • *Department of Diagnostic Radiology, Program in Quantitative Neuroscience with Magnetic Resonance, Magnetic Resonance Research Center, Yale University, School of Medicine, New Haven, CT*
AFONSO C. SILVA • *Cerebral Microcirculation Unit, Laboratory of Functional and Molecular Imaging, National Institute of Neurological Disorders and Stroke, Bethesda, MD*
ALLEN W. SONG • *Brain Imaging and Analysis Center, Duke University, Durham, NC*
YUL-WAN SUNG • *Laboratories for Sensibility and Welfare Research, Tohoku Fukushi University, Aoba, Sendai, Japan*
TAKAHIRO TAKANO • *Center for Aging and Developmental Biology, Department of Neurosurgery, University of Rochester Medical Center, Rochester, NY*
MANABU TANIFUJI • *Laboratory for Integrative Neural Systems, RIKEN Brain Science Institute, Saitama, Japan*
PASCALE TIRET • *Laboratory of Neurophysiology; Université Paris Descartes, INSERM U603, Paris, France*

TRONG-KHA TRUONG • *Brain Imaging and Analysis Center, Duke University, Durham, NC*

KAZUSHIGE TSUNODA • *Laboratory for Integrative Neural Systems, RIKEN Brain Science Institute, Saitama, Japan; Laboratory of Visual Physiology, National Institute of Sensory Organs, Tokyo, Japan*

KAMIL UGURBIL • *Center for Magnetic Resonance Research, Department of Radiology, University of Minnesota, School of Medicine, Minneapolis, MN*

XIAOHAI WANG • *Center for Aging and Developmental Biology, Department of Neurosurgery, University of Rochester Medical Center, Rochester, NY*

MARTY WOLDORFF • *Center for Cognitive Neuroscience, Duke University, Durham, NC*

DEJAN ZECEVIC • *Department of Cellular and Molecular Physiology, Yale University, School of Medicine, New Haven, CT*

NANYIN ZHANG • *Center for Magnetic Resonance Research, Department of Radiology, University of Minnesota, School of Medicine, Minneapolis, MN*

XIAOLIANG ZHANG • *Center for Magnetic Resonance Research, Department of Radiology, University of Minnesota, School of Medicine, Minneapolis, MN*

YI ZHANG • *Center for Magnetic Resonance Research, Department of Radiology, University of Minnesota, School of Medicine, Minneapolis, MN*

XIAO-HONG ZHU • *Center for Magnetic Resonance Research, Department of Radiology, University of Minnesota, School of Medicine, Minneapolis, MN*

Reviewers

BEAU M. ANCES, MD, PhD • *Department of Neurology, Washington University, St. Louis, MO 63110*
BRUNO AVERBECK, PhD • *Institute of Movement Neuroscience, Institute of Neurology, University College London, London WC1N 3BG, UK*
PETER A. BANDETTINI, PhD • *Laboratory of Brain and Cognition, NIMH, NIH, Bethesda, MD 20892*
DOUGLAS E. BEFROY, PhD • *Section of Endocrinology (& MRRC), Yale University School of Medicine, New Haven, CT 06520*
HAL BLUMENFELD, MD, PhD • *Departments of Neurology and Neurobiology, Yale University School of Medicine, New Haven, CT 06520*
FAWZI BOUMEZBEUR, PhD • *Neurospin / CEA / centre de Saclay, 91 191 Gif-sur-Yvette, Cedex, France*
JOSHUA C. BRUMBERG, PhD • *Psychology Department, Queens College, Flushing, NY 11367*
XAVIER GOLAY, PhD • *Laboratory of Molecular Imaging, Singapore Bioimaging Consortium, Singapore 138667*
PETER HERMAN, MD, PhD • *Program in Quantitative Neuroscience with Magnetic Resonance (& MRRC), Yale University School of Medicine, New Haven, CT 06520*
MATHIAS HOEHN, PhD • *In-vivo-NMR Laboratory, Max-Planck-Institute for Neurological Research, D-50931 Köln, Germany*
FAHMEED HYDER, PhD • *Program in Quantitative Neuroscience with Magnetic Resonance (& MRRC), Yale University School of Medicine, New Haven, CT 06520*
YOUNG RO KIM, PhD • *Center for Molecular Imaging, MGH, Harvard University, Charlestown, MA 02129-2060*
Hon C. Kwan, PhD • *Department of Physiology, University of Toronto, Toronto, Ontario M5S 1A8, Canada*
CHUNG-CHUAN LO, PhD • *Department of Neurobiology, Yale University School of Medicine, New Haven, CT 06520*
JED A. MELTZER, PhD • *Language Section, NIDCD, NIH, Bethesda, MD 20892*
ANNA WANG ROE, PhD • *Vision Research Center. Vanderbilt University, Nashville, TN 37240*
BASAVARAJU G. SANGANAHALLI, PhD • *Program in Quantitative Neuroscience with Magnetic Resonance (& MRRC), Yale University School of Medicine, New Haven, CT 06520*
JAMES R.A. SCHAFER, MD, PhD • *Department of Radiology, University of Michigan, Ann Arbor, MI 48109*

ROBERT G. SHULMAN, PhD • *Department of Diagnostic Radiology (& MRRC), Yale University School of Medicine, New Haven, CT 06520*
BOJANA STEFANOVIC, PhD • *Sunnybrook Health Sciences Center, Toronto, ON M4N 3M5, Canada*

Part I

Introduction

Chapter 1

Dynamic Imaging of Brain Function

Fahmeed Hyder

Abstract

In recent years, there have been unprecedented methodological advances in the dynamic imaging of brain activities. Electrophysiological, optical, and magnetic resonance methods now allow mapping of functional activation (or deactivation) by measurement of neural activity (e.g., membrane potential, ion flux, neurotransmitter flux), energy metabolism (e.g., glucose consumption, oxygen consumption, creatine kinase flux), and functional hyperemia (e.g., blood oxygenation, blood flow, blood volume). Properties of the glutamatergic synapse are used to model activities at the nerve terminal and their associated changes in energy demand and blood flow. This approach reveals that each method measures different tissue- and/or cell-specific components with characteristic spatiotemporal resolution. While advantages and disadvantages of different methods are apparent and often used to supersede one another in terms of specificity and/or sensitivity, no particular technique is *the* optimal dynamic brain imaging method because each method is unique in some respect. Since the demand for energy substrates *is* a fundamental requirement for function, energy-based methods may allow quantitative dynamic imaging in vivo. However, there are exclusive neurobiological insights gained by combining some of these different dynamic imaging techniques.

Key words: fMRI, glia, GABA, glutamate, glutamine, lactate, multi-modal, neuroimaging.

1. Introduction

The brain is a highly complex organ, both anatomically *(1)* and physiologically *(2)*, requiring an impressive arsenal of technological tools to study it. In recent years, neuroscientists and neurophysiologists have benefited from the emergence of several dynamic imaging techniques. Currently, a variety of electrophysiological, optical, and magnetic resonance methods allow in vivo probing of brain activities in terms of transients in neuronal activity as well as their fundamentally associated energetic and hyperemic events *(3)*. Since these functional imaging methods either measure directly or exploit relationships between electrical,

Fahmeed Hyder (ed.), *Dynamic Brain Imaging: Multi-Modal Methods and In Vivo Applications, vol. 489*

DOI 10.1007/978-1-59745-543-5_1 Springerprotocols.com

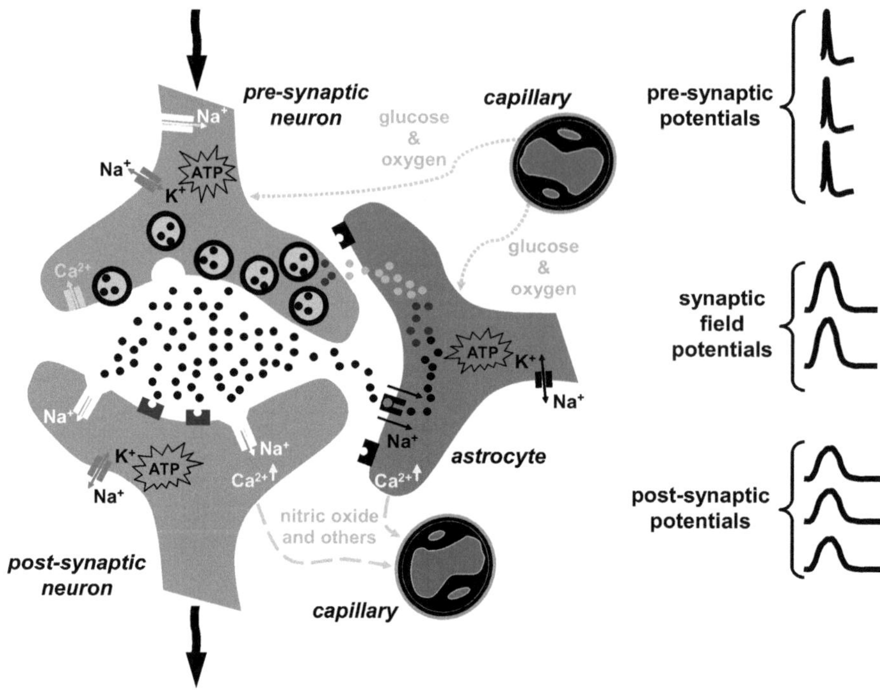

Fig. 1.1. Cytological association between microvasculature with neurons and astrocytes in the glutamatergic synapse.

metabolic, and/or hemodynamic changes, it is important to classify the underlying basic processes with appropriate spatial and temporal scales.

Functional integrity of the working brain is maintained by electrical communication amongst an enormous number of neurons with active partnership provided by astrocytes *(4, 5)*. Cytological association of neurons and astrocytes with the microvasculature (**Fig. 1.1**) provides the framework that links activities at the nerve terminal to energy demand *(6, 7)* and blood flow *(8, 9)*. In the mammalian cerebral cortex, glutamate is the major excitatory neurotransmitter, whereas γ-amino butyric acid (GABA) is its conjugate inhibitory partner. Together, they constitute nearly 90% of cortical neurons *(10)*. Glutamate metabolism plays a central role in both glutamatergic and GABAergic synapses because glutamate is a precursor of GABA and it is a constituent of both neurons and astrocytes *(11, 12)*. Thus, featuring properties of the glutamatergic system seem to be an appropriate starting point for this discourse.

2. Activities at the Nerve Terminal

Electrical communication between cells at the glutamatergic nerve terminal is characterized by 1–2 ms epochs of cellular discharge (or depolarization) which are followed by quiescent charging periods lasting a few ms *(2)*. Cellular excitability (or signaling) depends on ionic distribution across the cell membrane. The cell maintains a large difference in ionic distribution (~10:1) between the intra and extracellular compartments: higher concentrations of sodium (Na^+), calcium (Ca^{2+}), and chloride (Cl^-) ions outside and higher, but unbalanced, concentration of potassium (K^+) ions inside. Since the resting cell membrane is only marginally leaky to ions, passage of ions between intra and extracellular compartments is mediated predominantly by ion channels and pumps *(13)*. Both voltage- and ligand-gated ion channels allow ion movement down large chemical gradients, whereas ion pumps (e.g., Na^+/ K^+ ATPase) help restore the ionic distributions (back to that in the resting cell) by moving ions against large chemical gradients. Restoration of ion gradients (requiring energy input) is needed to keep neurons ready to discharge (or fire) whenever needed.

At rest, the cell membrane can be considered to be nearly impermeable to Na^+ ions with a membrane potential of approximately –70 mV (resting potential). Upon depolarization, however, the membrane becomes almost freely permeable to Na^+ ions and the potential is nearly reversed in an action potential (i.e., when the neuron has fired). Therefore, complete depolarization of the pre-synaptic axon (>1 mm long) can be approximated by increased permeability of Na^+ ions across the cell membrane. However, complementary roles of K^+ and Ca^{2+}voltage-gated ion channels are important. Delayed increase in K^+ conductance (in conjunction with delayed decrease in Na^+ conductance) causes the membrane potential of the pre-synaptic neuron to ultimately move back toward the resting value following depolarization *(14)*. Rapid pre-synaptic Ca^{2+} influx (~200 μM) within 1–2 ms after onset of depolarization triggers vesicles to release glutamate into the extracellular space *(15)*.

Glutamate discharge into the synaptic cleft (20 nm wide) is a vital step in transmission of the pre-synaptic signal on to post-synaptic elements *(2)*. Depolarization of the post-synaptic neuron is initiated by raising extracellular glutamate concentration (from nM to μM). Glutamate activation, mediated by ligand-gated ion channels on post-synaptic elements, occurs within 10 ms *(16)*. During transient binding of glutamate with these ionotropic receptors, Na^+ ions enter post-synaptic dendrites. The pyramidal dendritic branching patterns (spanning several mm) are thought to play a role in amplification of post-synaptic depolarization *(1, 2)*.

What is the fate of extracellular glutamate after pre-synaptic release? There is a large glutamate concentration gradient between pre-synaptic and extracellular compartments (mM vs. nM) *(17)*. Also, there is a high density of glutamate receptors on astrocytic endfeet *(13)*. Together, these favor extracellular; glutamate release followed by glial uptake of glutamate (which is co-transported with Na^+ ions). Astrocytic glutamate uptake prevents excitotoxicity, and in doing so, helps maintain glutamate homeostasis *(11, 12)*. Glutamate and glutamine are more concentrated respectively in neurons and astrocytes. Glutamate is used in astrocytes to synthesize glutamine which, in turn, is a precursor of glutamate synthesis in neurons. Re-establishment of glutamate reserves (requiring energy input) is essential for continuous signaling at the glutamatergic nerve terminal. The glutamate–glutamine cycle, therefore, is a necessary part of overall cellular excitability *(17)*, but more importantly, links membrane depolarization to synaptic activity *(11, 12)*. A prudent advisory, however, to the highlighted events at the nerve terminal is that synaptic activity is *not* exclusively mediated by digital (i.e., all or none) signaling alone *(2)*. Membrane potential changes can also be quite graded (and slow) to produce analog signals which may also influence activities at the nerve terminal *(18)*.

3. Energy Demand and Blood Flow

Signaling at the glutamatergic nerve terminal, spanning wide bandwidths *(19)*, depends on synchronized electrical as well as chemical events where neurons and astrocytes play complementary roles *(6, 7)*. Removal of any one step from the rest compromises function of the entire system *(11, 12)*. While most of the energy is expended for moving Na^+ and K^+ ions against large chemical gradients, a small but non-negligible fraction is needed for intracellular Ca^{2+} homeostasis and neurotransmitter recycling/synthesis/repackaging *(6, 7)*. Because activities at the nerve terminal are in continuous need of energy, demand for it *is* a fundamental requirement *(20)*.

In no other organ is the continuous energy supply more imperative than in the brain. In humans, the brain is merely 2% of the body's weight but it consumes more than 20% of the oxidative fuels in the entire body *(21)* and receives nearly 15% of the cardiac output *(22)* to supply nutrients (i.e., glucose and oxygen). Furthermore, endogenous energy reserves in the brain – glucose (1–3 mM *(23)*), oxygen (50–100 μM *(24)*), glycogen (2–4 mM *(25)*), and creatine (8–10 mM *(26)*) – are minimal *(27)*. Normal function, therefore, needs blood circulation to efficiently provide nutrients (and remove waste) throughout the brain *(28)*. Energy demand and blood flow are well correlated over different brain regions *(29)*.

The energetic costs for brain work are mainly met by ATP derived exclusively from glucose oxidation *(30)*. Glucose metabolism is limited by phosphorylation, not transport *(31)*.

Glucose is rapidly delivered *(32)* by a variety of transporters *(33)* located at the blood-brain barrier and plasma membranes and then broken down by non-oxidative (~5% ATP) and oxidative (~95% ATP) pathways to maintain ATP concentration (2–4 mM *(27)*). Some glucose is stored by glial cells as glycogen *(34)*. Because astrocytes lack the enzyme to generate glucose from glycogen, glucose-6-phosphate generated in glia may be transferred to neurons to serve as an additional energy buffer *(35)*. Total creatine, which represents the quantities of phosphocreatine (PCr) and creatine (Cr), can undergo phosphorylation–dephosphorylation reaction catalyzed by creatine kinase (i.e., ADP + PCr ↔ ATP + Cr). Therefore, PCr may provide an additional energy reserve when oxidative phosphorylation cannot maintain constant ATP supply *(36)*. These alternate energy reserves together can provide energy support for a short time (a couple of minutes) under ischemic conditions *(27)*. It is possible, however, that these extra non-oxidative pathways may provide faster ATP (in ms range) than from mitochondrial respiration *(35, 36)*.

To date, in spite of several decades of research, it remains unclear by what mechanism(s) nutrient supply (of glucose and oxygen) adjusts to changing energy demanded by neurons and astrocytes *(37)*. Many vasoactive agents have been implicated in mechanism(s) leading to functional hyperemia. Given the extremely low and high oxygen contents, respectively, in brain and blood (μM vs. mM *(24)*) but, at the same time, its ubiquitous need for oxidation of glucose (and other carbohydrates), should blood flow be tightly coupled to oxidative energy use, and thereby, suggest potential mechanism(s) for functional hyperemia *(38)*? Although theoretical *(39)* and experimental *(40)* studies show that these parameters are indeed well coupled, oxygen is not a candidate for a vasoactive agent because its excess has no impact on the hemodynamic response (within 500 ms) during sensory stimulation *(41)*. In other words, the system's use of energy substrates is ***only*** based on demand *(20)*, not availability *(42)*. Since suggestions of other agents (e.g., H^+, K^+, Ca^{2+}, adenosine) have lacked clear evidence of an impartial role with functional hyperemia *(37)*, new proposals *(43)* have shifted some attention to astrocytes as key participants, thereby revising the century-old neurovascular coupling idea *(44)* to include glia (i.e., "neurogliovascular" coupling).

Recent opinions *(5, 9, 12)* suggest that neuronal glutamate release not only induces metabolic responses in glial cells but it may also even trigger hemodynamic events through them. This seems plausible given that astrocytes are proximal to both cerebral microvasculature *(4, 5)* and glutamatergic and GABAergic neurons *(45, 46)*. Past experiments *(47)* suggested that astrocytes are activated by glutamate uptake (and increase glial glucose utilization as well as glial lactate production; see references 11 and 12 for details). Recent immunohistochemical and molecular studies *(48–51)* have identified receptors for a variety of neurotransmitters (including glutamate) on specific cellular components of the microvasculature. Furthermore, small increases (in nM range) in cytosolic Ca^{2+}, in both the post-synaptic neuron and astrocytes mediated respectively by activation of ionotropic *(52,53)* and metabotropic *(54,55)* glutamate receptors, may stimulate enzymes like nitric oxide synthase (and others) to generate strong vasoactive agents like nitric oxide (and others). Together, these new findings shed some preliminary *(56,57)* but not always complete *(58)* understanding about complexities of "neurogliovascular" coupling.

4. Functional Brain Imaging

The goal of *the* ideal functional imaging technique is to map task-induced changes in activity in the brain in vivo. The task may be sensory or cognitive and the extent of the activated region may be localized (as generally observed with sensory paradigms *(59)*) or delocalized (as often found with cognitive paradigms *(60)*). To what extent do different methods allow dynamic identification of task activated brain region(s)? The answer depends on the spatiotemporal resolution of the method. There are at least three relevant factors. Spatial coverage of the brain by the method is a practical limitation. Specificity depends on the physiological basis of the measurement. Sensitivity, if enhanced, can improve either the spatial or temporal resolution, but rarely both. Together, these factors describe the spatiotemporal resolution of the method (**Fig. 1.2**). Other (e.g., hardware) factors that affect spatiotemporal resolution of a given method are beyond the scope of discussion here.

Schematic illustration of spatial and temporal resolution ranges from a few experimental techniques is shown in **Figure 1.2** Since it is impractical to refer to the several dozen studies that were used to develop this figure, three reports *(61–63)* with similar comparisons are highlighted. But these may be questioned for several reasons. Techniques which did not reflect dynamic activity changes in their measurements (i.e., static or steady-state) were included with methods that were strictly designed for measuring transient signal changes. Methods specifically designed for use in humans were not separated from those applied in small animals. Similarly, differences between in vivo and in vitro methods were not distinguished. Thus the apparent superiority of one method versus another is debatable. Because there have also been technological advancements that influence

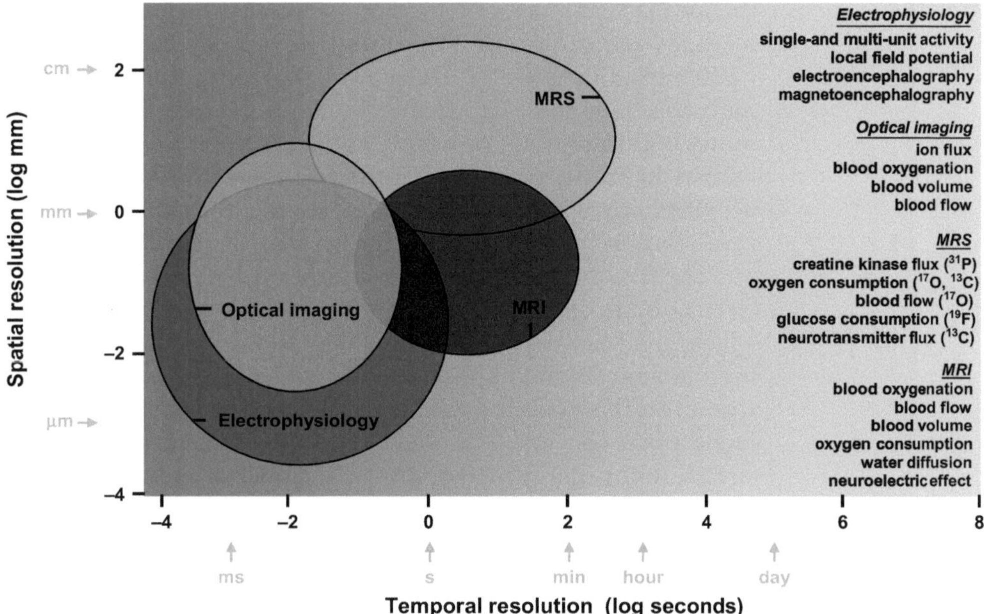

Fig. 1.2. Comparison of spatiotemporal resolutions of current methods used in neuroscience.

this type of cross comparison across methods since then, up-to-date results were compiled from in vivo dynamic imaging studies of primarily small animals. Unfortunately, these criteria excluded two dominant techniques in functional brain imaging: ^{14}C-deoxy-glucose (^{14}C-DG) autoradiography and positron emission tomography (PET).

^{14}C-DG, a glucose analog, measures glucose consumption by the amount of metabolically trapped ^{14}C-DG-6-phosphate throughout the brain *(64)*. However, about an hour is required for radioactivity to build up in the tissue. ^{14}C-DG autoradiography provides superior spatial resolution (50–100 µm) but it is a terminal experiment and therefore used only in animals.

PET, fashioned after the ^{14}C-DG method but designed for application to humans *(65)*, has several methods for functional imaging. Glucose consumption measurement is analogous to the ^{14}C-DG method except that ^{18}F-fluoro-deoxy-glucose (^{18}F-FDG) is the glucose analog for PET. ^{15}O-water is used to measure blood flow. This is done by injecting ^{15}O-water into the blood stream and then detecting the rate of appearance of ^{15}O-water in the brain tissue. Carbon monoxide with ^{11}C or ^{15}O, which mimics binding of oxygen to hemoglobin in red blood cells, is used to measure blood volume. The principle is that carbon monoxide is blood borne and, therefore, the detected radioactivity is only representative of the vascular compartment. The blood flow and blood volume methods in combination with inhaled ^{15}O-oxygen can be used to measure oxygen consumption by detecting the rate of metabolized ^{15}O-water in brain tissue. In all of these studies, the radioactive labels are at trace levels (μM or less). Most PET methods use radioactive isotopes that are relatively short-lived (<30 min), but as in ^{14}C-DG autoradiography, require long accumulation times and/or repeated scans (5–30 min) for radioactivity to build up. PET provides spatial resolution appropriate for human brain imaging (3–5 mm) and is well suited for clinical use. Recent advancements have improved PET resolution further (1–3 mm) to image brains of smaller animals *(66)*.

The techniques shown in **Figure 1.2** include a variety of electrophysiological, optical, and magnetic resonance methods which are now widely used in many neuroscience laboratories, both in animals and in humans. These methods dynamically measure everything from changes in neuronal activity (e.g., membrane potential, ion flux, neurotransmitter flux) to the coupled alterations in energy metabolism (e.g., glucose consumption, oxygen consumption, creatine kinase flux) and hemodynamics (e.g., blood oxygenation, blood flow, blood volume).

4.1. Electrophysiology

Electrophysiology deals with the study of the electrical properties of cells and tissues *(2)*. Classical electrophysiological methods, using different types of microelectrodes, measure changes in voltage (or current). Microelectrodes are inherently designed to cover a few specific regions because it is impractical to perforate very large areas of the brain. Microelectrode arrays cover slightly larger (e.g., 4×4 mm^2) superficial regions of the brain.

Intracellular recordings involve direct access into the intracellular milieu so that membrane potential can be measured with microelectrodes. The obvious advantage with intracellular recording is the unrivaled information about the individual cell, but the disadvantage is the difficulty in sampling large number of neurons. Extracellular recordings are made with the microelectrode tip (1–3 μm) situated next to cell bodies so that the extracellular voltage can be measured. Depending on the impedance and location of the electrode tip, extracellular recordings can capture the activity from either a single neuron (i.e., single-unit activity (SUA) recording) or from several neurons (i.e., multi-unit activity (MUA) recording). The clear benefit of extracellular recording is that several individual neurons can be assayed simultaneously. However, only the largest signals from neurons closest to the microelectrode tip can be reliably identified individually.

Since microelectrodes are designed to measure the high frequency action potentials (or spikes), the data are collected with high temporal resolution (< 100 μs). The high signal-to-noise ratio of action potentials can be sorted out from the lower amplitude signals (by spike sorting). The extracellular signal, however, is also susceptible to the slower waves representing the local field potential (LFP) which may arise from graded events at the nerve terminal as shown in **Figure 1.1**. Thus, appropriate filtering is applied to separate the MUA (10^2–10^4 Hz) from the LFP ($<10^2$ Hz) signals. SUA and MUA signals are typically representative of signals from neuron(s) in the microelectrode's vicinity (spanning 10–100 μm). LFP signals can integrate potentials over much larger distances (μm to mm) to represent the aggregate cellular activity in that region.

Both electroencephalography (EEG; *(67)*) and magnetoencephalography (MEG; *(68)*) allow mapping with much better brain coverage because the measurement devices nearly surround the entire head. EEG measures summed activity of post-synaptic currents, whereas MEG measures tiny magnetic fields (in fT range) produced by synchronized ionic currents flowing in dendrites. Since EEG and MEG signals originate from slightly different cortical locations and are acquired with different temporal resolutions (ms vs. μs), the types of information obtained vary. Because both EEG and MEG suffer from the inverse problem (i.e., difficulty localizing origin of signal), the spatial resolution suffers (mm to cm). EEG signals are susceptible to body movements and MEG signals from the brain sometimes compete with higher magnitude environmental noise. However, non-invasive use of EEG and MEG in humans is invaluable for basic science and clinical research.

4.2. Optical Imaging

Optical imaging techniques are based on the discovery *(69)* that optical properties of cells, and therefore tissue, change during activity. Sensitivity of all optical techniques is limited by the magnitude of the changed optical properties of the system. At present, four dynamic optical techniques dominate: optical intrinsic signal (OIS), voltage sensitive dye (VSD), near infra red spectroscopy (NIRS), and laser-Doppler flowmetry (LDF). These methods can be used to measure a variety of dynamic events in cells (e.g., ion flux) and vessels (e.g., blood oxygenation). OIS and VSD are primarily applied in animals because of the need for a window (e.g., 4×4 mm^2) on the skull and typically have high spatial resolution (10–50 μm). The data are limited to superficial regions of the brain because of limited light penetration in tissue (mm range). NIRS is a non-invasive technique and suited for use in humans. The spatial resolution of NIRS and LDF is limited by the distance between the emitting and detecting probes. Most of these techniques allow two dimensional images of brain activity with relatively high temporal resolution (10–100 ms).

OIS is perhaps the most widely used optical technique in neuroscience *(70)*, in part, because it does not require exogenous optical probes. However, the origin of the task-dependent changes in the intrinsic signal is complex. Changes in blood oxygenation (increase or decrease in deoxyhemoglobin), blood volume (vessel dilation, capillary recruitment, etc.), and light-scattering that accompany cell swelling with activation, all can affect the intrinsic signal in different ways and to various extents. Because these optical properties of the tissue vary in time, multiple illumination wavelengths (e.g., 600–630 nm for oximetry) are necessary to calculate the desired parameters. The intrinsic signals are based on small reflectance signal changes (0.1–1%) which peak shortly (2–3 s) after stimulus onset.

VSD uses small exogenous probes (weight <500 Da; length < 200 nm) that are designed to preferentially bind to the periphery of the cell membrane, and therefore, act as molecular transducers *(71)*. Changes in the membrane potential or ion concentration can then be transformed into an optical signal. There are two key factors to a successful dye used in VSD imaging. The optical yield (or contrast-to-noise ratio) when the molecule binds to the membrane is obviously critical. Another factor, which partly limits its application more widely, is successful delivery of the dye to the site of interest. Since the blood-brain barrier limits most foreign molecules from entering the extracellular milieu, direct injection of the dye into the site of interest is preferred. Because different events at the nerve terminal are linked to intracellular Ca^{2+} ion, dyes sensitive to Ca^{2+} voltage-gated ion channels have wide use in imaging activities, both neurons and astrocytes, with good temporal resolution ($<10^2$ ms). The VSD signal changes can be quite large (2–4%) which peak quite rapidly (<1 s) after stimulus onset.

NIRS is a spectroscopic method which utilizes the near infrared region (650–950 nm) of the electromagnetic spectrum to sense changes in the hemoglobin oxygenation state *(72)*. Because NIRS is non-invasive and has better tissue penetration (cm instead of mm) than visible light, the method has found wide use in both clinical monitoring of tissue oxygenation and functional imaging in humans and small animals *(73)*. Since near-infrared absorption spectra of oxyhemoglobin and deoxyhemoglobin allow separation of these two species, NIRS measurements commonly utilize three separate wavelengths to measure these components for blood oxygenation, and later sum them to get blood volume. The spatial resolution (in mm range) is lower than other optical methods because of the diffuse nature of photon migration through tissue which can affect sensitivity and quantitation of oxyhemoglobin and deoxyhemoglobin *(74)*. The NIRS signal changes are larger than intrinsic signal changes (>2%) and peak slowly (within 5 s) after stimulus onset.

LDF is an optical technique for measuring movement of red blood cells in the microvascular bed *(75)*. The principle is based on the "Doppler" frequency shift that arises in light that is scattered by moving red blood cells. By illuminating the tissue with single-frequency (low power) laser light and detecting the frequency distribution of the back-scattered light, red blood cell velocity and/or flux can be estimated. Since the LDF device is usually standardized with suspensions (e.g., latex spheres undergoing Brownian motion), it is a relative method and cannot easily provide absolute units of blood flow. LDF is a widely used technique in laboratories because of its minimal invasiveness and relatively high sensitivity for stimulation-induced blood flow/volume changes *(76)*. LDF has found popular applications in the clinical environment (e.g., neonatal care *(77)*) because it is quite simple to use. The LDF signal changes can be quite large (10–100%) and peak faster than NIRS signals (within 5 s) after stimulus onset.

4.3. Magnetic Resonance

Magnetic resonance (MR) is based upon the quantum mechanical properties of an atom's nucleus *(78)*. Nuclei that contain odd numbers of protons or neutrons have an intrinsic magnetic moment. Some MR visible and biologically relevant non-radioactive isotopes are ^{1}H, ^{13}C, ^{15}N, ^{17}O, ^{19}F, ^{23}Na, and ^{31}P. Sensitivity of MR detection for a given isotope depends on the physical intrinsic property of the isotope (i.e., the gyromagnetic ratio) and the abundance of the isotope prescribed by nature (can be enhanced by enrichment). The most sensitive and weak isotopes that are commonly used for in vivo MR experiments are ^{1}H and ^{13}C, respectively. ^{1}H has four times stronger intrinsic sensitivity than ^{13}C. Also ^{1}H is ~91% more naturally abundant than ^{13}C. Together, this makes ^{1}H orders of magnitude more sensitive than ^{13}C.

MR imaging (MRI) is primarily used in diagnostic imaging to map structure, but lately also function. MRI is predominantly based on ^{1}H signal from water. MRI data have good spatial resolution and the temporal resolution depends on the specific type of function being measured (see below). MR spectroscopy (MRS) is a powerful technique used to obtain rich chemical information about a wide range of biomolecules (other than water) based on their variable chemical shifts. MRS detection of an isotope depends on the experimental purpose (see below). MRS data can also be viewed in two or three dimensions but the images do not appear as crisp as MRI data because the MRS voxels are much larger. The lower spatial resolution of MRS (mm to cm) compared to MRI (μm to mm) is mainly because of the orders of magnitude of lower concentration of biomolecules (mM range) compared to water (>10 M) being detected. Since a wide range of MR methods can be used for neurobiological studies *(79)*, the few that meet the criteria of dynamic functional imaging in **Figure 1.2** are discussed.

4.3.1. MRS Methods

In the past, ^{31}P MRS had been mainly used to detect high energy phosphates (e.g., ATP, ADP) which lead to assessment of tissue pH *(80)*. Since these signals do not change significantly with modest functional challenges, ^{31}P MRS found little use in functional brain imaging. Recently, due to better sensitivity at high magnetic fields, ^{31}P MRS has been able to measure forward and backward fluxes of creatine kinase *(81)*. Although the temporal resolution of this approach is still low (seconds to minutes), there is great potential of its use because comparable techniques do not exist yet.

Similar to ^{31}P MRS, ^{17}O MRS was also demonstrated nearly two decades ago *(82)* and due to recent advancements, the method has been revitalized *(83)*. The ^{17}O MRS method for oxygen consumption follows almost the same approach as ^{15}O PET (i.e., inhale labeled-oxygen and detect labeled-water). However, a key difference between the two methods is that ^{17}O MRS can distinguish between labeled-water (very strong MRS signal) and labeled-oxygen (exceptionally weak MRS signal), whereas PET cannot. This allows for removal of a few assumptions for modeling oxygen consumption with ^{17}O MRS. The ^{17}O MRS method can also measure blood flow and follows the same principles as ^{15}O PET (i.e., inject labeled-water and detect loss of labeled-water).

Although ^{19}F is intrinsically almost as sensitive as ^{1}H and it is naturally 100% abundant *(78)*, there are no biomolecules in vivo (of relevance) containing fluorine. Therefore FDG, as in PET, has been applied with ^{19}F MRS for glucose consumption measurement *(84)*, but the MRS tracer is not radioactive. Furthermore, ^{19}F MRS can distinguish between FDG-6-phosphate created by metabolism of FDG from FDG itself, whereas ^{18}F PET cannot. Dynamic separation between the amount of FDG-6-phosphate and FDG helps quantify the glucose consumption rate. For ^{19}F MRS detection of these signals with modest signal-to-noise ratio, either very high FDG doses ($>$10 mM) have been used with acceptable acquisition times or very long acquisition times (hours) have been used with low FDG doses ($<$1 mM). More recently, due to superior sensitivity at high magnetic fields, ^{19}F MRS with reasonable spatial (mm range) and temporal resolution ($<$30 min) has been demonstrated using a low FDG dose *(85)*.

Glucose consumption measured by the FDG method (MRS or PET) has a non-oxidative and oxidative component, where the latter can be measured by ^{13}C MRS. Although ^{13}C detection sensitivity is quite low (see above), using ^{13}C-labeled exogenous biomolecules provides a viable alternative because of compartment/substrate specific information. ^{13}C MRS is a very powerful tool to study metabolism when it is used in combination with a ^{13}C-labeled substrate, such as [1-^{13}C]glucose *(86)*. Infusion of [1-^{13}C]glucose (into the blood stream) results in ^{13}C turnover into glucose, glutamate, glutamine, GABA, lactate, aspartate, etc, in brain tissue *(25, 42)*. Since ^{13}C turnover into these pools is time dependent, coupled differential equations can be used to estimate a range of fluxes (including neuronal/glial glucose oxidation, glutamate-GABA-glutamine cycling; see Refs. *(11, 17, 79)* for details. But this is a steady-state method where the measured turnover times are in the order of about half an hour (or slightly more). Recently, however, by dynamic nuclear polarization (DNP), the MR detection sensitivity can be artificially improved by orders of magnitude *(87)*. To date, this method (using injection of hyperpolarized ^{13}C or ^{15}N biomolecules) has primarily been used to demonstrate perfusion through organs with quite high temporal (within seconds) and spatial (below mm range) resolutions. Although technical challenges remain *(88)*, study of metabolism with ^{13}C and/or ^{19}F MRS in conjunction with DNP is quite promising.

4.3.2. MRI Methods

There are several hemodynamic-based functional MRI (fMRI) methods that are sensitive to functional hyperemic events. The blood oxygenation level dependent (BOLD) method *(89)* is the most popular, in part, due to its simplicity. Both BOLD and NIRS methods are sensitive to changing concentrations of oxyhemoglobin and deoxyhemoglobin (i.e., diamagnetic or red blood, and paramagnetic or black blood, respectively). With NIRS, both moieties are detected (by different absorption bands); with BOLD, the magnetic properties of blood changing with oxygenation influences the tissue water signal (i.e., less paramagnetic deoxyhemoglobin, higher MRI signal).

Just as hemoglobin is nature's endogenous MRI contrast agent to measure blood oxygenation changes, blood-borne exogenous MRI contrast agents are used to measure alterations in blood volume with almost the same assumptions (i.e., more paramagnetic agent, lower MRI signal) *(90)*. Principles of blood flow measured by MRI are similar to that in PET. Following magnetic labeling of arterial blood (water), subsequent dynamic MRI maps reflect the degree to which the label has decayed (by mixing with unlabeled water), thereby reflecting tissue perfusion *(91)*.

Since its discovery, the biophysical understanding of BOLD has improved significantly. Better specificity of the BOLD effect at higher magnetic fields *(92)* allows changes in oxygen consumption to be titrated out by combining the contributing changes in blood flow and blood volume with the BOLD signal *(93)*. At steady-state, the oxygen consumption predicted by BOLD calibration has been shown to be in good agreement with oxygen consumption measured by ^{13}C MRS *(94)* and which, in turn, concurs with ensemble neuronal activity measured by extracellular recordings *(95)*. Since these multi-modal MRI methods can be applied dynamically *(96)*, there is potential for transient energetics with higher temporal resolution (200–500 ms) *(97)*.

Recently, there has been a surge in fMRI that is based on non-hemodynamic events. Two prominent directions are effects of electrical currents directly or indirectly on the MRI signal *(98)* and task-related structural changes which may lead to alteration of diffusion of water molecules *(99)*. Both of these emerging fMRI approaches have potential to improve temporal resolution because the signals are not delayed by hemodynamics. However, the greater challenge is to enhance the magnitude of the non-hemodynamic effects (<1%) which are much smaller than hemodynamic-based fMRI (>3%).

5. Future

The brain is mapped by several dynamic imaging methods sensitive to a variety of events at the glutamatergic nerve terminal (**Fig. 1.1**). These methods span not only wide ranges of spatial and temporal resolutions (**Fig. 1.2**) where each technique has obvious limitations, but also definite advantages. No *one* method can cover the several orders of magnitude in temporal and spatial resolutions and at the *same* time capture the many cellular and vascular events. Any such promise *(100)* discounts potential reachable insights, even at present, gained by combining different techniques which complement each other *(101)*. The recent trend to combine fMRI *(102–105)* or optical imaging *(106–109)* with electrophysiology is in accord with this suggestion. Since energy is the currency of trade between cellular events and the substrates that those activities demand, dynamic functional imaging with energetics may be a crucial direction to pursue in the future *(79)*. In functional imaging, the qualitative way by which the baseline signal is conveniently differenced away to reveal stimulation-induced focal area(s) of interest ignored the high energy utilization in the resting brain *(6)* for a long time. Recent awareness *(6, 110)* of the high baseline activity, which slowly varies in time, has been investigated by multi-modal methods like fMRI *(111)*, optical imaging *(112)*, and electrophysiology *(113)*. The anatomical *(1)* and physiological *(2)* complexities of the brain, both at rest and during activation, reward an extraordinary set of dynamic imaging *(2, 3, 79)* and analytical *(114, 115)* tools.

Acknowledgements

This work was supported, in part, by NIH grants from NIMH (R01 MH-067528), NIDCD (R01 DC-003710), and NINDS (P30 NS-52519).

References

1. Shepherd GM (2004) *The Synaptic Organization of the Brain.* (Oxford University, New York, NY, USA)
2. Kandel ER, Schwartz JH, Jessell TM (2000) *Principles of Neural Science.* (McGraw-Hill, New York, NY, USA)
3. Huettel SA, Song AW, McCarthy G (2004) *Functional Magnetic Resonance Imaging* (Sinauer, Sunderland, MA, USA)
4. Jakovcevic D, Harder DR (2007) Role of astrocytes in matching blood flow to neuronal activity. Curr Top Dev Biol. 79:75–97
5. Iadecola C, Nedergaard M (2007) Glial regulation of the cerebral microvasculature. Nat Neurosci. 10:1369–1376
6. Shulman RG, Rothman DL (1998) Interpreting functional imaging studies in terms of neurotransmitter cycling. Proc Natl Acad Sci USA. 95:11993–11998
7. Attwell D, Laughlin SB (2001) An energy budget for signaling in the grey matter of the brain. J Cereb Blood Flow Metab. 21: 1133–1145
8. Attwell D, Iadecola C (2002) The neural basis of functional brain imaging signals. Trends Neurosci. 25:621–625
9. Drake CT, Iadecola C (2007) The role of neuronal signaling in controlling cerebral blood flow. Brain Lang. 102:141–152
10. Nicholls DG (1993) The glutamatergic nerve terminal. Eur J Biochem. 212:613–631
11. Hyder F, Patel AB, Gjedde A, Rothman DL, Behar KL, Shulman RG (2006) Neuronal-glial glucose oxidation and glutamatergic-GABAergic function. J Cereb Blood Flow Metab. 26:865–877
12. Bonvento G, Sibson N, Pellerin L (2002) Does glutamate image your thoughts? Trends Neurosci. 25:359–364
13. Meldrum BS (2000) Glutamate as a Neurotransmitter in the Brain: Review of Physiology and Pathology. J Nutr. 130(4S Suppl):1007S-1015S
14. Hodgkin AL, Huxley AF (1952) Propagation of electrical sickels along giant nerve fibres. Proc R Soc Lond B Biol Sci. 140: 177–183
15. Llinas R, Sugimori M, Silver RB (1992) Microdomains of high calcium concentration in a presynaptic terminal. Science. 256: 677–679
16. Dzubay JA, Jahr CE (1996) Kinetics of NMDA channel opening. J Neurosci. 16:4129–4134
17. Rothman DL, Sibson NR, Hyder F, Shen J, Behar KL, Shulman RG (1999) In vivo nuclear magnetic resonance spectroscopy studies of the relationship between the glutamate-glutamine neurotransmitter cycle and functional neuroenergetics. Philos Trans R Soc Lond B Biol Sci. 354:1165–1177
18. Marder E (2006) Neurobiology: Extending influence. Nature. 441:702–703
19. Buzsaki G, Draguhn A (2004) Neuronal oscillations in cortical networks. Science. 304:1926–1929
20. Magistretti PJ, Pellerin L, Rothman DL, Shulman RG (1999) Energy on demand. Science. 283:496–497
21. Sokoloff L (1991) "Relationship between functional activity and energy metabolism in the nervous system: Whether, where and why?" in *Brain Work and Mental Activity* (Eds, Lassen NA, Ingvar DH, Raichle ME, Friberg L). pp. 52–64 (Munksgaard, Copenhagen, Denmark)
22. Williams LR, Leggett RW (1989) Reference values for resting blood flow to organs of man. Clin Phys Physiol Meas. 10:187–217
23. Silver IA, Erecinska M (1994) Extracellular glucose concentration in mammalian brain: Continuous monitoring of changes during increased neuronal activity and upon limitation in oxygen supply in normo-, hypo-, and hyperglycemic animals. J Neurosci. 14:5068–5076
24. Herman P, Trubel HK, Hyder F (2006) A multiparametric assessment of oxygen efflux from the brain. J Cereb Blood Flow Metab. 26:79–91
25. Choi IY, Gruetter R (2003) In vivo ^{13}C NMR assessment of brain glycogen concentration and turnover in the awake rat. Neurochem Int. 43:317–322
26. Lin Y, Zhang YP, Xiao ZW, Li H, Shen ZW, Chen XK, Huang K, Wu RH (2006) Quantification of brain creatine concentration using PRESS sequence and LCModel: Comparison with HPLC method. Conf Proc IEEE Eng Med Biol Soc. 1: 1928–1931
27. Siesjo BK (1978) *Brain Energy Metabolism.* (Wiley, New York, NY, USA)
28. Trubel HK, Sacolick LI, Hyder F (2006) Regional temperature changes in the brain during somatosensory stimulation. J Cereb Blood Flow Metab. 26:68–78
29. Roland PE, Eriksson L, Stone-Elander S, Widen L (1987) Does mental activity change the oxidative metabolism of the brain? J Neurosci. 7:2373–2389
30. Ames A 3rd (2000) CNS energy metabolism as related to function. Brain Res Brain Res Rev. 34:42–68

31. Pardridge WM (1983) Brain metabolism: A perspective from the blood-brain barrier. Physiol Rev 63:1481–1535
32. Lund-Andersen H (1979) Transport of glucose from blood to brain. Physiol Rev. 59:305–352
33. Simpson IA, Carruthers A, Vannucci SJ (2007) Supply and demand in cerebral energy metabolism: The role of nutrient transporters. J Cereb Blood Flow Metab. 27:1766–1791
34. Magistretti PJ, Allaman I (2007) Glycogen: A Trojan horse for neurons. Nat Neurosci. 10:1341–1342
35. Shulman RG, Hyder F, Rothman DL (2001) Cerebral energetics and the glycogen shunt: Neurochemical basis of functional imaging. Proc Natl Acad Sci USA. 98:6417–6422
36. Shulman RG, Hyder F, Rothman DL (2001) Lactate efflux and the neuroenergetic basis of brain function. NMR Biomed. 14:389–396
37. Villringer A, Dirnagl U (1995) Coupling of brain activity and cerebral blood flow: Basis of functional neuroimaging. Cerebrovasc. Brain Metab Rev. 7:240–276
38. Buxton RB, Frank LR (1997) A model for the coupling between cerebral blood flow and oxygen metabolism during neural stimulation. J Cereb Blood Flow Metab. 17: 64–72
39. Hyder F, Shulman RG, Rothman DL (1998) A model for the regulation of cerebral oxygen delivery. J Appl Physiol. 85:554–564
40. Hyder F, Kennan RP, Kida I, Mason GF, Behar KL, Rothman DL (2000) Dependence of oxygen delivery on blood flow in rat brain: A 7 tesla nuclear magnetic resonance study. J Cereb Blood Flow Metab. 20:485–498
41. Wolf T, Lindauer U, Villringer A, Dirnagl U (1997) Excessive oxygen or glucose supply does not alter the blood flow response to somatosensory stimulation or spreading depression in rats. Brain Res. 761: 290–299
42. Shulman RG, Rothman DL, Behar KL, Hyder F (2004) Energetic basis of brain activity: Implications for neuroimaging. Trends Neurosci. 27:489–495
43. Lou HC, Edvinsson L, MacKenzie ET (1987) The concept of coupling blood flow to brain function: Revision required? Ann Neurol. 22:289–297
44. Roy CS, Sherrington CS (1890) On the regulation of the blood-supply of the brain. J Physiol. 11:85–108
45. Cauli B, Tong XK, Rancillac A, Serluca N, Lambolez B, Rossier J, Hamel E (2004) Cortical GABA interneurons in neurovascular coupling: Relays for subcortical vasoactive pathways. J Neurosci. 24:8940–8949
46. Rancillac A, Rossier J, Guille M, Tong XK, Geoffroy H, Amatore C, Arbault S, Hamel E, Cauli B (2006) Glutamatergic control of microvascular tone by distinct GABA neurons in the cerebellum. J Neurosci. 26: 6997–7006
47. Pellerin L, Magistretti PJ (1994) Glutamate uptake into astrocytes stimulates aerobic glycolysis: A mechanism coupling neuronal activity to glucose utilization. Proc Natl Acad Sci USA. 91:10625–10629
48. Cohen Z, Bouchelet I, Olivier A, Villemure JG, Ball R, Stanimirovic DB, Hamel E (1999) Multiple microvascular and astroglial 5-hydroxytryptamine receptor subtypes in human brain: Molecular and pharmacologic characterization. J Cereb Blood Flow Metab. 19:908–917
49. Elhusseiny A, Cohen Z, Olivier A, Stanimirovic DB, Hamel E (1999) Functional acetylcholine muscarinic receptor subtypes in human brain microcirculation: Identification and cellular localization. J Cereb Blood Flow Metab. 19:794–802
50. Gillard SE, Tzaferis J, Tsui HC, Kingston AE (2003) Expression of metabotropic glutamate receptors in rat meningeal and brain microvasculature and choroid plexus. J Comp Neurol. 461:317–332
51. Parfenova H, Fedinec A, Leffler CW (2003) Ionotropic glutamate receptors in cerebral microvascular endothelium are functionally linked to heme oxygenase. J Cereb Blood Flow Metab. 23:190–197
52. Garthwaite J, Charles SL, Chess-Williams R (1988) Endothelium-derived relaxing factor release on activation of NMDA receptors suggests role as intercellular messenger in the brain. Nature. 336:385–388
53. Ledo A, Barbosa RM, Gerhardt GA, Cadenas E, Laranjinha J (2005) Concentration dynamics of nitric oxide in rat hippocampal subregions evoked by stimulation of the NMDA glutamate receptor. Proc Natl Acad Sci USA. 102:17483–17488
54. Takano T, Tian GF, Peng W, Lou N, Libionka W, Han X, Nedergaard M (2006) Astrocyte-mediated control of cerebral blood flow. Nat Neurosci. 9:260–267
55. Zonta M, Angulo MC, Gobbo S, Rosengarten B, Hossmann KA, Pozzan T, Carmignoto G (2003) Neuron-to-astrocyte signaling is central to the dynamic control of brain microcirculation. Nat Neurosci. 6:43–50
56. Iadecola C (2004) Neurovascular regulation in the normal brain and in Alzheimer's disease. Nat Rev Neurosci. 5:347–360

57. Peppiatt C, Attwell D (2004) Neurobiology: Feeding the brain. Nature. 431:137–138
58. Mulligan SJ, MacVicar BA (2004) Calcium transients in astrocyte endfeet cause cerebrovascular constrictions. Nature. 431: 195–199
59. Logothetis NK (2003) The underpinnings of the BOLD functional magnetic resonance imaging signal. J Neurosci. 23:3963–3971
60. Raichle ME (2003) Functional brain imaging and human brain function. J Neurosci. 23:3959–3962
61. Churchland PS, Sejnowski TJ (1988) Perspectives on cognitive neuroscience. Science. 242:741–745
62. Cohen MS, Bookheimer SY (1994) Localization of brain function using magnetic resonance imaging. Trends Neurosci. 17: 268–277
63. Grinvald A, Hildesheim R (2004) VSDI: A new era in functional imaging of cortical dynamics. Nat Rev Neurosci. 5:874–885
64. Sokoloff L, Reivich M, Kennedy C, Des Rosiers MH, Patlak CS, Pettigrew KD, Sakurada O, Shinohara M (1977) The [^{14}C]deoxyglucose method for the measurement of local cerebral glucose utilization: Theory, procedure, and normal values in the conscious and anesthetized albino rat. J Neurochem. 28:897–916
65. Raichle ME (1988) "Circulatory and metabolic correlates of brain function in normal humans" in *Handbook of Physiology – The Nervous System V*, pp. 633–674 (Springer-Verlag, New York, NY, USA)
66. Chatziioannou AF (2002) Molecular imaging of small animals with dedicated PET tomographs. Eur J Nucl Med Mol Imaging. 29:98–114
67. Freeman WJ (2004) Origin, structure, and role of background EEG activity. Part 1. Analytic amplitude. Clin Neurophysiol. 115:2077–2088
68. Ioannides AA (2006) Magnetoencephalography as a research tool in neuroscience: State of the art. Neuroscientist. 12:524–544
69. Cohen LB (1973) Changes in neuron structure during action potential propagation and synaptic transmission. Physiol Rev. 53: 373–418
70. Roe AW (2007) Long-term optical imaging of intrinsic signals in anesthetized and awake monkeys. Appl Opt. 46:1872–1880
71. Baker BJ, Kosmidis EK, Vucinic D, Falk CX, Cohen LB, Djurisic M, Zecevic D (2005) Imaging brain activity with voltage- and calcium-sensitive dyes. Cell Mol Neurobiol. 25:245–282
72. Jobsis FF (1977) Noninvasive infrared monitoring of cerebral and myocardial oxygen sufficiency and circulatory parameters. Science. 198:1264–1267
73. Hoshi Y (2003) Functional near-infrared optical imaging: Utility and limitations in human brain mapping. Psychophysiology. 40:511–520
74. Boas DA, Dale AM, Franceschini MA (2004) Diffuse optical imaging of brain activation: Approaches to optimizing image sensitivity, resolution, and accuracy. Neuroimage. 23 Suppl 1:S275–S288
75. Stern MD (1975) In vivo evaluation of microcirculation by coherent light scattering. Nature. 254:56–58
76. Kida I, Maciejewski PK, Hyder F (2004) Dynamic imaging of perfusion and oxygenation by functional magnetic resonance imaging. J Cereb Blood Flow Metab. 24: 1369–1381
77. Sudikoff S, Banasiak K (1998) Techniques for measuring cerebral blood flow in children. Curr Opin Pediatr. 10:291–298
78. Abragam A (1961) *Principles of Nuclear Magnetism.* (Oxford University, Oxford, UK)
79. Shulman RG, Rothman DL (2004) *Brain Energetics & Neuronal Activity: Applications to fMRI and Medicine.* (Wiley, New York, NY, USA)
80. Ackerman JJ, Grove TH, Wong GG, Gadian DG, Radda GK (1980) Mapping of metabolites in whole animals by ^{31}P NMR using surface coils. Nature. 283:167–170
81. Du F, Zhu XH, Qiao H, Zhang X, Chen W (2007) Efficient in vivo ^{31}P magnetization transfer approach for noninvasively determining multiple kinetic parameters and metabolic fluxes of ATP metabolism in the human brain. Magn Reson Med. 57: 103–114
82. Fiat D, Dolinsek J, Hankiewicz J, Dujovny M, Ausman J (1993) Determination of regional cerebral oxygen consumption in the human: ^{17}O natural abundance cerebral magnetic resonance imaging and spectroscopy in a whole body system. Neurol Res. 15:237–248
83. Zhu XH, Zhang N, Zhang Y, Zhang X, Ugurbil K, Chen W (2005) In vivo ^{17}O NMR approaches for brain study at high field. NMR Biomed. 18:83–103
84. Nakada T, Kwee IL, Card PJ, Matwiyoff NA, Griffey BV, Griffey RH (1988) Fluorine-19 NMR imaging of glucose metabolism. Magn Reson Med. 6:307–313

85. Coman D, Sanganahalli BG, Cheng DW, McCarthy T, Rothman DL, Hyder F (2007) In vivo ^{19}F CSI of 2-fluoro-2-deoxy-D-glucose and 2-fluoro-2-deoxy-D-glucose-6-phosphate in rat brain. Proc Inter Soc Magn Reson Med. 1:577
86. Morris P, Bachelard H (2003) Reflections on the application of ^{13}C-MRS to research on brain metabolism. NMR Biomed. 16: 303–312
87. Golman K, Ardenkjaer-Larsen JH, Petersson JS, Mansson S, Leunbach I (2003) Molecular imaging with endogenous substances. Proc Natl Acad Sci USA. 100:10435–10439
88. Klomp DW, Renema WK, van der Graaf M, de Galan BE, Kentgens AP, Heerschap A (2006) Sensitivity-enhanced ^{13}C MR spectroscopy of the human brain at 3 Tesla. Magn Reson Med. 55:271–278
89. Ogawa S, Lee TM, Kay AR, Tank DW (1990) Brain magnetic resonance imaging with contrast dependent on blood oxygenation. Proc Natl Acad Sci USA. 87: 9868–9872
90. Kennan RP, Scanley BE, Innis RB, Gore JC (1998) Physiological basis for BOLD MR signal changes due to neuronal stimulation: separation of blood volume and magnetic susceptibility effects. Magn Reson Med. 40:840–846
91. Williams DS, Detre JA, Leigh JS, Koretsky AP (1992) Magnetic resonance imaging of perfusion using spin inversion of arterial water. Proc Natl Acad Sci USA. 89:212–216
92. Ugurbil K, Adriany G, Andersen P, Chen W, Garwood M, Gruetter R, Henry PG, Kim SG, Lieu H, Tkac I, Vaughan T, Van De Moortele PF, Yacoub E, Zhu XH (2003) Ultrahigh field magnetic resonance imaging and spectroscopy. Magn Reson Imaging. 21:1263–1281
93. Hoge RD, Pike GB (2001) Oxidative metabolism and the detection of neuronal activation via imaging. J Chem Neuroanat. 22:43–52
94. Hyder F, Kida I, Behar KL, Kennan RP, Maciejewski PK, Rothman DL (2001) Quantitative functional imaging of the brain: Towards mapping neuronal activity by BOLD fMRI. NMR Biomed. 14: 413–431
95. Smith AJ, Blumenfeld H, Behar KL, Rothman DL, Shulman RG, Hyder F (2002) Cerebral energetics and spiking frequency: The neurophysiological basis of fMRI. Proc Natl Acad Sci USA. 99:10765–10770
96. Kida I, Rothman DL, Hyder F (2007) Dynamics of changes in blood flow, volume, and oxygenation: Implications for dynamic functional magnetic resonance imaging calibration. J Cereb Blood Flow Metab. 27: 690–696
97. Sanganahalli BG, Herman P, Hyder F (2007) Transient energetics from fMRI: Single event to block design paradigms. J Cereb Blood Flow Metab. BP47–6W
98. Bandettini PA, Petridou N, Bodurka J (2005) Direct detection of neuronal activity with MRI: Fantasy, possibility, or reality? Appl Magn Reson. 29:65–88
99. Le Bihan D (2007) The 'Wet Mind': Water and functional neuroimaging. Phys Med Biol. 52:R57–R90
100. Jasanoff A (2007) Bloodless FMRI. Trends Neurosci. 30:603–610
101. Hyder F (2004) Neuroimaging with calibrated fMRI. Stroke. 35(11 Suppl 1): 2635–2641
102. Shmuel A, Augath M, Oeltermann A, Logothetis NK (2005) Negative functional MRI response correlates with decreases in neuronal activity in monkey visual area V1. Nat Neurosci. 9:569–577
103. Maandag NJ, Coman D, Sanganahalli BG, Herman P, Smith AJ, Blumenfeld H, Shulman RG, Hyder F (2007) Energetics of neuronal signaling and fMRI activity. Proc Natl Acad Sci USA. 104:20546–20551
104. Sharon D, Hamalainen MS, Tootell RB, Halgren E, Belliveau JW (2007) The advantage of combining MEG and EEG: Comparison to fMRI in focally stimulated visual cortex. Neuroimage. 36:1225–1235
105. Stefanovic B, Schwindt W, Hoehn M, Silva AC (2007) Functional uncoupling of hemodynamic from neuronal response by inhibition of neuronal nitric oxide synthase. J Cereb Blood Flow Metab. 27: 741–754
106. Devor A, Dunn AK, Andermann ML, Ulbert I, Boas DA, Dale AM (2003) Coupling of total hemoglobin concentration, oxygenation, and neural activity in rat somatosensory cortex. Neuron. 39:353–359
107. Sheth SA, Nemoto M, Guiou M, Walker M, Pouratian N, Toga AW (2004) Linear and nonlinear relationships between neuronal activity, oxygen metabolism, and hemodynamic responses. Neuron. 42: 347–355
108. Jones M, Hewson-Stoate N, Martindale J, Redgrave P, Mayhew J (2004) Nonlinear coupling of neural activity and CBF in rodent barrel cortex. Neuroimage. 22:956–965
109. Chen-Bee CH, Agoncillo T, Xiong Y, Frostig RD (2007) The triphasic intrinsic signal: Implications for functional imaging. J Neurosci. 27:4572–4586

110. Fox MD, Raichle ME (2007) Spontaneous fluctuations in brain activity observed with functional magnetic resonance imaging. Nat Rev Neurosci. 8:700–711
111. Biswal B, Yetkin FZ, Haughton VM, Hyde JS (1995) Functional connectivity in the motor cortex of resting human brain using echo-planar MRI. Magn Reson Med. 34:537–541
112. Arieli A, Sterkin A, Grinvald A, Aertsen A (1996) Dynamics of ongoing activity: Explanation of the large variability in evoked cortical responses. Science. 273:1868–1871
113. Leopold DA, Murayama Y, Logothetis NK (2003) Very slow activity fluctuations in monkey visual cortex: Implications for functional brain imaging. Cereb Cortex. 13: 422–433
114. Mitra PP, Pesaran B (1999) Analysis of dynamic brain imaging data. Biophys J. 76:691–708
115. Eke A, Herman P, Kocsis L, Kozak LR (2002) Fractal characterization of complexity in temporal physiological signals. Physiol Meas. 23:R1–R38

Chapter 2

Fractal Characterization of Complexity in Dynamic Signals: Application to Cerebral Hemodynamics

Peter Herman, Laszlo Kocsis, and Andras Eke

Abstract

We introduce the concept of spatial and temporal complexity with emphasis on how its fractal characterization for 1D, 2D or 3D hemodynamic brain signals can be carried out. Using high-resolution experimental data sets acquired in animal and human brain by noninvasive methods – such as laser Doppler flowmetry, laser speckle, near infrared, or functional magnetic resonance imaging – the spatiotemporal complexity of cerebral hemodynamics is demonstrated. It is characterized by spontaneous, seemingly random (that is disorderly) fluctuation of the hemodynamic signals. Fractal analysis, however, proved that these fluctuations are correlated according to the special order of self-similarity. The degree of correlation can be assessed quantitatively either in the temporal or the frequency domain respectively by the Hurst exponent (H) and the spectral index (β). The values of H for parenchymal regions of white and gray matter of the rat brain cortex are distinctly different. In human studies, the values of β were instrumental in identifying age-related stiffening of cerebral vasculature and their potential vulnerability in watershed areas of the brain cortex such as in borderline regions between frontal and temporal lobes. Biological complexity seems to be present within a restricted range of H or β values which may have medical significance because outlying values can indicate a state of pathology.

Key words: Complexity, fractals, laser Doppler flowmtery, laser speckle imaging, MRI, cerebral blood flow.

1. Introduction

The human brain represents the highest degree of organization of matter and is often referred to as being "hypercomplex" *(1)*. The brain's hypercomplexity does not only reflect its assembly consisting of astronomical number of neurons *(2)* capable of operating in concert to produce various functions of the nervous system, but also highlights the fact that traditional approaches and available

Fahmeed Hyder (ed.), *Dynamic Brain Imaging: Multi-Modal Methods and In Vivo Applications, vol. 489*

DOI 10.1007/978-1-59745-543-5_2 Springerprotocols.com

knowledge are insufficient to comprehend the laws governing the development of its complex structures and the operation of its functions therein.

The essence of somewhat simpler systems, known as "complex", still cannot be revealed from the behavior of their isolated components because these components do have interactions with one another, and the overall behavior of this complex system is in fact determined by the very ways its elements are coupled. Even a few tightly coupled elements can produce very complex dynamics known as "chaos" on purely deterministic ground *(3)*. The degree of freedom in these chaotic systems is low, suggesting that a few coupled differential equations govern the complex behavior *(4)*. When a complex system is assembled from a large number of components, these cannot possibly be tightly coupled. Hence the degree of freedom for the system is rendered to be high. The complexity of this system is manifested in a peculiar pattern of spatial and/or temporal correlation known as self-similar or fractal behavior.

Here, we overview the application of the fractal approach on one such complex system, the cerebral hemodynamics. We briefly introduce the fractal properties, and show how they relate to complex systems. Because many aspects of complexity are present in hemodynamics, we will demonstrate that to extract these features from such signals requires an array of measuring and analytical tools. We give an overview of fractal descriptors for the 1 dimensional signal up to the 4 dimensions of time and space.

2. Fractal Characterization

No straightforward axiom exists to determine if an observed object is fractal or not, hence this decision can only be made based on identifying the presence of its fractal properties *(5)*. These are fundamental, interrelated features like the self-similarity (or self-affinity), the power-law scaling relationship of features, scale invariance, scaling range and fractal (non-integer) dimension of the object. (For a more detailed explanation see ref. *(6)*).

2.1. Self-similarity

Pieces of a fractal object when enlarged are similar to larger pieces or to that of the whole. We can classify the self-similar properties from two points of view: *1. Similarity.* If the pieces are identical, rescaled replica of each other, the fractal is exact. When the similarity is present only in between statistical populations of observed data of a given feature assessed at different scales, the fractal is referred to as statistical; *2. Scaling.* If the scaling is uniform in all directions (isotropic), then the fractal is self-similar. If the scaling is anisotropic, i.e. in one direction, the proportions between the enlarged pieces are different from those in the other direction, then the fractal is self-affine. This distinction is, however, often smeared and for the purpose of being more expressive, self-similarity is used when indeed self-affinity is meant *(7)*.

2.2. Power Law Scaling Relationship, Fractal Dimension

The power law scaling relationship gives the mathematical basis for calculating a fractal parameter. It can mathematically be derived from self-similarity *(8)*. Briefly, when a quantitative property, q, is measured on scale s, its value will be dependent on s according to the following scaling relationship:

$$q = f(s) \tag{2.1}$$

For fractals, q does not converge if s is decreasing, but, instead exhibits a power law scaling relationship with respect to s, whereby with decreasing s, it increases (because newer details become visible) without any limit.

$$q = ps^{\varepsilon} \tag{2.2}$$

where p is a factor of proportionality (prefactor) and ε is a negative number, the scaling exponent. The value of ε can be easily determined as the slope of the linear regression fitted to the data pairs on the plot of log q versus log s:

$$\log\ q = \log\ p + \varepsilon \log\ s. \tag{2.3}$$

The scaling exponent (or its simple derivative) is itself the most essential fractal parameter, the fractal dimension. Because its value usually is non-integer, this is the eponym of fractals (i.e., fractus = broken) *(9)*. Instead of the fractal dimension, another fractal descriptor, the Hurst exponent, is widely used *(8)* because the most essential scaling properties of *temporal* signals can be compared on a scale of 0 to 1. The fractal dimension can describe the highly dimensional structure of the process; its value can vary between 0 and theoretically infinitum, however, in practice rarely higher than 3. If the scaling exponent is described in the frequency domain of the signal, then it is referred to as spectral index (β). The spectral index is, by convention, the negative value of the scaling exponent.

2.3. Scale Invariance and Scaling Range

The most remarkable visual feature of a fractal object is its scale invariance, which refers to the fact that the physical size of the object cannot be judged by its perceived image. Mathematically speaking, the ratio of two estimates of q measured at two different scales, s_1 and s_2, q_2/q_1 depends only on the ratio of scales (relative scale), s_2/s_1, and not directly on the absolute scale of observations, s_1 or s_2

$$q_2/q_1 = ps_2^{\varepsilon}/ps_1^{\varepsilon} = (s_2/s_1)^{\varepsilon} \tag{2.4}$$

For exact fractals, s_2/s_1 is a discrete variable, while for statistical fractals, like those in nature, s_2/s_1 may change in a continuous fashion while not violating the validity of the equation (2.4). For statistical fractals, the two sides of the equation are equal only in distribution (note that $=_d$ should be used instead of $=$ here)

$$q_2/q_1 =_d ps_2^{\varepsilon}/ps_1^{\varepsilon} =_d (s_2/s_1)^{\varepsilon} \tag{2.5}$$

For natural fractals, scale invariance holds only for a restricted range of the absolute scale *(10)*. The upper limit of validity s_{max}, for natural fractals falls into the range of the size of the structure itself, likewise the lower limit, s_{min}, falls into the dimensions of the smallest structural elements (i.e. there are no more newer details when the resolution is increased further). The scaling range, SR, is then given in decibels.

$$\mathrm{SR} = \log_{10}(s_{max}/s_{min}). \tag{2.6}$$

3. Complex Systems

A complex system has a great deal of structural and/or functional redundancies and which, in turn, lend it a robust behavior, where quite a few of its elements may be dropped without compromising the structural or functional integrity of the complex system. In the absence of such robust behavior, the system can still be regarded as "complicated." In this system, however, excluding one key element can bring the whole system to a halt.

Another key behavior of a complex system is that it can adapt to changing conditions often in producing a non-linear response, a very hard to predict behavior. The description of a complex system's structure and function requires special tools such as those used in chaos and/or fractal theories *(8)*. Chaos theory describes the non-linear dynamics emerging from coupled differential equations and predicts the behavior of the system in Newtonian space which is characterized by a fractal attractor *(4)*. The fractal approach is aimed at identifying the presence of self-similarity in the design of complex structures or in the dynamics of complex temporal processes, where both manifestations are correlation of patterns *(11)*. Complex systems in nature can be viewed as implementations of nature's "blueprints" employing fractal spatiotemporal dynamics, keeping in mind that fractal dynamics is only a model which can indeed describe many complex systems of nature. In living systems, a certain degree of complexity is what is considered normal *(8)*. Processes upsetting this particular degree of complexity typically lead to pathology and can be categorized as disease states *(8)*.

Sampled structures or processes of living biological systems produce spatial and/or temporal signals. Depending on the spatiotemporal resolution of the observation(s) and the presence of intrinsic or extrinsic factors altering the observed signal(s), the dynamics within the system can be revealed. Almost all biological signals can be shown to be dynamic if the spatial or temporal window of the observation is properly chosen. For example, blood flow measured in large brain areas, under resting conditions for a minute or two, due to its minute fluctuations, can be regarded as a temporal signal without significant dynamics. However, resting blood flow measured in "micro" areas, as measured by laser speckle imaging *(12)* and as shown in **Fig. 2.1**, can exhibit great spatial and temporal variations and hence should be regarded as a dynamic signal.

Emergence of dynamics in biological behavior is not only a function of the scale of observation but is fundamentally determined by the energy flux through the observed system (**Fig. 2.2**) *(13)*. At the low end of the energy flux, the system tends to equilibrate at a non-zero steady level of operation or decay to zero level (i.e., no operation). The higher the energy flux, the more pronounced is the dynamic character of the signal, eventually reaching a state of utmost complexity known as pure random behavior. In between these two extreme states (i.e., equilibrium and random), the physiological dynamics can be found surrounded by a range of pathologic states.

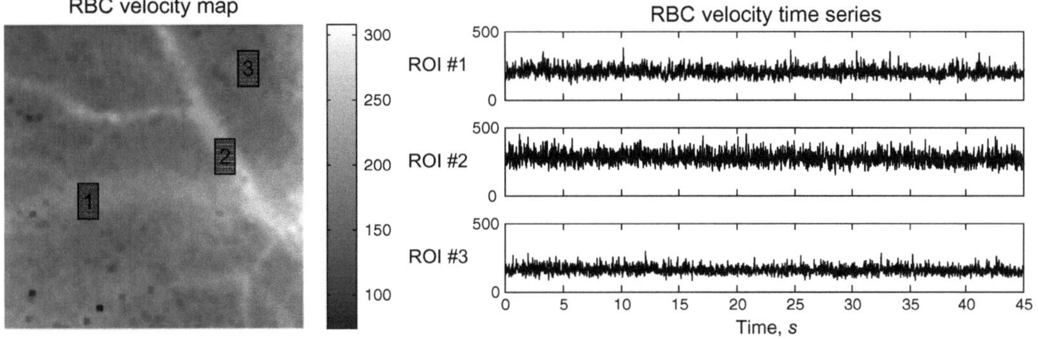

Fig. 2.1. *Spatial distribution of microflow across the brain cortex and its temporal dynamics for three regions of interest.* Laser speckle flow imaging was used to map red blood cell perfusion at 252 × 252 voxel resolution for an area of 1 mm^2 in the brain cortex of an anesthetized rat. Intensity coded data are plotted on the left, while time series of spontaneous perfusion fluctuation is shown for a pial vein (1) artery (2) and parenchymal area (3). An apparent heterogeneity, randomness dominates these signals, lending them a complex appearance.

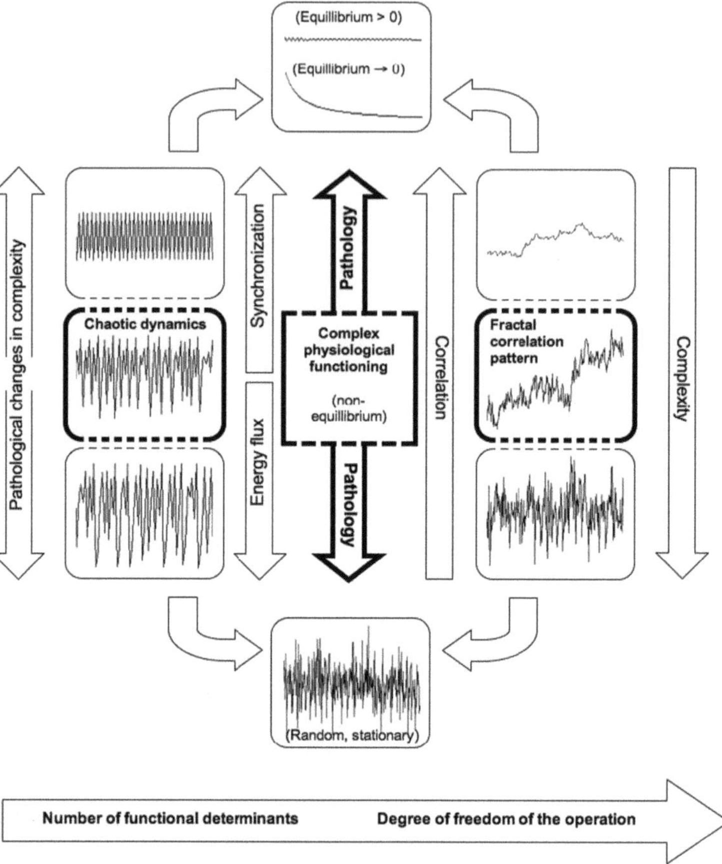

Fig. 2.2. *Conceptual representation of dynamic properties of temporal biological signals and their complexity with reference to ranges of normal and abnormal biological function.* Note that truly random dynamics is the highest degree of complexity which is not typical to the normal, physiological function, just as well as one that either decays to zero or exhibits a level with no variations in time. From this latter state, increasing energy flux takes the system to higher levels of complexity. If the number of determinants is small, the complex dynamics can likely be chaotic, while in most biological systems, due to the very large number of simultaneously operating factors, the dynamics can likely be of a temporal fractal, which shows self-similar pattern of fluctuation as observed across a range of time scales. Physiological dynamics are to be found within a restricted range of dynamic complexity. Disease states are typically associated with a disruption of complexity in directions of decreasing and increasing dynamics.

Our task at revealing the significance of dynamics in physiology and pathology should begin with the understanding of how complexity in these signals can be quantitatively characterized. Two major trends of analysis (**Fig. 2.2**) have emerged in this respect in the past decades: One (chaos theory) that can be applied to a deterministic system with a few control parameters in action, the other applicable to a system in which the dynamics result from the simultaneous impact of a large number of operating factors producing patterns of correlation (fractal concept) *(6)*. Since most of the biological systems rely on a concerted action of large number of factors (degrees of freedom), we focus on fractal analysis of dynamic signals.

4. Basic Requirements of Experimental Methods

In order to capture complex dynamics in time and in space, the process needs to be sampled at a proper rate and resolution *(6, 14)*. Non-invasive experimental methods should be preferred in that their use does not disturb the delicate patterns in vivo. To capture the complex patterns, these tools should produce spatial and/or temporal data sets of high resolution. Tools of data analysis of complexity cannot compensate for an inadequate spatial and/or temporal sampling. In our experimental work, having kept pace with advancing technologies, optical imaging (laser Doppler flowmetry (LDF), laser speckle imaging (LSI), near infrared spectroscopy (NIRS)) and functional magnetic resonance imaging (fMRI) proved adequate in this regard.

The LDF and LSI methods monitor the blood perfusion in the brain tissue. Both methods rely on the frequency analysis of the back scattered laser light: LDF monitor evaluates the Doppler shift of the photons, whereas LSI analyzes the blurring effect of the photons' interference pattern called "speckle". NIRS and fMRI measure the intensity changes of the photon flux at given wavelengths, which are in the visible and radiofrequency ranges of the electromagnetic energy spectrum. All of these methods yield large data sets as required by the fractal analysis.

The detailed description of our animal and human experiments are found elsewhere *(15–18)*. In brief, in the animal studies, all rats were anesthetized and artificially ventilated. In the optical studies, the animal's head was fixed in a stereotaxic frame, the scalp was removed and the skull was thinned to transparency. The probe of the LDF monitor (Model MBF3D, Moor Instruments, Millwey, Axminster, Devon, UK) and an custom made LSI system *(12)* detected cerebral hemodynamics from the parietal somatosensory cortex. In the fMRI experiments, the blood oxygenation level dependent (BOLD) signal was collected on a 9.4T Bruker horizontal-bore spectrometer (Billerica, MA) with ^{1}H resonator surface coil (1.4 cm diameter) using a echo planar imaging (EPI) sequence, where the repetition delay and gradient echo time were 200 and 15 ms, respectively. The coronal slice (2 mm) slice thickness was selected at the level of the bregma. The NIRS experiments were conducted in awake human volunteers

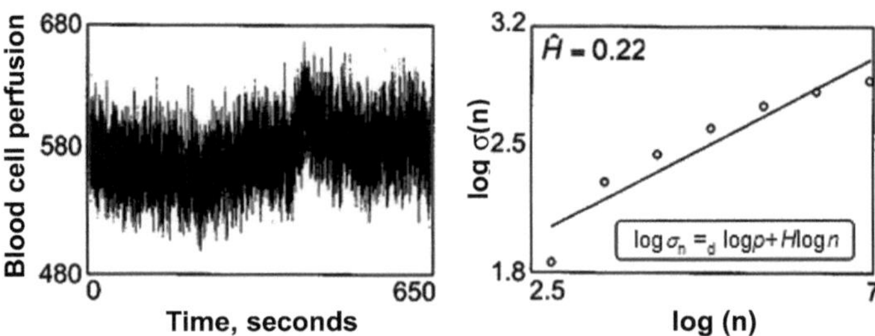

Fig. 2.3. *The concept of fractal analysis as demonstrated on a time series of blood perfusion sampled from the brain cortex by laser Doppler flowmetry.* A descriptive statistical parameter, the standard deviation, σ, of the time series data is calculated for a set of observation window, *n*. Log $\sigma(n)$ is plotted as a function of log(*n*) and the fractal descriptor is found according to the model of the scaled windowed variance method as the regression slope across the plotted data.

who provided their written consent to the study. The single and multiple probes of the spectroscope (NIRO-500, Hamamatsu, Japan) and the imager (courtesy of Dr. Briton Chance, University of Pennsylvania, PA, U.S.A.), respectively, were placed on the forehead of the volunteers seated in an armchair.

The aim of fractal analysis is to describe, in quantitative terms, the spatial or temporal correlation in the signals. Most of the tools have a descriptive statistical underpinnings configured according to the fractal concept aimed at identifying the presence of self-similarity, as a basic fractal property, in the statistical distribution of the data set *(6, 8)*. The novelty of the fractal approach is that traditional statistical tools are used in a novel manner to calculate descriptive statistical parameters in recursion for a range of scales of observation (scaling range) and thus, the power-law relationship between the scales and the observed descriptive statistical parameters (see insert in **Fig. 2.3**), becomes assessable by fitting across the data pairs (**Fig. 2.3**).

5. Pitfalls in the Fractal Analysis

It is crucial to realize that one cannot apply any one of the fractal tools to an available signal without an assessment of the signal variance. Fractal signals fall in two categories based on their variance *(6, 14)*. If variance is time dependent, the signal is referred to as nonstationary, while if it is not, it is called stationary. However, the visual appearance of these signal types can be very different in a certain resolution: the fluctuating nonstationary signal can wander from its trend line (e.g., see signal in **Fig. 2.3**), while the stationary signal generally fluctuates around a steady level (e.g., see signals in **Fig. 2.1**). A precise mathematical analysis is required to distinguish between these two signal types *(14)*. In the former case, if self-similarity and a related fractal property given by the power-law relationship can be demonstrated in the time domain, the signal is termed fractional Brownian motion (fBm). Its difference signal is fractional Gaussian noise (fGn), a stationary time series *(14)*. Based on these two types of signal classification, the fractal analytical tools can be split into three categories. Those that can be applied either to fGn (dispersional, autocorrelation, Hurst's rescaled range, etc.; see below) or fBm (scaled windowed variance, averaged wavelet coefficient, etc.; see below) only, and a third one that can handle both signal classes (spectral, detrended fluctuation, etc.; see below). If this correspondence between signal classification and compatible tool of analysis is disregarded or not fully appreciated, the fractal estimate obtained by the analysis can end up being correct or meaningless by chance *(6, 14)*.

Even if the proper tool of analysis is rationally selected, considerations have to be given to the nature and degree of error in the fractal estimate (Hurst exponent, *H*; spectral index, β; fractal dimension, *D*). The performance of the analytical methods can be evaluated on synthetic, ideal fractal signals generated in chosen length and at a preset degree of fractal correlation corresponding to a particular level of *H*. The numerical testing shown in **Fig. 2.4** utilized the method of Davies and Harte (DHM) *(19)* to generate fGn signals that were converted to their fBm counterparts by summation *(14)*. These signals of 2^8–2^{18} in length were analyzed by their class specific methods Disp (dispersional) *(20)*,

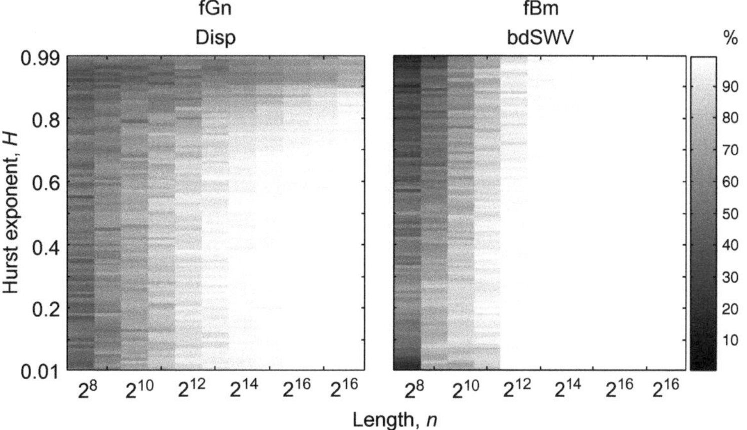

Fig. 2.4. *Evaluation of two robust fractal methods, the dispersion analysis, Disp, and the bridge detrended scaled windowed variance analysis as a function of signal length.* Note that even though these methods are considered robust, their performance strongly depends on the actual length of the time series data and the degree of fractal correlation in the data set as expressed by the actual level of the Hurst exponent, *H*. For further details, see text.

bdSVW (bridge detrended scaled windowed variance) *(21)* . The intensity-coded parameter is the percent estimates falling into the range of $H_{true} \pm 0.05$. The light areas on the plot designate the range of signal lengths where reliable values of H were obtained from the respective methods. It should be noted, however, that dynamic signals present themselves as complex only if the fractal methods for their analysis assume proper application of a complex protocol, as suggested earlier.

6. Temporal (1 Dimensional) Fractal Analysis

As first choice of applying fractal analysis to characterize temporal complexity of a dynamically fluctuating physiological parameter, we chose the cortical perfusion signal obtained by LDF from the brain cortex of anesthetized rats (**Fig. 2.3**) *(14, 16, 22)*. The dynamics in this signal is present at rest without any change in the functional state of the brain, and in spite of its random appearance, proved representative of the self-similar order of a temporal fractal. When the cerebrovascular system is challenged by graded arterial hypotension, the fractal correlation in the signal was found to persist *(16)*. In spite of the challenge in perfusion pressure, the fractal correlation of presumably the flowmotion pattern remains stable as demonstrated by Hurst exponents falling into the narrow range of 0.25–0.29. This can be taken as an indication of a strong self-organization *(23, 24)* of the regional flowmotion emerging from rhythms intrinsic to the vascular segments and proved quite immune to strong drop in perfusion pressure.

The presence of a fractal, spontaneous, hemodynamic fluctuation in the brain cortex was later demonstrated in humans *(25)*. Cerebral blood volume (CBV) was found to fluctuate in a manner dependent on age and gender alike *(15)* (**Fig. 2.5**). The motivation behind these studies was to demonstrate – using a noninvasive measurement of a cardinal cerebrovascular parameter (CBV by a single NIRS probe) and the above described set of fractal tools – whether age-related stiffening of the cerebral vessels would have any significant influence on the fluctuation patterns thought to be impacted by vasomotion and/or flowmotion within the arterial tree. An important finding of this study was that a gradual shift in this pattern as characterized by the spectral index began to develop in the young adult male, while this tendency was found absent in the age-matched female groups of the premenopausal female. However, when the female hormonal protection of the cerebral vasculature was presumably removed by menopause, the spectral index emerged to be more powerful (**Fig. 2.5**) *(15)*.

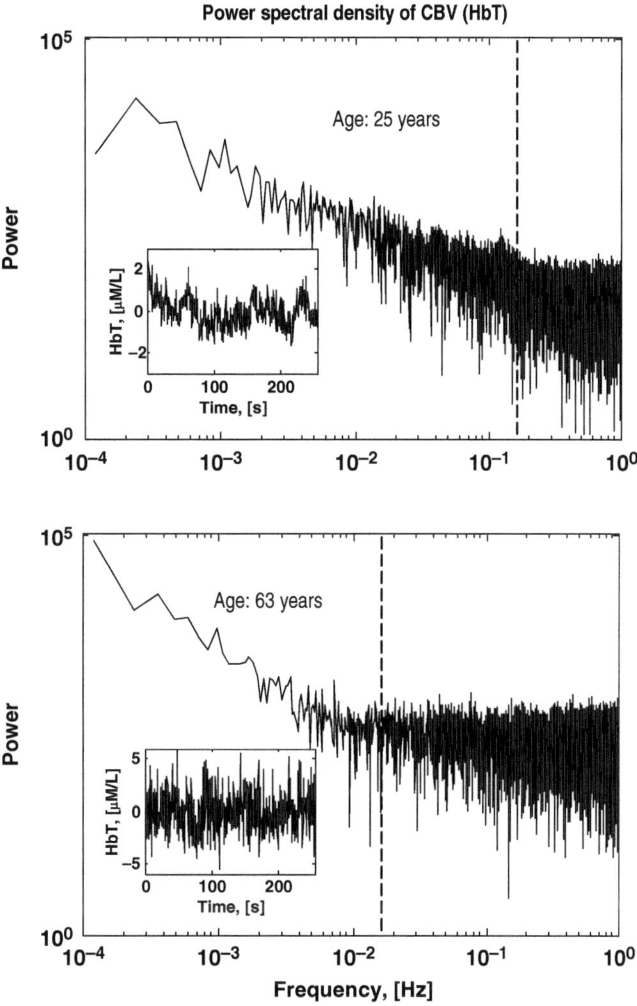

Fig. 2.5. *Cerebral blood volume (CBV) fluctuation patterns (inserts) and their corresponding bimodal spectra in a young and an old woman.* The correlated, i.e. self-similar, fractal pattern is seen in the low frequencies with $\beta = 1.27$ and 1.55, respectively, while an uncorrelated, noisy dynamics dominates above this frequency range with $\beta \approx 0$. The aging process increases the noisy domination in CBV dynamics by shifting the range of the noisy dynamic from 0.146–1 Hz to 0.014–1 Hz.

Given that low and very low frequency oscillations can likely be associated with segmental vasomotion rhythms, alterations observed in the CBV fluctuation spectrum via its spectral index or specific frequencies deviating from the fractal pattern can be indicative of a cerebral vascular segment(s) undergoing stiffening. The process of vessel wall stiffening may arise from loss of segmental control as manifested by the altered segmental rhythms.

7. Spatial (2 Dimensional) Surface Mapping of Fractal Parameter

The LDF and NIRS optical probes used in the animal and human studies *(14–16, 22, 25)* (**Figs. 2.3** and **2.5**, respectively) did not allow for sampling spatially mapped temporal complexity in these dynamic signals. Therefore, more advanced techniques were needed to gain access to the spatial dimension. Flow imaging with LSI *(12)* was chosen to acquire high resolution maps of red cell perfusion fluctuation in the brain cortex of anesthetized rats and an NIRS imager enabled us to map CBV fluctuations in 16 regions over the forehead and temporal regions in humans *(26)*. Both of these optical approaches generated surface maps of temporal dynamics. Due to photon diffusion geometry in the brain cortex and the particular imaging geometry of LSI, a single voxel represents a tissue volume of ~100 mm^3, while that in the NIRS imaging raster corresponds to ~3 cm^3. Our earlier findings demonstrated that fluctuations in perfusion or vascular volume are a fundamental and robust property of the cerebrovascular system *(15)*. **Figures 2.6** and **2.7** show topographic examples of temporal dynamics in the brain cortex where particular "micro" (**Fig. 2.6**) or "macro" (**Fig. 2.7**) region may have a temporal fractal correlation pattern of its own, different from its neighbors.

When spatial resolution of mapping is high as with LSI, these differences create a functional map of cerebrovascular microhemodynamics *(18)*, where arterial, parenchymal and venous segments of the circulation can be easily discerned from the parametric images of the Hurst exponent and, in addition, otherwise

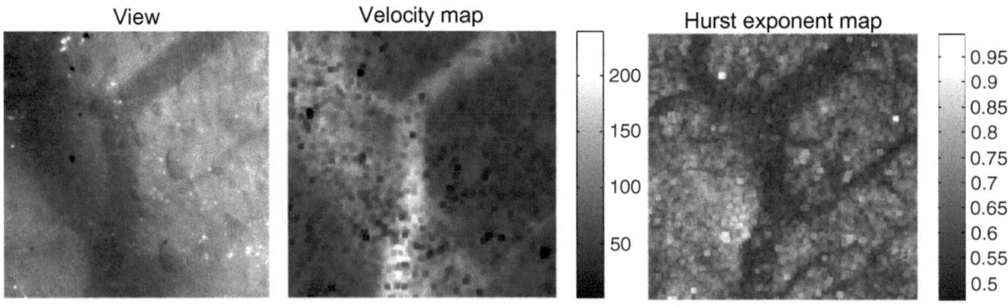

Fig. 2.6. *Spatially mapped temporal complexity of microflow dynamics over the rat brain cortex.* Exemplary parametric maps are shown for spatial intensity (*left*), average microflow velocity (*middle*), and the Hurst exponents, *H*, calculated for microflow fluctuations (*right*) in the rat brain cortex. The speckle contrast images (4096 images/150 s) were collected from the rat parietal cortex with a Coolsnap CF camera with 256×256 resolution (voxel size: 16 μm^2) at 655 nm. A 5×5 kernel was used to calculate flow velocity and the fractal parameter map was calculated from a stack of the 4096 flow velocity images. Note that the spatially mapped temporal fractal descriptor, *H*, forms a very articulate structural map of the pial and intraparenchymal circulatory networks indicating that vasomotion/flowmotion patterns captured in the corresponding perfusion time series are different and characteristic to these thoroughfare channels.

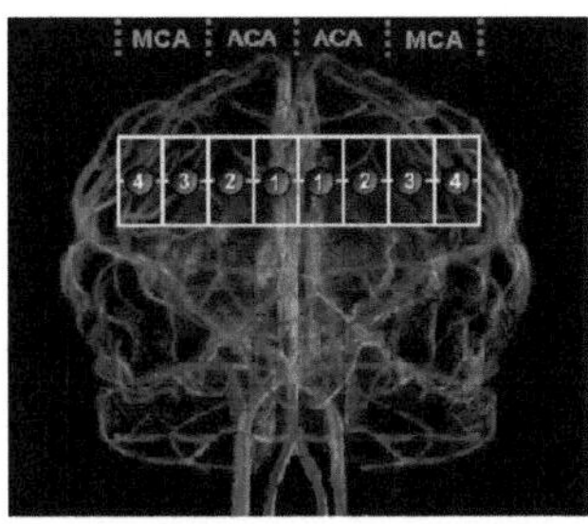

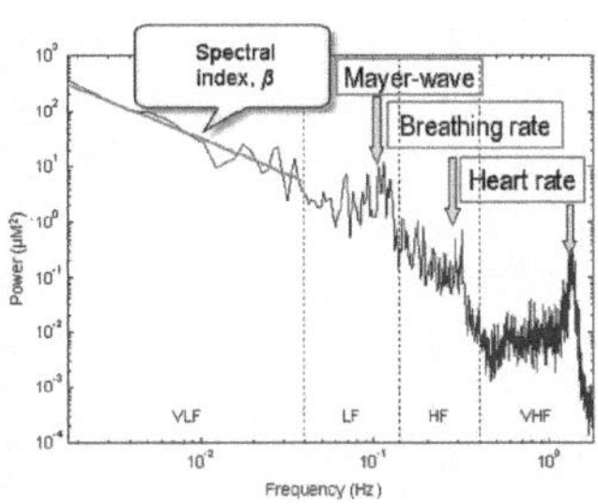

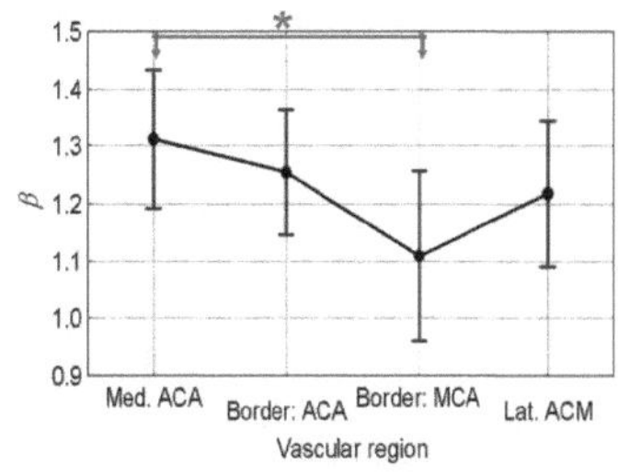

Fig. 2.7. *Spatially mapped temporal complexity of CBV dynamics over the human brain cortex.* CBV fluctuations were mapped by NIRS across the anterior cerebral and middle cerebral arterial territories (*left panel*) in humans in a square array of 16 regions. The total hemoglobin concentration, which is proportional to CBV, was measured on the forehead of 12 young (males 21.5±0.8, females 22.3±1.4 years old) volunteers, in sedentary position for 20 minutes. Data from the 16 channels of the NIRS LED imager were aggregated into four groups (medial ACA (anterior cerebral artery), lateral ACA, medial MCA (middle cerebral artery) and lateral MCA). Spectral analysis revealed peaked (heart rate, breathing rate, and Mayer waves) and self-similar frequency components in the signal (*middle*). The latter was found in the very low frequency range (VLF) and was assessed by fitting for the regression slope through the VLF range of the spectral estimates. The spectral index was found significantly lower for the MCA borderline zone indicative of a noisier fluctuation pattern equivalent to a less rhythmic vasomotion of the monitored small arteries along the borderline MCA arterial tree. (This study has been carried out with a significant contribution from Ms. Andrea Mile as a part of her student research project in Dr. Eke's laboratory.)

hidden parenchymal thoroughfare channels can also be revealed (**Fig. 2.6**). All this is possible as a result of a quantitative evaluation of temporal complexity of spatially mapped temporal data sets. One reasonable explanation of these findings is that the Hurst exponents within the extra- and the intraparenchymal vessels (pial arteries and veins, and capillaries, respectively) reflect two distinct flow dynamics. While flow in the arteries and veins is being influenced by rapid acceleration and deceleration resulting in anticorrelated patterns in the capillaries red blood cells flow in a single file ("plug flow") resulting in a strong correlation of hemodynamic events.

Regional dynamic signals such as those mapped across the frontotemporal regions of the human brain cortex (**Fig. 2.7**) do contain a wide range of frequencies *(15, 27, 28)*. Fluctuations within any given range of these frequencies (very high frequency, VHF; high frequency, HF; low frequency, LF; very low frequency, VLF) can be associated with an underlying energy source driving the fluctuations: heart rate at ~1.2 Hz, breathing rate at ~0.3 Hz, and Mayer-waves at ~0.11 Hz (**Fig. 2.7**, middle panel). Given the fact that these frequency ranges are not produced by the vasomotion activity of the monitored cerebral vessels, they need to be excluded from the fractal analysis if the intention is to characterize the complexity due to local factors such as vasomotion *(29)*. The negative of regression slope (spectral index, β) fitted across the VLF range of the CBV fluctuations is a measure of the fractal correlation in the signal. A smaller β value indicates the presence of a less correlated temporal fractal in any particular region. The

NIRS optode array was placed such that the optical signals were sampled from the frontal and temporal lobes which are respectively supplied by the anterior (ACA) and medial (MCA) cerebral arteries (**Fig. 2.7**, left panel). This arrangement allowed for mapping across the borderline zones in between adjacent lobes known to be susceptible to hypotensive provocation *(30)*, and ischemia, both having severe clinical consequences *(31, 32)*. As seen in the right panel of **Fig. 2.7**, β is significantly lower within the borderline MCA territory than in the surrounding areas of the brain cortex *(32)*. This regional difference in temporal complexity is interpreted as a manifestation of an altered vasomotion activity of the small arteries within the borderline MCA territory which is less synchronized than in the neighboring regions. A less synchronized vasomotion results in a less efficient "arterial pump", which otherwise could aid local perfusion when perfusion pressure drops to critically low levels and this can lead to susceptibility of these borderline zones to ischemic insults – a well known clinical scenario *(30, 31)*.

8. Towards 4 Dimensional (Spatiotemporal) Description of Brain Hemodynamics

Our optical approaches to assess dynamic complexity utilizing surface measurements by LDF, LSI, and NIRS imaging did not allow resolving the cerebrovascular dynamics in deeper regions of the brain. For this purpose, fMRI was pursued (**Fig. 2.8**) *(17)*. Out of the various MRI modalities, the gradient echo or BOLD weighted EPI maps were analyzed. Coronal scanning across the rat brain was carried out under normoxic, resting conditions when the main determinant of the signal is local perfusion or fluctuations in tissue oxygenation due to perfusion. As observed by the Hurst exponent parametric map (**Fig. 2.8**, right panel), similar to the optical modalities, the BOLD signal also shows spontaneous fluctuations at rest across the brain. H is around 0.5 indicative of an uncorrelated pattern (random) within the white matter where blood flow is lower than in gray matter, while H is much higher in the cerebral cortex and in deep nuclear regions where blood flow is much higher. It is worth noting, that the ranges of H are comparable within the cortical microareas of the brain, using either the optical imaging or magnetic resonance modalities. Both of these signal modalities are of a correlated fGn temporal fractal with $H \sim 0.7$. Spatial differences in β may have inference to regionally different metabolic activities, via differences in regional neuronal firing rates mapped into the regional spectra of the metabolically driven BOLD fluctuations.

Optical imaging and magnetic resonance are both suitable to collect high definition spatiotemporal data sets (volumes) from

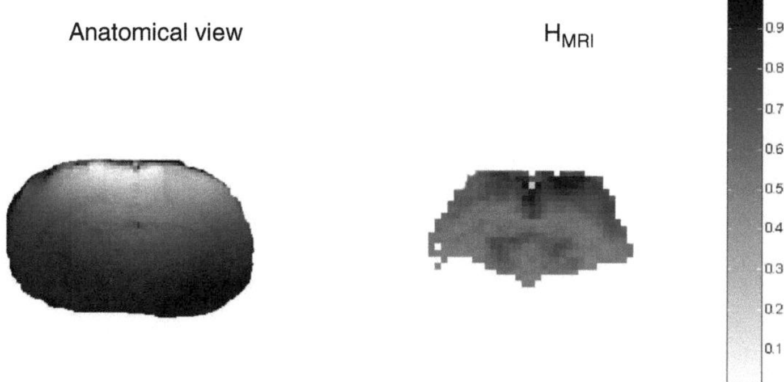

Fig. 2.8. *Spatially mapped temporal complexity of BOLD related signal across a coronal section of the rat brain.* The anatomical view of the brain section is seen on the left while the corresponding Hurst exponent map calculated from a set of gradient echo T_2-weighted EP images is shown on the right along with an intensity coded bar of *H*. The experiment was conducted on a 9.4T spectrometer (Bruker, Billerica, MA) using a ^{1}H resonator/surface-coil radio-frequency probe *(36)*. Gradient echo EPI data were acquired with $TR = 0.2$ s. The images were collected in matrix 64×64 spatial resolutions; the slice thickness was 2 mm, the volume of one voxel is ~0.15 μL. The slice position was selected at the level of Bregma. Similar to the signals shown in **Fig. 2.1**, these signals are also of fGn class. A value of $H < 0.5$ indicates the presence of an anticorrelated signal, $H = 0.5$ that of a pure random signal (i.e., no correlation), while $H > 0.5$ is associated with a correlated signal where a temporal event shows a given degree of dependence on values preceding it. Notice that BOLD related signal fluctuations are correlated in those areas (cortex, thalamus) where blood flow is high.

the brain (**Fig. 2.9**). However, for technical reasons, this is more easily achievable with optical imaging than fMRI. Optical reflectance imaging offers ways of mapping microregional blood flow, and blood volume from superficial layers of such volume in the brain cortex *(33, 34)*. When feature extraction is performed on the map of the latter, the pial network can be traced and its 2D complexity assessed by the calculation of box dimension *(35)* (**Fig. 2.9**, right panel). The same approach can be applied in 3D also to quantify the multi-dimensional complexity and its dynamics. In this respect, lessons learned from studies aimed at defining signal properties and evaluating fractal tools of analysis in the 2D spatial *(35)* and 1D temporal and frequency domains can be useful *(6, 14)*. Some questions that might be worth pursuing: What voxel size can be considered adequate given current findings with the time series data? Should anisotropy be considered as a complicating factor, and if so, how can its treatment be devised? Which of the known and tested fractal tools can be considered as candidates in 3D assessment of spatial complexity? Should one attempt to model 3D data sets similarly to the dichotomous fGn/fBm model of time series in order to enhance selection criteria for the analysis? These and other issues should certainly be made the subject of future research before their detailed use in animal or human experimentation begins.

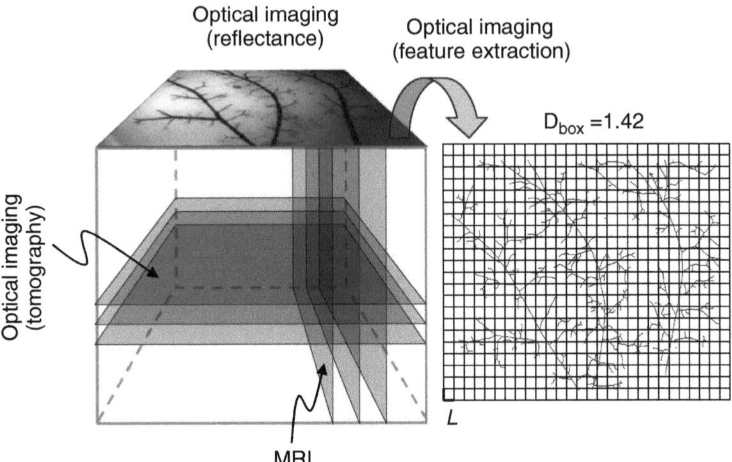

Fig. 2.9. ***Methodological possibilities in exploring 3D spatial and temporal complexity in the brain.*** The optical and MRI approaches can be instrumental in acquiring high resolution spatiotemporal data sets within the brain cortex (*left panel*) for biological functions, such as supply networks of arteries and dynamic distribution of flow arriving through these networks to a 3D array of tissue microareas. Surface reflectance imaging with feature extraction has proven adequate in assessing spatial complexity of the pial arterial network in 2D in the rat (*right panel*). This network proved a spatial fractal that distributes blood flow along the surface of the brain cortex via a self-similar pattern of vascular arborization with a box dimension *(35)*, D_{box}=1.42. This approach can be taken further into the depth of the brain applied to high-definition 3D data sets of structure and associated perfusion pattern in time. From a combination of the characterization of complexity in spatial and temporal dynamics of tissue perfusion, a better understanding of how global blood supply and local demand for adequate perfusion is matched can emerge.

9. Summary

We have presented a overview of the fractal characterization of spatial and temporal hemodynamic signals obtained from the mammalian brain using non-invasive methods with reasonable high temporal and/or spatial resolutions. We demonstrated that the fractal analysis of optical (LDF, LSI, NIRS) and fMRI data can capture the spatiotemporal complexity of cerebral hemodynamics. Fractal analysis proved that the seemingly random fluctuations are correlated according to the special order of self-similarity. We used the Hurst exponent (H) and the spectral index (β) to characterize the degree of correlation *(6, 14)*.

We showed in a rat model that the cortical perfusion is an anti-persistent fractal process, which remains very stable during hypotension *(16)* thus indicating strong self-organization *(23, 24)* of regional flowmotion emerging from intrinsic segmental flowmotion patterns. The capillary (parenchymal) and the small vessel blood flow can be discerned based on their respective fractal properties *(18)*. Moreover, H values for white and gray matters and within parenchymal regions of the rat brain cortex were distinctly different *(17)*.

In our human studies, fractal analysis revealed age and gender related stiffening of the cerebral vasculature *(15)*. We found that the hemodynamics of the borderline regions in between the frontal and temporal lobes (i.e. between the vascular territories of the ACA and MCA) has a fractal pattern which is noisier than their neighbors' suggesting that vascular segments in these areas contain less efficient "arterial pumps". This can potentially render these regions more vulnerable to hypoperfusion.

It seems that normal, physiological biological complexity presents itself within a restricted range of H or β values, while their values outside this range can indicate disease states or progress towards pathology.

Acknowledgments

The authors gratefully acknowledge the contribution of Ms. Andrea Mile to the human LED Imager study and Drs. Fahmeed Hyder, Ikuhiro Kida and Basavaraju G. Sanganahalli to the MRI study. This work was supported by the Hungarian Research Foundation (OTKA) by its grants T016953, T34122 and the High Performance Computing of the Hungarian National Information Infrastructure Development Program.

References

1. Norris, V., Cabin, A. and Zemirline, A. (2005) Hypercomplexity. *Acta Biotheor* **53**, 313–330.
2. Krone, G., Mallot, H., Palm, G. and Schuz, A. (1986) Spatiotemporal receptive fields: A dynamical model derived from cortical architectonics. *Proc R Soc Lond B Biol Sci* **226**, 421–444.
3. Grassberger, P. and Procaccia, I. (1983) Measuring the strangeness of strange attractors. *Physica D* **9**, 189–208.
4. Kaplan, D. and Glass, L. (1997) *Understanding Nonlinear Dynamics*. Springer-Verlag, New York.
5. Falconer, K. (1990) *Fractal geometry: Mathematical Foundations and Applications*. Wiley, Chichester, New York.
6. Eke, A., Herman, P., Kocsis, L. and Kozak, L. R. (2002) Fractal characterization of complexity in temporal physiological signals. *Physiol Meas* **23**, R1–38.
7. Beran, J. (1994) *Statistics for Long Memory Processes*. Chapman and Hall, New York.
8. Bassingthwaighte, J., Liebovitch, L. and West, B. (1994) *Fractal Physiology*. Oxford University Press, New York, Oxford.
9. Mandelbrot, B. (1983) *The Fractal Geometry of Nature*. WH Freeman, San Francisco.
10. Avnir, D., Biham, O., Lidar, D. and Malcai, O. (1998) Is the geometry of nature fractal? *Science* **279**, 39–40.
11. Mandelbrot, B. (1985) Self-affine Fractals and Fractal Dimension. *Phys Scripta* **32**, 257–260.
12. Dunn, A. K., Bolay, H., Moskowitz, M. A. and Boas, D. A. (2001) Dynamic Imaging Of Cerebral Blood Flow Using Laser Speckle. *J Cereb Blood Flow Metab* **21**, 195–201.
13. Eke, A. (2003) *Fractal, chaos, physiological complexity*. In: *Studia Physiologica* (Series Ed.: A. Juhasz-Nagy) **13**,1–157, Scientia Kiado, Budapest.
14. Eke, A., Herman, P., Bassingthwaighte, J. B., Raymond, G. M., Percival, D. B., Cannon, M., Balla, I. and Ikrenyi, C. (2000) Physiological

time series: Distinguishing fractal noises from motions. *Pflugers Arch* **439**, 403–415.

15. Eke, A., Herman, P. and Hajnal, M. (2006) Fractal and noisy CBV dynamics in humans: Influence of age and gender. *J Cereb Blood Flow Metab* **26**, 891–898.
16. Herman, P. and Eke, A. (2006) Nonlinear analysis of blood cell flux fluctuations in the rat brain cortex during stepwise hypotension challenge. *J Cereb Blood Flow Metab* **26**, 1189–1197.
17. Herman, P., Kida, I., Sanganahalli, B., Hyder, F. and Eke, A. (2005) Fractal correlation structure in fMRI data of rat brain. *J Cereb Blood Flow Metab* **25**, S379.
18. Herman, P., Kocsis, L., Portöro, I. and Eke, A. (2007) *Heterogenous response in CBF during autoregulation: A non-invasive laser speckle study in the rat brain cortex.* Brain'07. The 23rd International Symposium on Cerebral Blood Flow, Metabolism and Function, Osaka, Japan.
19. Davies, R. B. and Harte, D. S. (1987) Test for Hurst effect. *Biometrika* 74, 95–101.
20. Bassingthwaighte, J. B. and Raymond, G. M. (1995) Evaluation of the dispersional analysis method for fractal time series. *Ann Biomed Eng* **23**, 491–505.
21. Cannon, M., Percival, D. B., Caccia, D., Raymond, G. M. and Bassingthwaighte, J. B. (1997) Evaluating scaled windowed variance methods for estimating the Hurst coefficient of time series. *Physica A* **241**, 606–626.
22. Eke, A., Herman, P., Bassingthwaighte, J. B., Raymond, G. M., Balla, I. and Ikrenyi, C. (1997) Temporal fluctuations in regional red blood cell flux in the rat brain cortex is a fractal process. *Adv Exp Med Biol* **428**, 703–709.
23. Turcotte, D. L., Malamud, B. D., Guzzetti, F. and Reichenbach, P. (2002) Self-organization, the cascade model, and natural hazards. *Proc Natl Acad Sci USA* **99**, 2530–2537.
24. Waliszewski, P. (2005) A Principle Of Fractal-Stochastic Dualism And Gompertzian Dynamics Of Growth And Self-Organization. *Biosystems* **82**, 61–73.
25. Eke, A. and Herman, P. (1999) Fractal analysis of spontaneous fluctuations in human cerebral hemoglobin content and its oxygenation level recorded by NIRS. *Adv Exp Med Biol* **471**, 49–55.
26. Chance, B., Anday, E., Nioka, S., Zhou, S., Hong, L., Worden, K., Li, C., Murray, T., Ovetsky, Y., Pidikiti, D. and Thomas, R. (1998) A novel method for fast imaging of brain function, non-invasively, with light. *Opt. Express* **2**, 411–423.
27. Intaglietta, M. (1990) Vasomotion and flowmotion – physiological mechanisms and clinical evidence. *Vasc Med Rev* **1**, 101–112.
28. Schroeter, M. L., Schmiedel, O. and von Cramon, D. Y. (2004) Spontaneous low-frequency oscillations decline in the aging brain. *J Cereb Blood Flow Metab* **24**, 1183–1191.
29. Nilsson, H. and Aalkjaer, C. (2003) Vasomotion: Mechanisms and physiological importance. *Mol Interv* **3**, 79–89, 51.
30. Miklossy, J. (2003) Cerebral hypoperfusion induces cortical watershed microinfarcts which may further aggravate cognitive decline in Alzheimer's disease. *Neurol Res* **25**, 605–610.
31. Jorgensen, L. and Torvik, A. (1969) Ischemic cerebrovascular diseases in an autopsy series. 2. Prevalence, location, pathogenesis, and clinical course of cerebral infarcts. *J Neurol Sci* **9**, 285–320.
32. Yong, S. W., Bang, O. Y., Lee, P. H. and Li, W. Y. (2006) Internal and cortical border-zone infarction: Clinical and diffusion-weighted imaging features. *Stroke* **37**, 841–846.
33. Eke, A. (1993) Multiparametric imaging of microregional circulation over the brain cortex by video reflectometry. *Adv Exp Med Biol* **333**, 183–191.
34. Eke, A., Hutiray, G. and Kovach, A. G. (1979) Induced hemodilution detected by reflectometry for measuring microregional blood flow and blood volume in cat brain cortex. *Am J Physiol* **236**, H759–768.
35. Herman, P., Kocsis, L. and Eke, A. (2001) Fractal branching pattern in the pial vasculature in the cat. *J Cereb Blood Flow Metab* **21**, 741–753.
36. Hyder, F., Kida, I., Behar, K. L., Kennan, R. P., Maciejewski, P. K. and Rothman, D. L. (2001) Quantitative functional imaging of the brain: Towards mapping neuronal activity by bold fMRI. *NMR Biomed* **14**, 413–431.

Part II

Optical Imaging

Chapter 3

Wide-Field and Two-Photon Imaging of Brain Activity with Voltage- and Calcium-Sensitive Dyes

Ryota Homma, Bradley J. Baker, Lei Jin, Olga Garaschuk, Arthur Konnerth, Lawrence B. Cohen, Chun X. Bleau, Marco Canepari, Maja Djurisic and Dejan Zecevic

Abstract

This chapter presents three examples of imaging brain activity with voltage- or calcium-sensitive dyes. Because experimental measurements are limited by low sensitivity, the chapter then discusses the methodological aspects that are critical for optimal signal-to-noise ratio. Two of the examples use wide-field (1-photon) imaging and the third uses two-photon scanning microscopy. These methods have relatively high temporal resolution ranging from 10 to 10,000 Hz.

The three examples are the following: (1) Internally injected voltage-sensitive dye can be used to monitor membrane potential in the dendrites of invertebrate and vertebrate neurons in in vitro preparations. These experiments are directed at understanding how individual neurons convert the complex input synaptic activity into the output spike train. (2) Recently developed methods for staining many individual cells in the mammalian brain with calcium-sensitive dyes together with two-photon microscopy made it possible to follow the spike activity of many neurons simultaneously while in vivo preparations are responding to stimulation. (3) Calcium-sensitive dyes that are internalized into olfactory receptor neurons in the nose will, after several days, be transported to the nerve terminals of these cells in the olfactory bulb glomeruli. There, the population signals can be used as a measure of the input from the nose to the bulb.

Three kinds of noise in measuring light intensity are discussed: (1) Shot noise from the random emission of photons from the preparation. (2) Extraneous (technical) noise from external sources. (3) Noise that occurs in the absence of light, the dark noise. In addition, we briefly discuss the light sources, the optics, and the detectors and cameras.

The commonly used organic voltage and ion sensitive dyes stain all of the cell types in the preparation indiscriminately. A major effort is underway to find methods for staining individual cell types in the brain selectively. Most of these efforts center around fluorescent protein activity sensors because transgenic methods can be used to express them in individual cell types.

Fahmeed Hyder (ed.), *Dynamic Brain Imaging: Multi-Modal Methods and In Vivo Applications, vol. 489*

DOI 10.1007/978-1-59745-543-5_3 Springerprotocols.com

Key words: Two-photon microscopy, one-photon (wide-field) microscopy, voltage-sensitive dyes, calcium-sensitive dyes, neural networks.

1. Introduction

An optical measurement of brain activity using a molecular probe can be beneficial in a variety of circumstances. An obvious advantage of any imaging method is the possibility of simultaneous measurements from many locations. This is especially important in the study of the nervous system where many parts of an individual cell, or many cells, or many regions, are simultaneously active. In addition, optical recording offers the possibility of recording from processes which are too small or fragile for electrode recording.

An important advantage of optical methods for measuring activity is their relatively high spatial resolution. In one of the examples presented below, the resolution is on the order of a few microns; in another, it is easy to distinguish the activity of individual cells. However, optical recordings are generally limited to areas of the brain that can be visualized and, in these areas, limited to a recording depth of about 0.5 mm.

Using a molecular probe has advantages (and disadvantages) over intrinsic signals. For optical imaging of membrane potential, the use of molecular probes has resulted in greatly increased signal size in comparison to intrinsic signals. Furthermore, with the introduction of protein activity sensors, there is now the possibility for cell type specificity. In addition, many of the molecular probes are fast with a time resolution in microseconds or milliseconds but others, especially the protein-based biosensors, are substantially slower. Finally, there are novel probes that are sensitive to various ions and metabolic intermediates. With these advantages comes the responsibility for demonstrating that pharmacological and photodynamic effects are minimal (see below).

1.1. Voltage-Sensitive Dyes

Several different optical properties of membrane-bound organic dyes are sensitive to membrane potential including fluorescence, absorption, dichroism, birefringence, Förster resonance energy transfer (FRET), non-linear second harmonic generation, and resonance Raman absorption. Similarly, another group of organic dyes are sensitive to ion concentrations. However, because the vast majority of applications have involved fluorescence or absorption measurements, these will be the only subjects of this review.

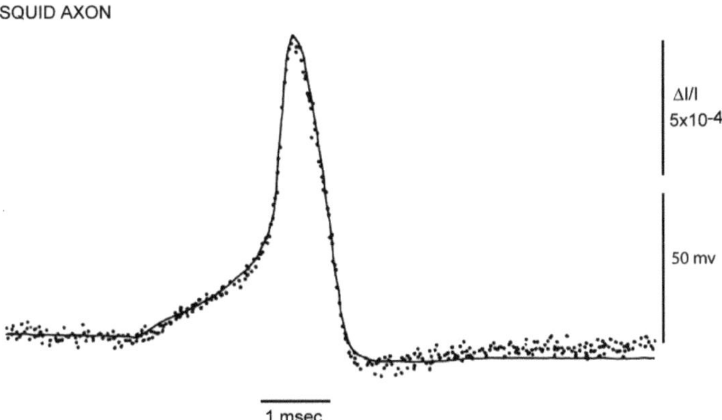

Fig. 3.1. Changes in absorption (dots) of a giant axon stained with a merocyanine dye, XVII (**Fig.** 3.2), during a membrane action potential (smooth trace) recorded simultaneously. The change in absorption and the action potential had the same time course. In this and subsequent figures, the size of the vertical line represents the stated value of the fractional change in intensity of transmitted (ΔI/I), or fluorescent (ΔF/F) light. The response time constant of the light measuring system was 35 μsec; 32 sweeps were averaged. Modified from Ross et al. *(59).*

The voltage-sensitive dye signals described in this chapter are "fast" signals *(1)* that are presumed to arise from membrane-bound dye; they follow changes in membrane potential with a time course that is rapid compared to the rise time of an action potential. **Figure 3.1** illustrates the kind of result that is used to define a voltage-sensitive dye. In a model preparation, the giant axon from a squid, these optical signals are fast, following membrane potential with a time constant of <10 μsec *(2)* and their size is linearly related to the size of the change in potential (e.g. *(3)*). Thus, these dyes provide a direct, fast, and linear measure of the change in membrane potential of the stained membranes.

Several voltage-sensitive dyes (e.g., **Fig. 3.2A**) have been used to monitor changes in membrane potential in a variety of preparations. This figure illustrates four different chromophores (the merocyanine dye, XVII, was used for the measurement illustrated in **Fig. 3.1**). For each chromophore, approximately 100 analogues have been synthesized in an attempt to optimize the signal-to-noise ratio that can be obtained in a variety of preparations. (This screening was made possible by synthetic efforts of three laboratories: Jeff Wang, Ravender Gupta and Alan Waggoner then at Amherst College; Rina Hildesheim and Amiram Grinvald at the Weizmann Institute; and Joe Wuskell and Leslie Loew at the University of Connecticut Health Center.) For each of the four chromophores illustrated in **Fig. 3.2**, there were 10 or 20 dyes that gave approximately the same signal size on squid axons *(3)*. However, dyes that had nearly identical signal size on squid axons could have very different responses in other

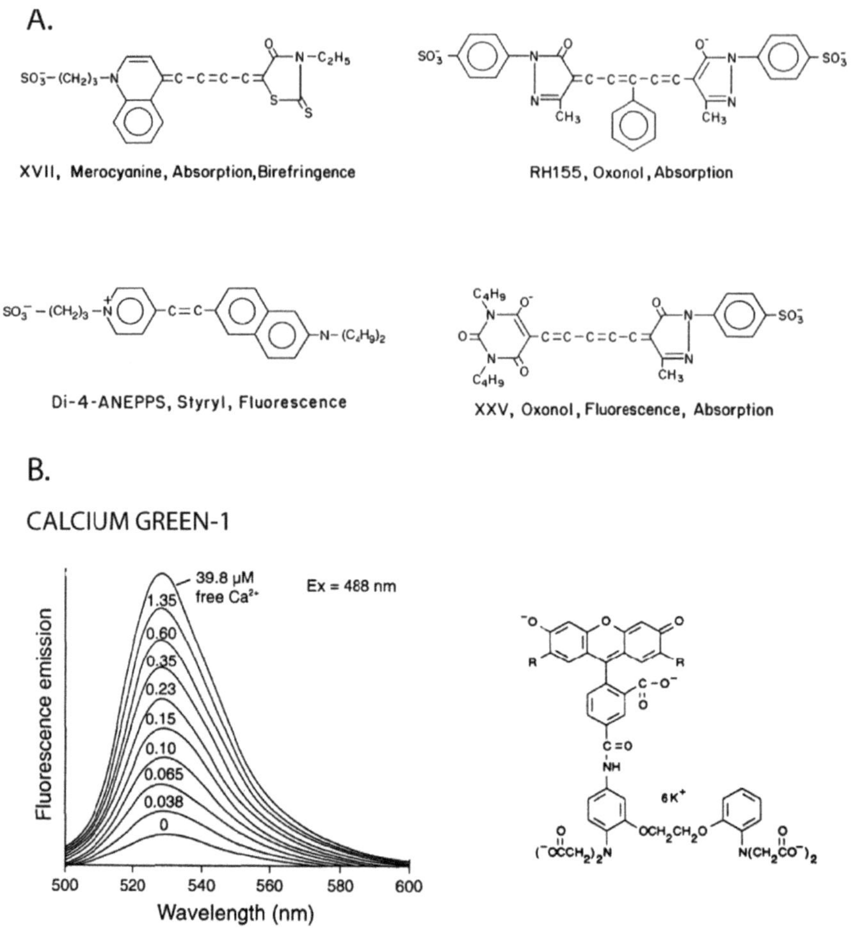

A.

XVII, Merocyanine, Absorption, Birefringence

RH155, Oxonol, Absorption

Di-4-ANEPPS, Styryl, Fluorescence

XXV, Oxonol, Fluorescence, Absorption

B.

CALCIUM GREEN-1

Fig. 3.2. A. Examples of four different chromophores that have been used to monitor membrane potential. The oxonol dye, RH155, and its analog RH482 (not shown), are commercially available as NK3041 and NK3630 from Hayashibara Biochemical Laboratories Inc./Kankoh-Shikiso Kenkyusho Co. Ltd., Okayama, Japan. The oxonol, XXV (WW781) and styryl, di-4-ANEPPS, are available commercially as dye R-1114 and D-1199 from Invitrogen Molecular Probes, Junction City, OR. Merocyanine dyes, including XVII (WW375, NK2495), and its analog, NK2761, need negotiation with Hayashibara Biochemical Laboratories. **B**. The chemical structure (*right*) and the emission spectrum as a function of calcium concentration of Calcium Green-1. The conjugate with the 10 kD dextran is available commercially from Molecular Probes as C-3713. Data taken from the Handbook of Fluorescent Probes and Research Chemicals, sixth edition, Molecular Probes.

preparations, and thus, tens of dyes usually have to be tested to obtain the largest possible signal. A common problem was that the dye did not penetrate through connective tissue or along intercellular spaces to the membrane of interest.

The following rules-of-thumb seem to be useful: First, each of the chromophores is available with a fixed charge which is either a quaternary nitrogen (positive) or a sulfonate (negative). Generally, the positively charged dyes have given larger signals in vertebrate preparations. Second, each chromophore is available with carbon chains of various lengths. The more hydrophilic dyes (methyl or ethyl) work best if the dye has to penetrate through a compact tissue (vertebrate brain) or needs to travel a long distance in a dendritic tree.

A number of studies have been conducted to determine the molecular mechanisms that result in potential-dependent optical properties. The available evidence supports three different mechanisms (for different dyes): Dipole rotation *(5)*, electrochromism *(2)*, and a potential sensitive monomer-dimer equilibrium *(4)*. In many cases, it has been possible to show that pharmacological and photodynamic effects are small (e.g. *(6–12)*).

1.2. Calcium-Sensitive Dyes

Figure 3.2B shows the chemical structure of a calcium-sensitive dye, Calcium Green-1, together with a plot of the fluorescence spectrum as a function of the free calcium concentration. This dye signal reaches 50% of its maximum at a calcium concentration of about 0.2 μM. In contrast to the voltage-sensitive dyes, the calcium dyes are located intracellularly. The dye is presumed to be in the cytoplasm and to report changes in the calcium concentration in the cytoplasm, although some of the dye may be in other intracellular compartments. These dyes are slower to respond and, because they also act as buffers of calcium, the calcium signal in the presence of dye may substantially outlast the change in calcium concentration that would occur in the absence of the dye *(13, 14)*.

We begin with examples obtained from measurements addressing three quite different neurobiological problems. In one example where the light level was low, the camera was a fast, back-illuminated, CCD camera with 80×80 pixel spatial resolution. In the second example, cellular resolution could only be obtained using two-photon scanning microscopy. In the third example, a slower CCD camera with 256×256 pixel spatial resolution was used. The optical signals in the three examples are not large, they represent fractional changes in light intensity ($\Delta I/I$) of from 5×10^{-3} to 5×10^{-1} and have modest signal-to-noise ratios. Nonetheless, they can be measured with an acceptable signal-to-noise ratio after attention to details of the measurement which are described in the second part of the chapter.

Figure 3.3 illustrates the three qualitatively different applications in neurobiology where imaging has been useful. First (left panel), in order to know how a neuron integrates its synaptic input into its action potential output, one needs to be able to measure membrane potential everywhere where synaptic input occurs and at the places where spikes are initiated. Second (middle panel), in order to understand how a nervous system generates a behavior, it is important to measure the action potential activity of many of the participating neurons. Third (right panel), responses to sensory stimuli and generation of motor output in the vertebrate brain are often accompanied by synchronous

PARTS OF A NEURON

Many Detectors - One neuron

Potential changes in dendrites.
Microinject dye: stain one neuron.

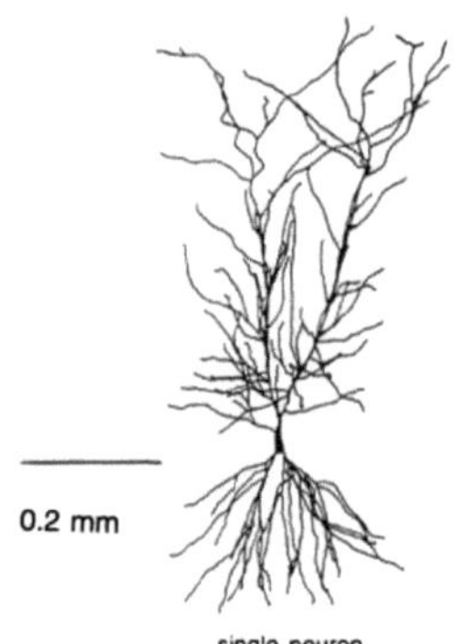

INDIVIDUAL NEURONS

2-photon microscopy

Follow spike activity of many individual neurons.

POPULATION SIGNALS

One Detector - Many Neurons

Signals are population average.
Vertebrate brain.
Bath applied dye. **Transported dye.**

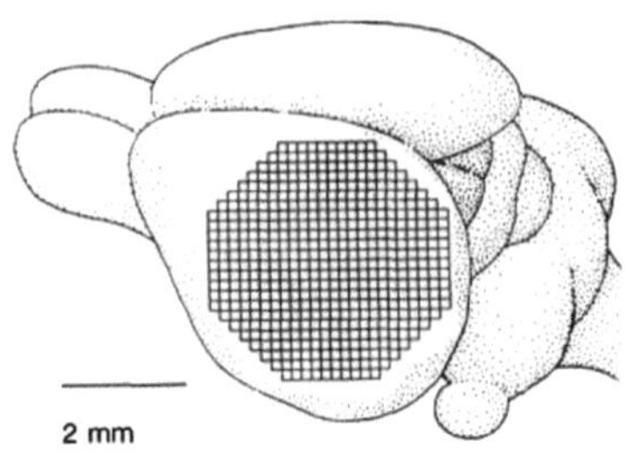

Fig. 3.3. Schematic drawings of the three kinds of measurements described as examples. *Left*, an individual cortical hippocampal CA1 pyramidal cell. Each pixel of an 80×80 pixel CCD camera would receive light from a small part of the dendrite, axon, or cell body of the neuron. An optical measurement of membrane potential would provide important information about how the neuron converts its synaptic input into its spike output. *Middle*, a drawing of cells in a vertebrate cortex. 2-photon imaging allows imaging of the fluorescence of many individual neurons stained with a calcium sensitive dye. *Right*, a vertebrate brain with a superimposed pixel array. Each pixel of the array would receive light from thousands of cells and processes. The signal would be the population average of the change in membrane potential in those cells and processes.

activation of many neurons in wide-spread brain areas; optical imaging allow simultaneous measurement of population signals from many areas. In these three instances, optical recordings have provided kinds of information about the function of the nervous system that were previously unobtainable.

2. Three Examples

2.1. Processes of an Individual Neuron (Fig. 3.3, Left Panel)

Understanding the biophysical properties of single neurons and how they process information is fundamental to understanding how the brain works. At present, however, the detailed functional structure of nerve cells is not fully understood. Part of the explanation for this incomplete understanding is that neurons are exceedingly complex. It is widely recognized that dendritic membranes of many vertebrate CNS neurons contain active conductances such as voltage-activated Na^+, Ca^{2+} and K^+ channels (e.g. *(15–20)*). An important consequence of active dendrites is that regional electrical properties of branching neuronal processes will be extraordinarily complex, dynamic, and, in the general case, impossible to predict in the absence of detailed measurements. To obtain such a measurement, one would, ideally, like to be able to monitor, at multiple sites, subthreshold events as they travel from the sites of origin on neuronal processes and summate at particular locations to influence action potential initiation. This can be achieved by using voltage-imaging. Recently, the sensitivity of intracellular voltage-sensitive dye technique has been dramatically improved allowing direct recording of subthreshold and action potential signals from the neurites of individual neurons *(10–12, 21–24)*.

2.1.1. Dye Screening

We carried out experiments to develop a methodology for optical monitoring of the generation and spread of subthreshold synaptic potentials and action potentials in neuronal processes using selective staining of individual invertebrate and vertebrate neurons by intracellular application of membrane impermeant voltage-sensitive dyes *(25)*. First, we tested analogues of the most successful absorption dyes known from extracellular application because absorption dyes were substantially better in terms of photodynamic damage as compared to fluorescence dyes. A total of 30 dyes were tested. The best results were obtained with two positively-charged pyrazolone-oxonol dyes (designated JPW1177 and JPW1245), and two negatively-charged merocyanine dyes (designated WW375 and JPW1124). However, the signal-to-noise ratio for the best dyes in these series varied from 1 to 4 in single trials. We concluded that this mode was probably not sensitive enough to allow monitoring neuronal signals from distal processes.

Fluorescence measurements are more effective than absorption measurements in the situation where the image of the object (e.g. thin process) is much smaller than the size of a photodetector *(6, 26)*. We found that JPW1114 was better than the three other fluorescent styryl molecules we tested (RH461, RH437 and JPW1063) as determined by signal-to-noise ratio. There was an improvement in sensitivity (as defined by the signal-to-noise ratio) by a factor of about 50 over previously reported (e.g. *(27, 28)*) absorption and fluorescence signals from neuronal processes stained by either intra- or extracellular application of dyes *(25)*. The obtained sensitivity depends on using fluorescence measurements and selective staining of only one neuron within a ganglion *(29)*, and, on finding a dye that gives relatively large signals when applied internally. The fluorescent voltage-sensitive styryl dye JPW1114 (also called di-2-ANEPEQ), is a doubly positively charged analog of the ANEPPS series of lipophilic voltage sensitive dyes that is still sufficiently water soluble to be used for microinjection. JPW1114 (available from Invitrogen - Molecular Probes Inc. as D6923), together with its close analog JPW3028 (di-1-ANEPEQ), is presently the optimal voltage indicator for intracellular application in recording from neuronal processes of individual neurons in brain slices *(12, 23)*.

2.1.2. Methods

We use a stationary stage upright microscope equipped with two camera ports. One port has a standard high spatial resolution CCD camera for infrared differential interference contrast video-microscopy. The other camera port has a fast data acquisition camera with relatively low spatial resolution (80 × 80 pixels) but outstanding dynamic range (14 bits; NeuroCCD-SM, RedShirtImaging LLC, Decatur, GA). The analysis and display of data are carried out using the NeuroPlex program (RedShirtImaging) written in IDL (ITT Visual Information Solutions, Boulder, Colorado) running on a Windows XP computer. The brain slice is placed on the stage of the microscope and the fluorescent image of the stained cell projected by a water immersion objective via an optical coupler (0.09–0.11 X; RedShirtImaging) onto the CCD chip. A 250 W xenon, short-gap, arc lamp (Osram, XBO 250 W/CR ORF) powered by a low ripple power supply (Model 1700XT/A, Opti-Quip, Highland Mills, NY) is the source of excitation light.

With the above dye, the best signals are obtained using an excitation interference filter of 520±45 nm, a dichroic mirror with the central wavelength of 570 nm, and a 610 nm barrier filter (a Schott RG610). With the sensitivity of the intracellular voltage-sensitive dyes used (1–6% ΔF/F per 100 mV in recording from dendritic processes), relatively good signal-to-noise ratios can be obtained in single-trial recordings with modest spatial averaging (4–10 pixels) from dendritic regions that are about 300 μm away from the soma. Modest signal-averaging (4 trials) is often used to improve the signal-to-noise ratio further.

In experiments where calcium and voltage imaging is carried out from the same neuron, Ca^{2+} signals are recorded at the frame rate of 500 Hz using an excitation interference filter of 380±15 nm, a dichroic mirror with central wavelength of 400 nm and a 450 nm barrier filter. Calcium fluorescence transients are expressed as $\Delta F/F=(F_0-F)/F_0$, where F is the stimulated fluorescence and F_0 is the resting fluorescence. The value of F is calculated after subtracting auto-fluorescence determined for an area far from the loaded dendrite. Due to uneven illumination of the field of view and due to differences in auto-fluorescence between different parts of the slice, background fluorescence values vary between different regions on the same slice typically by a factor of up to 3. However, when signals from the same dendritic regions are compared under different conditions, possible errors in calculating ΔF/F introduced by uncertainties about actual contribution of auto-fluorescence to the resting light intensity can be neglected. We did not detect any toxicity of $[Ca^{2+}]_i$ indicator bis-fura-2; loading of neurons did not have detectable pharmacological effects (toxicity of the indicator dye in the absence of excitation light) on the electrical behavior of neurons. During calcium imaging, the exposure of the preparation to the excitation light was limited to the recording periods of 400 ms.

Even after 50–100 recording trials, typically collected for one experiment, no change has been found in the amplitude or the time course of optical or electrical signals due to photodynamic damage (toxicity induced by the interaction of the indicator dye with high intensity light).

The voltage-sensitive dyes, JPW 1114 or JPW 3028, have little or no pharmacological effect when applied at functional concentrations to both invertebrate *(22, 25)* and vertebrate neurons *(11, 12, 21, 23, 24)*. We also demonstrated, for invertebrate neurons, rat neocortical layer V and hippocampal CA1 pyramidal neurons, and mitral cells of the olfactory bulb, that photodynamic damage during optical recording was not present if the exposures of the dendritic arbor to excitation light were kept relatively short (100 ms) and if successive trials were separated by dark intervals lasting several minutes *(10, 12, 21, 23, 24)*.

2.1.3. Example Results

The current sensitivity of voltage imaging allowed experiments in which subthreshold, synaptic potentials were monitored at the site of origin, in the dendritic tuft of mitral cells in the olfactory bulb of the rat. **Figure 3.4-I** illustrates presently available signal-to-noise ratio obtained at the desired spatial resolution. Our results showed consistently that the highest sensitivity was achieved in recording from thin terminal branches in the mitral cell tuft, likely due to the favorable surface-to-volume ratio of these structures. The effect is based on the fact that, in addition to the plasma membrane, the cytoplasm and the internal membranes (e.g. ER membranes, mitochondria) of the stained cell also contain the dye. Dye bound to internal membranes does not contribute to the signal because it is not in the membrane that changes potential. This internal dye, nonetheless, contributes to the resting fluorescence and degrades the fractional change in light intensity ($\Delta F/F$). In addition, the higher sensitivity of recording from the tuft is accompanied by an improved signal-to-noise ratio (for the same $\Delta F/F$) based on a large membrane surface area of the terminal dendritic arborization. Larger membrane surface area results in the larger amount of the membrane bound dye and higher light intensity; higher light intensity, in turn, results in the higher signal-to-noise ratio under our experimental conditions (dominant noise is statistical, shot noise). This is a fortunate feature of voltage imaging because thin dendritic branches receive all of the excitatory synaptic inputs in mitral cells, are not accessible to direct electrical measurement, and are likely to be a key compartment for signal integration.

Figure 3.4-I illustrates the spatial resolution and the sensitivity of voltage imaging from small dendritic structures (single pixels received light from 4 × 4 μm areas in the object plane).

I

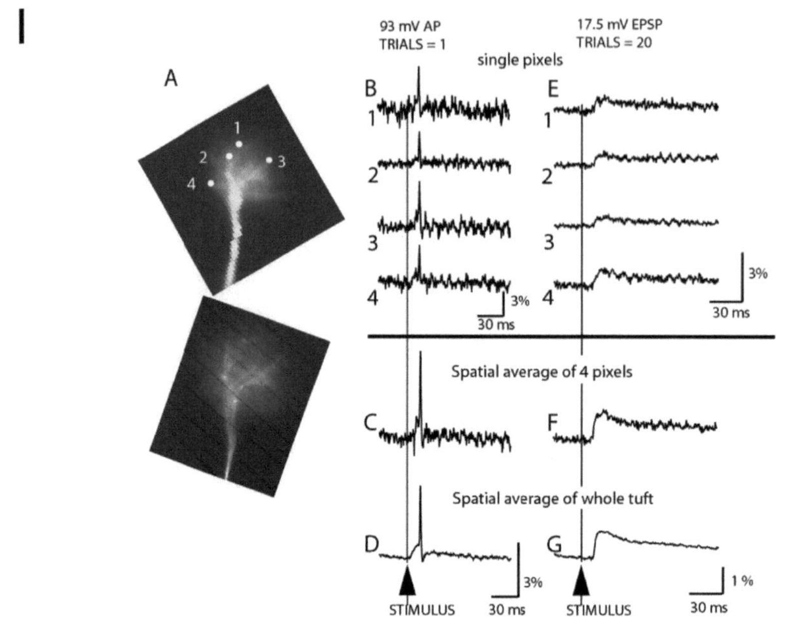

II

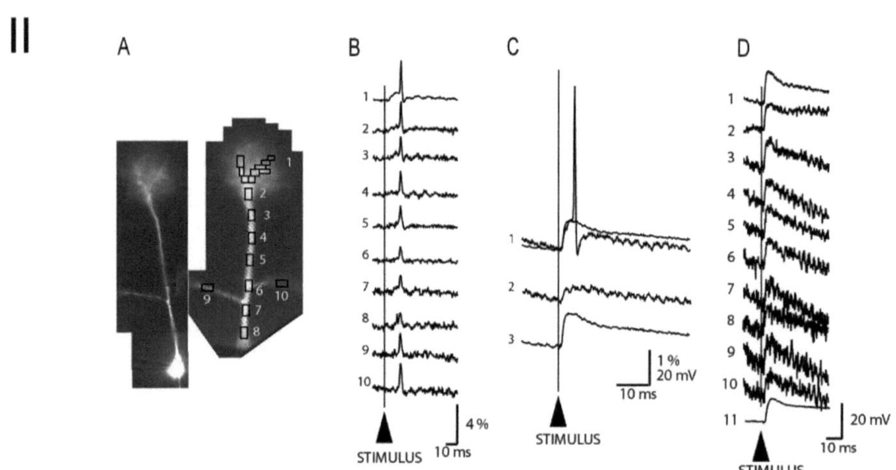

Fig. 3.4. Spatial resolution and sensitivity of voltage-imaging from neuronal processes. (**I**) (**A**) Low resolution (*upper*) and high resolution (*lower*) image of the terminal dendritic tuft. (**B**) Recordings of action potential signals from 4 individual pixels/locations on the dendritic tuft in single-trial measurements (no averaging). (**C**) The signal-to-noise ratio with spatial averaging of 4 pixels/locations from B. (**D**) The signal-to-noise ratio with spatial averaging of the entire tuft area. (**E**) An evoked EPSP recorded from 4 individual locations on the dendritic tuft; 20 trials averaged. (**F**) Spatial average of 4 pixels shown in E. (**G**) Spatial average of the entire tuft area. (**II**) Attenuation of the EPSP along the primary dendrite. (**A**) High and low resolution image of a mitral cell. (**B**) Single-trial optical recordings of action potential signals from ten different regions on the primary and oblique dendrites. The rectangles indicate pixels that were averaged to obtain traces 1–10. The amplitude of the action potential measured from the soma by a patch-electrode was 93 mV (resting potential-to-peak). The amplitude of optical signals ($\Delta F/F$) corresponding to a spike of constant size varied with the location *(1–10)* due to unequal sensitivity of the optical measurement from different locations. The sensitivity profile, determined from action potential measurements, provides a calibration for the conversion of optical signals to membrane potential changes. In the measurements shown, the calibration factors (from the tuft to the most proximal part of the primary dendrite) are: 1; 0.74; 0.56; 0.63; 0.53; 0.38; 0.40; 0.32 and 0.47 for the left and 0.59 for the right oblique

The images of the dendritic tuft of a mitral cell (upper image – low resolution; lower image – high resolution) are shown in (A). Single pixel recordings of action potential signals from four individual locations on the dendritic tuft in single-trial measurements (no averaging) are shown in B. Action potential signals from individual locations could easily be resolved. An improvement in the signal-to-noise ratio with minimal spatial averaging (signals from the four pixels shown in B is illustrated in panel C). A dramatic improvement with more extensive spatial averaging (the whole tuft; 184 pixels) made the optical recording shown in D appear similar in signal-to-noise ratio to electrode measurements. Measurements of the same type were carried out during an evoked synaptic potential at the dendritic tuft (**Fig. 3.4-I, E, F and G**). Because excitatory post-synaptic potential (EPSP) was about 5 times smaller in amplitude than an action potential, we averaged 20 individual trials to obtain a signal-to-noise ratio similar to the one corresponding to action potentials (the signal-to-noise ratio increases with the square root of the number of averages). Panels E, F and G of **Figure 3.4-I** again illustrate the striking improvement in the sensitivity of recording with both spatial averaging and averaging of individual repetitive events. This sensitivity was routinely obtained from mitral cells that were positioned relatively close to the surface of the slice (not deeper than 70 μm) and that had its primary dendrite and tuft close to the focal plane.

Figure 3.4-II illustrates a representative example from a series of optical measurements designed to determine the characteristics of the evoked EPSP at the site of origin and its attenuation along the primary dendrite. This information was not possible to obtain before using electrode measurements because of the small diameter of terminal dendritic branches. First, a spike was elicited and optically recorded from multiple sites in a single-trial measurement to serve as a calibration standard (**Fig. 3.4-II B**, traces 1–10). The calibration of optical data from multiple sites in terms of membrane potential requires a voltage

Fig. 3.4. (continued) branch. (**C**) Calibration of optical signals (ΔF/F) in terms of membrane potential (mV). All traces represent the average output of the same group of 35 pixels that receive light from the dendritic tuft (rectangles over the tuft in A). Twenty trials were averaged to improve the signal-to-noise ratio. Trace 1 shows an optical signal corresponding to an action potential of 93 mV used as a calibration standard. Trace 2 is a subthreshold EPSP signal evoked by olfactory nerve stimulation and calibrated to be 9 mV in amplitude at the site of origin (tuft). Trace 3 shows a threshold EPSP signal recorded from the tuft after the action potential was blocked by intracellular application of QX-314. In the measurement shown in trace 3, the stimulus delivered to the olfactory nerve was identical to the one applied in the measurement shown in trace 1. The EPSP signal in trace 3 (also superimposed over trace1) overlaps closely with the local response preceding a spike in trace 1 indicating that the spike is eliminated by QX-314 while the synaptic potential was unchanged. (**D**) The amplitude of EPSP signals on a voltage scale at 10 recording sites. The calibration of optical signals shows that EPSP at location 8, only 15 μm away from soma, is 13.2 mV in amplitude. Electrical recording from the soma (trace 11 in (**D**)) was similar (12.5 mV). Traces *(9, 10)* are signals from left and right oblique dendritic branches.

signal that has a known amplitude at all recording locations on neuronal processes *(29)*. In a general case, such a signal is not available. However, the action potential amplitude in mitral cells is constant along the entire length of the primary dendrite, as established by direct electrical measurements *(30–32)* and, thus, can be used as a calibration signal.

A typical threshold EPSP recorded from the dendritic tuft is shown in **Fig. 3.4-II** C. Trace 1 shows the EPSP that was suprathreshold for spike initiation in the control solution and an EPSP alone after spike was blocked with QX-314 to facilitate averaging synaptic potentials. The amplitude and the shape of the EPSP were mostly unaltered by the drug. The EPSP was monitored optically and calibrated in terms of membrane potential along a 260 μm primary dendrite. The optical signals are shown on a voltage scale in **Fig. 3.4-II** D. The recordings from 8 locations on the primary dendrite (traces 1–8), including the terminal tuft, and two sites on oblique dendrites (traces 9 and 10) are shown together with the electrical recording from the soma (trace 11). The calibration of the EPSP in the tuft revealed that the peak amplitude was 18 mV. The EPSP amplitude in the soma, determined from optical data, was 13.2 mV (panel D, trace 8). The direct electrical recording of the EPSP in the soma (panel D, trace 11) showed amplitude of 12.5 mV. This result indicates that the error in calibrating optical signals can be very small. A series of measurements of this type provided information about the characteristics of the EPSPs at the synaptic sites and its attenuation along the primary dendrite. On average, the EPSP attenuated by 35 ± 9% in 300 μm. The mean apparent "length constant" in the primary dendrites of mitral cells was unusually long, corresponding to approximately 1200 μm *(24)*.

It is often useful to monitor and correlate regional dendritic membrane potential signals and the associated local calcium transient, the two signaling pathways that control a variety of functions in individual neurons. We combined voltage imaging and Ca^{2+} imaging from the dendritic arbor of individual neurons loaded with two indicators. In this approach, changes in membrane potential and in $[Ca^{2+}]i$ are recorded sequentially from the dendritic tree of the same neuron as fractional changes in fluorescence intensity using two different filter sets *(12)*. **Figure 3.5** is an example of typical measurements carried out to correlate *Vm* signals with corresponding changes in $[Ca^{2+}]_i$ during the EPSP–AP pairing protocol that typically induces long-term potentiation (LTP) in hippocampal CA1 pyramidal neurons. *Vm* signals and $[Ca^{2+}]i$ transients were monitored during three stimulation protocols (APs alone, EPSPs alone, paired stimulation) from multiple sites on a dendritic tree. The recordings from one location on an oblique dendrite close to a stimulating electrode revealed a marked localized supralinear increase in the $[Ca^{2+}]_i$ signal evoked by pairing a train of EPSPs with a burst of two backpropagated

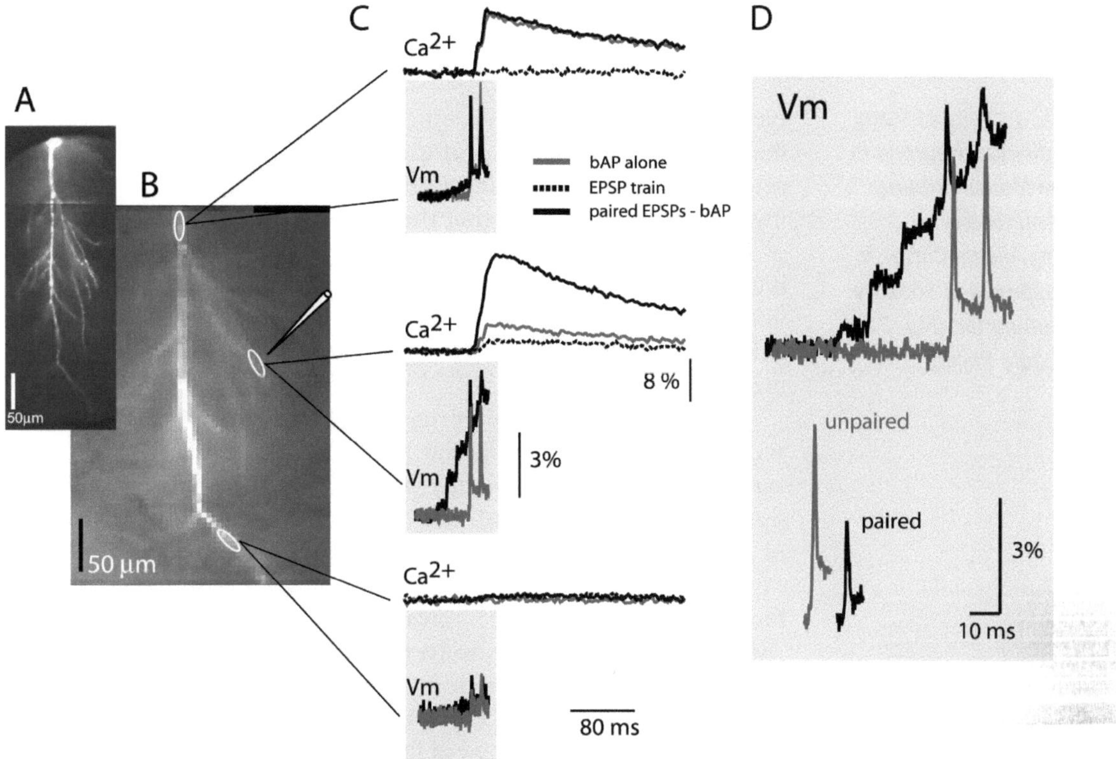

Fig. 3.5. Comparison of *Vm* and $[Ca^{2+}]_i$ signals from same dendritic locations. (**A**) A composite image of a neuron (voltage-sensitive dye excitation). (**B**) Low resolution images (bis-fura-2 excitation) of the dendritic arbor in recording position. Representative recording locations marked by white ovals; the position of the extracellular electrode shown schematically. (**C**) Vm signals and $[Ca^{2+}]_i$ signals related to backpropagated action potentials (bAP, gray traces), EPSP train (dashed traces, shown only for $[Ca^{2+}]_i$ signals), and paired EPSP–AP activity (dark traces). Vm signals (100 ms) during paired activity are superimposed with the recordings of bAP signals alone. Calcium recordings (400 ms) are shown above Vm traces – signals during three stimulation protocols (bAPs, EPSP train, paired) are superimposed. (**D**) Vm recordings of bAPs on the expanded time scale superimposed on the same baseline to show region-specific changes in the peak membrane depolarization during paired activity (upper traces). Traces are shifted horizontally with respect to each other for clarity. Lower traces: The first spike signals (unpaired and paired) aligned to indicate the difference in baseline-to-peak amplitude (EPSP signal subtracted).

APs (**Fig. 3.5C**; middle set of traces). At the same time, a dramatic reduction in the baseline-to-peak local AP amplitude was observed during the paired stimulation. The reduction in the size of the backpropagated AP during paired activity is illustrated in **Fig. 3.5D**. The spike signals corresponding to backpropagated APs evoked alone and during paired activity are superimposed on the baseline membrane potential before the EPSP train (upper traces) and on the baseline membrane potential immediately preceding the spikes evoked during the last two EPSPs in a train (lower traces). Clearly, a supralinear increase in $[Ca^{2+}]_i$ that serves as a trigger for LTP induction did not require boosting in bAP baseline-to-peak amplitude. This initial result was followed by a series of experiments carried out to characterize the relationship between electrical signals and related $[Ca^{2+}]_i$ transients during

EPSP-AP pairing activity pattern that induces persistent changes in synaptic efficacy *(12)*.

2.2. Activity of Many Individual Neurons in the Mouse Primary Somatosensory/Visual Cortex in Response to Sensory Stimulation (Fig. 3.3, Middle Panel)

Nervous systems are made up of large numbers of neurons and, many of these are active during the generation of behaviors. The original motivation for developing optical methods for monitoring activity was the hope that they could be used to record activity of many neurons simultaneously during behaviors *(33)*. Obtaining information about the activity of many cells is essential for understanding the roles of the individual neurons in generating behavior and for understanding how nervous systems are organized.

One simple and widely used method to monitor neuronal activity relies on imaging the neuron's intracellular Ca^{2+} concentration. Indeed, in living cells, most depolarizing electrical signals are associated with Ca^{2+} influx caused by the activation of different types of voltage-gated Ca^{2+} channels, abundantly expressed in the nervous system *(34, 35)*. Such signals are often further amplified by Ca^{2+} release from intracellular Ca^{2+} stores *(36)*. The easiest technique to monitor activity of many individual neurons by imaging their intracellular Ca^{2+}concentration uses a membrane permeant acetoxymethyl (AM) ester form of a Ca^{2+} indicator dye. Such dyes were first introduced by R.Y. Tsien *(37)* and widely used ever since. For the vertebrate brain in vivo, the method allowing imaging of neural ensembles with single cell resolution was introduced by Stosiek et al. *(38)*. The technique was originally developed for imaging layer 2/3 neurons in the mouse cortex (**Fig. 3.6**), and was successfully applied later to stain dif-

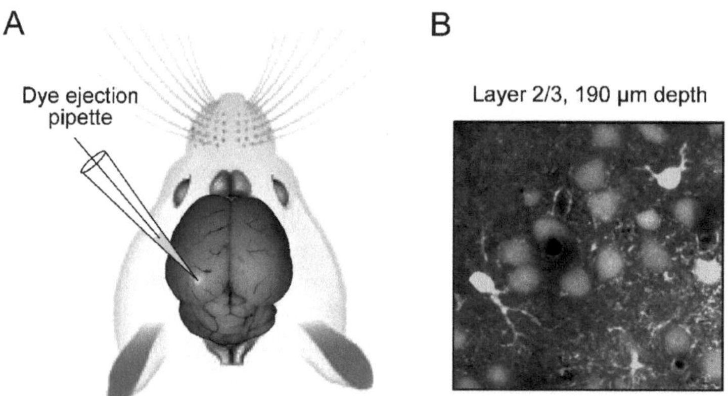

Fig. 3.6. A, a schematic drawing illustrating in vivo multi cell bolus loading of cortical neurons with a calcium indicator dye. The membrane-permeant dye is pressure-injected from the patch pipette into the extracellular space. Subsequently, it diffuses into the cells where it is deesterified by intracellular esterases. B, an example of image quality. The cells in layer 2/3 of the primary visual cortex were stained using multi cell bolus loading and visualized using two-photon imaging.

ferent neuronal tissues in a variety of species from lower vertebrates to mammals *(39–45)*. This technique utilizes a "bolus" injection of a membrane-permeant Ca^{2+} indicator dye into the extracellular space. Briefly, a pipette containing the AM form of an indicator dye is inserted into the tissue of interest and approximately 400 fl of the dye-containing solution is pressure-ejected at the desired depth. We typically use patch-like pipettes with a pipette resistance of 6–9 MOhm when filled with the pipette solution *(38)*. For the detailed step-by-step description of the technique (including the list of required equipment as well as tip and troubleshooting table), we refer the reader to Garaschuk et al. *(112)*. When injected into the brain parenchyma, the dye diffuses into the cells of interest where it is hydrolyzed by intracellular esterases. The remaining dye is rapidly removed from the extracellular space by microcirculation *(46)*. The technique accommodates any AM indicator dye of interest (also Ca^{2+} insensitive dyes like, for example, Calcein AM). Ca^{2+}indicator dyes successfully used so far include: Fura-2 AM, Fura-PE3 AM, Fura Red AM, Indo-1 AM, Calcium Green-1 AM, Oregon Green 488 BAPTA-1 AM, Fluo-4 AM, and Magnesium Green AM.

2.2.1. Example Results

Figure 3.7 illustrates Ca^{2+} transients evoked in layer 2/3 cortical neurons in response to sensory stimulation. In **A**, neurons in the barrel cortex were activated by the air-puff induced movement of the majority of whiskers on the contralateral side of the mouse's snout. In **B**, cells in the primary visual cortex were activated by repetitive light flashes. Note that not all cells in **B** are responding to this kind of stimulus (compare cells 1, 4 with cells 2, 3). The figure also illustrates two different approaches used for recording stimulus-induced Ca^{2+} transients. In **A**, Ca^{2+} transients were recorded at 200 Hz sampling rate using line scan mode. While providing "real time" temporal resolution, this technique reduces the spatial dimension of the recordings to a single line (*scan line* in **A**). In **B**, the area containing 4 neighboring neurons was imaged at frame rate of ~20 Hz. Note that although the individual cell bodies can be clearly distinguished, the multi cell bolus loading technique does not provide imaging contrast sufficient to resolve individual dendritic processes.

2.3. Population Signals from Olfactory Receptor Neuron Terminals in the Olfactory Bulb Glomeruli (Fig. 3.3, Right Panel)

A very large number (~10,000) of olfactory receptor neurons send their axons to each glomerulus in the olfactory bulb. These nerve terminals are small, ~1 μm in diameter, far below the spatial resolution obtainable using wide-field microscopy. Thus, wide-field imaging from the glomerulus will result in a population signal, a signal that is the population average of what is happening in all 10,000 terminals. Selective labeling of olfactory receptor neuron nerve terminals with a calcium-sensitive dye was used to visualize the spatial-temporal patterns in the input from the nose to

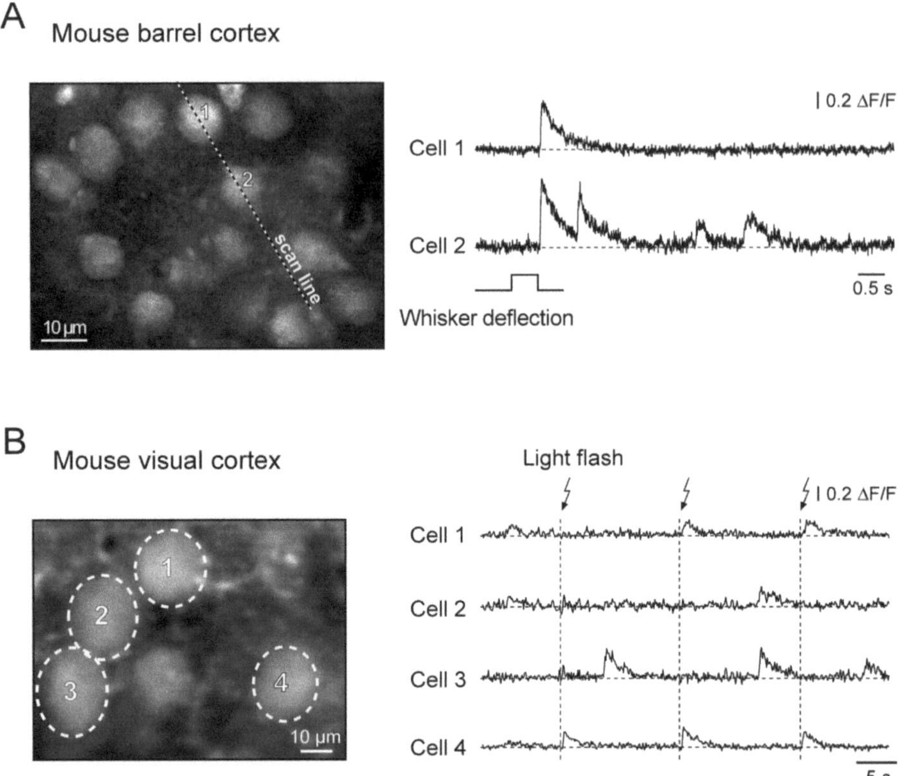

Fig. 3.7. Sensory-driven Ca^{2+} transients in individual cortical neurons. A, individual neurons in the mouse barrel cortex (*left*) and the corresponding Ca^{2+} transients (*right*) evoked by whisker deflection. Whiskers at the contralateral side of the snout were moved by air puffs. The transients were recorded using the line-scan mode (5 ms/line; the position of the scanned line is indicated). B, individual neurons in the mouse visual cortex (*left*) and the corresponding Ca^{2+} transients (*right*) evoked by brief consecutive light flashes. A is reproduced, with permission from National Academy of Sciences USA, copyright 2003, from Stosiek et al. *(38)*, B is reproduced, with permission from Springer-Verlag copyright 2006 from Garaschuk et al. *(112)*.

the dorsal surface of the olfactory bulb. This area encompassed approximately 150 glomeruli. While functional imaging methods such as fMRI, 2-deoxyglucose, and intrinsic imaging reveal patterns of glomerular activity (e.g., *(47–49)*), with these methods, it is difficult to determine whether the signals come from presynaptic terminals and/or from post-synaptic juxtaglomerular and/or mitral/tufted processes in the glomerulus. In selected instances, labeling neurons with voltage- or calcium-sensitive dyes via retrograde or anterograde transport allows selective monitoring of activity in defined neuronal populations (e.g., *(50–52)*). This approach was first developed for the olfactory receptor neurons by Friedrich and Korsching *(53)* in the zebrafish and later adapted for use in the mouse *(54)*.

2.3.1. Dye Screening

We tested one calcium-sensitive dye, Calcium Green-1 (Invitrogen-Molecular Probes, Eugene OR), that was not dextran conjugated. While it did label receptor neurons in the olfactory epithelium, we did not detect labeling in the neuron terminals in the olfactory bulb. Thus, our choice of calcium dyes was limited to those that can be obtained as dextran conjugates. We tried both Calcium Green-1 dextran and Fluo-4-dextran *(51)*. In our hands, labeling was more reliable and the fluorescence signals were larger with Calcium Green-1 dextran. We tested both the 3,000 kD and the 10,000 kD dextran conjugates of Calcium Green-1. No clear difference was observed. Yaksi and Friedrich *(55)* used rhod-dextran but noted that the signal-to-noise ratio was lower than that obtained with the green calcium indicators.

2.3.2. Methods for the In Vivo Mouse Preparation

Because dextran-conjugated dyes are membrane impermeant, loading olfactory receptor neurons with Calcium Green-1 dextran requires treatment with a permeabilizing agent. Friedrich and Korsching *(53)* found that coapplication of Calcium Green-1 dextran with a dilute solution of Triton-X 100 detergent was an effective method for loading zebrafish olfactory receptor neurons.

For imaging, mice were anesthetized with pentobarbital (50 mg/kg, i.p.). A double tracheotomy was performed so that an artificial sniff paradigm using the upper tracheotomy tube allowed for precise control of odorant access to the nasal cavity. This helped to ensure a fixed and rapid onset of the signal which was important when multiple trials were averaged. The mice breathed freely through the lower tracheotomy tube. The dorsal surface of one olfactory bulb was illuminated with 480±25 nm light using a 150 W Xenon arc lamp and 515 nm long-pass dichroic mirror, and fluorescence emission above 530 nm was collected (**Fig. 3.8**, right panel). Images were acquired and digitized with an 80×80 or 256×256 pixel CCD camera (NeuroCCD-SM or NeuroCCD-SM256; RedShirtImaging LLC, Decatur, GA) and stored on disk at a 25 or 32.25 Hz frame rate. Fluorescence was imaged using a 10.5×, 0.2 N.A. objective (spatial resolution, 20 μm per pixel assuming no scattering or out-of-focus signals) or a 14×, 0.4 N.A. objective (15 μm per pixel resolution). The olfactometer *(56)* was an improved version of the one used in Lam et al. *(57)*.

While odorant-evoked signals were detected in single trials (e.g. **Fig. 3.9C**), we typically collected, then averaged, responses of two to eight consecutive odorant presentations in order to improve the signal-to-noise ratio and to obtain a measure of trial-to-trial variability. To avoid habituating the response, we waited a minimum of 60 s between trials. The primary source of extrinsic noise was movement associated with respiration and heartbeat. The noise was largest in regions adjacent to major blood vessels, and so pixels overlying these regions were sometimes removed from the data set (omitted) prior to analysis. Occasional trials with widespread artifactual signals (primarily due to movement) were

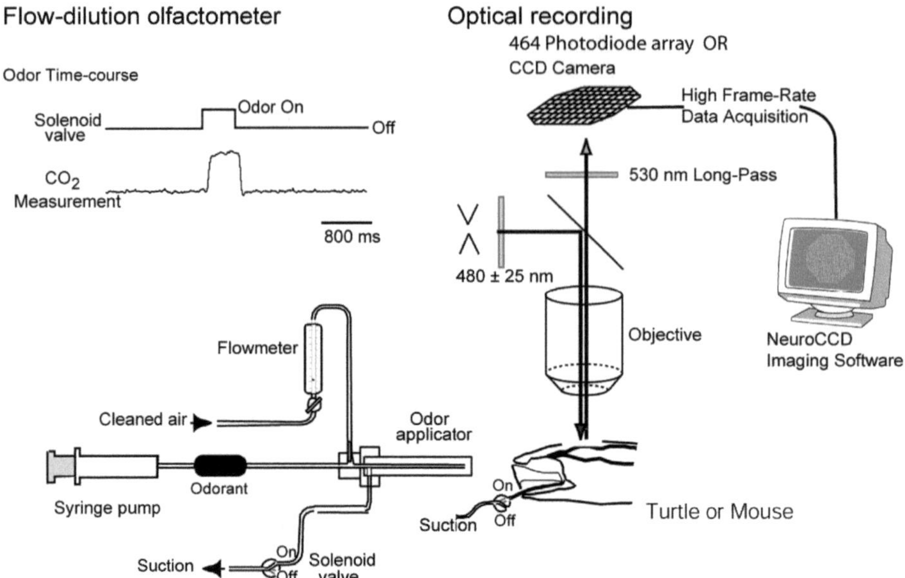

Fig. 3.8. *Left Panel*: *Top*: Time course of the odor output from the olfactometer measured by monitoring the CO_2 in the carrier gas. The upper trace shows the time-course of the command pulse delivered to the suction solenoid of the outer barrel of the odor applicator. The lower trace is the output of the CO_2 detector probe. There is a delay of about 100 ms between the command pulse and the arrival of the pulse at the CO_2 detector. The odor-pulse is approximately square shaped. *Bottom*: Schematic diagram of the olfactometer. Compressed air was cleaned, hydrated and then mixed with room air saturated with odorant vapor in the odor applicator. The flow rates of the air and the odorant vapor were controlled by a flow-meter and a syringe pump respectively. The odor-applicator had two barrels; the outer one was normally under suction to remove the odor. Turning-off of the suction to the outer barrel releases odorant from the end of the applicator. *Right Panel*: Schematic diagram of the optical imaging apparatus. The olfactory bulb was illuminated using a 100 W tungsten halogen lamp or a xenon arc lamp. The incident light passed through a heat filter and a 480 $\pm$ 30 nm band-pass interference filter and was reflected onto the preparation by a long-pass dichroic mirror (510 nm). The image of the preparation was formed by a 10.5× or 14× objective lens onto an 80×80 CCD camera after passing through a 530 nm long-pass secondary filter. The secondary filter is needed to block reflected incident wavelengths that are transmitted by the dichroic mirror. (D. Vucinic, L. B. Cohen, S. Kosmidis, unpublished.)

discarded. After averaging, data from each pixel were temporally filtered with a 1–2 Hz low-pass Gaussian and a 0.017 Hz high-pass digital RC filter.

To correct for unequal labeling of glomeruli, the signal from each pixel was divided by its resting fluorescence obtained at the beginning of each trial. A significant part of the resting fluorescence arises from dye in axons running along the surface of the bulb prior to entering in glomeruli. Because there is no detected increase in calcium in axons *(58)*, this correction is only partially successful. To construct the spatial maps of input to the bulb, response amplitudes for each pixel were measured by subtracting the temporal average of frames in a time window just preceding the stimulus from a temporal average centered at approximately the peak of the response.

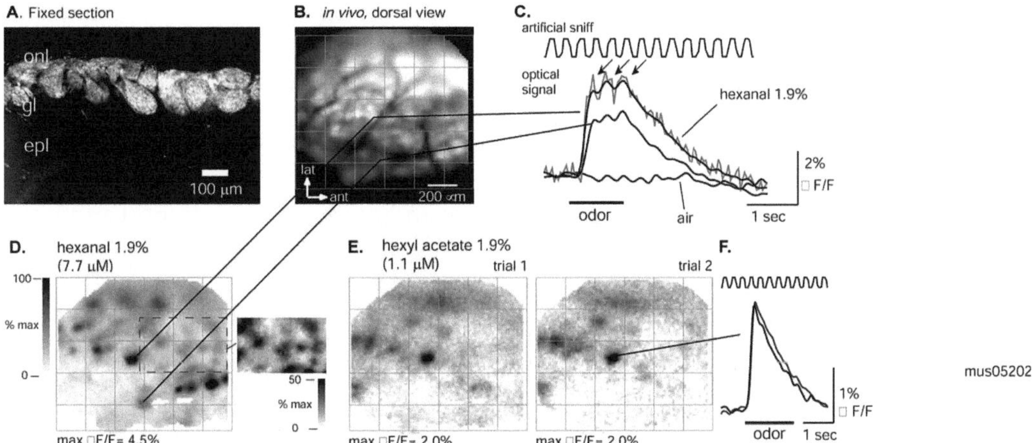

Fig. 3.9. Imaging mouse olfactory receptor neuron activation after in vivo loading with Calcium Green-1 dextran. (**A**) Confocal image of a section through one olfactory bulb fixed 5 days after loading. Fixable rhodamine dextran (10 kD) was used instead of Calcium Green dextran to preserve labeling in fixed tissue. Glomeruli are strongly labeled. There is no evidence of transsynaptic labeling. onl, olfactory nerve layer; gl, glomerular layer; epl, external plexiform layer. (**B**) Resting Calcium Green-1 dextran fluorescence imaged in vivo, 7 days after loading. The image was contrast-enhanced to emphasize individual glomeruli. Blood vessels appear as dark lines. The saturated regions in the upper right are from olfactory nerve bundles that obscure underlying glomeruli. Lines originate from two glomeruli whose responses are shown in (**C**) and (**D**). (**C**) 1.9% hexanal evoked rapid (~200 ms rise-time) increases in fluorescence in the two glomeruli indicated in (**B**) (*lines*). Each trace shows the optical signal measured from one pixel and from a single trial after band-pass filtering from 0.017 to 2 Hz. (**D**) Gray-scale map of the evoked signal for the trial shown in (*C*), showing foci of fluorescence increases. A region of the map normalized to 50% of the maximum signal (*inset*) shows additional smaller-amplitude foci. (**E**) The spatial distribution and amplitude of the signals was consistent across repeated odorant presentations and was different for different odorants (D–E). Redrawn from Wachowiak and Cohen *(54)*.

2.3.3. Example Results

Calcium Green-1 dextran loading resulted in labeling of olfactory receptor axon terminals in the olfactory bulb (**Fig. 3.9A, B**). We imaged odorant responses from the dorsal olfactory bulb of anesthetized mice 4–8 days after loading. Odorant presentation evoked rapid (100–500 ms rise-time) increases in fluorescence of up to 9% $\Delta F/F$ (**Fig. 3.9C**). Spatial maps of the response amplitude measured from each pixel showed well-defined foci of fluorescence increases (**Fig. 3.9D**), often corresponding to individual glomeruli visible from the resting fluorescence (**Fig. 3.9B**). Odorant-evoked signals showed a range of amplitudes in different glomeruli. Using an expanded gray scale (**Fig. 3.9D**, inset), it is clear that even smaller signals have glomerular localization. The spatial distribution and amplitude of the signals were consistent across repeated odorant presentations (**Fig. 3.9E**) and were different for different odorants (**Fig. 3.9D–E**).

The noise in this measurement of fluorescence from transported dye is also consistent with expectations from a calculation of the shot-noise (see below). In measurements made at a frame rate of 100 Hz, the number of photoelectrons per pixel per millisecond was approximately 2×10^4. Because we digitally low-pass filtered the data at 2 Hz, the effective sample period is 500 msec and thus the number of photons/sample period is 10^7. The shot noise in this measurement should then be approximately 3×10^{-4} of the resting intensity. Consistent with this prediction, the noise in the measurements shown in **Fig. 3.9C** is less than 2×10^{-3} of the resting intensity. In preliminary measurements, we found that the shot noise and noise from respiration and heart beat were similar in magnitude.

The spatial resolution shown in **Fig. 3.9**, D-E is on the order of 20 μm, far better than might have been anticipated from the measurements illustrated in **Fig. 3.12**. However, both factors that could contribute to blurring are minimized in the measurements shown in **Fig. 3.9**. First, scattering will be lower because the Calcium Green-1 dextran is only in the outer two layers of the olfactory bulb. Second, out-of-focus signals will be small because the glomerular layer is only 100 μm thick.

Because many (up to 50) maps of this sort can be obtained from each preparation, considerable information has been obtained about the input from the nose to the bulb. However, it remains a challenge to determine the map of the output of the bulb carried by the mitral/tufted cells. If both the input and the output were known, we would have a strong clue about the function of this first way station in processing olfactory information.

3. Methodological Considerations

The three examples given above involved fractional intensity changes and signal-to-noise ratios that are not large. To measure these signals, the noise in the measurements had to be a substantially smaller fraction of the resting intensity. In the sections that follow, some of the considerations necessary to achieve such a low noise are outlined.

3.1. Signal Type

Sometimes it is possible to decide in advance which kind of optical signal will give the best signal-to-noise ratio, but in other situations, an experimental comparison is necessary. The choice of signal type often depends on the optical characteristics of the preparation. Extrinsic birefringence signals are relatively large in preparations that, like axons, have a cylindrical shape and radial optic axis *(59)*. However, in preparations with spherical symmetry (e.g., cell soma), the birefringence signals in adjacent quadrants will cancel *(60)*. Thick preparations (e.g. mammalian cortex) also dictate the choice of signal. In this circumstance, transmitted light measurements are not easy (a subcortical implantation of a light guide would be necessary), and the small size of the absorption signals that are detected in reflected light *(59, 61)* meant that fluorescence or reflectance would be optimal *(62)*. Fluorescence signals have most often been used in monitoring membrane potential and calcium concentration from tissue-cultured neurons. Both fluorescence and absorption have been used in measurements from ganglia and brain slices. Fluorescence or reflectance has always been used in measurements from intact brains.

3.2. Amplitude of the Voltage or Calcium Change

Both the signals are often presented as a fractional intensity change ($\Delta I/I$). These signals give information about the time course of the potential or calcium concentration change but no direct information about the absolute magnitude. However, in some instances, approximate estimations can be obtained. For example, the size of the optical signal in response to a sensory stimulus can be compared to the size of the signal in response to an epileptic event *(62)*. Another approach is the use of ratiometric measurements at two independent wavelengths *(63, 64)*. However, to determine the amplitude of the voltage or calcium change from a ratio measurement, one must know the fraction of the fluorescence that results from dye in the expected location, i.e. bound to active *versus* inactive membranes for voltage-sensitive dyes, or dye free in the axoplasm *versus* dye bound to protein or in intracellular compartments for calcium dyes. These requirements are only approximately met in special circumstances. For voltage-sensitive dyes, the best calibration is an electrode measurement of membrane potential. One case where this was possible was described in the first example presented above.

4. Measuring Technology

4.1. Noise

4.1.1. Shot Noise

The limit of accuracy with which light can be measured is set by the shot noise arising from the statistical nature of photon emission and detection. Random fluctuations in the number of photons emitted (and measured) per unit time occur; the root-mean-square (RMS) deviation in the number emitted is the square root of the average number. In a shot noise limited measurement, the signal-to-noise ratio is directly proportional to the square root of the number of measured photons and inversely proportional to the square root of the bandwidth of the photodetection system *(65, 66)*. The basis for the square root dependence on intensity is illustrated in **Figure 3.10**. In 10A, the result of using a random number table to distribute 20 photons into 20 time windows is shown. In 10B, the same procedure was used to distribute 200 photons into the same 20 bins. Relative to the average light level, there is more noise in the top trace (20 photons) than in the bottom trace (200 photons). On the right side of **Fig. 3.10**, the measured signal-to-noise ratios are listed; the improvement from

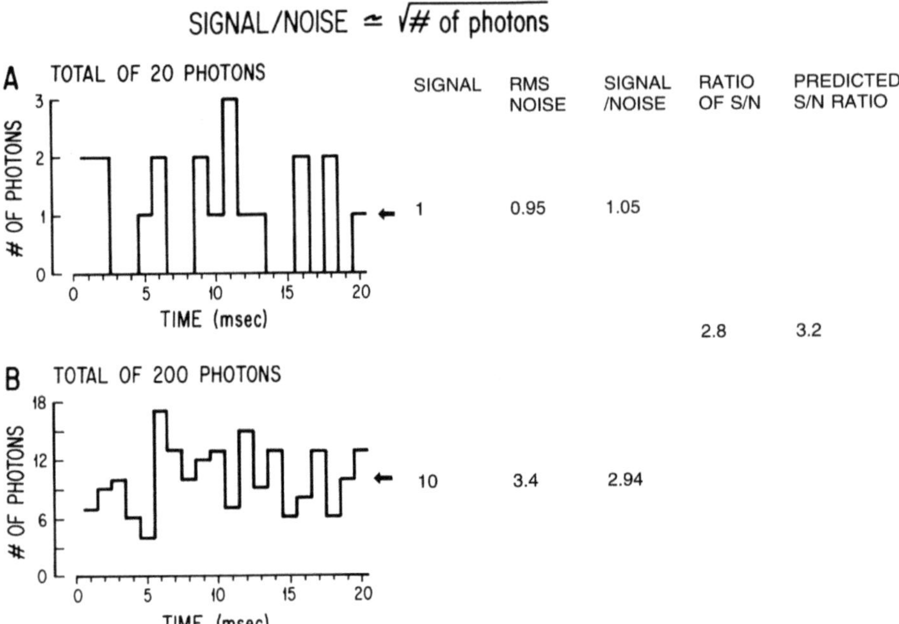

Fig. 3.10. Plots of the results of using a table of random numbers to distribute 20 photons (*top*, **A**) or 200 photons (*bottom*, **B**) into 20 time bins. The result illustrates the fact that when more photons are measured, the signal-to-noise ratio is improved. On the right, the signal-to-noise ratio is measured for the two results. The ratio of the two signal-to-noise ratios was 0.43. This is close to the ratio predicted by the relationship that the signal-to-noise ratio is proportional to the square root of the measured intensity.

A to B is similar to that expected from the square-root relationship. This square-root relationship is indicated by the dotted line in **Fig. 3.11** which plots the light intensity divided by the noise in the measurement versus the light intensity. In a shot-noise limited measurement, improvement in the signal-to-noise ratio can only be obtained by (i). Increasing the illumination intensity. (ii) Improving the light-gathering efficiency of the measuring system or (iii) Reducing the bandwidth.

4.1.1.1. The Optimum Signal to Noise Ratio in a Wide-Field Measurement

A tungsten filament lamp emits an average of 10^{16} photons/ms and the root-mean-square (RMS) deviation in the number emitted is the square root of this number or 10^8 photons/ms. However, because only a small fraction of the photons will be measured, a signal-to-noise ratio of 10^8 (see above) cannot be achieved. A partial listing of the light losses follows. A 0.9-NA lamp collector lens would collect 0.1 of the light emitted by the source. Only 0.2 of that light is in the visible wavelength range; the remainder is infrared (heat). Limiting the incident wavelengths to those which have the signal means that only 0.1 of the visible light is used. Thus, the light reaching the preparation might typically be reduced to 10^{13} photons/ms. If the light-collecting system that forms the image has high efficiency e.g., in an absorption measurement with matched condenser and

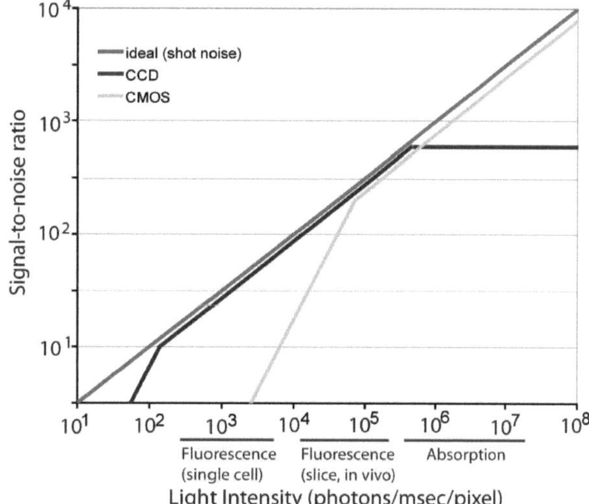

Fig. 3.11. The ratio of light intensity divided by the noise in the measurement as a function of light intensity in photons/ms/pixel. The theoretical optimum signal-to-noise ratio (medium line) is the shot-noise limit. Two camera systems are shown, a 10 kHz frame rate, 128×128 pixel, CMOS camera and a back-illuminated, 2 kHz frame rate, 80×80 pixel, CCD camera. The CMOS camera (*light line*) provides a near optimal signal-to-noise ratio at higher light intensities while the CCD camera (*dark line*) is better at lower intensities. The approximate light intensity per detector in fluorescence measurements from a single neuron, fluorescence measurements from a slice or in vivo preparation, and in absorption measurements from a ganglion or a slice is indicated along the x axis. The signal-to-noise ratio for the CMOS camera falls away from the ideal at low intensities because of dark noise. The lower dark noise of the cooled CCD allows it to function at the shot-noise limit at lower intensities until read noise dominates at about 10^2photons/pixel/msec. The CCD camera saturates at intensities above 5×10^5 photons/ms/pixel.

objective numerical apertures, about 10^{13} photons/ms will reach the image plane. (In a fluorescence measurement, there will be much less light measured because 1. only a fraction of the incident photons are absorbed by the fluorophores, 2. only a fraction of the absorbed photons appear as emitted photons, and 3. only a fraction of the emitted photons are collected by the objective.) If the camera has a quantum efficiency of 1.0, then, in absorption, a total of 10^{13} photoelectrons/ms will be measured. With a camera of 10,000 pixels, there will be 10^9 photoelectrons/ms/pixel. The shot noise will be $10^{4.5}$ photoelectrons/ms/pixel; thus, the very best that can be expected is a noise that is $10^{-4.5}$ of the resting light (a signal-to-noise ratio of ~90 db). The extra light losses in a fluorescence measurement would further reduce the maximum obtainable signal-to-noise ratio.

One way to compare the performance of different camera systems for wide-field imaging and to understand their deviations from optimal (shot-noise limited) is to determine the light intensity divided by the noise in the measurement and plot that versus the number of photons reaching each pixel of the camera. The dotted line in **Fig. 3.11** is the plot for an ideal camera. At high light intensities, this ratio is large and thus small changes in intensity can be detected. For example, at 10^{10} photons/ms, a fractional intensity change of 0.1% can be measured with a signal-to-noise ratio of 100. On the other hand, at low intensities, the ratio of intensity divided by noise is small and only large signals can be detected. For example, at 10^4 photons/msec, the same fractional change of 0.1% can be measured with a signal-to-noise ratio of 1 only after averaging 100 trials.

In addition, **Fig. 3.11** compares the performance of two particular camera systems, a CMOS camera (solid lines) and a back-illuminated cooled CCD camera (dashed lines), with the shot noise ideal. A 128×128 pixel CMOS camera approaches the shot-noise limitation over the range of intensities from 10^5 to 10^9 photons/ms/pixel. This is the range of intensities obtained in absorption measurements and fluorescence measurements using bath applied dye on in vitro slices and intact brains. On the other hand, the cooled 80×80 pixel CCD camera approaches the shot noise limit over the range of intensities from 10^2 to 5×10^5 photons/ms. This is the range of intensities obtained from fluorescence experiments on individual cells and neurons. In the discussion that follows, we will indicate the aspects of the measurements and the characteristics of the two camera systems which cause them to deviate from the shot noise ideal. The two camera systems we have used to illustrate in **Fig. 3.11** have excellent dark noise and saturation characteristics; other cameras would be dark noise limited at higher light intensities and would saturate at lower intensities.

4.1.1.2. The Optimum Signal-to-Noise Ratio in Two-Photon Scanning Microscopy Measurements

Similar considerations apply to two-photon measurements. Because two-photon excitation will only occur with the nearly synchronous arrival of two low-energy photons, excitation is proportional to the square of light intensity and only very high intensity sources achieve significant excitation. In practice, this restricts the light source to pulsed lasers with very narrow pulses. In the experiments illustrated in Example 2 (above), the laser intensity was not limiting because the targeted cells were only some 150–250 μm below the surface of the preparation and losses in focal intensity from light scattering were not large. For cells as deep as mitral cells (~400 μm below the surface), the loss of the excitation light due to photon dispersion within the tissue becomes significant. In addition, in the neural tissue, the quality of in-depth imaging often suffers from (i) inhomogeneity of the refractive indices within the tissue (e.g. blood vessels, clusters of cell bodies acting as microlenses *(67)*), (ii) light absorption by the hemoglobin, (iii) generation of out-of-focus fluorescence while increasing laser power etc. For a thorough description of approaches used to improve the imaging depth as well as image quality, we refer the reader to the recent review of Helmchen and Denk *(67)*.

4.1.2. Extraneous (Technical) Noise

A second type of noise, termed extraneous or technical noise, is more apparent at high light intensities or at high spatial resolution. At high light intensities, the sensitivity to extraneous noise is high because the fractional shot noise and dark noise are low. One type of extraneous noise is caused by fluctuations in the output of the light source (see below). Two other sources of extraneous noise are vibrations and movement of the preparation.

4.1.2.1. Vibrational Noise in Wide-Field Measurements

A number of precautions for reducing vibrational noise have been described *(68, 69)*. The pneumatic isolation mounts on many vibration isolation tables are more efficient in reducing vertical vibrations than in reducing horizontal movements. There has been a series of progressively more successful (and expensive) solutions for vibration problems. A simple and inexpensive remedy is air-filled, soft rubber tubes (Newport Corp, Irvine, CA). For more severe vibration problems, Minus K Technology (Inglewood, CA) sells vibration isolation tables with very low resonant frequencies. The Halcyonics Micro 60 (Menlo Park, CA) is an active (piezoelectric drivers) isolator and can defeat airborne vibrations as well as those transmitted through the "ground" (Brian Salzberg, personal communication). Nevertheless, it has been difficult to reduce vibrational noise to less than 10^{-5} of the total light. With this amount of vibrational noise, increases in measured intensity beyond 10^{10} photons/ms would not improve the signal-to-noise ratio. For this reason and because of well size limitations, the performance of the CMOS system reaches a ceiling (**Fig. 3.11**, solid line).

4.1.2.2. Vibrational Noise in Two-Photon Measurements

Using a Newport Corporation Research Grade vibration isolation table reduces the vibrational noise so that it is less than the shot noise.

4.1.2.3. Preparation Movement

Preparation movement is often the limiting noise factor in in vivo measurements. This is true in wide-field measurements because the light intensity can be relatively high, reducing the shot noise to a level that is smaller than the movement noise. It is true in two-photon measurements because the spatial resolution and contrast are high and thus the sensitivity to movement is also high. The movement artifacts in vivo usually consist of irregular movements of the entire animal (less of a problem in anesthetized preparations) as well as of heart beat- and breathing-related vibrations. The higher frequency vibrations are usually the result of heart beat pulsation and are more profound in regions with a high density of blood vessels. The heart beat- and breathing-related artifacts significantly increase their amplitudes when the skull and the *dura mater* are removed. (The multi cell bolus loading technique does not require dura removal.) The stability of recordings also depends on the diameter of the craniotomy. Thus, openings larger than 1 mm in diameter are often accompanied by large movement artifacts occurring at the heart beat frequency. Under such circumstances, the vibration noise can be reduced by covering the skull opening with 2% agarose and a glass coverslip *(70)*.

4.1.3. Dark Noise/Read Noise

Dark noise will degrade the signal-to-noise ratio at low light levels.

4.1.3.1. Wide-Field Imaging

Because of its electronic characteristics and because it is cooled, the CCD camera dark noise is substantially lower than that of the CMOS system. The CCD camera has read noise of 10e-rms at 1 kfps, low in an absolute sense. The larger dark noise in the CMOS camera accounts for the fact that the break in the curve in **Fig. 3.11** is substantially to the right of the intensity where the CCD camera becomes read-noise limited (10^2 photons/pixel/msec).

4.1.3.2. Two-Photon Imaging

The dark noise of photomultipliers (anode dark current) critically depends on supply voltage. At low-to-medium supply voltages, anode dark current is relatively low, but it increases dramatically at high supply voltages (usually higher than 700–800 V). To minimize dark noise, it is therefore advisable to conduct measurements at lowest PMT gain (e.g. lowest supply voltage) provided that the preparation tolerates the required increase in the intensity of excitation light. As a rule-of-thumb, the illumination intensity has to be as high and the corresponding PMT gain as low as possible to provide recording conditions just below the dye bleaching threshold.

4.2. Light Sources

4.2.1. Wide-Field Imaging

Three kinds of sources have been used. Tungsten filament lamps are a stable source, but their intensity is relatively low, particularly at wavelengths less than 480 nm. Arc lamps are somewhat less stable but can provide higher intensity. Laser illumination can provide even more intense illumination.

4.2.2. Tungsten Filament Lamps

It is not difficult to provide a power supply stable enough so that the output of the bulb fluctuates by less than 1 part in 10^5. In absorption measurements, where the fractional changes in intensity are relatively small, only tungsten filament sources have been used. On the other hand, fluorescence measurements often have larger fractional changes that will better tolerate light sources with systematic noise, and the measured intensities are lower, making improvements in signal-to-noise ratio from brighter sources attractive.

4.2.3. Arc Lamps

Cairn Research Ltd (Faversham, UK) provides xenon power supplies, lamp housings, and arc lamps with noise that is in the range of 1 part in 10^4. A 150 watt lamp yielded 2–3 times more light at 520 ± 45 nm than a tungsten filament bulb. The extra intensity is especially useful for fluorescence measurements from processes of single neurons where the light intensity is low and the dark noise is a problem. In that situation, the signal-to-noise ratio will improve linearly with intensity.

4.2.4. Lasers

Laser illumination can provide the highest illumination intensity but avoiding photodynamic damage requires careful attention to minimizing illumination duration. The interference from speckle noise can be eliminated by reducing the beam coherence (Dejan Zecevic and Thomas Knopfel, personal communication).

4.2.5. Two-Photon Imaging

Two-photon imaging requires the use of a specific type of light source, pulsed lasers. These lasers use either a Titanium- sapphire (Ti:Sapphire) or a Nd:YAG (neodymium-doped yttrium aluminium garnet) oscillator. The Ti:Sapphire lasers provide wide tuning range (from 690 to 1080 nm) whereas Nd:YAG lasers typically operate at a wavelength of 1064 nm (although they can be also made to lase at their non-principal wavelengths of 946, 1320 or 1123 nm). At present, mostly Ti:Sapphire lasers are used for two-photon imaging of living tissues *(67, 71, 72)*.

4.3. Optics

4.3.1. Numerical Aperture

The need to maximize the number of measured photons has been a dominant factor in the choice of optical components. In wide-field epifluorescence, both the excitation light and the emitted light pass through the objective, and the intensity reaching the photodetector is proportional to the *fourth* power of numerical aperture *(73)*. In two-photon imaging, only the collection of emitted light is affected by numerical aperture and thus the intensity reaching the photodetector is proportional to the square of numerical aperture. Clearly, numerical aperture is an important consideration in the choice of lenses. However, direct comparison of the intensity reaching the image plane has shown that the light collecting efficiency of an objective is not completely determined by the stated magnification and NA. In wide-field imaging, differences of a factor of five between lenses of the same specification have been observed.

4.3.2. Back-Aperture of the Objective

Because of the strong light scattering within living tissue, in vivo two-photon imaging critically depends on detection of scattered photons. Under this condition, the size of back aperture of the objective (but also the size of all other apertures within the detection pathway) becomes important. Ideally, one would like to employ an objective with high NA, large field of view (low magnification) and large back aperture. Such objectives recently became available from different manufacturers (Olympus, Nikon, Zeiss, etc.).

4.3.3. Depth of Focus in Wide-Field Imaging

Salzberg et al. *(68)* determined the effective depth of focus for a 0.4 NA objective lens by recording an optical signal from a neuron when it was in focus and then moving the neuron out of focus by various distances. They found that the neuron had to be moved 300 μm out of focus to cause a 50% reduction in signal size. Using 0.5 NA optics, Kleinfeld and Delaney *(74)* found that 100 μm out of focus led to a reduction in signal size of 50%.

4.3.4. Light Scattering and Out-of-Focus Light in Wide-Field Imaging

Light scattering can limit the spatial resolution of an optical measurement. **Figure 3.12** illustrates the results of experiments carried out on tissue from the salamander olfactory bulb. The top section indicates that when no tissue is present, essentially all of the light (750 nm) from a small spot falls on one detector. The bottom section illustrates the result when a 500 μm-thick slice of olfactory bulb is present. The light from the small spot is spread to about 200 μm. Mammalian cortex appears to scatter more than the salamander olfactory bulb. Thus, light scattering will cause considerable blurring of signals in vertebrate preparations. Presumably, this effect will be more severe at lower wavelengths because scattering by nervous tissue is inversely related to wavelength squared *(75)*.

A second source of blurring is signal from regions that are out of focus. For example, if the active region is a cylinder (a column) perpendicular to the plane of focus, and the objective is focused at the middle of the cylinder, then the light from the focal plane will have the correct diameter at the image plane. However, the light from the regions above and below is out of focus and will have a diameter that is too large. The middle section of **Fig. 3.12** illustrates the effect of moving the small spot of light 500 μm out of focus. The light from the small spot is spread to about 200 μm. Thus, in preparations with considerable scattering or with out-of-focus signals, the actual spatial resolution is likely to be limited by the preparation and not by the number of pixels in the imaging device.

4.3.5. Confocal and Two-Photon Microscopes

The confocal microscope *(76)* substantially reduces both the scattered and out-of-focus light by using a pinhole in the detection pathway. Two-photon microscopes also reduce out-of-focus light as well as out-of-focus phototoxicity and photobleaching *(77)*. In this case, however, the excitation of the fluorophore is restricted to the focal plane. With both types of microscope, one can obtain

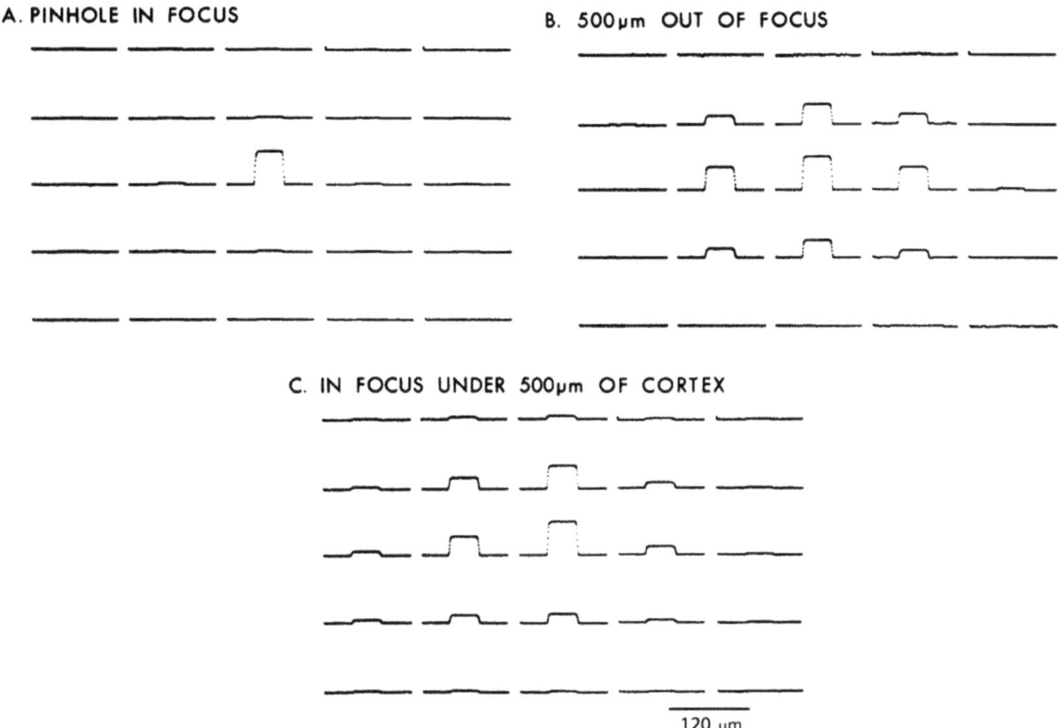

Fig. 3.12. Effects of focus and scattering on the distribution of light from a point source onto a camera. The data were acquired using a 464 element photodiode array. (**A**) A 40 µm pinhole in aluminum foil covered with saline was illuminated with light at 750 nm. The pinhole was in focus. More than 90% of the light fell on one detector. (**B**) The stage was moved downward by 500 µm. Light from the out-of-focus pinhole was now seen on several detectors. (**C**) The pinhole was in focus but covered by a 500 µm slice of salamander cortex. Again the light from the pinhole was spread over several detectors. A 10 × 0.4 NA objective was used. Kohler illumination was used before the pinhole was placed in the object plane. The recording gains were adjusted so the largest signal in each of the three trials would be approximately the same size in the figure. Redrawn from Orbach and Cohen *(61)*.

images from intact vertebrate preparations with much better spatial resolution than is achieved with wide-field microscopy. The two-photon microscope has been successfully used to monitor changes in calcium concentration inside small processes of neurons *(78)* and, in many individual cells, after bulk loading the AM ester calcium dyes (*(38)*; Example 2 above). Using a scanned line confocal microscope from Prairie Technologies (Middletown, WI), Jeff Magee and collaborators (personal communication) were able to measure voltage-sensitive dye signals from hippocampal cell dendrites at a frame rate of 3 kfps. Slower voltage-sensitive dye signals were measured confocally much earlier *(79)*.

4.4. Cameras

4.4.1. Quantum Efficiency

Because the signal-to-noise ratio in a shot noise limited measurement is proportional to the square root of the number of photons converted into photoelectrons, quantum efficiency is important. Silicon photodiodes and CCD and CMOS cameras can have quantum efficiencies approaching the ideal (1.0) at wavelengths where most dyes absorb or emit light (500–900 nm). In contrast, only specially chosen vacuum photocathode devices (phototubes, photomultipliers, or image intensifiers) have a quantum efficiency as high as ~0.4. A relatively low fill factor and front illumination reduce the quantum efficiency of CMOS cameras to about 50%. This reduction will reduce the signal-to-noise ratio in a shot-noise limited measurement by a factor of about 1.4. Thus, in a low light level shot-noise limited situation, a CCD camera will have a larger signal-to-noise ratio. Photographic film has a much smaller quantum efficiency, <0.01 *(80)*.

4.4.2. EM-CCD Cameras

These cameras have on-chip multiplication and should lead to better signal-to-noise performance at very low light levels. However, the multiplication process adds noise (a factor of 1.4) and some existing chips are even noisier than expected from the factor of 1.4. If an ordinary CCD has a read noise of 10e-, then an ideal EM-CCD camera will have a better signal-to-noise ratio only at light levels less than 100 photons/pixel/frame. The light level achieved in the experiments illustrated in Example 1, where there is relatively little dye in small distal processes, is ~1000 photons/pixel/msec. Thus, in this measurement and all measurements with even higher light intensity, the CCD will have a better signal-to-noise ratio than an EM-CCD.

5. Future Directions

5.1. Organic Dyes

Because the light measuring apparatus is already reasonably optimized for all of the above applications, any improvement in sensitivity will need to come from the development of better dyes and/or investigating signals from additional optical properties of the dyes.

5.2. Organic Voltage-Sensitive Dyes

The voltage-sensitive dyes in **Fig. 3.2** and the vast majority of those synthesized are of the general class named polyenes *(81)*, a group that was first used to extend the wavelength response of photographic film. It is possible that improvements in signal size can be obtained with new polyene dyes (see *(4, 5)* for a discussion of maximum possible fractional changes in absorption and fluorescence). On the other hand, the maximum fractional change has not increased in recent years (L. B. Cohen, A. Grinvald, K. Kamino, L. Loew, and B. M. Salzberg, personal communication), and most improvements (e.g. *(9, 25, 26, 51)*) have involved synthesizing analogues that work well on new preparations.

The best of the organic dyes have fluorescence changes of 10–20%/100 mv in situations where the staining is specific to the membrane whose potential is changing *(26, 82, 83)*. Gonzalez and Tsien *(84)* introduced a new scheme for generating voltage-sensitive signals using two chromophores and fluorescence resonance energy transfer (FRET). While these fractional changes were also in the range of 10%/100 mv, more recent results are about 30% (Gonzalez and Tsien, personal communication). However, in order to achieve fast response times (> 100 Hz), one of the chromophores must be very hydrophobic and, as a result, does not penetrate into brain tissue. Thus far, it has not been possible to measure fast signals with a fast pair of dyes in intact tissues (Gonzalez and Tsien; Obaid and Salzberg; personal communication), although impressive results have been obtained where speed was not critical *(85, 86)*.

Membrane potential changes the non-linear second harmonic generation from styryl dyes *(87–89)*. Large (50%) fractional changes were measured. But, because the number of detected photons is small, the signal-to-noise ratios remain substantially smaller than that obtained with wide-field measurements.

5.3. Neuron-Type Specific Staining

An important new direction is the development of methods for neuron-type specific staining which would make it possible to determine the role of specific neuron types in generating the input-output function of a brain region. Three quite different approaches have been tried.

5.3.1. Directed Transport

The use of anterograde and retrograde dye transport has resulted in specific staining of olfactory receptor neuron terminals (Friedrich and Korsching *(53)*; Example 3) and motor neurons in embryonic chick and in lamprey spinal cords *(51)*. This method depends on finding a location where only one cell type is present that has a process leading to the brain area of interest. In lamprey experiments, spike signals from individual neurons were sometimes measured *(90)*. Further effort at optimizing this staining procedure is needed.

5.3.2. Staining Dependent on β-galactosidase Expression

The use of cell-type specific staining developed for fluorescein by Nirenberg and Cepko *(91)* might be extended to ion-sensitive or voltage-sensitive dyes.

5.3.3. Genetically Encoded Activity Sensors

5.3.3.1. Voltage Sensors

Siegel and Isacoff *(92)* constructed a genetically encoded combination of a potassium channel and green fluorescent protein. When expressed in a frog-oocyte, this molecule, FlaSh, had a (relatively slow; few hundred millisecond) voltage-dependent signal with a fractional fluorescence change of 5%. More recently, Ataka and Pieribone *(93)* have developed a similar construct, SPARC, with very rapid kinetics. Unfortunately, in mammalian cells, both of these FP-voltage sensors are only minimally expressed in the extracellular membrane *(94)*. They appear to be retained in the endoplasmic reticulum. Thomas Knopfel and collaborators *(95)* investigated a new class of voltage- sensitive proteins, the voltage-sensitive phosphatase from *Ciona intestinalis (96)*. They replaced the phosphatase domain with a FRET pair and this protein, VSFP2.1, both trafficked well to the external membrane of hippocampal neurons and had a voltage-dependent response. VFSP2.1 was faster than FlaSh but slower than SPARC. Future efforts will be needed to improve the response time and signal size of VFSP2.1 and to test the construct in transgenic systems.

5.3.4. Calcium Sensors

There exists a rather large family of genetically encoded fluorescent Ca^{2+} indicator proteins. Until recently, all protein calcium sensors employed Ca^{2+} binding to calmodulin (CaM) and Ca^{2+}-dependent interaction of calmodulin and the CaM binding peptide M13 as a calcium sensor. The CaM-M13 complex was then attached to green fluorescent protein or one of its variants to generate a single-wavelength Ca^{2+} indicator *(97–102)*. Alternatively, the CaM-M13 complex was sandwiched between CFP (cyan fluorescent protein) and YFP (yellow fluorescent protein) to generate ratiometric Ca^{2+} sensors like the "Cameleons" *(103, 104)*. Calmodulin-based calcium sensors function well in invertebrates and lower vertebrates (for review see refs. in Miyawaki *(104)*), but show rather poor performance in mammals *(105–108)*. Significant improvement of protein calcium sensor performance in mammalian neurons came from the use of alternative Ca^{2+}-binding protein Troponin C (TnC; *(109–111)*). The members of the TnC family have relatively high dynamic range (4-fold increase in signal strength upon changing Ca^{2+} concentration from 0 to 10 mM), rather fast rise and decay times *(108)* and they respond linearly to an increase in Ca^{2+} concentration within the physiological activity range *(102, 111)*. TnC-based indicators can be expressed transgenically in mice and allow in vivo imaging of neural function with single-cell and even subcellular resolution *(111)*. While recent developments significantly improved the properties of protein calcium sensors, when compared to organic Ca^{2+} indicators, the protein sensors are still inferior when Ca^{2+} sensitivity and/or dynamic range of the indicator are considered (see Table 1 in Garaschuk et al. *(108)*).

Taken together, optical recordings already provide unique insights into brain activity and organization. Improvements in sensitivity or selectivity would make these methods even more powerful.

Acknowledgments

The authors are indebted to their collaborators Vicencio Davila, Amiram Grinvald, Kohtaro Kamino, Les Loew, Bill Ross, Brian Salzberg, Dejan Vucinic, Alan Waggoner, Matt Wachowiak, and Jian-young Wu for numerous discussions about optical methods. The experiments carried out in our laboratories were supported by NIH grants DC05259 and NS42739, Deutsche Forschungsgemeinschaft (SFB 391 and SFB 596) and the Bundesministerium für Bildung und Forschung (NGFN-2).

References

1. Cohen L.B., Salzberg B.M. (1978). Optical measurement of membrane potential. Rev Physiol Biochem Pharmacol, 83, 35–88.
2. Loew L.M., Cohen L.B., Salzberg B.M., Obaid A.L., Bezanilla F. (1985). Charge-shift probes of membrane potential. Characterization of aminostyrylpyridinium dyes on the squid giant axon. Biophys J, 47, 71–77.
3. Gupta R.K., Salzberg B.M., Grinvald A., Cohen L.B., Kamino K., Lesher S., Boyle M.B., Waggoner A.S., Wang C.H. (1981). Improvements in optical methods for measuring rapid changes in membrane potential. J. Membr. Biol., 58, 123–137.
4. Waggoner A.S., Grinvald A. (1977). Mechanisms of rapid optical changes of potential sensitive dyes. Annu NY Acad Sci, 303, 217–241.
5. Fromherz P., Dambacher K.H., Ephardt H., Lambacher A., Muller C.O., Neigl R., Schaden H., Schenk O., Vetter T. (1991). Fluorescent dyes as probes of voltage transients in neuron membranes: Progress report. Ber. Bunsenges. Phys. Chem., 95, 1333–1345.
6. Cohen L.B., Lesher, S. (1986). Optical monitoring of membrane potential: methods of multisite optical measurement. Soc Gen Physiol Ser, 40, 71–99.
7. Grinvald A., Frostig R.D., Lieke E., Hildesheim R. (1988) Optical imaging of neuronal activity. Physiological Reviews. 68: 1285–1366.
8. Nakashima M., Yamada S., Shiono S., Maeda M., Satoh F. (1992) 448-detector Optical Recording System: Development and application to *Aplysia* gill-withdrawal reflex. IEEE Trans Biomed Eng. 39: 26–36.
9. Momose-Sato Y., Sato K., Sakai T., Hirota A., Matsutani K., Kamino, K. (1995). Evaluation of optimal voltage-sensitive dyes for optical measurement of embryonic neural activity. J Memb. Biology, 144, 167–176.
10. Zecevic D. (1996). Multiple spike-initiation zones in single neurons revealed by voltage-sensitive dyes. Nature. 381: 322–325.
11. Palmer L.M., Stuart G.J. (2006) Site of action potential initiation in layer 5 pyramidal neurons. J Neurosci. 2006 Feb 8;26(6):1854–63.
12. Canepari M., Djurisic M., Zecevic D. (2007) Dendritic signals from rat hippocampal CA1 pyramidal neurons during coincident pre- and post-synaptic activity: a combined voltage- and calcium-imaging study. J Physiol. 580:463–484.
13. Neher E., Augustine G.J. (1992) Calcium gradients and buffers in bovine chromaffin cells. J Physiol. 450:273–301.
14. Helmchen F., Imoto K., Sakmann, B (1996) Ca^{2+} buffering and action potential-evoked $Ca^{2}+$ signaling in dendrites of pyramidal neurons. Biophys. J. 70:1069–1081.
15. Stuart G.J., Sakmann B. (1994). Active propagation of somatic action potentials into neocortical pyramidal cell dendrites. Nature 367, 69–72.
16. Spruston N., Schiller Y., Stuart G., Sakmann B. (1995). Activity-dependent action potential invasion and calcium influx into hippocampal CA1 dendrites. Science 268, 297–300.
17. Magee J.C., Johnston D. (1995). Synaptic activation of voltage-gated channels in the dendrites of hippocampal pyramidal neurons. Science, 268, 301–304.
18. Magee J.C., Christofi G., Miyakawa H., Christie B., Lasser-Ross N., Johnston D. (1995). Subthreshold synaptic activation of voltage-gated calcium channels mediate a localized calcium influx into dendrites

of hippocampal pyramidal neurons. J. Neurophysiol., 74, 335–324.
19. Stuart G.J., Hausser M. (2001) Dendritic coincidence detection of EPSPs and action potentials. Nat. Neurosci. 4:63–71.
20. Berger T., Larkum M.E., Luscher H.R. (2001) High I(h) channel density in the distal apical dendrite of layer V pyramidal cells increases bidirectional attenuation of EPSPs. J Neurophysiol 85: 855–868.
21. Antic S., Major G., Zecevic D. (1999) Fast optical recording of membrane potential changes from dendrites of pyramidal neurons. J. Neurophysiol., 82, 1615–1621.
22. Antic S., Wuskell J.P., Loew L., Zecevic D. (2000). Functional profile of the giant metacerebral neuron of *Helix aspersa*: Temporal and spatial dynamics of electrical activity in situ. J Physiol (Lond) *527*,55–69.
23. Antic S. (2003). Action potentials in basal and oblique dendrites of rat neocortical pyramidal neurons. J Physiol 550, 35–50.
24. Djurisic M., Antic S., Chen, W-r., Zecevic D. (2004) Voltage imaging from dendrites of mitral cells: EPSP attenuation and spike trigger zones. J. Neuroscience. 24, 6703–6714.
25. Antic S., Zecevic D. (1995). Optical signals from neurons with internally applied voltage- sensitive dyes. J Neuroscience, 15, 1392–1405.
26. Grinvald A., Hildesheim R., Farber I.C., Anglister L. (1982). Improved fluorescent probes for the measurement of rapid changes in membrane potential. Biophys J, 39, 301–308.
27. Ross W.N., Krauthamer V. (1984) Optical measurements of potential changes in axons and processes of neurons of a barnacle ganglion. J Neurosci 4: 659-672.
28. Grinvald A., Salzberg B. M., Lev-Ram V., Hildesheim R. (1987). Optical recording of synaptic potentials from processes of single neurons using intracellular potentiometric dyes. Biophys J. 51, 643–651.
29. Zecevic D., Antic S. (1998) Fast optical measurement of membrane potential changes at multiple sites on an individual nerve cell. Histochem J., 30:197–216.
30. Bischofberger J., Jonas P. (1997) Action potential propagation into the presynaptic dendrites of rat mitral cells. J Physiol (Lond) 504: 359–65.
31. Chen W.R., Midtgaard J., Shepherd G.M. (1997). Forward and backward propagation of dendritic impulses and their synaptic control in mitral cells. Science 278, 463–467.
32. Christie J.M., Westbrook G.L. (2003) Regulation of backpropagating action potentials in mitral cell lateral dendrites by A-type potassium currents. J Neurophysiol 89: 2466–2472.
33. Davila H.V., Salzberg B.M., Cohen L.B., Waggoner A.S. (1973). A large change in axon fluorescence that provides a promising method for measuring membrane potential. Nature, New Biol., 241, 159–160.
34. Tsien R.W., Tsien R.Y. (1990). "Calcium channels, stores, and oscillations." Annu. Rev. Cell Biol. 6: 715–760.
35. Berridge M.J., Lipp P., Bootman M.D. (2000). "The versatility and universality of calcium signaling." Nat. Rev. Mol. Cell. Biol. 1: 11–21.
36. Verkhratsky A. (2005). "Physiology and pathophysiology of the calcium store in the endoplasmic reticulum of neurons." Physiol. Rev. 85: 201–279.
37. Tsien R.Y. (1981). "A non-disruptive technique for loading calcium buffers and indicators into cells." Nature 290: 527–528.
38. Stosiek C., Garaschuk O., Holthoff K., Konnerth A. (2003) In vivo two-photon calcium imaging of neuronal networks. Proc. Natl. Acad. Sci. U S A. 100:7319–7324.
39. Brustein E., Marandi N., Kovalchuk Y., Drapeau P., Konnerth A. (2003). "'In vivo' monitoring of neuronal network activity in zebrafish by two-photon Ca^{2+} imaging." Pflugers Arch. 446: 766–773.
40. Nimmerjahn A., Kirchhoff F., Kerr J.N.D., Helmchen F. (2004). "Sulforhodamine 101 as a specific marker of astroglia in the neocortex in vivo." Nature Methods 1: 31–37.
41. Kerr J.N., Greenberg D., Helmchen F. (2005). "Imaging input and output of neocortical networks in vivo." Proc. Natl. Acad. Sci. U S A 102: 14063–14068.
42. Li J., Mack J.A., Souren M., Yaksi E., Higashijima S., Mione M., Fetcho J.R., Friedrich R.W. (2005). "Early development of functional spatial maps in the zebrafish olfactory bulb." J. Neurosci. 25: 5784–5795.
43. Niell C.M., S.J. Smith (2005). "Functional imaging reveals rapid development of visual response properties in the zebrafish tectum." Neuron 45: 941–951.
44. Ohki K., Chung S., Ch'ng Y.H., Kara P., Reid R.C. (2005). "Functional imaging with cellular resolution reveals precise microarchitecture in visual cortex." Nature 433: 597–603.
45. Sullivan M.R., Nimmerjahn A., Sarkisov D.V., Helmchen F., Wang S.S. (2005). "In vivo calcium imaging of circuit activity in cerebellar cortex." J. Neurophysiol. 94: 1636–1644.

46. Garaschuk O., Milos R.I., Grienberger C., Marandi N., Adelsberger H., Konnerth A. (2006). "Optical monitoring of brain function in vivo: From neurons to networks." Pflugers Arch. 453: 385–396.
47. Xu F., Kida I., Hyder F., Shulman R. (2000) Assessment and discrimination of odor stimuli in rat olfactory bulb by dynamic functional MRI. Proc Natl Acad Sci USA 97:10601–10606.
48. Woo C.C., Hingco E.E., Johnson B.A., Leon M. (2007) Broad activation of the glomerular layer enhances subsequent olfactory responses. Chemical Senses, 32: 51–55.
49. Rubin B., Katz L. (1999) Optical imaging of odorant representations in the mammalian olfactory bulb. Neuron 23:499–511.
50. O'Donovan M.J., Sholomenko S.Ho.G., Yee W. (1993). Real-time imaging of neurons retrogradely and anterogradely labeled with calcium-sensitive dyes. J. Neuroscience Methods, 46, 91–106.
51. Tsau Y., Wenner P., O'Donovan M.J., Cohen L.B., Loew L.M., Wuskell, J.P.. (1996). Dye screening and signal-to-noise ratio for retrogradely transported voltage-sensitive dyes. J. Neuroscience Methods, 70, 121–129.
52. Kreitzer A.C., Gee K.R., Archer E.A., Regehr W.G. (2000) Monitoring presynaptic calcium dynamics in projection fibers by in vivo loading of a novel calcium indicator. Neuron. 27, 25–32.
53. Friedrich R.W., Korsching S.I., (1997) Combinatorial, and chemotropic odorant coding in the zebrafish olfactory bulb visualized by optical imaging. Neuron, 18, 737–752.
54. Wachowiak M., Cohen L.B., (2001) . Representation of odorants by receptor neuron input to the mouse olfactory bulb. Neuron, Neuron, 32: 725–737.
55. Yaksi E., Friedrich R.W. (2006) Reconstruction of firing rate changes across neuronal populations by temporally deconvolved Ca^{2+} imaging. Nat. Methods. 5:377–383.
56. Vucinic D., Cohen L.B., and Kosmidis E.K. (2004), Presynaptic center-surround inhibition shapes sensory input to the mouse olfactory bulb. J. Neurophysiology, 2006 95:1881–1887.
57. Lam Y.-W., Cohen L.B., Wachowiak M., Zochowski M.R., (2000), Odors elicit three different oscillations in the turtle olfactory bulb. J. Neuroscience, 20:749–762.
58. Wachowiak M., Cohen L.B., (1999), Presynaptic inhibition of primary olfactory afferents mediated by different mechanisms in the lobster and turtle. J. Neuroscience, 19, 8808–8817.
59. Ross W.N., B.M. Salzberg L.B. Cohen A. Grinvald H.V. Davila A.S. Waggoner, Wang C.H. (1977). Changes in absorption, fluorescence, dichroism, and birefringence in stained giant axons : Optical measurement of membrane potential. J Memb. Biol, 33, 141–183.
60. Boyle M.B., Cohen L.B. (1980). Birefringence signals that monitor membrane potential in cell bodies of molluscan neurons. Fed Proc, 39, 2130.
61. Orbach H.S., Cohen L.B. (1983). Optical monitoring of activity from many areas of the in vitro and in vivo salamander olfactory bulb: A new method for studying functional organization in the vertebrate central nervous system. J Neuroscience, 3, 2251–2262.
62. Orbach H.S., Cohen L.B., Grinvald A. (1985). Optical mapping of electrical activity in rat somatosensory and visual cortex. J Neuroscience, 5, 1886–1895.
63. Grynkiewicz G., Poenie M., Tsien R.Y. (1985) A new generation of Ca^{2+} indicators with greatly improved fluorescence properties. J Biol Chem. 260:3440–3450.
64. Gross E., Bedlack R.S., Loew L.M. (1994). Dual-wavelength ratiometric fluorescence measurements of the membrane dipole potential. Biophysical J. 67, 208–216.
65. Braddick H.J.J. (1960). Photoelectric photometry. Rep Prog Physics, 23, 154–175.
66. Malmstadt H.V., Enke C.G., Crouch S.R., Harlick G.. (1974). Electronic Measurements for scientists, Benjamin, Menlo Park, CA.
67. Helmchen F., Denk W. (2005). "Deep tissue two-photon microscopy." Nat. Methods 2: 932–940.
68. Salzberg B.M., Grinvald A., Cohen L.B., Davila H.V., Ross W.N. (1977). Optical recording of neuronal activity in an invertebrate central nervous system: Simultaneous monitoring of several neurons. J Neurophysiol, 40, 1281–1291.
69. London J.A., Zecevic D., Cohen, L.B.. (1987). Simultaneous optical recording of activity from many neurons during feeding in *Navanax.* J Neurosci, 7, 649–661.
70. Svoboda K., Denk W., Kleinfeld D., Tank D.W. (1997). "In vivo dendritic calcium dynamics in neocortical pyramidal neurons." Nature 385: 161–165.
71. Denk W., Svoboda K. (1997). "Photon upmanship: Why multiphoton imaging is more than a gimmick." Neuron 18: 351–357.
72. Svoboda K., Yasuda R. (2006). "Principles of two-photon excitation microscopy and its

applications to neuroscience." Neuron 50: 823–839.
73. Inoue S. (1986) Video Microscopy. Plenum Press, New York. p 128.
74. Kleinfeld D., Delaney K.R.. (1996). Distributed representation of vibrissa movement in the upper layers of somatosensory cortex revealed with voltage-sensitive dyes. J Comparative Neurology, 375, 89–108.
75. Cohen L.B., Keynes R.D. (1971). Changes in light scattering associated with the action potential in crab nerves. J. Physiol (Lond), 212, 259–275.
76. Petran M., Hadravsky M. (1966). Czechoslovakian patent 7720.
77. Denk W, Strickler J.H., Webb W.W. (1990). "Two-photon laser scanning fluorescence microscopy." Science 248: 73–76.
78. Yuste R., W. Denk. (1995). Dendritic spines as basic functional units of neuronal integration. Nature 375, 682–684.
79. Loew L.M. (1993). Confocal microscopy of potentiometric fluorescent dyes. Methods Cell Biol 38, 195–209.
80. Shaw R. (1979). Photographic detectors. Appl Optics Optical Eng, 7, 121–154.
81. Hamer F.M. (1964). The Cyanine Dyes and Related Compounds, Wiley, New York.
82. Loew L.M., Cohen L.B., Dix J., Fluhler E.N., Montana V., Salama G., Wu J.Y. (1992). A napthyl analog of the aminostyryl pyridinium class of potentiometric membrane dyes shows consistent sensitivity in a variety of tissue, cell, and model membrane preparations. J Memb. Biology, 130, 1–10.
83. Rohr S., Salzberg B.M. (1994). Multiple site optical recording of transmembrane voltage in patterned growth heart cell cultures: assessing electrical behavior, with microsecond resolution, on a cellular and subcellular scale. Biophys J. 67, 1301–1315.
84. Gonzalez J.E., Tsien R.Y. (1995). Voltage sensing by fluorescence energy transfer in single cells. Biophysical J, 69, 1272–1280.
85. Cacciatore T.W., Brodfuehrer P.D., Gonzalez J.E., Jiang T., Adams S.R., Tsien R.Y., Kristan W.B. Jr., Kleinfeld D. (1999) Identification of neural circuits by imaging coherent electrical activity with FRET-based dyes. Neuron. 23: 449–459.
86. Briggman K.L., Kristan W.B. (2006) Imaging dedicated and multifunctional neural circuits generating distinct behaviors. J Neurosci. 26:10925–10933.
87. Bouevitch O., Lewis A., Pinevsky I., Wuskell J., Loew L. (1993). Probing membrane potential with non-linear optics. Biophysical J, 65, 672–679.
88. Millard A.C., Jin L., Wuskell J.P., Boudreau D.M., Lewis A., Loew L.M. (2005) Wavelength- and time-dependence of potentiometric non-linear optical signals from styryl dyes. J Membr Biol. 208:103–111.
89. Dombeck D.A., Sacconi L., Blanchard-Desce M., Webb W. W. (2005) Optical recording of fast neuronal membrane potential transients in acute mammalian brain slices by second-harmonic generation microscopy. J Neurophysiol. 94:3628–3636.
90. Hickie C., Wenner P., O'Donovan M., Tsau Y., Fang J., Cohen L.B. (1996). Optical monitoring of activity from individual and identified populations of neurons retrogradely labeled with voltage-sensitive dyes. Abs Soc Neuroscience, 22, 321.
91. Nirenberg S., Cepko C. (1993). Targeted ablation of diverse cell classes in the nervous system in vivo. J Neuroscience, 13, 3238–3251.
92. Siegel M.S., Isacoff E.Y., (1997) A genetically encoded optical probe of membrane voltage. Neuron 19:735–41.
93. Ataka K., Pieribone V.A. (2002) A genetically-targetable fluorescent probe of channel gating with rapid kinetics. Biophysical J., 82, 509–516.
94. Baker B.J., Lee H., Pieribone V.A., Cohen L.B., Isacoff E.Y., Knopfel T., Kosmidis E.K. (2007) Fluorescent protein voltage sensors exhibit low plasma membrane expression in mammalian cells. J. Neuroscience Methods, 161: 32–38.
95. Dimitrov D., He Y., Mutoh H., Baker B.J., Cohen L. Akemann W., and Knopfel T. (2007) Engineering and characterization of an enhanced fluorescent protein voltage sensor. PLOS One, 2(5); e440.
96. Murata Y, Iwasaki H, Sasaki M, Inaba K, Okamura Y (2005) Phosphoinositide phosphatase activity coupled to an intrinsic voltage sensor. Nature 435: 1239–1243.
97. Baird G.S., Zacharias D.A., Tsien R.Y. (1999). "Circular permutation and receptor insertion within green fluorescent proteins." Proc. Natl. Acad. Sci. U S A 96: 11241–11246.
98. Nagai T., Sawano A., Park E.S., Miyawaki A. (2001). "Circularly permuted green fluorescent proteins engineered to sense Ca^{2+}." Proc. Natl. Acad. Sci. U S A 98: 3197–3202.
99. Nakai J., Ohkura M., Imoto K. (2001). "A high signal-to-noise Ca^{2+} probe composed of a single green fluorescent protein." Nat. Biotechnol. 19: 137–141.
100. Zhang J., Campbell R.E., Ting A.Y., Tsien R.Y. (2002). "Creating new fluorescent

probes for cell biology." Nat Rev Mol Cell Biol 3(12): 906–918.

101. Ohkura M., Matsuzaki M., Kasai H., Imoto K., Nakai J. (2005). "Genetically encoded bright Ca^{2+} probe applicable for dynamic Ca^{2+} imaging of dendritic spines." Anal. Chem. 77: 5861–5869.
102. Reiff D.F., Ihring A., Guerrero G., Isacoff E.Y., Joesch M., Nakai J., Borst A. (2005). "In vivo performance of genetically encoded indicators of neural activity in flies." J. Neurosci. 25: 4766–4778.
103. Miyawaki A., Llopis J., Heim R., McCaffery J.M., Adams J.A., Ikura M., Tsien R.Y. (1997). "Fluorescent indicators for Ca^{2+} based on green fluorescent proteins and calmodulin." Nature 388: 882–887.
104. Miyawaki A. (2005). "Innovations in the imaging of brain functions using fluorescent proteins." Neuron 48: 189–199.
105. Hasan M.T., Friedrich R.W., Euler T., Larkum M.E., Giese G., Both M., Duebel J., Waters J., Bujard H., Griesbeck O., Tsien R.Y., Nagai T., Miyawaki A., Denk W. (2004). "Functional fluorescent Ca^{2+} indicator proteins in transgenic mice under TET control." PLoS Biol. 2: 763–775.
106. Nagai T., Yamada S., Tominaga T., Ichikawa M., Miyawaki A. (2004). "Expanded dynamic range of fluorescent indicators for $Ca^{2+}$by circularly permuted yellow fluorescent proteins." Proc. Natl. Acad. Sci. U S A 101: 10554–10559.
107. Diez-Garcia J., Matsushita S., Mutoh H., Nakai J., Ohkura M., Yokoyama J., Dimitrov D., Knopfel T. (2005). "Activation of cerebellar parallel fibers monitored in transgenic mice expressing a fluorescent Ca^{2+} indicator protein." Eur. J. Neurosci. 22: 627–635.
108. Garaschuk O., Griesbeck O., Konnerth A. (2007). "Troponin C-based biosensors: A new family of genetically encoded indicators for in vivo calcium imaging in the nervous system." Cell Calcium. in press.
109. Heim N. and O. Griesbeck (2004). "Genetically encoded indicators of cellular calcium dynamics based on troponin C and green fluorescent protein." J. Biol. Chem. 279: 14280–14286.
110. Mank M., Reiff D.F., Heim N., Friedrich M.W., Borst A., Griesbeck O. (2006). "A FRET-based calcium biosensor with fast signal kinetics and high fluorescence change." Biophys. J. 90: 1790–1796.
111. Heim N., Garaschuk O., Friedrich M.W., Mank M., Milos R.I., Kovalchuk Y., Konnerth A., Griesbeck O. (2007). "Improved calcium imaging in transgenic mice expressing a troponin-C based biosensor." Nature Methods 4: 127–129.
112. Garaschuk O., Milos R.I., Konnerth A. (2006). "Targeted bulk-loading of fluorescent indicators for two-photon brain imaging in vivo." Nat. Prot. 1: 380–386.

Chapter 4

Two-Photon Imaging of Capillary Blood Flow in Olfactory Bulb Glomeruli

Pascale Tiret, Emmanuelle Chaigneau, Jérôme Lecoq, and Serge Charpak

Abstract

Two-photon laser scanning microscopy (TPLSM) is an efficient tool to study cerebral blood flow (CBF) and cellular activity in depth in the brain. We describe here the advantages and weaknesses of the olfactory bulb as a model to study neurovascular coupling using TPLSM. By combining intra- and extracellular recordings, TPLSM of CBF in individual capillaries, local application of drugs, we show that odor triggers odorant-specific and concentration-dependent increases in CBF. We also demonstrate that activation of neurons is required to trigger blood flow responses.

Key words: Calcium, neuronal activity, odor, neurovascular coupling.

1. Introduction

In humans, most non-invasive brain imaging techniques assess neuronal activation from measurement of hemodynamic changes. However, the molecular and cellular mechanisms underlying the neurovascular coupling are not fully understood (for review see *(1–5)*). If there is a general consensus that neuronal activation is tightly correlated to cerebral blood flow (CBF) responses, the requirement of postsynaptic activation in the generation of vascular responses is still under debate *(6–11)* (for review see *(5, 12, 13)*). In vitro brain preparations may seem ideal preparations to investigate this question and studies using them have shown several molecular mechanisms involved in the regulation of arteriolar diameter by glial cells and neurons *(14–20)*. However, these preparations have also raised issues such as the absence

Fahmeed Hyder (ed.), *Dynamic Brain Imaging: Multi-Modal Methods and In Vivo Applications, vol. 489*

DOI 10.1007/978-1-59745-543-5_4 Springerprotocols.com

of adequate smooth muscle resting tone, the disruption of cellular networks, the use of abnormal glucose and oxygen levels, issues that certainly affect neurovascular coupling. Therefore, in vitro results require to be validated in vivo with the same temporal and spatial resolution, a challenge that can now be achieved using new imaging tools. We describe here the advantages of combining two-photon laser scanning microscopy and well defined neuronal networks such as olfactory bulb glomeruli, in order to investigate the role of postsynaptic neuronal activation in triggering vascular responses.

2. The Glomerular Module as a Model to Study Neurovascular Coupling

Deciphering the nature of neurovascular coupling in the brain requires the choice of a region with the following properties: It should be activated with physiological stimuli, its anatomical organization should allow to determine which cell type and number are activated upon stimulation as well as which capillaries supply the activated cells. Under these conditions, one can realistically envisage to understand the entire chain of events that links transmitter release to local cerebral blood flow (CBF) changes. At first glance, the rodent olfactory bulb seems to be an ideal model. As the cortex, it is a dorsal region easily accessible to imaging and electrophysiological recordings. Because of its anatomical organization, it is functionally organized in modules: Olfactory receptor neurons (ORN), expressing a given odorant receptor type, send their axons to the bulb where they converge onto only 2 topographically fixed glomeruli *(21, 22)*, each one located on one side of a plane of symmetry that separates each bulb in two *(23)*. Thus, at low concentration, one may expect an odorant to activate only 2 distant "symmetric" glomeruli and 2 different odorants to activate different glomeruli (*see* **Fig. 4.1**). However, several imaging methods have revealed that odorants activate small glomeruli ensembles. These methods include: (i) Blood oxygenation level dependent functional MRI *(24–27)*, (ii) Optical imaging of calcium *(28–30)* voltage *(31)* and intrinsic signals *(32–37)*, (iii) Electrophysiological recordings *(38–40)*, (iv) Measurements of 2-deoxyglucose consumption *(41, 42)*. Thus, if diluting odorants does reduce the number of activated glomeruli, the limitation of odorant receptor specificity, the nature of the signals detected as well as the detection thresholds of these methods do not allow to ensure that only the 2 "symmetric" glomeruli are activated in an olfactory bulb. Still, this region remains a good model to study neurovascular coupling for several other reasons: (i) It is the first relay of olfaction in the central nervous system and, as such, it is weakly influenced by anesthesia in comparison to cortical areas.

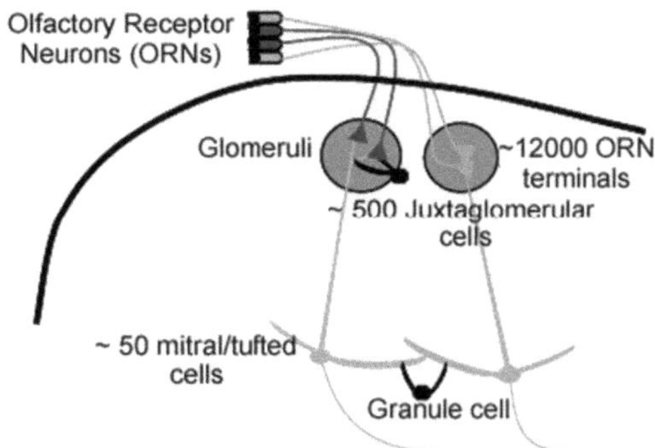

Fig. 4.1. Synaptic organization of the olfactory bulb. Terminals from ORNs expressing different odorant receptors converge into different glomeruli.

(ii) The glomerular synaptic connectivity is rather well known with about 12,000 ORN terminals converging and releasing glutamate onto ~500 juxtaglomerular cells and only ~50 output (mitral/tufted) cells. It should be thus possible to quantify precisely the relationship between changes in glomerular CBF and mitral cell output. (iii) The glomerular module offers the possibility to record simultaneously blood flow and neuronal activity at the cellular level by using two-photon laser scanning microscopy.

3. Two-Photon Laser Scanning Microscopy (TPLSM), Strengths and Limitations

It took less than a decade between the demonstration of two-photon excitation *(43)* and its first application to record neuronal activity *(44)* and blood flow *(45)* in vivo. We first used TPLSM to analyze mitral cell responses to odor in the anesthetized rat. We could demonstrate that several types of calcium signals were evoked by odor, involving the resting membrane potential, postsynaptic potentials and backpropagating spikes *(46, 47)*. Simultaneously, we investigated which optical and biological parameters limit TPLSM in depth. Because the principle of TPLSM has been reviewed several times, we will just mention several technical points important to keep in mind when imaging in depth using standard optics (without using gradient refractive index lenses) *(48, 49)* or optical fibers *(50, 51)*. Light scattering limits both the incident infrared light and the collected fluorescent light, the number of ballistic (unscattered) photons diminishing in an exponential fashion with the thickness of brain tissue. This has obvious implications for in vivo imaging *(52)*:

3.1. Incident Light

The incident light scattering length increases with the wave length. As a result, in order to improve imaging in depth, it may be worthwhile to excite a chromophore at the longest wavelength possible (900–1000 nm instead of 800 nm) even though such wavelength does not necessarily give the best two-photon absorption cross-section. (i) The incident light scattering length also decreases with the age of brain tissue. We found that it is divided by two in the adult brain (3-month old) in comparison to 2-week old brains. A good compromise is to use young adults in which most neurophysiological processes are matured and in which the scattering length has intermediate values. (ii) Imaging at large depth raises a problem of fluorescence background: the high incident light intensity required to excite the chromophore in depth becomes sufficient to excite chromophores at the surface or just below, even if the beam is not focussed. As a result, the autofluorescence or the fluorescence generated in cells or vessels on the beam path may generate a background that decreases the signal-to-noise ratio of the signal of interest. Finally, the best is to choose the incident light experimentally i.e. to vary and choose the incident beam wavelength that gives the best signal-to-noise ratio at the depth of interest. (iii) Imaging at large depth can also be improved by redistributing the laser power with a regenerative amplifier: the average power is kept constant while the laser pulse energy is increased and the repetition rate decreased *(53, 54)*. Note that the photodamage caused by such pulses in vivo still need to be investigated. (iv) An additional problem worth mentioning concerns light aberrations resulting from refraction index changes at the surface and in the tissue. These aberrations reduce the resolution and efficiency of the incident light. They can partially be corrected with adaptive optics *(55)* but it has not been attempted in the mammalian brain in vivo.

3.2. Collection of Fluorescence Light

If it is acknowledged that high numerical aperture objectives are best for fluorescence collection, it is necessary that the angular acceptance of the microscope detection system matches that of the objective. Practically, the detection system must be as close as possible to the objective back aperture. The gain in fluorescence collection is then particularly interesting at large depth *(52)*.

4. Capillary Distribution in the Superficial Olfactory Bulb Layers

As shown in the cortex *(45)*, an I.V. injection of fluorescein- or texas red dextran easily reveals the vascular architecture of the dorsal OB. In order to classify the vessel types according to the different OB superficial layers, we labeled glomeruli by intranasal injections of Oregon green- or Texas red dextrans (PM 10,000) 1–5 days prior to microcirculation experiments *(28)*. Note that fluorescence from nasal and vessel dyes were acquired on two “green” and “red” channels (separated using a dichroic mirror with a cut-off wavelength of 560 nm). TPLSM revealed fluorescent axons in the olfactory nerve layer, converging terminals in precisely delineated glomeruli. This allowed us to show that, in the rat, capillaries were exclusively located below the olfactory nerve layer, i.e. in the glomerular layer and below *(56)*. Their presence was tightly linked to the sites of synaptic interactions and there was no clear organization within the glomerular layer, capillaries often crossing the juxtaglomerular zones between two glomeruli. The capillary network thus does not present an obvious differentiation in close relationship with the glomerular organization. In addition, because sensory axons converging on a single glomerulus outnumber principal cells by a factor of several hundred and because action potentials are predicted to consume a major part of the energy budget during signaling *(57)*, our result raises the question of how much energy is consumed in the olfactory nerve layer and how it is distributed to olfactory nerve axons. Glucose consumption is significant in the olfactory nerve layer *(41, 42, 58)*. However, our preliminary data based on oxygen measurements suggest that oxygen is much more consumed in glomeruli than in the nerve layer during brief odor stimulations. What is thus possible is that few axons are activated during odor stimulation and that part of the oxygen is brought by diffusion from arteries and arterioles *(59)*. Still, because in some cases, few, if any, vessels were apparent in the nerve layer within a 400–400 μm column, it is probable that oxygen is supplied by diffusion from the glomerular layer. Hence, the question of the energy distribution and consumption in the nerve layer remains open.

5. Blood Flow in Individual Glomerular Capillaries at Rest and During Odor Stimulation

Three blood flow parameters can be extracted from line scan measurements in individual capillaries, instantaneous red blood cell (inst.RBC) flow; inst.RBC velocity and inst.RBC linear density (*see* **Figs. 4.1 and 4.2**).

At rest, all parameters are variable and fluctuate in time. Variations of inst.RBC flow result principally from changes in the inst.RBC linear density rather than from changes in inst.RBC velocity. In the rat, we usually have to screen 5–20 odorant molecules in order to obtain a response in a given glomerulus, even though the same region is targeted in the dorsal bulb. Note that this "efficiency" depends on the anesthetics used. In the majority of cases, odor induces an increase in RBC flow that results from an increase in RBC velocity (**Fig. 4.1** right). In about a quarter of the cases, an increase in RBC linear density is also observed. As expected from neuronal

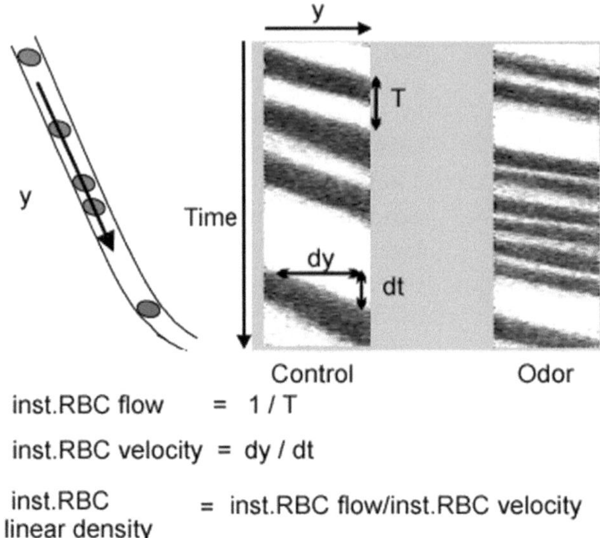

Fig. 4.2. Longitudinal line scans through a capillary reveals red blood cells (RBCs) as oblique shadows. RBC velocity (dy/dt) increases during odor (*right*).

responses, vascular responses are concentration-dependent and odor-specific. In our initial study *(56)*, we measured blood flow in contiguous glomeruli and observed that vascular responses seemed glomerular-specific. However, because optical functional imaging of calcium *(28, 29)*, voltage *(31)* and intrinsic signals *(32–37)* have demonstrated that several clustered glomeruli are often activated by an odor and because we did not map responses in all contiguous glomeruli, it remained unclear whether vascular responses are regulated at the level of a single glomerulus.

6. Vascular Responses Involve Several Clustered Glomeruli in Most Cases

Solving the problem (of clustered activation of glomeruli) requires to find the ideal condition where only one glomerulus is activated. However, this condition is impossible to ascertain using either TPLSM imaging of blood flow at the level of a single capillary or other optical imaging techniques. The first approach is so precise that it would require to screen one capillary per glomerulus over a large portion of the dorsal olfatory bulb. Other optical techniques detect signals generated by the activation of neuronal populations even though activation of a few cells may be sufficient to trigger vascular responses. We thus recently used an alternative approach which relies on the pharmacological inactivation of synaptic transmission or even of neuronal activity of all the neurons belonging to a single glomerular module *(60)*. To our surprise, such glomerular inactivation (with pressure applications of NBQX + DAP5 or tetrotoxin) did not block and even did not affect vascular responses even though local neuronal responses were abolished . This implies that in our experimental conditions, more than one glomerulus are usually activated by the odor. Silencing the recorded glomerulus does not affect a vascular response involving several clustered glomeruli.

7. Vascular Responses Require Postsynaptic Activation of Glomerular Neurons

The difficulty to activate a single glomerulus precludes the use of intraglomerular pharmacology to assert or exclude the role of postsynaptic activation in the generation of vascular responses. What is necessary is to block glutamatergic synaptic transmission in all glomeruli from the rat dorsal olfactory bulb. We recently superfused glutamate antagonists onto the bulb surface, i.e. omitting the application of agar while imaging blood flow. Imaging blood flow requires brain movements inferior to 1–2 μm and we obtained stable measurements of RBC flow in only 1/3 experiments. High concentrations of NBQX and D-AP5 decreased both neuronal and vascular responses by about 40 % (**Fig. 4.3**).

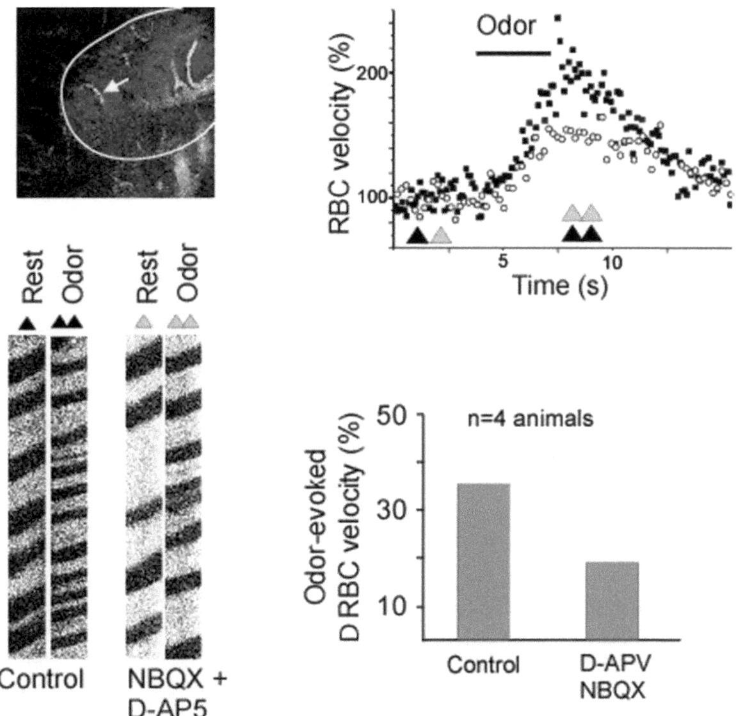

Fig. 4.3. Superfusion of glutamate antagonists decreases odor-evoked vascular responses (rat, top left, the arrow points to the recorded capillary).

These experiments were however extremely difficult to perform and raised several problems: (i) The effect of drugs was unpredictable, sometimes rapidly reversible due to drug washout by blood flow, (ii) High concentrations of NBQX decreased the signal-to-noise ratio of fluorescent signals and occasionally decreased or blocked basal blood flow. We repeated these superfusion experiments in the mouse in which the olfactory nerve layer is thin as opposed to rat's, favoring the drug access to the glomerular layer. We used G-CaMP2 transgenic mice *(61)* which express G-CaMP2 fluorescence in mitral cells and thus strongly improve our capacity to detect responding glomeruli. In G-CaMP2 mice, superfusion of glutamate receptor antagonists strongly reduced or even fully blocked both postsynaptic Ca^{2+} and vascular responses. These results allowed us to conclude that activation of postsynaptic glutamate receptor plays a major role in triggering the vascular response. It is thus possible that intrinsic signals observed in the presence of glutamate antagonists *(11)* reflect changes in light scattering involving either olfactory nerve terminals or astrocytes.

8. Concluding Remarks

Can we eliminate a possible involvement of glial cells in glomerular vascular responses? Our data show that direct activation of glial cells by the glutamate release from olfactory nerve terminals is not sufficient to trigger vascular responses. However, blocking postsynaptic neuronal responses also blocked dendritic release. Is it thus conceivable that dendritic release from mitral cells or from some juxtaglomerular cell type activates glial cells and a downstream cascade leading to vasodilation. The question of a "disynaptic" activation of glial cells remains open. Still, the regulation of neurovascular coupling seems different in the olfactory bulb and in the neocortex. *(62)*

In the course of our investigations, our data questionned the use of local field potential (LFP) responses as reliable markers of neuronal activity to study neurovascular coupling *(60)*. We initially characterized which component of the odor-evoked LFP response was generated locally, within glomerular boundaries. We placed extracellular fluorescent recording pipettes in the center of glomeruli under TPLSM "visual" control and found that odor evoked various types of LFP responses with several components. However, one of them was exclusively glomerulus-specific: It consisted of "rapid" negativities that were locked to the 2 Hz respiration frequency, disappeared at the glomerular border and were sensitive to intraglomerular NBQX or TTX. As suggested by Rall and Shepherd decades ago *(63)*, we demonstrated that LFP negativities were most likely generated by mitral/tufted cell EPSPs.

The problem arose because the "rapid" negativities could be fully blocked by intraglomerular NBQX or TTX without affecting vascular responses. This meant that under physiological conditions where several spatially distinct neuronal populations are activated, i.e. several glomeruli, local blow flow could be independent of the glomerular activity recorded locally with a single electrode. This problem will be particularly important when the distribution of the activated brain loci is unknown or when the spatial resolution of the imaging technique used to measure CBF includes several activated loci.

A very recent study confirmed our results that local intraglomerular blockade of glutamate ionotropic antagonists did not blocked odor-evoked vascular responses and even increased them by decreasing presynaptic inhibition. More importantly, it also showed that local activation of astrocytic metabotropic glutamate receptors and glutamate transporters can trigger functional hyperemia independently of neuronal activation *(64)*. How can these results be reconciled with the fact that, as in cerebellum, topical application of NBQX and D-AP5 decreased (rat) or blocked (mice) odor-evoked functional hyperemia? One possibility is that the experimental conditions required to achieve topical applications alter some mechanisms of neurovascular coupling. A second one is that we used higher odor concentrations, inducing postsynaptic responses that trigger additional regulatory mechanisms of vascular responses *(65)*? To conclude, in the bulb as in other brain regions, several pathways most probably participate to functional hyperemia.

References

1. Villringer, A. & Dirnagl, U. Coupling of brain activity and cerebral blood flow: Basis of functional neuroimaging. *Cerebrovasc. Brain Metab Rev.* 7, 240–276 (1995).
2. Magistretti, P.J., Pellerin, L., Rothman, D.L., & Shulman, R.G. Energy on demand. *Science* **283**, 496–497 (1999).
3. Attwell, D. & Iadecola, C. The neural basis of functional brain imaging signals. *Trends Neurosci.* **25**, 621–625 (2002).
4. Logothetis, N.K. & Pfeuffer, J. On the nature of the BOLD fMRI contrast mechanism. *Magn. Reson. Imaging* **22**, 1517–1531 (2004).
5. Lauritzen, M. Reading vascular changes in brain imaging: Is dendritic calcium the key? *Nat. Rev. Neurosci.* **6**, 77–85 (2005).
6. Mathiesen, C., Caesar, K., Akgoren, N., & Lauritzen, M. Modification of activity-dependent increases of cerebral blood flow by excitatory synaptic activity and spikes in rat cerebellar cortex. *J. Physiol.-Lond.* **512**, 555–566 (1998).
7. Yang, G. & Iadecola, C. Activation of cerebellar climbing fibers increases cerebellar blood flow: Role of glutamate receptors, nitric oxide, and cGMP. *Stroke* **29**, 499–507 (1998).
8. Nielsen, A.N. & Lauritzen, M. Coupling and uncoupling of activity-dependent increases of neuronal activity and blood flow in rat somatosensory cortex. *J. Physiol.-Lond.* **533**, 773–785 (2001).
9. Hoffmeyer, H.W., Enager, P., Thomsen, K.J., & Lauritzen, M.J. Nonlinear neurovascular coupling in rat sensory cortex by activation of transcallosal fibers. *J. Cereb. Blood Flow Metab*(2006).
10. Gsell, W. *et al.* Differential effects of NMDA and AMPA glutamate receptors on functional magnetic resonance imaging signals and evoked neuronal activity during forepaw

stimulation of the rat. *J. Neurosci.* **26**, 8409–8416 (2006).

11. Gurden, H., Uchida, N., & Mainen, Z.F. Sensory-evoked intrinsic optical signals in the olfactory bulb are coupled to glutamate release and uptake. *Neuron* **52**, 335–345 (2006).
12. Logothetis, N.K. The Underpinnings of the BOLD Functional Magnetic Resonance Imaging Signal. *J. Neurosci.* **23**, 3963–3971 (2003).
13. Iadecola, C. & Nedergaard, M. Glial regulation of the cerebral microvasculature. *Nat Neurosci* **10**, 1369–1376 (2007).
14. Zonta, M. *et al.* Neuron-to-astrocyte signaling is central to the dynamic control of brain microcirculation. *Nat. Neurosci.* **6**, 43–50 (2003).
15. Cauli, B. *et al.* Cortical GABA interneurons in neurovascular coupling: Relays for subcortical vasoactive pathways. *J. Neurosci.* **24**, 8940–8949 (2004).
16. Mulligan, S.J. & MacVicar, B.A. Calcium transients in astrocyte endfeet cause cerebrovascular constrictions. *Nature* **431**, 195–199 (2004).
17. Filosa, J.A., Bonev, A.D., & Nelson, M.T. Calcium dynamics in cortical astrocytes and arterioles during neurovascular coupling. *Circ. Res.* **95**, e73–e81 (2004).
18. Metea, M.R. & Newman, E.A. Glial cells dilate and constrict blood vessels: A mechanism of neurovascular coupling. *J. Neurosci.* **26**, 2862–2870 (2006).
19. Filosa, J.A. *et al.* Local potassium signaling couples neuronal activity to vasodilation in the brain. *Nat. Neurosci.* **9**, 1397–1403 (2006).
20. Rancillac, A. *et al.* Glutamatergic Control of Microvascular Tone by Distinct GABA Neurons in the Cerebellum. *J. Neurosci.* **26**, 6997–7006 (2006).
21. Vassar, R. *et al.* Topographic organization of sensory projections to the olfactory bulb. *Cell* **79**, 981–991 (1994).
22. Ressler, K.J., Sullivan, S.L., & Buck, L.B. Information coding in the olfactory system: Evidence for a stereotyped and highly organized epitope map in the olfactory bulb. *Cell* **79**, 1245–55 (1994).
23. Nagao, H., Yoshihara, Y., Mitsui, S., Fujisawa, H., & Mori, K. Two mirror-image sensory maps with domain organization in the mouse main olfactory bulb. *Neuroreport* **11**, 3023–3027 (2000).
24. Yang, X. *et al.* Dynamic mapping at the laminar level of odor-elicited responses in rat olfactory bulb by functional MRI. *Proc. Natl. Acad. Sci. U. S. A* **95**, 7715–7720 (1998).
25. Xu, F., Kida, I., Hyder, F., & Shulman, R.G. Assessment and discrimination of odor stimuli in rat olfactory bulb by dynamic functional MRI. *Proc. Natl. Acad. Sci. U. S. A* **97**, 10601–10606 (2000).
26. Kida, I., Xu, F., Shulman, R.G., & Hyder, F. Mapping at glomerular resolution: fMRI of rat olfactory bulb. *Magn Reson. Med.* **48**, 570–576 (2002).
27. Schafer, J.R., Kida, I., Xu, F., Rothman, D.L., & Hyder, F. Reproducibility of odor maps by fMRI in rodents. *Neuroimage.* **31**, 1238–1246 (2006).
28. Wachowiak, M. & Cohen, L.B. Representation of odorants by receptor neuron input to the mouse olfactory bulb. *Neuron* **32**, 723–735 (2001).
29. Spors, H., Wachowiak, M., Cohen, L.B., & Friedrich, R.W. Temporal dynamics and latency patterns of receptor neuron input to the olfactory bulb. *J. Neurosci.* **26**, 1247–1259 (2006).
30. Wachowiak, M., Denk, W., & Friedrich, R.W. Functional organization of sensory input to the olfactory bulb glomerulus analyzed by two-photon calcium imaging. *Proc Natl Acad Sci U S A* **101**, 9097–9102 (2004).
31. Spors, H. & Grinvald, A. Spatio-temporal dynamics of odor representations in the mammalian olfactory bulb. *Neuron* **34**, 301–315 (2002).
32. Rubin, B.D. & Katz, L.C. Optical imaging of odorant representations in the mammalian olfactory bulb. *Neuron* **23**, 499–511 (1999).
33. Uchida, N., Takahashi, Y.K., Tanifuji, M., & Mori, K. Odor maps in the mammalian olfactory bulb: Domain organization and odorant structural features. *Nat. Neurosci.* **3**, 1035–1043 (2000).
34. Belluscio, L. & Katz, L.C. Symmetry, stereotypy, and topography of odorant representations in mouse olfactory bulbs. *J. Neurosci.* **21**, 2113–2122 (2001).
35. Luo, M. & Katz, L.C. Response correlation maps of neurons in the mammalian olfactory bulb. *Neuron* **32**, 1165–1179 (2001).
36. Meister, M. & Bonhoeffer, T. Tuning and topography in an odor map on the rat olfactory bulb. *J. Neurosci.* **21**, 1351–1360 (2001).
37. Yuan, Q., Harley, C.W., McLean, J.H., & Knopfel, T. Optical imaging of odor preference memory in the rat olfactory bulb. *J. Neurophysiol.* **87**, 3156–3159 (2002).
38. Imamura, K., Mataga, N., & Mori, K. Coding of odor molecules by mitral/tufted cells in rabbit olfactory bulb. I. Aliphatic compounds. *J. Neurophysiol.* **68**, 1986–2002 (1992).
39. Katoh, K., Koshimoto, H., Tani, A., & Mori, K. Coding of odor molecules by mitral/tufted cells in rabbit olfactory bulb. II. Aromatic

compounds. *J. Neurophysiol.* 70, 2161–2175 (1993).
40. Yokoi, M., Mori, K., & Nakanishi, S. Refinement of odor molecule tuning by dendrodendritic synaptic inhibition in the olfactory bulb. *Proc Natl Acad Sci U S A* **92**, 3371–5 (1995).
41. Stewart, W.B., Kauer, J.S., & Shepherd, G.M. Functional organization of rat olfactory bulb analysed by the 2-deoxyglucose method. *J. Comp Neurol.* **185**, 715–734 (1979).
42. Johnson, B.A., Woo, C.C., & Leon, M. Spatial coding of odorant features in the glomerular layer of the rat olfactory bulb. *J. Comp Neurol.* %20;393, 457–471 (1998).
43. Denk, W., Strickler, J.H., & Webb, W.W. Two-photon laser scanning fluorescence microscopy. *Science* **248**, 73–76 (1990).
44. Svoboda, K., Denk, W., Kleinfeld, D., & Tank, D.W. In vivo dendritic calcium dynamics in neocortical pyramidal neurons. *Nature* **385**, 161–165 (1997).
45. Kleinfeld, D., Mitra, P.P., Helmchen, F., & Denk, W. Fluctuations and stimulus-induced changes in blood flow observed in individual capillaries in layers 2 through 4 of rat neocortex. *Proc Natl Acad Sci U S A* **95**, 15741–15746 (1998).
46. Charpak, S., Mertz, J., Beaurepaire, E., Moreaux, L., & Delaney, K. Odor-evoked calcium signals in dendrites of rat mitral cells. *Proc Natl Acad Sci U S A* **98**, 1230–1234 (2001).
47. Debarbieux, F., Audinat, E., & Charpak, S. Action potential propagation in dendrites of rat mitral cells in vivo. *J. Neurosci.* **23**, 5553–5560 (2003).
48. Jung, J.C., Mehta, A.D., Aksay, E., Stepnoski, R., & Schnitzer, M.J. In vivo mammalian brain imaging using one- and two-photon fluorescence microendoscopy. *J. Neurophysiol.* **92**, 3121–3133 (2004).
49. Levene, M.J., Dombeck, D.A., Kasischke, K.A., Molloy, R.P., & Webb, W.W. In vivo multiphoton microscopy of deep brain tissue. *J. Neurophysiol.* **91**, 1908–1912 (2004).
50. Gobel, W., Kerr, J.N., Nimmerjahn, A., & Helmchen, F. Miniaturized two-photon microscope based on a flexible coherent fiber bundle and a gradient-index lens objective. *Opt. Lett.* **29**, 2521–2523 (2004).
51. Helmchen, F., Fee, M.S., Tank, D.W., & Denk, W. A miniature head-mounted two-photon microscope: High-resolution brain imaging in freely moving animals. *Neuron* **31**, 903–912 (2001).
52. Oheim, M., Beaurepaire, E., Chaigneau, E., Mertz, J., & Charpak, S. Two-photon microscopy in brain tissue: Parameters influencing the imaging depth. *J. Neurosci. Methods* **111**, 29–37 (2001).
53. Beaurepaire, E., Oheim, M., & Mertz, J. Ultra-deep two-photon fluorescence excitation in turbid media. *Opt. Commun* **188**, 25–29 (2001).
54. Theer, P., Hasan, M.T., & Denk, W. Two-photon imaging to a depth of 1000 microm in living brains by use of a Ti:Al2O3 regenerative amplifier. *Opt. Lett.* **28**, 1022–1024 (2003).
55. Rueckel, M., Mack-Bucher, J.A., & Denk, W. Adaptive wavefront correction in two-photon microscopy using coherence-gated wavefront sensing. *Proc. Natl. Acad. Sci. U. S. A* **103**, 17137–17142 (2006).
56. Chaigneau, E., Oheim, M., Audinat, E., & Charpak, S. Two-photon imaging of capillary blood flow in olfactory bulb glomeruli. *Proc Natl Acad Sci U S A* **100**, 13081–13086 (2003).
57. Attwell, D. & Laughlin, S.B. An energy budget for signaling in the grey matter of the brain. *J. Cereb. Blood Flow Metab* **21**, 1133–1145 (2001).
58. Sharp, F.R., Kauer, J.S., & Shepherd, G.M. Local sites of activity-related glucose metabolism in rat olfactory bulb during olfactory stimulation. *Brain Res.* **98**, 596–600 (1975).
59. Tsai, A.G. *et al.* Microvascular and tissue oxygen gradients in the rat mesentery. *Proc. Natl. Acad. Sci. U. S. A* **95**, 6590–6595 (1998).
60. Chaigneau, E. *et al.* The relationship between blood flow and neuronal activity in the rodent olfactory bulb. *J. Neurosci.* **27**, 6452–6460 (2007).
61. Diez-Garcia, J. *et al.* Activation of cerebellar parallel fibers monitored in transgenic mice expressing a fluorescent Ca2+ indicator protein. *Eur. J. Neurosci.* **22**, 627–635 (2005).
62. Takano, T. *et al.* Astrocyte-mediated control of cerebral blood flow. *Nat. Neurosci.* **9**, 260–267 (2005).
63. Rall, W. & Shepherd, G.M. Theoretical reconstruction of field potentials and dendrodendritic synaptic interactions in olfactory bulb. *J. Neurophysiol.* **31**, 884–915 (1968).
64. Petzold, G.C., Albeanu, D.F., Sato, T.F., & Murthy, V.N. Coupling of neural activity to blood flow in olfactory glomeruli is mediated by astrocytic pathways. Neuron **58**, 897–910 (2008).
65. Shepherd, G.M. & Charpak, S. The olfactory glomerulus: a model for neuro-glio-vascular biology. Neuron **58**, 827–829 (2008).

Chapter 5

Astrocytic Calcium Signaling: Mechanism and Implications for Functional Brain Imaging

Xiaohai Wang, Takahiro Takano, and Maiken Nedergaard

Abstract

Astrocytes are electrically non-excitable cells that, on a slow time scale of seconds, integrate synaptic transmission by dynamic increases in cytosolic Ca^{2+}. A number of groups have recently shown that astrocytic Ca^{2+} signaling regulates vascular tones and that astrocytes play a central role in functional hyperemia by Ca^{2+}-dependent release of Prostaglandin E2 (PGE_2). Astrocytes are, however, not simple detectors of excitatory transmission, since a number of neuromodulator and hormones trigger elevations in astrocytic Ca^{2+} independently of synaptic transmission. Furthermore, astrocytes exhibit ex vivo intrinsic Ca^{2+} excitability, or spontaneous increases in Ca^{2+} that are not triggered by receptor activation. The notion that astrocytes can regulate vascular tone independently of synaptic transmission challenges the notion that changes in the blood oxygenation level dependent (BOLD) signal is directly proportional to neuronal activity and may thus require a reevaluation of the large body of data accumulated using functional magnetic resonance imaging (fMRI).

Key words: Photolysis, 2-photon imaging, functional brain imaging.

1. Introduction

Until about two decades ago, astrocytes were regarded as non-excitable supporting cells in the brain, lacking membrane properties required to generate action potentials. The view of astrocytes as simple supporting cells was questioned in the early 1990s, with the demonstration that glutamate can trigger Ca^{2+} oscillations among cultured glial cells often resulting in propagating intercellular Ca^{2+} waves *(1)*. The notion that astrocytes can actively propagate Ca^{2+} waves among each other led to the studies of Ca^{2+} signaling between astrocytes and neurons in mixed cultures. These studies revealed that astrocytic Ca^{2+} waves

Fahmeed Hyder (ed.), *Dynamic Brain Imaging: Multi-Modal Methods and In Vivo Applications, vol. 489*

DOI 10.1007/978-1-59745-543-5_5 Springerprotocols.com

modulate neuronal cytosolic Ca^{2+}, large spike-like increases in neuronal Ca^{2+} levels, suggesting that astrocytes participate more directly in neurotransmission than previously recognized *(2, 3)*. Bidirectional communication between neurons and astrocytes was later demonstrated in acute brain slices *(4–7)* and the isolated retina preparation *(8)*. Although bidirectional communication has been studied for more than a decade, it was first recently demonstrated that astrocytes are activated during sensory stimulation in vivo *(9)*. This was an important step, because activation of astrocytes in previous ex vivo experiments often required excessive stimulation, including high intensity electrical stimulation *(10)*, mechanical distortion *(11)*, traumatic injury *(12)*, agonist application *(13)* and laser light *(14)*. Understanding the role of astrocytic Ca^{2+} signaling in the intact brain has direct implication for functional brain imaging, since new lines of evidence suggest that astrocytes play a key role in functional hyperemia (reviewed in *(15)*). The basis of blood oxygen level dependent (BOLD) functional magnetic resonance imaging (fMRI), and positron emission tomography (PET) functional imaging, as well as intrinsic signal optical imaging, is that neural activity is coupled to increases in local blood flow *(16)*. However, the mechanisms linking neuronal activity to increases in local perfusion remain poorly understood. Since the first description by Golgi at the end of 1800s that astrocytic endfeet contact arterioles and capillaries in the brain, it has been speculated that astrocytes have the potential to modulate the cerebral blood flow. Indeed, several in vitro studies highlighted the importance of neuron-to-glia signaling in vessel diameter *(10, 17–20)*. Although different mechanisms have been proposed for the function of astrocytes in control of microcirculation, a common conclusion is that elevations in astrocytic cytosolic Ca^{2+} levels are necessary and sufficient to cause changes in the diameter of arterioles. Yet, an essential question to be addressed is whether these astrocytic Ca^{2+} transients occur in vivo during physiological neural activation.

We have, in recent work, addressed the signaling pathway from neurons to astrocytes in live adult mice using 2-photon laser scanning microscopy. We analyzed changes in neuronal local field potentials (LFPs) and astrocytic Ca^{2+} signaling evoked by whisker stimulation in barrel cortex layer 2 of adult mice. We found that whisker stimulation promptly evoked astrocytic Ca^{2+} responses in vivo. The slow onset ($\sim$3 s) and long time course ($\sim$20 s) of the astrocytic Ca^{2+} signaling compared with the properties of LFPs suggested that temporal summation of electrical events with a long time constant determined the development of the astrocytic Ca^{2+} elevation. We found that the changes of astrocytic Ca^{2+} signaling were a direct function of the summated LFPs amplitude, which was commonly used to predict the amplitude of increases in cerebral blood flow changes during sensory stimulation *(21, 22)*.

Our recent observation on astrocytic Ca^{2+} signaling will here be discussed in the context of functional brain imaging.

2. Astrocytes Can Sense and Modulate Synaptic Activity

2.1. Astrocytes Increase Ca^{2+} in Response to Receptors Activation

Astrocytes make contact with most synapses in the CNS and respond to synaptic activity with increases in cytosolic Ca^{2+} ex vivo *(4, 23)*. Although astrocytes are electrically non-excitable, they express receptors for most neurotransmitters *(24)*, including glutamate *(1)*, GABA *(4)*, norepinephrine *(25)* and acetylcholine *(26)*. Retinal Müller cells and astrocytes also respond to synaptic release of ATP by activation of purinergic receptors *(14)*. The types of neurotransmitter receptors found on astrocytes in situ and in vivo are mainly metabotropic receptors, which link to the second messenger system including, activation of phospholipase C, adenyl cyclase and production of IP3, Ca^{2+} and cAMP *(27, 28)* (**Fig. 5.1**).

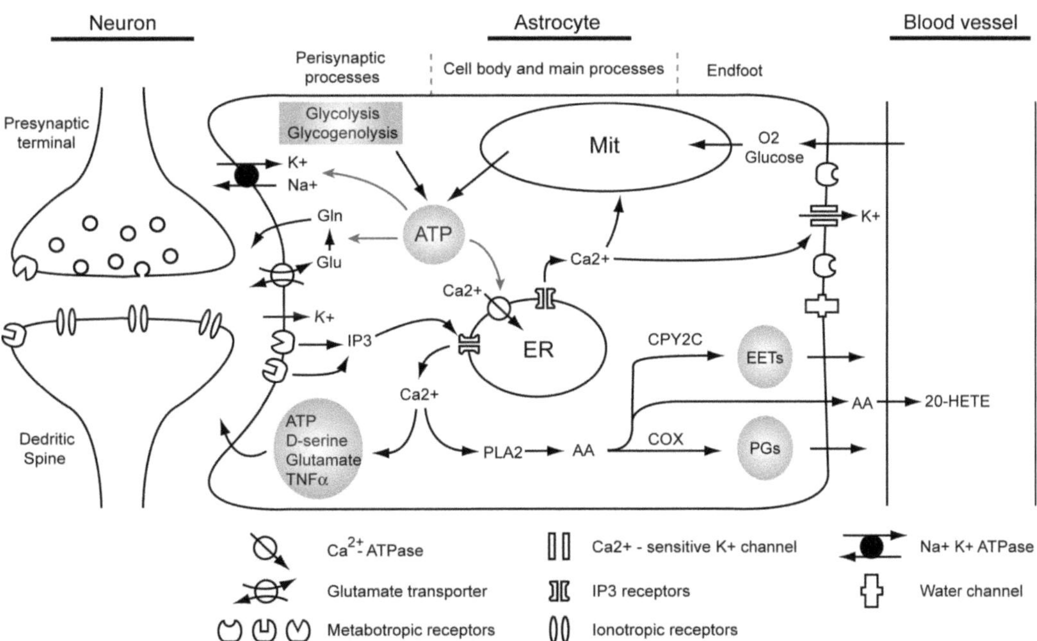

Fig. 5.1. Proposed mechanisms that link synaptic activity, cerebral vasculature regulation and astrocytic energy metabolism with astrocytic Ca^{2+}signaling. Synaptic released neurotransmitters activate metabotropic receptors in astrocytic perisynaptic processes, causing elevation of intracellular Ca^{2+} through a phospholipase C-dependent pathway. Increases in astrocytic Ca^{2+} are associated with release of gliotransmitters which feed back to the synaptic activity. Astrocytic Ca^{2+} signaling may also propagate to the astrocytic endfoot, releasing vasoactive agents through phospholipase A2(PLA2)-dependent pathway as well as activation of Ca^{2+}-sensitive K^+ channels. Furthermore, Ca^{2+} is a well established stimulator of oxidative-phosphorylation which accelerates ATP production from mitochondria (Mit). Due to the small diameter of perisynaptic processes and the absence of mitochondria in these structures, ATP can also be generated through glycolysis and glycogenolysis pathways particularly in astrocyte perisynaptic processes. The major energy consuming processes in astrocyte include converting glutamate (Glu) to glutamine (Gln) after taken up by glutamate transporters, uptake of K^+ from extracellular space through Na^+-K^+ ATPase and restoring astrocytic cytosolic Ca^{2+} concentration by activating both plasma membrane and endoplasmic reticulum (ER) Ca^{2+} -ATPase.

Several lines of evidence indicate that metabotropic glutamate receptors (mGluR) play a key role in the mobilization of intracellular Ca^{2+} stores in response to synaptic release of glutamate *(5, 6)*. In culture, astrocytes respond to glutamate with oscillatory increases in Ca^{2+} and intercellular Ca^{2+} waves propagation *(1)*. Glutamate released during neuronal activity may reach astrocytic receptors through synaptic "spillover" *(29, 30)* or ectopic release *(31)*. In addition, neurotransmitters may directly activate astrocytes through volume transmission *(32)*.

2.2. Modulatory Role of Gliotransmitters Released by Astrocytes

Although we know that astrocytes sense the neurotransmitters in extracellular space, we know very little about the consequences of astrocytic receptors activation. In vitro preparations have shown that astrocytic Ca^{2+} signaling is associated with release of gliotransmitters. Several of the best studied gliotransmitters include glutamate, ATP, D-serine, TNF-alpha, and arachidonic acid metabolites *(3, 10, 33–35)* (**Fig. 5.1**). Thus, by sensing the extracellular neurotransmitter concentration and increasing intracellular Ca^{2+}, astrocytes may actively participate in the information processing in the normal brain.

3. Interface Between Astrocytic Endfeet and the Vasculatures

Astrocytes send processes to contact not only synapses but also the vasculature forming specialized endfoot structures in the astrocyte-vasculature interface (AVI). A variety of channels and receptors required for interaction between astrocytes and vasculatures are clustered in the astrocytic endfoot processes. For instance, purinergic receptors P2Y2 and P2Y4 as well as gap junctions composed of Cx43 have been shown to be preferentially expressed in the astrocytic endfeet *(36)*. Accordingly, direct application of ATP, electrical stimulation and uncaging evoked astrocytic Ca^{2+} waves can propagate along the AVI *(17, 36)*. Increases of astrocytic Ca^{2+} in the endfeet can activate Ca^{2+} – sensitive K^+ channels, which is also abundantly expressed in the endfeet *(37)*, and subsequently dilate intracerebral arterioles by increasing extracellular K^+ concentration in the AVI *(19, 38)* (**Fig. 5.1**).

4. In Vivo Imaging of Astrocytic Ca2+ Signaling by 2-Photon Laser Scanning Microscopy (2PLSM)

Due to the technical challenges, imaging astrocytic Ca^{2+} signaling in vivo was only possible after the introduction of 2PLSM *(39)* and the in vivo Ca^{2+} indicator loading techniques *(40, 41)*.

The invention of 2PLSM by Webb's group in 1990 *(39)* has tremendously improved our understanding of neurobiological phenomena in both normal and diseased brain. With two-photon excitation (2PE), it is possible to image deep brain tissue in the live animals with single cell spatial resolution *(42)*. We have used 2PLSM because 2PE provides several key advantages compared with single photon excitation (1PE). First, because of the localized excitation in 2PE, all emission fluorescent photons contribute to the useful signals. In contrast, a pinhole has to be used to block the emission light from the unfocused plan with 1PE. Therefore, 2PE has much higher efficiency than 1PE. Second, since only the fluophores in the focused point are activated by 2PE, whereas the entire light pathway is activated during 1PE, phototoxicity and photobleaching are greatly reduced in 2PE. This is especially important for time-lapse imaging of live tissue. Moreover, the excitation wavelength used in 2PE (740~900 nm) penetrates tissue better than 1PE (350~550 nm), which is essential for in vivo imaging of live animals.

Although blind intracellular recordings *(43)* have been widely used for loading Ca^{2+} indicators into neuronal structures under 2PLSM, delivering Ca^{2+} indicators into single astrocyte through recording pipette in live animals is still challenging. However, recent studies demonstrated that astrocyte population can be labeled by either local dye injection *(40)* or pial surface dye application *(41)*. Together with the discovery of specific astrocyte fluorescent indicator sulforhodamine 101 *(44)*, these pioneer studies enable, for the first time, the observation of astrocytic Ca^{2+} activity in response to physiological stimulation in the live animals *(9)* (**Fig. 5.2**).

We found that astrocytic Ca^{2+} signals can be clearly detected from pial surface to ~300 μm deep in adult mouse primary somatosensory cortex (barrel cortex) following surface loading with fluorescent Ca^{2+} indicators. This penetration depth correlates with the layer1 and part of the layer 2/3 of mouse barrel cortex.

4.1. Animal Preparation

Adult FVB or C57BL/6 mice (~8 weeks old) were used for in vivo imaging primarily because application of fluorescent Ca^{2+} indicators on mouse pial surface produced satisfactory astrocyte loading in vivo. Same surface loading procedure only works on immature rats up to ~3 weeks old (unpublished observations). Animals were anesthetized with intraperitoneal injection of ketamine (0.12 mg/g) and xylazine (0.01 mg/g). Depth of anesthesia was monitored by field potential recording and hind limb pinch withdrawal reflexes and kept constant at stage III/3 *(45)* with supplemental doses of anesthetics. Tracheotomy was performed and the mice were intubated and artificially ventilated with a small animal ventilator (SAAR-830, CWE). The

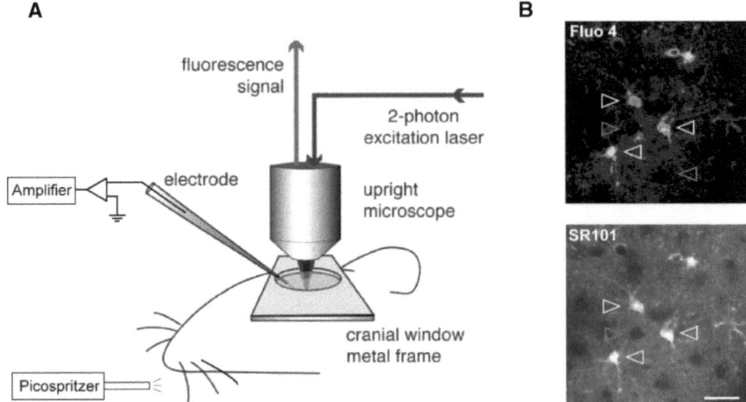

Fig. 5.2. Diagrams of experimental setup. (**A**) A cranial window was opened on the primary somatosensory cortex. A metal frame was glued to the skull to prevent the movement of the brain. Whisker stimulation was performed in the contra lateral side of the animal's snout by air puffs which were generated from a high pressure nitrogen tank and control by picospritzer. A plastic tube connected to the picospritzer was placed ~2 cm away from the C6 whisker in a rostral-caudal direction to deliver the high pressure air. Electrical field potential recording and 2-photon Ca^{2+} imaging were performed simultaneously during whisker stimulation. Agarose (1%) was poured onto the pial surface before the coverglass was mounted to the metal frame to minimize the brain pulsation. (**B**) Dual labeling of barrel cortex with Ca^{2+} indicator fluo-4 AM (*top*) and astrocyte specific marker SR101 (*bottom*). Only SR101 positive astrocytes were labeled with fluo-4 AM (*white arrowheads*). Neurons appear as dark, round areas without either fluo-4 AM or SR101 labeling (*gray arrowheads*). The image shown was located 120 μm below the pial surface. Scale bar, 30 μm.

respiration rate was set at ~100/min; inspiration time, 0.3 s; and the tidal volume, ~0.2 ml. Lung pressure was carefully monitored and maintained at lower than 10 cm H2O to prevent lung injury. Blood gas was tested by taking arterial blood through a femoral artery catheter and pCO_2, pO_2 and pH were analyzed in microsamples (Rapidlab 248, Bayer, sample size 40 μl). Blood samples were taken every 1–2 h and totally 3–4 times from each animal. Experiments were completed only if physiological variables remained within the normal limits. The normal limits for pCO_2 were set at 30–45 mm Hg; pO_2, 80–115 mm Hg; and pH, 7.35–7.45. To image astrocytic Ca^{2+} signals in mouse barrel cortex, a cranial window (2–3 mm in diameter) was prepared centered at 0.5 mm posterior to the bregma and 3.5 mm lateral from midline. To stabilize the brain during imaging, a custom-made metal plate was glued to the skull with dental acrylic cement. Dura matter was removed for surface loading of fluorescent indicators. Body temperature was monitored by a rectal probe and maintained at 37°C by a heating blanket (BS4, Harvard Apparatus).

4.2. Ca^{2+} Indicators Loading

The major difference comparing surface bulk loading and local dye injection is that the former loading procedure only labels astrocyte structures due to the uptake by astrocyte limiting membranes that completely cover the pial surface *(41)*, whereas the latter protocol loads both astrocytes and neurons *(40,46)*. For the surface bulk loading, Fluo-4 AM (0.5–1 mM) was dissolved in dimethylsulphoxide (DMSO) with 20% pluronic acid and mixed in artificial cerebrospinal fluid (aCSF) containing 126 mM NaCl, 2.5 mM KCl, 1.25 mM NaH2PO4, 2 mM MgCl2, 2 mM CaCl2, 10 mM glucose and 26 mM NaHCO3 (pH 7.4), gassed with 95% O2 and 5% CO2 at 37°C *(47)*. aCSF containing the dye (10–15 μl) was dropped on the pial surface, then the cranial window was covered by a small piece of parafilm to prevent the dye from drying. After a 45-min incubation, the exposed brain was washed for 15 min with aCSF without dye. In selected experiments, the exposed cortex was also loaded with the astrocyte-specific fluorescent indicator sulforhdamine 101 (SR101, 10 μM) *(44)* for 10 min after fluo-4 AM loading (**Fig. 5.2**). To minimize brain pulsation, 0.9% NaCl containing 1% agarose (37°C) was poured on the cranial window and the glass coverslip was glued to the metal plate with dental acrylic cement. A small opening between coverglass and metal plate was left for inserting recording electrode into the cortex (**Fig. 5.2**).

Serial images from pial surface to barrel cortex layer2/3 (~300 μm deep) revealed that fluo-4 AM signals strictly co-localized with SR101 staining (**Fig. 5.2**). Astrocytes appear as bright fluorescent cells with multiple processes. Astrocyte end-feet are clearly labeled which outline the vasculatures. In contrast, neuronal cell bodies appear as dark, round shaped areas which can be detected at ~90 μm deep in the barrel cortex, owing to their lack of uptake of both fluo-4 AM and SR101(**Fig. 5.2**).

4.3. Two-Photon Imaging

A custom-built microscope attached to Tsunami/Millinium laser (10W, SpectraPhysics) and a scanning box (FV300, Olympus) using Fluoview software and a 20X (0.9 NA, Olympus) objective was used for in vivo imaging. Excitation wavelength was in the range of 820~840 nm for both fluo-4 AM and SR101 imaging simultaneously. Two-channel detection of emission wavelength was achieved by using a 565 nm dichroic mirror (Chroma) and two external photomultiplier tubes. A 525/40 bandpass filter (Chroma) was used to detect fluo-4 AM emission light, and a 620/60 bandpass filter (Chroma) was used to detect SR101 signals (**Fig. 5.2**). For image collection, ~1 s was needed to record a single frame of image at 512×512 pixels resolution. Time-lapse images of astrocytic Ca2+ signaling were collected every 1~3 s. Normally, a sampling interval of 3 s was found sufficient to detect whisker stimulation evoked astrocytic Ca^{2+} elevation, and this low sampling rate was used to avoid the photodamage. The two-photon laser power was carefully adjusted according to the depth of imaging in the brain, since the photodamage induced by high power laser can cause astrocytic Ca^{2+} oscillation in vivo *(9)*. The average laser power that was used for detecting astrocytic Ca^{2+} signaling in barrel cortex layer 2 was less than 30 mV under the objective.

4.4. Whisker Stimulation

Whiskers were stimulated by air puffs *(48)*. Briefly, whiskers on one side of the snout were trimmed to a 10-mm length. Air puffs consisted of pulses of compressed air, which were generated from a high pressure nitrogen tank and controlled by picospritzer (General Valve) at a pressure of 10 psi and delivered by a thick wall plastic tubing (4.5 mm outer-diameter, 2.5 mm inner-diameter). The tube was placed parallel to the left side of the mouse snout and 10–20 mm away from the C6 whisker. The direction of air flow was from rostral to caudal along the whisker row. This protocol will stimulate most of the vibrissae on one side of the whisker pad (at least three rows of whiskers were consistently activated by air puffs) (**Fig. 5.2**). The frequency of air puffs was set to 1, 3, 5, 7 or 10 Hz and controlled by Master-8 (A.M.P.I). Pulse width in all experiments was 10 ms. The duration of stimulation was adjusted from 1 to 60 s in different experiments. Single whisker stimulation was not attempted at this point, but in future studies, it will be intriguing to deflect a principle whisker using mechanic stimulation *(49)* and image the astrocyte Ca^{2+} responses in its related barrel column and the nearby columns.

4.5. Electrophysiological Recording

LFP recordings were obtained from layer 2 of barrel cortex (100–150 μm below the pial surface) by a patch pipette (TW100F-4, WPI; outer diameter, 1.0 mm; inner diameter, 0.75 mm; tip diameter, ~3 μm; tip resistance 3–5 MΩ), containing 0.2% Texas red–dextran (77,000 molecular weight, Sigma) in aCSF. LFP signals were amplified using Axopatch 700B and pCLAMP 8.2 program with DigiData 1332A interface (Axon Instruments). The signals were bandpass filtered at (1–100 Hz) and digitized at 10 kHz.

5. Properties of Astrocytic Ca^{2+} Signaling Evoked by Whisker Stimulation

Astrocytes in barrel cortex layer 2/3 of adult mouse were labeled with Ca^{2+} indicator fluo4-AM. Recording electrode was inserted into the field of imaging to detect LFP activity (**Fig. 5.3A**). Although increased neuronal activity in layer 2/3 of barrel cortex can be detected as early as 10–12 ms after whisker deflection *(50)*, increases in astrocyte somatic Ca^{2+} were delayed by ~3 s after the onset of whisker stimulation *(9)*. The onset of the increase in astrocytic Ca^{2+} was defined as the earliest time point at which the

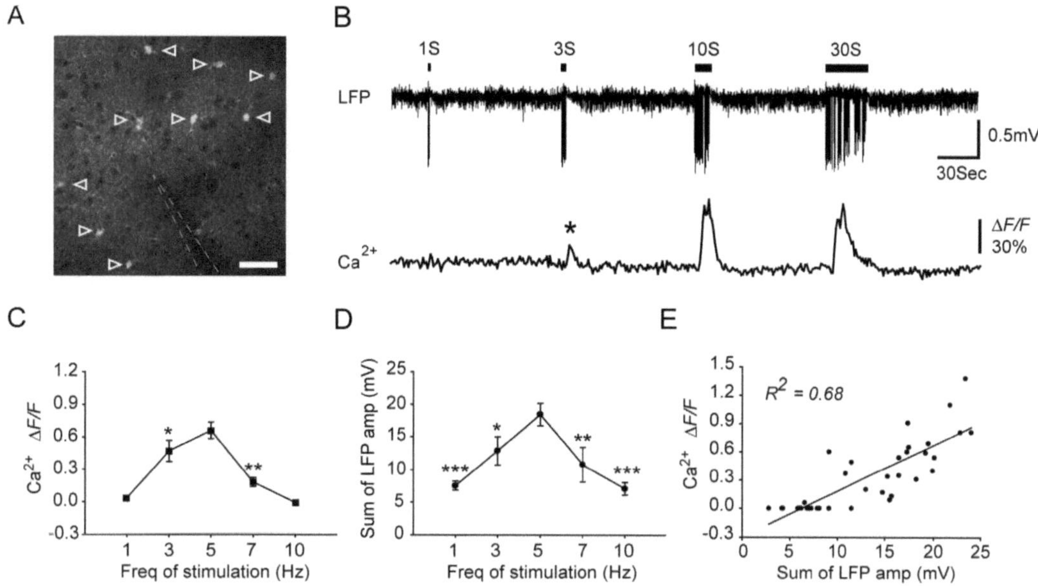

Fig. 5.3. Whisker stimulation evokes astrocytic Ca^{2+} increases in mouse barrel cortex. (**A**) Two-photon fluorescence image taken from layer 2/3 of mouse barrel cortex (~150 μm below the pial surface), showing that astrocytes were labeled with Ca^{2+} indicator fluo-4 AM (*open arrowheads*). Neuronal cell bodies appear as dark, round area, due to their lack of uptake of fluo-4 AM. The location of recording electrode is indicated by white dashed line. Scale bar, 30 μm. (**B**) 5 Hz whisker stimulation induced astrocytic Ca^{2+} increases as a function of the stimulation duration. Upper trace, LFP recorded from recording electrode shown in panel A. Lower trace, averaged relative fluorescent change $\Delta F/F$ in the astrocytic cell bodies indicated in panel A. (**C**) Mean increase in fluo-4 emission after 9 s of different frequencies of stimulation. (**D**) Summed LFP amplitude within the first 9 s of whisker stimulation as a function of the frequency of stimulation. Data are presented as mean ± s.e.m. * $P<0.05$, ** $P<0.01$ and *** $P<0.001$ compared with 5 Hz whisker stimulation induced responses. One-way analysis of variance with Dunnett's test. (**E**) Correlation between the summed LFP amplitude within the first 9 s and the mean increase in fluo-4 emission 9 s after the onset of stimulation ($R^2=0.68$, $P<0.001$).

changes of somatic fluorescent signal reached 3 standard deviation of baseline fluorescent activity. Notably, 1-s stimulation failed to evoke astrocytic Ca^{2+} signaling, whereas 3-s stimulation was sufficient to induce astrocytic Ca^{2+} elevation (asterisk) (**Fig. 5.3B**).

In addition, the amplitude of astrocytic Ca^{2+} increases was similar in response to 10-s and 30-s stimulation, while the duration of astrocytic Ca^{2+} elevation was longer during 30-s stimulation. This evidence indicates that the amplitude of astrocytic Ca^{2+} elevation was mainly determined by the neuronal activity during the first ~10 s of stimulation (**Fig. 5.3B**). Most astrocytes showed a single increase in Ca^{2+} in response to 1 min of whisker stimulation (72 cells from 13 mice), whereas others showed a second Ca^{2+} increase (11 cells from 13 mice).

In addition to the effects of the duration of stimulation on astrocytic Ca^{2+} elevations, astrocytic Ca^{2+} responses are also a function of the frequency of stimulation. During 1 Hz whisker stimulation, only 3 of 37 astrocytes tested in 9 mice showed an increase in Ca^{2+} *(9)*. Rapid sensory adaptation, or a decrease in LFP amplitude within 1 s, occurred at 10 Hz stimulation. Accordingly, 10 Hz whisker stimulation did not trigger astrocytic Ca^{2+} responses *(9)*. Additional experiments using 3 and 7 Hz whisker stimulation confirmed that astrocytic Ca^{2+} responses peaked at 5 Hz and decreased at both lower and higher frequencies (**Fig. 5.3C**). Since astrocytic Ca^{2+} signaling was delayed by ~3 s compared with neuronal field potential signals, there must be an accumulating effect of neuronal activity on the astrocytic Ca^{2+} elevations. To test the idea that astrocytic Ca^{2+} increases are a function of the intensity of local synaptic input, we calculated the summed LFP amplitude during the first 9 s of stimulation. As an index of total neuronal activity within a given time window, the sum of LFPs has also been widely used to positively correlate neuronal activity with cerebral blood flow (CBF) *(21)*. Notably, the summed LFP amplitudes peaked at 5 Hz during the first 9 s of stimulation (**Fig. 5.3D**) and had a strong correlation with astrocyte somatic Ca^{2+} signaling ($R^2 = 0.68$, $P < 0.001$; **Fig. 5.3E**)

6. Implications for BOLD Signaling

6.1. Astrocytic Ca^{2+} Elevations Contribute to Cerebral Blood Flow Regulation

The basis for a positive BOLD signal is the increase in blood flow and volume that occur during neural activity. As discussed above, new development has shown that astrocytes can mediate functional hyperemia by release of vasoactive agents, indicating that astrocytic Ca^{2+} signaling may be both necessary and sufficient for changes in the BOLD signal *(10, 17–19, 51)*. The initial studies on astrocytic control of the microvasculature were performed in slice preparations. One line of work showed that astrocytic Ca^{2+} increases are linked to activation of Ca^{2+} - sensitive phospholipase A2 (PLA2), which, in turn, stimulate the production of arachidonic acid (AA). Several AA metabolites are powerful vasoactive compounds. For example, the products of cyclooxygenase (COX) induced vessel dilation in both brain slice and in vivo preparations *(10)*. Furthermore, the cytochrome P450 epoxygenase pathway produces epoxyeicosatrienoic acids (EETs), which mediated vessel dilation in retina *(18)*, whereas astrocytic Ca^{2+} signaling triggered vessel dilation through activation of Ca^{2+} -sensitive K^+ channels in the astrocyte endfeet in cortical slices *(19)*. In addition, two lines of work have suggested that astrocytic Ca^{2+} signaling in both retina and acute hippocampal slices is linked to production of the vasoconstrictor 20-hydroxyeicosatetraenoic acid, 20-HETE *(17, 18)* (**Fig. 5.1**). However, a major drawback of studying blood flow regulation in ex vivo preparation is the loss of pressure-induced vasculature tone. Our in vivo experiments, employing targeted photolysis of caged Ca^{2+} in astrocytic

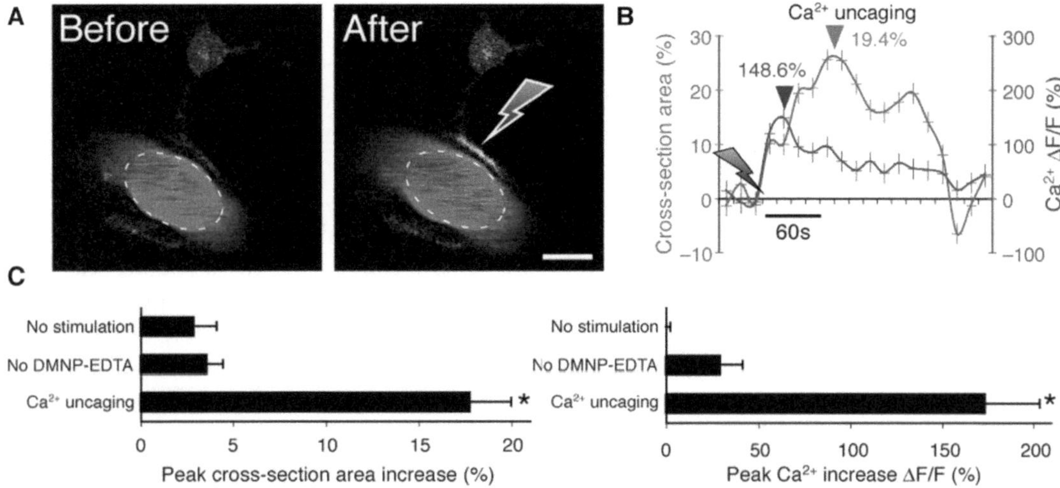

Fig. 5.4. Vasodilation triggered by uncaging (**A**) *Upper panel*: Two-photon images of a vascular astrocyte, which displayed Ca^{2+} increases in the endfoot following photolysis of caged Ca2+. Astrocytes were loaded with the Ca^{2+} indicator dye, rhod-2/am and DMNP-EDTA/am, whereas the vasculature was stained with FITC-dextran. Gray arrow indicates the position of photostimulation. Scale bar, 10 μm. (**B**) Time-course tracings show that photostimulation caused a rapid Ca^{2+} increase and arterial vasodilation. Arrows indicate time of photolysis stimulation. Dark arrowhead indicates the peak of Ca^{2+} elevation, and light arrowhead indicates peak dilation. (**C**) Summary histograms of maximum arterial cross-section area (*left panel*) and maximum Ca^{2+} increase (*right panel*) without photostimulation, with photostimulation but omitting DMNP-EDTA/am loading, and with DMNP-EDTA loading and photolysis. Maximum increase within 1 min after photostimulation was measured relative to baseline before stimulation. Mean $\pm$ s.e.m. * $P < 0.01$ compared to no-stimulation, Tukey-Kramer.

endfeet, indicated that the primary mechanism by which astrocytes mediated vessel dilation is through release of PGE2 in a COX-1-dependent pathway *(51)* (**Fig. 5.4**).

Interestingly, several lines of work have documented that the changes in blood flow and BOLD signal detected during brain activation have a stronger correlation with synaptic inputs compared with spiking activity *(52, 53)*. Indeed, local spiking activity does not appear to play a central role in functional hyperemia *(53)*. Thus, activity-dependent vascular signals reflect excitatory synaptic activity or synaptic release of glutamate *(54, 55)*. Similarly, astrocytes respond to the synaptic inputs, or glutamate released from synaptic terminal during sensory stimulation, and local spiking activity has no effect upon astrocytic Ca^{2+} elevation *(9)*. Thus, astrocytic Ca^{2+} signaling may represent a signaling pathway that locally integrates synaptic inputs and controls the microvasculature. Astrocytes detect presynaptic neurotransmitters release and display increases in intracellular Ca^{2+}. In turn, astrocytic Ca^{2+} signaling activates PLA2 resulting in release of vasoactive COX-1 metabolic products. Other pathways, such as NO, adenosine and pH may in parallel contribute to functional hyperemia. Several of these pathways may be similar to glutamatergic pathways which require astrocytes as a necessary intermediary linking synaptic activity to vasodilatation.

6.2. Astrocytic Ca^{2+} Signaling May Increase Oxidative Metabolism

Glutamate transporters (GluTs), which are mainly expressed in the perisynaptic processes of astrocytes, play an important role in removing the glutamate released during synaptic transmission preventing the extracellular glutamate concentration from reaching excitotoxic concentrations *(56)*. Glutamate uptake is an energy requiring process and stimulates the glycolytic metabolism in cultured astrocytes *(57, 58)*. The main energy consuming processes associated with glutamate transport into astrocytes include extrusion of Na^{+} and conversion of glutamate to glutamine *(59)*. In the model proposed by Pellerin and Magistretti, astrocytes utilize ATP produced from glycolytic glucose metabolism and lactate is transported to neurons for oxidative metabolism *(57, 58)*. Because astrocytes express the full machinery for oxidative-phosphorylation *(60)*, it is somewhat surprising that glutamate uptake increases glycolysis rather than oxidative metabolism *(57)*. A possible explanation is that the perisynaptic processes of astrocytes are essential devoid of most organelles (**Fig. 5.1**). The fine processes of astrocyte covering synapses have a small diameter (~0.3 μm) and are almost devoid of mitochondria *(61)*. Therefore, the glycolytic metabolism of glucose, most likely occur in these fine structures of astrocytes which are heavily engaged in glutamate uptake *(59, 62)*. Does oxidative metabolism in the astrocytic cell bodies contribute to increased ATP production during periods of intense glutamate uptake? Astrocytes contain a large number of mitochondria and the mitochondria volume fraction in the somatic areas of neurons and astrocytes are directly comparable *(60, 61)*.

It is in this regard important to note that mGluRs are expressed in the astrocyte perisynaptic processes in both rodent and primate *(63, 64)*. Ex vivo experiments have shown that electrical stimulation of Schaffer collaterals elicits mGluRs-dependent Ca^{2+} elevation in hippocampal astrocytes *(5, 6)*. Depending on the intensity of electrical stimulation, astrocytic Ca^{2+} signaling can be detected either in the perisynaptic microdomains or propagate intracellularly into the soma. Furthermore, activation of astrocytic mGluRs contributes to increases of astrocytic Ca^{2+} signaling during sensory stimulation in vivo *(9)*. Astrocytic Ca^{2+} increases are first initiated in the perisynaptic processes and propagate within a time frame of 0.5–2 s to the soma triggering a 50–300% fold increase in cytosolic Ca^{2+} *(9)*. Although little is known about the role of Ca^{2+} signaling as a modulator of glycolytic metabolism, it has been established that elevations in mitochondria Ca^{2+} are strong stimulator of oxidative-phosphorylation and accelerate ATP production *(65)*. Therefore, we proposed that astrocytic Ca^{2+} signaling initiated in the perisynaptic processes following propagation into mitochondria rich somatic regions may stimulate the oxidative-phosphorylation. ATP produced by mitochondria in the soma may in turn diffuse back into the perisynaptic processes and supplement ATP produced by local glycolysis during periods of intense glutamate uptake. Thus, a possible function of astrocytic Ca^{2+} signaling evoked during synaptic activity may be to link energy metabolism in processes and cell body by activating oxidative-phosphorylation in mitochondria located distant to the perisynaptic processes.

6.3. Energy Consumption by Astrocytes During Neural Activation

Several reviews have proposed that astrocytes only account for 5% of the total energy budget consumed in rodents and about 6% in humans *(66)*. These estimations were based on calculation of known properties of channels and synapses in neurons and astrocytes *(66)*. In contrast, studies using in vivo nuclear magnetic resonance spectroscopy indicate that astrocytes contribute to 15% to 30% of total oxidative metabolism in the brain *(67, 68)*. The prior calculations may neglect many of the energy requiring functions of protoplasmic astrocytes *(66)*. In addition to glutamate uptake, multiple ATP consuming processes are activated in astrocytes during neural activation. For instance, restoring astrocytic Ca^{2+} concentrations following mGluRs mediated Ca^{2+} mobilization requires activation of both plasma membrane Ca^{2+}-ATPase and ER Ca^{2+}-ATPase *(69)*. Furthermore, astrocytes buffer at least, in part, extracellular K^+ by active uptake of K^+ through Na^+-K^+-ATPase. Astrocytes are the principle cell type responsible for water homeostasis and clearance of tissue swelling during episodes of intense synaptic activity likely requires increased metabolism *(70)*. Moreover, Ca^{2+} dependent activation of PLA2 results in release of arachidonic acid, which in turn requires new lipid synthesis. In addition to this energy requiring supportive function of astrocytes, astrocytic Ca^{2+} signaling likely evokes downstream effects that increase energy demands of both neurons and other non-neuronal cells. Intercellular astrocytic Ca^{2+} signaling is mediated by ATP release from astrocytes *(71)*. ATP is, in addition to its function as an energy metabolite, an important gliotransmitter that triggers Ca^{2+} signaling in surrounding astrocytes *(33)*, neurons *(72)*, microglia *(73)* and smooth muscle cells *(19)*. Normalization of Ca^{2+} in all of these cell types occurs through activation of Ca^{2+}-ATPases. Conversely, hydrolysis of extracellular ATP results in accumulation of adenosine, which reduces release of glutamate and thereby dampens energy demands. The notion that astrocytic Ca^{2+} signaling is an energy requiring process initiated during synaptic activity finds support in the observation that inhibition of postsynaptic activity by application of CNQX/APV and TBOA failed to block changes in NADH signaling evoked during intense electrical stimulation in brain slice preparation *(74)*. Glycolytic metabolism was shown not to play an essential role in the CNQX/APV resistant NADH changes *(74)*, indicating that astrocytic Ca^{2+} increases most significantly affect oxidative- phosphorylation.

Of interest is the observation that the various tasks of astrocytes may preferentially activate oxidative or glycolytic metabolism. In a recent study using 2-photon NADH imaging, we found that cortical spreading depression was associated with a sharp decrease in NADH signaling reflecting increased oxidation in perivascular endfeet. The decrease in NADH occurred during the repolarization phase during which plasma membrane pumps are engaged in normalization of ion gradients *(75)*.

7. Concluding Remarks

Deciphering the metabolic responses of astrocytes in response to neural activation is likely to add a new dimension to our understanding of BOLD signal changes during fMRI. A question for future research is whether astrocytic Ca^{2+} signaling is necessary and sufficient for increases in local blood flow similar to what have been demonstrated in mice using targeted photolysis of caged Ca^{2+} *(51)*.

Acknowledgments

This work was supported by NINDS/NIH NS030007, NS038073, NS50315.

References

1. Cornell-Bell, A.H., et al., *Glutamate induces calcium waves in cultured astrocytes: long-range glial signaling.* Science, 1990. **247**(4941): pp. 470–3.
2. Nedergaard, M., *Direct signaling from astrocytes to neurons in cultures of mammalian brain cells.* Science, 1994. **263**(5154): pp. 1768–71.
3. Parpura, V., et al., *Glutamate-mediated astrocyte-neuron signalling.* Nature, 1994. **369**(6483): pp. 744–7.
4. Kang, J., et al., *Astrocyte-mediated potentiation of inhibitory synaptic transmission.* Nat Neurosci, 1998. **1**(8): pp. 683–92.
5. Porter, J.T. and K.D. McCarthy, *Hippocampal astrocytes in situ respond to glutamate released from synaptic terminals.* J Neurosci, 1996. **16**(16): pp. 5073–81.
6. Pasti, L., et al., *Intracellular calcium oscillations in astrocytes: A highly plastic, bidirectional form of communication between neurons and astrocytes in situ.* J Neurosci, 1997. **17**(20): pp. 7817–30.
7. Pascual, O., et al., *Astrocytic purinergic signaling coordinates synaptic networks.* Science, 2005. **310**(5745): pp. 113–6.
8. Newman, E.A. and K.R. Zahs, *Modulation of neuronal activity by glial cells in the retina.* J Neurosci, 1998. **18**(11): pp. 4022–8.
9. Wang, X., et al., *Astrocytic Ca(2+) signaling evoked by sensory stimulation in vivo.* Nat Neurosci, 2006. **9**(6): pp. 816–23.
10. Zonta, M., et al., *Neuron-to-astrocyte signaling is central to the dynamic control of brain microcirculation.* Nat Neurosci, 2003. **6**(1): pp. 43–50.
11. Araque, A., et al., *Calcium elevation in astrocytes causes an NMDA receptor-dependent increase in the frequency of miniature synaptic currents in cultured hippocampal neurons.* J Neurosci, 1998. **18**(17): pp. 6822–9.
12. Rzigalinski, B.A., et al., *Intracellular free calcium dynamics in stretch-injured astrocytes.* J Neurochem, 1998. 70(6): pp. 2377–85.
13. Fellin, T., et al., *Neuronal synchrony mediated by astrocytic glutamate through activation of*

extrasynaptic NMDA receptors. Neuron, 2004. **43**(5): pp. 729–43.
14. Newman, E.A., *Calcium increases in retinal glial cells evoked by light-induced neuronal activity.* J Neurosci, 2005. **25**(23): pp. 5502–10.
15. Haydon, P.G. and G. Carmignoto, *Astrocyte control of synaptic transmission and neurovascular coupling.* Physiol Rev, 2006. **86**(3): pp. 1009–31.
16. Fox, P.T. and M.E. Raichle, *Focal physiological uncoupling of cerebral blood flow and oxidative metabolism during somatosensory stimulation in human subjects.* Proc Natl Acad Sci U S A, 1986. **83**(4): pp. 1140–4.
17. Mulligan, S.J. and B.A. MacVicar, *Calcium transients in astrocyte endfeet cause cerebrovascular constrictions.* Nature, 2004. **431**(7005): pp. 195–9.
18. Metea, M.R. and E.A. Newman, *Glial cells dilate and constrict blood vessels: A mechanism of neurovascular coupling.* J Neurosci, 2006. **26**(11): pp. 2862–70.
19. Filosa, J.A., et al., *Local potassium signaling couples neuronal activity to vasodilation in the brain.* Nat Neurosci, 2006. **9**(11): pp. 1397–403.
20. Peppiatt, C.M., et al., *Bidirectional control of CNS capillary diameter by pericytes.* Nature, 2006. **443**(7112): pp. 700–4.
21. Mathiesen, C., K. Caesar, and M. Lauritzen, *Temporal coupling between neuronal activity and blood flow in rat cerebellar cortex as indicated by field potential analysis.* J Physiol, 2000. **523** Pt 1: pp. 235–46.
22. Ngai, A.C., et al., *Frequency-dependent changes in cerebral blood flow and evoked potentials during somatosensory stimulation in the rat.* Brain Res, 1999. **837**(1–2): pp. 221–8.
23. Charles, A.C., *Glia-neuron intercellular calcium signaling.* Dev Neurosci, 1994. **16**(3–4): pp. 196–206.
24. Porter, J.T. and K.D. McCarthy, *Astrocytic neurotransmitter receptors in situ and in vivo.* Prog Neurobiol, 1997. **51**(4): pp. 439–55.
25. Duffy, S. and B.A. MacVicar, *Adrenergic calcium signaling in astrocyte networks within the hippocampal slice.* J Neurosci, 1995. **15**(8): pp. 5535–50.
26. Perea, G. and A. Araque, *Properties of synaptically evoked astrocyte calcium signal reveal synaptic information processing by astrocytes.* J Neurosci, 2005. **25**(9): pp. 2192–203.
27. Verkhratsky, A., R.K. Orkand, and H. Kettenmann, *Glial calcium: Homeostasis and signaling function.* Physiol Rev, 1998. 78(1): pp. 99–141.
28. Volterra, A. and J. Meldolesi, *Astrocytes, from brain glue to communication elements: The revolution continues.* Nat Rev Neurosci, 2005. **6**(8): pp. 626–40.
29. Bergles, D.E. and C.E. Jahr, *Synaptic activation of glutamate transporters in hippocampal astrocytes.* Neuron, 1997. **19**(6): pp. 1297–308.
30. Porter, J.T. and K.D. McCarthy, - *Astrocytic neurotransmitter receptors in situ and in vivo.* 1997. **51**: p. 455.
31. Matsui, K. and C.E. Jahr, *Ectopic release of synaptic vesicles.* Neuron, 2003. **40**(6): pp. 1173–83.
32. Zoli, M., et al., *The emergence of the volume transmission concept.* Brain Res Brain Res Rev, 1998. **26**(2–3): pp. 136–47.
33. Cotrina, M.L., J.H. Lin, and M. Nedergaard, *Cytoskeletal assembly and ATP release regulate astrocytic calcium signaling.* J Neurosci, 1998. **18**(21): pp. 8794–804.
34. Schell, M.J., M.E. Molliver, and S.H. Snyder, *D-serine, an endogenous synaptic modulator: Localization to astrocytes and glutamate-stimulated release.* Proc Natl Acad Sci U S A, 1995. **92**(9): pp. 3948–52.
35. Beattie, E.C., et al., *Control of synaptic strength by glial TNFalpha.* Science, 2002. **295**(5563): pp. 2282–5.
36. Simard, M., et al., *Signaling at the gliovascular interface.* J Neurosci, 2003. **23**(27): pp. 9254–62.
37. Price, D.L., et al., *Distribution of rSlo Ca2+-activated K+ channels in rat astrocyte perivascular endfeet.* Brain Res, 2002. **956**(2): pp. 183–93.
38. Filosa, J.A., A.D. Bonev, and M.T. Nelson, *Calcium dynamics in cortical astrocytes and arterioles during neurovascular coupling.* Circ Res, 2004. **95**(10): pp. e73–81.
39. Denk, W., J.H. Strickler, and W.W. Webb, *Two-photon laser scanning fluorescence microscopy.* Science, 1990. **248**(4951): pp. 73–6.
40. Stosiek, C., et al., *In vivo two-photon calcium imaging of neuronal networks.* Proc Natl Acad Sci U S A, 2003. **100**(12): pp. 7319–24.
41. Hirase, H., et al., *Calcium dynamics of cortical astrocytic networks in vivo.* PLoS Biol, 2004. **2**(4): pp. E96.
42. Svoboda, K. and R. Yasuda, *Principles of two-photon excitation microscopy and its applications to neuroscience.* Neuron, 2006. **50**(6): pp. 823–39.
43. Svoboda, K., et al., *Spread of dendritic excitation in layer 2/3 pyramidal neurons in rat barrel cortex in vivo.* Nat Neurosci, 1999. **2**(1): pp. 65–73.
44. Nimmerjahn, A., et al., *Sulforhodamine 101 as a specific marker of astroglia in the neocortex in vivo.* Nat Methods, 2004. **1**(1): pp. 31–7.

45. Friedberg, M.H., S.M. Lee, and F.F. Ebner, *Modulation of receptive field properties of thalamic somatosensory neurons by the depth of anesthesia.* J Neurophysiol, 1999. **81**(5): pp. 2243–52.
46. Kerr, J.N., D. Greenberg, and F. Helmchen, *Imaging input and output of neocortical networks in vivo.* Proc Natl Acad Sci U S A, 2005. **102**(39): pp. 14063–8.
47. Kang, J., et al., *Astrocyte-mediated potentiation of inhibitory synaptic transmission.* Nat Neurosci, 1998. **1**(8): pp. 683–92.
48. Sosnik, R., S. Haidarliu, and E. Ahissar, *Temporal frequency of whisker movement. I. Representations in brain stem and thalamus.* J Neurophysiol, 2001. **86**(1): pp. 339–53.
49. Pinto, D.J., J.C. Brumberg, and D.J. Simons, *Circuit dynamics and coding strategies in rodent somatosensory cortex.* J Neurophysiol, 2000. **83**(3): pp. 1158–66.
50. Petersen, C.C., A. Grinvald, and B. Sakmann, *Spatiotemporal dynamics of sensory responses in layer 2/3 of rat barrel cortex measured in vivo by voltage-sensitive dye imaging combined with whole-cell voltage recordings and neuron reconstructions.* J Neurosci, 2003. **23**(4): pp. 1298–309.
51. Takano, T., et al., *Astrocyte-mediated control of cerebral blood flow.* Nat Neurosci, 2006. **9**(2): pp. 260–7.
52. Logothetis, N.K. and B.A. Wandell, *Interpreting the BOLD signal.* Annu Rev Physiol, 2004. **66**: pp. 735–69.
53. Mathiesen, C., et al., *Modification of activity-dependent increases of cerebral blood flow by excitatory synaptic activity and spikes in rat cerebellar cortex.* J Physiol, 1998. **512** (Pt 2): pp. 555–66.
54. Peppiatt, C. and D. Attwell, *Neurobiology: Feeding the brain.* Nature, 2004. **431**(7005): pp. 137–8.
55. Lauritzen, M., *Reading vascular changes in brain imaging: Is dendritic calcium the key?* Nat Rev Neurosci, 2005. **6**(1): pp. 77–85.
56. Bergles, D.E., J.S. Diamond, and C.E. Jahr, *Clearance of glutamate inside the synapse and beyond.* Curr Opin Neurobiol, 1999. **9**(3): pp. 293–8.
57. Pellerin, L. and P.J. Magistretti, *Food for thought: Challenging the dogmas.* J Cereb Blood Flow Metab, 2003. **23**(11): pp. 1282–6.
58. Pellerin, L. and P.J. Magistretti, *Glutamate uptake into astrocytes stimulates aerobic glycolysis: A mechanism coupling neuronal activity to glucose utilization.* Proc Natl Acad Sci U S A, 1994. **91**(22): pp. 10625–9.
59. Dienel, G.A. and N.F. Cruz, *Nutrition during brain activation: Does cell-to-cell lactate shuttling contribute significantly to sweet and sour food for thought?* Neurochem Int, 2004. **45**(2–3): pp. 321–51.
60. Pysh, Y. and T. Khan, *Variation in mitochondrial structure and content of neurons and neuroglia in rat brain: An electron microscopic study.* Brain Research, 1972. **36**(1): pp. 1–18.
61. Peters, A., L.P. Sandford, and H.D. Webster, *Fine structure of the Nervous System: Neurons and Their Supporting Cells.* 3rd edition ed. 1991: Oxford University Press, Oxford.
62. Hertz, L., L. Peng, and G. Dienel, *Energy metabolism in astrocytes: High rate of oxidative metabolism and spatiotemporal dependence on glycolysis/glycogenolysis.* Journal of Cerebral Blood Flow & Metabolism, 2007. **27**: pp. 219–249.
63. van den Pol, A.N., C. Romano, and P. Ghosh, *Metabotropic glutamate receptor mGluR5 subcellular distribution and developmental expression in hypothalamus.* J Comp Neurol, 1995. **362**(1): pp. 134–50.
64. Munoz, A., X.B. Liu, and E.G. Jones, *Development of metabotropic glutamate receptors from trigeminal nuclei to barrel cortex in postnatal mouse.* J Comp Neurol, 1999. **409**(4): pp. 549–66.
65. Brookes, P.S., et al., *Calcium, ATP, and ROS: A mitochondrial love-hate triangle.* Am J Physiol Cell Physiol, 2004. **287**(4): pp. C817–33.
66. Attwell, D. and S. Laughlin, *An Energy Budget for Signaling in the Grey Matter of the Brain.* Journal of Cerebral Blood Flow and Metabolism, 2001. **21**: pp. 1133–1145.
67. Lebon, V., et al., *Astroglial Contribution to Brain Energy Metabolism in Humans Revealed by 13C Nuclear Magnetic Resonance Spectroscopy: Elucidation of the Dominant Pathway for Neurotransmitter Glutamate Repletion and Measurement of Astrocytic Oxidative Metabolism.* J Neurosci, 2002. **22**(5): pp. 1523–1531.
68. Oz, G., et al., *Neuroglial Metabolism in the Awake Rat Brain: CO2 Fixation Increases with Brain Activity.* The Journal of Neuroscience, 2004. **22**(50): pp. 11273–11279.
69. Kahlert, S. and G. Reiser, *Requirement of Glycolytic and Mitochondrial Energy Supply for Loading of Ca2+ Stores and InsP3-Mediated Ca2+ Signaling in Rat Hippocampus Astrocytes.* J Neurosci Res, 2000. **61**: pp. 409–420.
70. Feustel, P.J., Y. Jin, and H.K. Kimelberg, *Volume-regulated anion channels are the predominant contributors to release of excitatory amino acids in the ischemic cortical penumbra.* 2004. **35**: p. 1168.
71. Arcuino, G., et al., *Intercellular calcium signaling mediated by point-source burst release of*

ATP. Proc Natl Acad Sci U S A, 2002. **99**(15): pp. 9840–5.

72. Zhang, J.M., et al., *ATP released by astrocytes mediates glutamatergic activity-dependent heterosynaptic suppression.* Neuron, 2003. **40**(5): pp. 971–82.
73. Davalos, D., et al., *ATP mediates rapid microglial response to local brain injury in vivo.* Nat Neurosci, 2005. **8**(6): pp. 752–8.
74. Brennan, A.M., J.A. Connor, and C.W. Shuttleworth, *NAD(P)H fluorescent transients after synaptic activity in brain slices: Predominant role of mitochondrial function.* J Cereb Blood Flow Metab, 2006. **26**(11): pp. 1389–406.
75. Takano, T., et al., *Cortical spreading depression causes and coincides with tissue hypoxia.* Nat Neurosci, 2007. **10**(6): pp. 754–62.

Chapter 6

Using the Light Scattering Component of Optical Intrinsic Signals to Visualize In Vivo Functional Structures of Neural Tissues

Uma Maheswari Rajagopalan, Kazushige Tsunoda, and Manabu Tanifuji

Abstract

Visualization of changes in reflected light from *in vivo* brain tissues reveals spatial patterns of neural activity. An important factor which influences the degree of light reflected includes the change in light scattering elicited by neural activation. Microstructures of neural tissues generally cause light scattering, and neural activities are associated with some changes in the microstructures. Here, we show that the optical properties unique to light scattering enable us to visualize spatial patterns of retinal activity non-invasively (FRG: functional retinography), and resolve functional structures in depth (fOCT: functional optical coherence tomography).

Key words: Intrinisic signal imaging, OCT, optical coherence tomography, light scattering.

1. Introduction

In 1986, Blasdel and Salame visualized orientation columns in monkey visual cortex *in vivo* by staining the brain surface with a voltage-sensitive dye and observed spatial patterns of absorption changes elicited during visual stimuli *(1)*. After this initial finding, it has been shown that functional structures can also be visualized intrinsically, by measuring changes in light reflection without the need for dyes *(2)*. This technique, the measurement of intrinsic reflection changes elicited by neural activation, is called optical intrinsic signal imaging (OISI), and is widely used to map cortical functional structures in neural tissues of living animals *(3–9)*. Because the intrinsic signals were originally found as byproduct

Fahmeed Hyder (ed.), *Dynamic Brain Imaging: Multi-Modal Methods and In Vivo Applications, vol. 489*

DOI 10.1007/978-1-59745-543-5_6 Springerprotocols.com

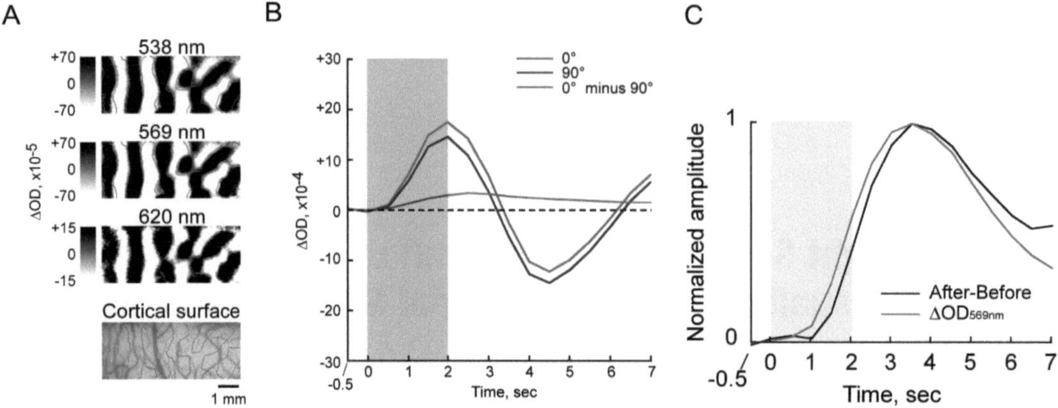

Fig. 6.1. Orientation columns visualized by OISI at different wavelengths. Typically in visual cortex, intrinsic signals consist of stimulus-specific and -nonspecific components. Orientation columns are reflected in the stimulus-specific component that is a local modulation of a stimulus-nonspecific component. In (A), we subtracted activation patterns obtained by one stimulus (90 deg. orientation) from the other (0 deg. orientation) to remove stimulus-nonspecific components. Please note that, depending on the size of the functional structures of interest, stimulus specificity of synaptic inputs, and spread of intrinsic signals, distinction between stimulus-specific and nonspecific components may not be required (for example, see **Figs. 6.3 and 6.4**). Columnar patterns obtained at different wavelengths are nearly the same as indicated by contours in red drawn for columnar patterns obtained at 620 nm. Although it is not shown here, similar patterns of functional structures were also observed using infrared light. Bottom image shows vessel patterns of the exposed cortical surface where the above recordings were made. (B) Time courses of stimulus-nonspecific (red and blue lines) and stimulus-specific (green line) components obtained at 620 nm. Increase of ΔOD corresponds to increase of deoxyhemoglobin. The visual stimulus was given from 0 to 2 s. (C) Time course of stimulus-nonspecific components obtained at 570 nm (Green line). The black line indicates time course of blood volume changes estimated by subtraction between time courses obtained before and after the extrinsic absorption dye infusion into the blood stream (see Fukuda et al., 2005 for details). The visual stimulus was given from 0 to 2 s. (*See* Color Plate)

of dye-sensitive absorption changes, the sources of these signals have since been debated.

One unique feature of OISI is that functional structures produce nearly identical spatial activation patterns at a wide range of imaging wavelengths (**Fig. 6.1A**). The dominant sources of intrinsic signals, however, differ from wavelength to wavelength. At least three components are involved in intrinsic signals, and all of them are secondarily elicited by neural activations. These components are changes in the oxygenation levels of hemoglobin, changes in blood volume within neural tissues, and changes in microstructures of neural tissues.

Oxy- and deoxygenated hemoglobin in blood vessels are the major light absorbing molecules in neural tissues at around 500–650 nm. Particularly, changes in the oxygenation level of hemoglobin are considered the major source of intrinsic signals around 610 nm because light absorption is dominated by deoxyhemoglobin at this wavelength. Vanzetta and Grinvald recorded changes in tissue oxygen tension by measuring changes in phosphorescence of an oxygen-sensitive dye *(10)*. They found that oxygen tension initially decreases within 2 s after stimulus onset, and then increases beyond the baseline oxygen tension during persistent stimulation. This time course of changes in oxygen tension was quite consistent with the time course of intrinsic signals (stimulus-nonspecific component) observed at 620 nm (**Fig. 6.1B**). This result provides good evidence that the deoxygenation level of hemoglobin is one of the sources of intrinsic signals.

A second source of intrinsic optical signals stems from changes in blood volume within neural tissues. One of the isosbestic points of oxy- and deoxy-hemoglobin absorption is located at 570 nm within the major absorption spectrum band. Thus, the intrinsic signals at this wavelength should be dominated by changes in blood volume in tissue blood vessels. Several studies provide supporting evidence for the involvement of blood volume changes in intrinsic signals *(11, 12)*. For example, infusion of extrinsic absorption dye into the blood stream increases absorption changes elicited by neural activation at the wavelength specific for that dye, and the time course of the signal was nearly identical to the time course of intrinsic signals recorded at 570 nm, where blood volume changes seem to dominate (**Fig. 6.1C**) *(12)*.

Microstructures of neural tissue, such as intricate subcomponents of neurons, multiple types of glial cells, and collection of blood vessels of various sizes generally cause scattering of light that penetrates the neural tissues. If neural activities are associated with changes in these microstructures, then light scattering changes from intrinsic signals could be a component for neuroimaging.

Changes in tissue light scattering is indeed considered a source of intrinsic signals since intrinsic signals are also observed at the wavelengths outside of the major band of the hemoglobin absorption spectrum. The light scattering component of the intrinsic signal has unique properties that do not exist in the absorption of oxy- and deoxy-hemoglobin. First, we can detect light scattering changes at a wide range of wavelengths, including infrared light, where hemoglobin absorption is minimal. Using light scattering in the infrared offers several advantages. Infrared light permits visualization of functional signals from tissues sensitive to visible light, such as the retina, as well as detection of signals from deeper structures because of the increased light penetration in the infrared light range compared to the visible range. Second, we can detect light scattering changes not only through changes in light reflection but also through phase-sensitive detection such as optical coherence tomography (OCT), which enables us to resolve functional structures in greater depth. Finally, light scattering changes may have faster time courses than the other intrinsic signals, allowing us to resolve neural events with higher temporal resolution than signals originating from hemodynamics. Nevertheless, firm evidence for the contribution of the light scattering changes has not been demonstrated until recently. Here, we will focus on the light scattering component of intrinsic signals. In particular, we will show evidence for the involvement of light scattering changes in intrinsic signals, applications of OISI to functional imaging of the retina using the light scattering component, and depth-resolved functional imaging with OCT.

2. OISI for Functional Imaging from Retina (Functional Retinography)

With the increasing number of people suffering from vision-threatening retinal diseases, such as age-related macular degeneration, there is an urgent need for the development of objective methods to measure retinal function, by which functional disorder can be detected before symptomatic or structural changes occur. The distribution of retinal responsiveness could not be adequately mapped by conventional techniques, such as electroretinogram (ERG). Therefore, we have developed a recording system for measuring flash-evoked intrinsic signals from the macaque retina *(13, 14)*.

We have applied OISI to macaque retina, and successfully demonstrated the topography of cone- and rod-induced neural function by measuring the light reflectance changes following flash stimulus. Because of the differences in anatomical structures, the properties of intrinsic signals in retina are quite different from those in the cerebral cortex. Here, we will focus on the light reflectance changes in two discrete regions in the posterior retina: fovea and posterior retina apart from the fovea (perifoveal regions).

The ocular fundus of Rhesus monkey under anesthesia was monitored via a modified fundus camera equipped with a CCD camera (**Fig. 6.2**). The intrinsic signals evoked by white diffuse flash stimuli were calculated by dividing the averaged images obtained after the flash by those obtained before the flash *(13, 14)* (see details in Appendix 1). With an observation light wavelength of 630 nm, the light reflectance from the fovea increased (retinal image became brighter), whereas, the light reflectance from the perifoveal regions decreased (retinal image became darker) following a flash stimulus (**Fig. 6.3A**). The increase in light reflectance following a flash in the foveal region was attributed to the bleaching of photopigments by visible light *(15–18)*. On the other hand, the decrease in light reflectance in the perifoveal region was attributed to the hemoglobin-related reflectance changes or tissue light scattering changes, which can be observed in the cerebral cortex *(13)*. The pseudocolor map (**Fig. 6.3A**) shows that the signals in the foveal and perifoveal

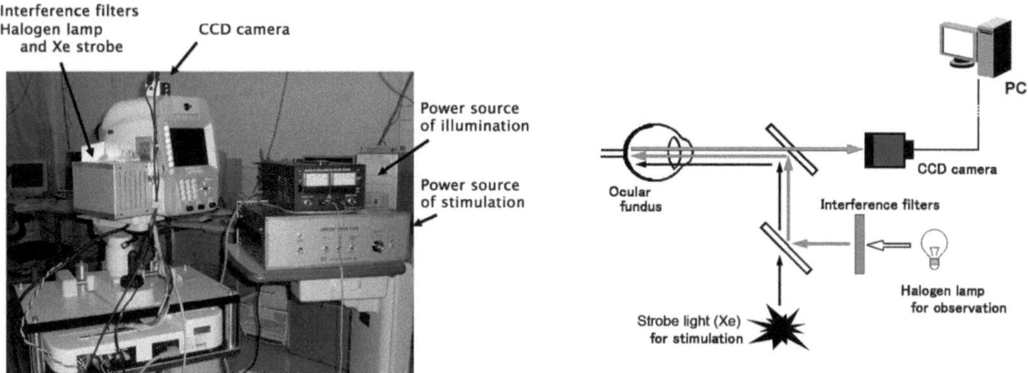

Fig. 6.2. Overall view of the intrinsic signals imaging system for retina (left) and schematic drawing of the experimental setup (right). Throughout the recording trial, the fundus was continuously illuminated with observation light through one of the bandpass filters. (*See* Color Plate)

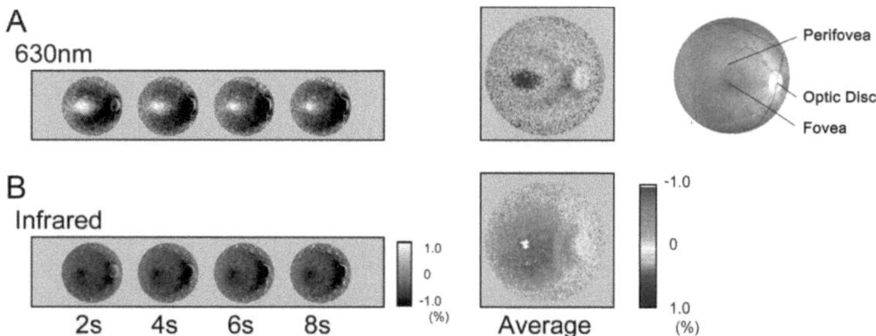

Fig. 6.3. Time courses of two-dimensional images of retina showing light reflectance changes following a flash stimulus observed with 630 nm (**A**) and infrared (**B**) light, measured in the posterior pole region of normal retina. The averaged reflectance changes during the first second after the flash are shown in pseudocolor maps on the right. Color indicates relative light reflectance changes from pre-stimulus level. Fundus photograph of a normal macaque retina is shown in the right. (*See* Color Plate)

regions have different polarities. In the foveal region, the contribution of brightening by photopigment bleaching is greater than that of darkening by light scattering changes.

In retinal optical imaging, we need to take into account two points when visible light is used for OISI: (1) Visible light for observation evokes neural activity and the reflectance can be changed without giving flash stimuli. The baseline retinal reflectance is thus unstable during recording.(2) Following bleaching of photopigments under visible light, the foveal reflectance is dramatically increased (retinal image becomes brighter) and this bleaching-related reflectance change has an opposite polarity to conventional intrinsic signals, which are commonly observed as decreases in light reflectance (retinal image becomes darker).

However, with infrared observation light (840–900 nm), the results differ. The whole posterior region of the retina became darker after the stimulus and the pseudocolor map shows that the most prominent decrease in light reflectance was at the center (**Fig. 6.3B**). The response topography under the light-adapted condition demonstrated a steep peak of darkening at the fovea, together with a gradual decrease of signal intensity away from the fovea toward the periphery (**Fig. 6.4A**). This is consistent with the topography of psychophysical cone sensitivity in normal human subjects measured with bright background *(19–21)*. If the optical signal reflects the central accumulation of cone photoreceptors under the light-adapted condition, the local retinal region with high rod density, called the 'rod ring', should also be activated under the dark-adapted condition. Following forty minutes of dark adaptation, the intrinsic signal showed additional peaks along the circular region surrounding the macula at the eccentricity of the optic disk, together with the central peak at the fovea (**Fig. 6.4B**) *(22–24)*.

In order to ensure that the intrinsic signals reflect the neuronal activities in retina, the local optical signal values under light-adapted condition were compared with the local neuronal

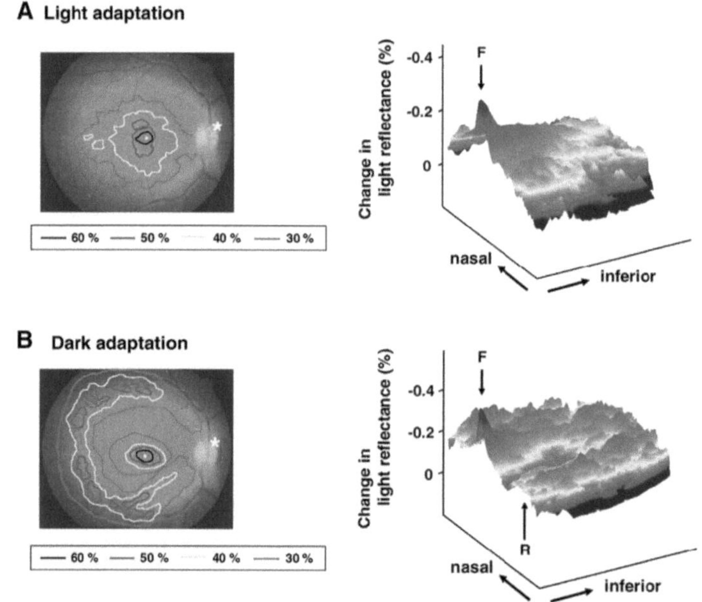

Fig. 6.4. Response topography of normal retina under light-adapted (**A**) and dark-adapted (**B**) conditions. The foveal center is indicated by a white dot and the optic disk is indicated by an asterisk. Regions with 60%, 50%, 40% and 30% of the peak signal intensity value at the fovea were outlined by different colors. Pseudocolor topographic maps of light reflectance changes in the inferior retina, profiled along the horizontal meridian, are shown on the right. The location of the fovea is indicated by F, and the crest of 'rod ring' is indicated by R. (*See* Color Plate)

activities, measured electrophysiologically by multifocal ERG *(13)*. Multifocal ERG (VERIS, EDI, San Mateo, California) measures the local neuronal activity in multiple small regions in the posterior retina, when presented (on a TV monitor) with achromatic flicker which alternates independently between black and white in individual hexagonal segments *(25)*. The correlation was evaluated between the amplitudes of the focal ERG responses and optical signals at the corresponding locations, from the same eye. In order to artificially reduce the function of the photoreceptor layer of the retina, the upper half of the left fundus of a monkey was locally coagulated with Argon Laser photocoagulation (**Fig. 6.5A**). The electrophysiological responses in the damaged region were reduced accordingly (**Fig. 6.5B**). In intrinsic signal imaging, the damaged region also showed smaller amount of light reflectance decrease (**Fig. 6.5C**). The light reflectance decrease and the multifocal ERG signal showed statistically significant positive correlation ($r^2 = 0.79$, $p < 0.001$, $n = 45$) (**Fig. 6.5D**). The

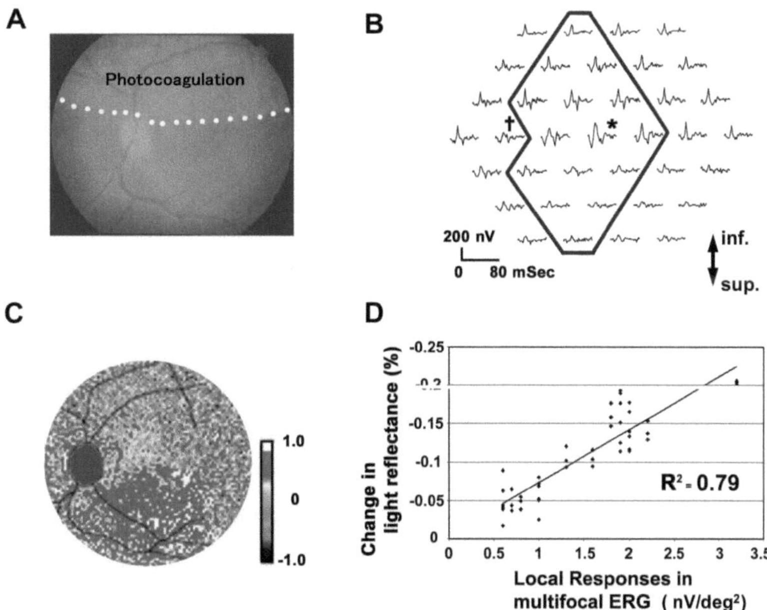

Fig. 6.5. (**A**) Photograph of the monkey's retina artificially damaged by Argon Laser Photocoagulation. The upper half of posterior retina was densely coagulated, sparing the macular area. (**B**) Array of 37 local responses of multifocal ERG, taken from the fundus in (**A**). ERGs with $*$ and † indicate the location of macula and optic disk, respectively. Note that the responses of the intact retina in the lower half region are inversely shown in the upper half of the arrays. (**C**) pseudocolor map of flash-evoked intrinsic signals measured with infrared light. Red color indicates light reflectance decrease (*darkening*) from pre-stimulus level. Flash-evoked darkening could not be observed in the damaged region (*upper half*). (**D**) Correlation between the change in light reflectance decrease and the focal responses in ERG at the corresponding retinal location under infrared light. Correlation coefficient: $r^2 = 0.79$ ($p < 0.001$, n=45). (*See* Color Plate)

results of optical imaging and electrophysiological measurement were well correlated not only in their response amplitudes but also in the spatial location of reduced responses: the border between normal and reduced response regions in both measurements corresponded to the border between normal site and photocoagulation site.

Interestingly, with infrared observation, the time course of the intrinsic signals evoked by a brief flash stimulus was different for different regions of the ocular fundus. Representative time courses of flash-evoked response at the foveal and perifoveal regions under the dark-adapted condition are shown in **Fig. 6.6**. The reflectance changes at the fovea were rapid and reached a negative peak (darkening) within 100 to 200 ms following the flash. The darkening then gradually returned toward the pre-stimulus baseline. The signals in the perifoveal regions (3°–12°) were composed of both fast and slow components. The time courses of the intrinsic signals of the perifoveal regions were approximately the same and distinct from the foveal response: the light reflectance decreased rapidly within 100 ms (flexural point), and then gradually decreased to reach a trough. The light reflectance at the fovea did not decrease following the initial negative peak.

The fovea is a central region (300 μm in diameter) in the posterior retina that is composed of cone photoreceptors and is free of capillaries and middle or inner layer structures *(26)*. The perifoveal region has both cone and rod photoreceptors, and its

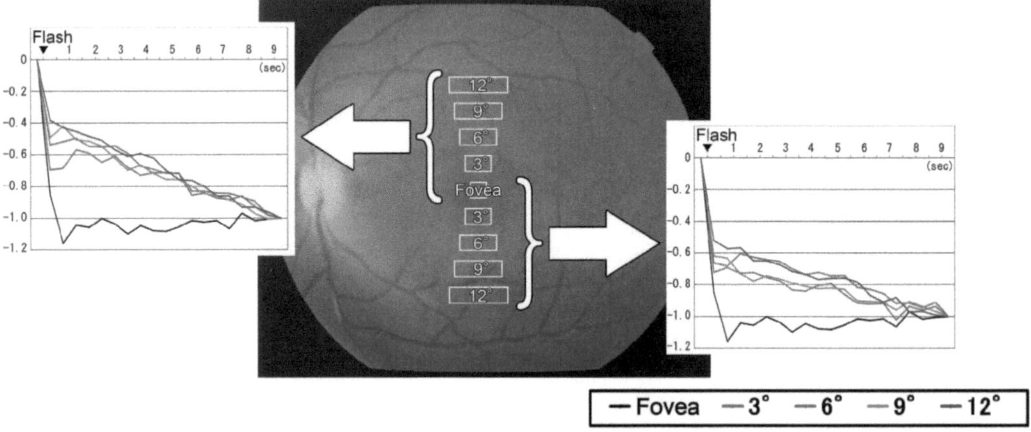

Fig. 6.6. Time courses of light reflectance changes in a single trial following a diffuse flash, measured at the fovea and different regions within twelve degrees superior or inferior to the fovea. Amplitudes are indicated as values relative to the light reflectance changes at the end of each trial (1.0). The four regions tested in each quadrant are indicated as distances from the fovea (3,6,9 and 12°). (*See* Color Plate)

inner and middle layers are nourished by retinal vessels and capillaries. We believe that the light scattering changes following activation of the cone photoreceptors are probably the source of the fast intrinsic signals observed at the fovea because the foveal avascular region is not subject to changes in hemoglobin concentration or blood volume following neural activation *(14)*. The rapid darkening observed at the perifoveal regions may also be derived from light scattering changes because, under infrared light, the change in the optical signal due to deoxygenated hemoglobin concentration is thought to be much smaller than that from tissue light scattering *(27)*. The light scattering changes following a flash are thought to be derived from microstructural changes in the outer segment disks, membrane hyperpolarization, cell swelling, and changes in the composition of the inter-photoreceptor matrix. Recent functional OCT studies using blood-free slice preparations *(28,29)* or in vivo retina *(30)* showed that the reflectance in the photoreceptor layer is strongly modulated by neural activation followed by microscopic morphological changes.

As for the sources of the slow signals in the perifoveal regions, our recent studies have suggested that the light scattering changes due to blood flow changes in capillaries are the main contributors to changes in reflectance *(42)*. Direct measurement of blood flow with laser Doppler flowmetry has shown that a flash stimulus evokes a slow increase in blood flow only in the perifoveal regions and its time course exactly matched that of the slow component of the intrinsic signals we described above. A fast increase in blood flow, however, has not been observed either in foveal or perifoveal regions.

We have shown that the light scattering change, which is independent of blood oxygenation level, also correlates well with neuronal activity and can be used for mapping neural function. We believe that this fast scattering signal is of great value for mapping neuronal activity because it may have better spatial and temporal resolutions than the blood-flow- or blood oxygenation related signals.

3. Optical Coherence Tomography (OCT) for Functional Imaging Resolved in Depth (Functional OCT)

In conventional OISI with CCD cameras, the measured reflected light is actually the integrated signals over depths determined by the collection optics. Hence potential variations in functional organization across depth may go undetected. Optical coherence tomography (OCT) is an optical imaging technique that has the potential to show reflectivity at specific depths because the method is a sensitive measure of refractive index variations across depth *(31,32)*. In this technique, light from a low-coherence light source is focused onto the tissue and reflectivity of the internal microstructures at different depths is measured by an interferometer, thus providing a map of the structural profile of the tissue. As described in the previous section on conventional OISI, in addition to the oxy- and deoxy-hemoglobin related absorption changes, scattering changes also contribute to the intrinsic signal measured with OCT. Scattering changes could result from changes in the size of the scatterer or the density of the scatterer, or both. During neural activation, secondary physiological structural changes such as those in photoreceptors described above *(13)*, capillary dilation *(12,33)*, change in the density of red blood cells *(34)* and swelling of glial cells *(35)* can occur *(36)*. We expect that the changes in scattering characteristics would result in an activity-dependent reflectivity change, and that the sensitivity of OCT to refractive index changes would make it theoretically sensitive to the scattering changes such as those happening during neural activation. We refer to this technique as functional OCT (fOCT) and its signal as fOCT signal.

To demonstrate the potential of OCT in functional studies, we used primary visual cortex (V1) of cats to confirm that detection of a stimulus-specific reflectivity change is feasible *(36, 37)*. The reliability of the technique was demonstrated by comparison with results of conventional OSIS and multi-unit activity recorded electrophysiologically. Recently, supporting evidence for the potential of OCT in functional studies has been reported in squid *(38)* and in retina *(28–30)*.

4. Brief Introduction for Optical Coherence Tomography (OCT)

Figure 6.7A shows a simple schematic of the principles of an interferometer: a broad-spectrum light source is divided by a half mirror into two beams, one illuminating the reference mirror and the other illuminating a turbid medium such as the cortex. The light reflected back from the reference mirror and the cortex are recombined at the half mirror to reach the detector. The reflected light beams would interfere only if their total light path length difference (Lr-Ls) is within the coherence length of the source, or, in other words, if the light reaching the detector has temporal correlation. The extent of temporal correlation is determined by autocorrelation of the source, and can be described in terms of the spectral width (Proportional to $\lambda_0^2/\Delta\lambda$) of the source, where λ_0 is the central mean wavelength and $\Delta\lambda$ the spectral width. For the case of the light source used in OCT, the coherence length is on the order of a few μm to a few tens of μm. So, by having a mechanism to move the reference mirror, it is possible to dissect

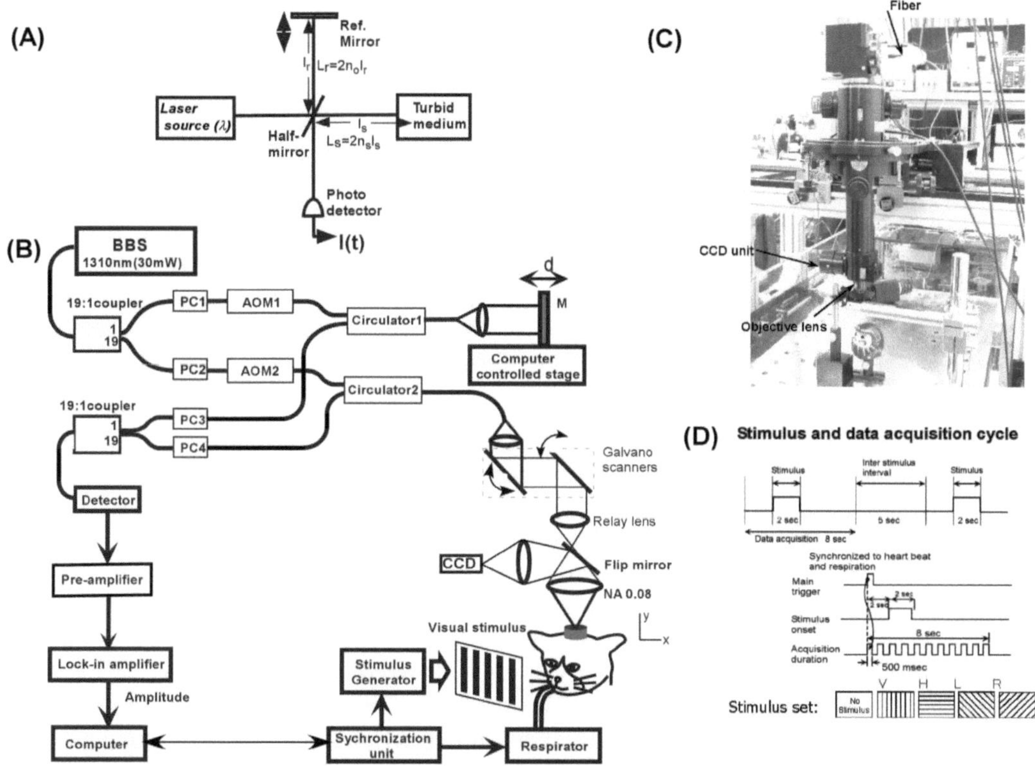

Fig. 6.7. A schematic of the basic principle of OCT (**A**) and the experimental system used (**B**) along with a picture of the probe unit (**C**) and a schematic of the scanning paradigm (**D**). In the figure (**B**), the abbreviations denote: BBS - Broad Band Source, AOM- Acousto-Optic Modulator, PC - Polarization Controller, M – Mirror and O – Objective lens. (*See* Color Plate)

the cortex optically and obtain depth-resolved reflectivity maps. The experimental system used in our studies and other details are described in Appendix 2.

5. Functional Imaging with OCT (fOCT)

Prior to doing functional imaging with OCT, we performed in vivo optical intrinsic signal imaging with the exposed cortical surface of cat visual cortex (**Fig. 6.8A**) at a wavelength of 607 nm. The stimulus set was identical to the one used in fOCT. It consisted of four differently oriented gratings and a blank screen used as a control (see detailed protocol in Appendix 2). **Figure 6.8B** shows the thresholded difference maps obtained when horizontal and vertical grating visual stimuli were presented to the cat. Dark and bright regions indicate the activated regions for horizontal and vertical gratings, respectively.

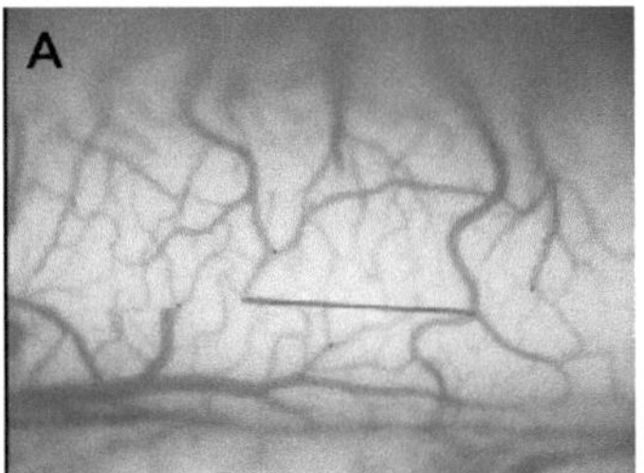

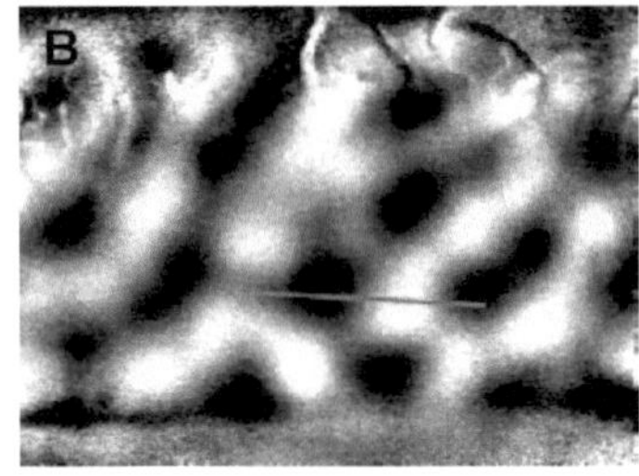

Fig. 6.8. (**A**) Exposed cortical surface of cat visual cortex with (**B**) a thresholded activation map overlay . *Dark* and *light* patches represent the activated regions for horizontal and vertical gratings, respectively, and *green lines* indicate the region of OCT scans. (*See* Color Plate)

An OCT *x-z* scan was then conducted across the green line indicated in **Figure 6.8**. **Figure 6.9A** shows the results of the OCT structural *x-z* image obtained. The cortical surface border has been drawn manually and the warm-colored regions indicate the scattering centers within the cortex. Here, the light beam was adjusted to be incident normal to the cortical surface and at a position of interest in relation to the cortical surface. The depth profile showing the intensity variation as a function of depth is shown in **Fig. 6.9B**. **Figure 6.9C** shows the calculated fOCT map obtained as a difference of the fOCT maps collected for horizontal and vertical grating stimuli obtained across the green line of **Fig. 6.8** (see Appendix 2 for calculation used to extract

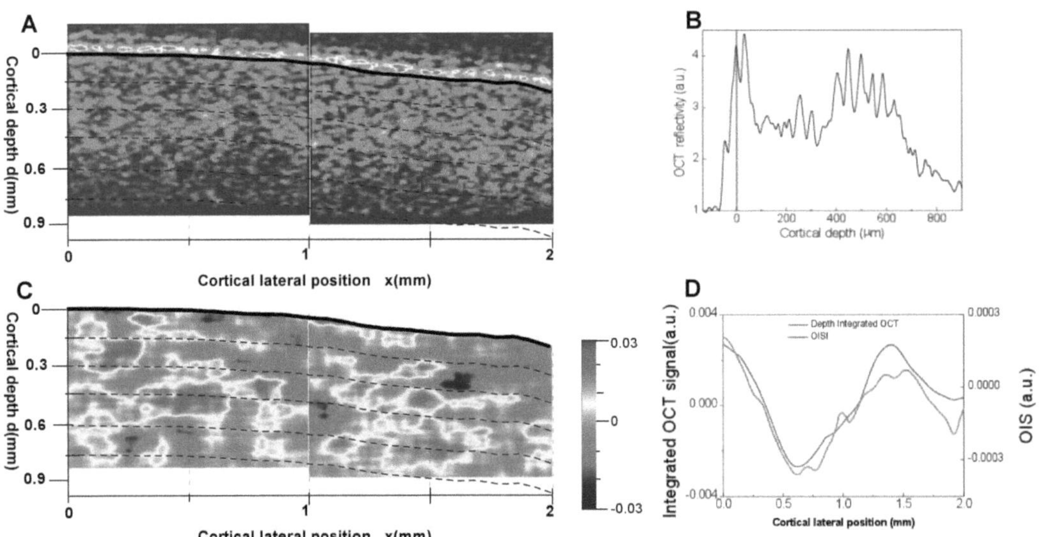

Fig. 6.9. (**A**) OCT scan obtained across the line indicated in Fig. 6.8 with (**B**) a typical depth reflectivity profile from (**A**) and the corresponding (**C**) Functional OCT map and (**D**) Consistency of OISI result with the integrated result of fOCT. In (**C**), *red* and *blue* patches represent the activated regions for horizontal and vertical gratings, respectively. In (**D**) green line indicates the variation of OISI across the line indicated in Fig. 6.8B. while the *red line* obtained by calculating the functional signal from integrating the OCT scans across the full scanned depth range of Fig. 6.9A. (*see* Color Plate)

functional signals). Here, red and blue patches indicate the activation for horizontal and vertical grating stimuli, respectively. In order to clarify the reliability of maps, a pixel-by-pixel comparative t-test was done for the horizontal against the vertical grating stimuli. The tests revealed that the results obtained were statistically significant to within a 5% tolerance limit. Further, a blank minus blank map did reveal a flat distribution without any patches.

From the fOCT map, we can make the following inferences:

1. There is a discrete distribution of activation patches across depth, which is stimulus specific.
2. In a very superficial region (100–200 μm under cortical surface), there are no activation patches indicating layer 1 where neurons are scarce.
3. In the region deeper than 100–200 μm, there exist several localized patches across depth showing no regular structure.
4. The localized patches extend up to the measured depth of approximately 1 mm. The presence of orientation columns in primary visual cortex is well known *(39)*. This indicates that, across depth, there might exist a continuous cylindrical organization more complex than the commonly understood columnar organization.

As the maps revealed by fOCT suggest a surprisingly discrete distributed columnar organization, it becomes necessary to validate the technique. In the present measurements, there are two potential issues that need to be considered:

1. Although we could observe a functional signal that arises as a result of scattering change due to neural activation, it is still not possible to specify the origin regarding the exact nature of the scattering changes such as glial swelling or capillary dilation. One way to resolve this issue may be to increase the spatial resolution of the technique, a possibility that is technically feasible. But even with increased resolution, allowing us to visualize the details of the structural organization, it is still not clear whether the reflectivity changes could provide enough contrast to resolve finer details such as neuronal cells, glial cells and blood vessels etc. To resolve such structural differences, one way is by specifically attaching contrast agents such as gold particles to specific structures such as neurons. We previously proposed a method of increasing the reflectivity in OCT *(40)* by introducing properly sized gold particles. With such specific labeling, we may be able to increase the reflectivity from specific structures, which would, in turn, enable us to specify the origin of scattering change.
2. Another issue is that the site of localized scattering change may not correspond to the site of neural activity. To address this problem, we conducted electrophysiology recordings to measure the neuronal activity at the sites revealed by fOCT discussed in the following section.

6. Reliability of the Signal as a Measure of Localization of Neural Activity

As the correlation of OISI maps with neural activity is already well established *(3, 4, 41)*, a comparison of fOCT with OISI is a reasonable validation approach. We compared the intensity variation of an OISI map across the scanned line (**Fig. 6.8**) with the integrated profile of fOCT. In order to calculate the integrated profile, all the pixels across the *z*-direction were summed to obtain a profile that varies only across *x*; then the differential OCT at each *x* was calculated. **Figure 6.9D** shows the result of such a comparison with the red line indicating the integrated result and green line indicating the OIS intensity variation. A clear and remarkable agreement between the profiles can be seen. This indicates that fOCT signal is indeed correlated with OISI and thus correlated with the neural activity. Experiments were done in 5 cats and showed a good correlation between the integrated fOCT signal and the intensity profiles obtained from intrinsic maps. Correlation coefficients vary in the range of 0.3–0.9. **Table 6.1** gives the results of correlation coefficients obtained from a cat from different scan positions. However, a general caveat is that there are many differences between OISI and fOCT, including illumination and detection geometries, wavelengths used and the origin of the signals. In fOCT, as near-infrared wavelength is used, oxy- and deoxy-hemoglobin have almost the same absorption coefficients and hence the absorption changes are minimal. The main source of the signal is thought to be scattering changes. Nevertheless, the OISI results corresponded very well with the integrated profiles of fOCT.

Table 6.1
The correlation coefficients between the profiles obtained by the OISI intensity profile and the corresponding depth-integrated fOCT intensity profile obtained from 6 different scan positions of a single cat

Scan position index	Correlation for 0–90°	Correlation for 45–135°
Position 1	0.58	0.65
Position 2	0.32	0.29
Position 3	0.24	0.39
Position 4	0.89	0.77
Position 5	0.407	0.539
Position 6	0.566	0.559

Next, to investigate what the patches in fOCT maps represent (i.e., whether they represent a localization in the distribution of neurons themselves or a localization of the secondary mechanisms that are behind the origin of the functional signals), we recorded multi-unit activities (MUA) in the scanned region to determine the correlation with the distribution of strength of neuronal activity. MUA recordings were done with glass-coated tungsten electrodes. **Figure 6.10A** shows a representative example of the variation of MUA with respect to the cortical depth for four different grating orientations. As seen from the figure, the stimulus selectivity across a single track is retained. However, the evoked response at different depths is not of the same magnitude and it varies across depth with the response being the largest for a depth of 100–400 μm. This implies that, across the depth, all neurons do not behave in the same way and they have individual characteristics as would be expected from an ensemble of neurons. We would like to emphasize this variation in the evoked response and we expect this variation could be detected by the proposed fOCT measurement.

Figure 6.10B shows an example of the result of a comparison of the fOCT depth profile with the MUA profile. Variation of the fOCT profile generally agrees with the non-uniformity of evoked response across different depths. This suggests that the functional signal measured by fOCT may be correlated with the spiking activity of neurons. A calculation of the correlation coefficient revealed that the profiles are well interrelated. Results of correlation obtained from six different scan positions and three different cats are presented in **Table 6.2A and B**, respectively. Except for two cases, the correlation coefficient was fairly high, indicating that fOCT could indeed measure functional maps.

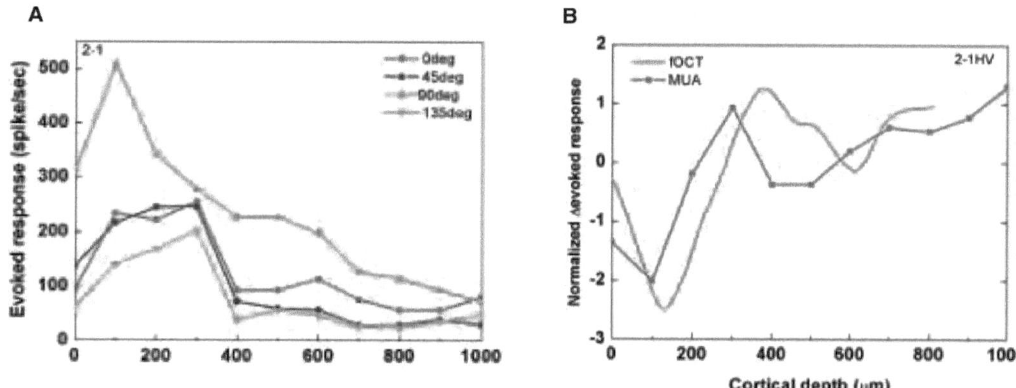

Fig. 6.10. An example of (**A**) MUA-evoked responses as a function of depth obtained from a single track for different orientation stimuli and (**B**) a comparison of difference of the evoked MUA response (shown in red) with the fOCT profile (shown in green) as a function of depth for the difference of 0° minus 90° orientation stimulus. (*See* Color Plate)

Table 6.2
Cross-correlation coefficients obtained from (A) different tracks of a single averaged x–z scan and (B) population result from three different cats

	Correlation between fOCT and unit profiles (A)	
Site #	**Correlation coeff. for 0–90 deg**	**Correlation coeff. for 45–135**
#1	0.94	0.66
#2	0.809	0.764
#3	0.908	0.926
#4	0.852	0.899
#5	0.348	0.374
#6	0.751	0.365
(B)		
Cat #	# of sites that show significant correlation ($p<0.005$)	
C-22	4/8	
C-30	9/11	
C-36	5/11	

7. fOCT – Future Prospects

It should be mentioned that in the comparison study discussed in the previous section, different methods of OISI, fOCT and MUA were conducted independently to be compared later. However, there could be some spatial ambiguity of 100 μm or more in such comparisons. For more reliable comparisons, it may be necessary to conduct simultaneous measurements of OISI, fOCT and MUA. In addition, recently, we have started the implementation of a high speed Fourier domain OCT system that would largely improve the sensitivity of OCT signal enabling a clearer picture of the columnar organization of orientation columns in cat visual cortex.

Recently, we have also started applying fOCT to the study of the rat olfactory system and our preliminary results show clear odor-dependent responses. We therefore expect this method to provide novel insights regarding the response distribution in granule cell layers that receive input from olfactory glomeruli and lie deeper from superficial regions of the bulb.

Appendix 1: Technical Details of Functional Retinography

The monkeys were anesthetized by a mixture of 70% N_2O, 30% O_2 supplemented with 1.0–1.5% of isoflurane and were paralyzed with vecuronium bromide (0.1–0.2 mg/ kg/ hour). Before the recordings, the pupils were fully dilated with topical tropicamide (0.5%) and phenylephrine hydrochloride (0.5%). A modified digital fundus camera system (NM-1000, Nidek, Aichi, Japan) was used to observe and measure the light reflectance changes from the ocular fundus. The fundus images were recorded with a CCD camera (PX-30BC, Primetech Engineering, Tokyo, Japan), and the images were digitized with an IBM/PC-compatible computer equipped with a video frame grabber board (Corona II, Matrox, Quebec, Canada: gray level resolution, 10 bits; spatial resolution, 640 x 480; temporal resolution, 1/30 s). The camera was focused on the macular vessels, and the area recorded covered 45 degrees, which included the macula, superior and inferior vascular arcades, and the optic disk.

The fundus was continuously monitored with light from a halogen lamp filtered through a red (610–650 nm) or infrared (840–900 nm) interference filter. Each recording trial consisted of three hundred video frames collected at 30 frames/sec for a total recording time of 10 s. A Xenon flash (duration: 1 msec) was given to the whole posterior pole of the ocular fundus, 500 msec after the initiation of data acquisition. The flash intensity measured at the cornea was 1.54×10^2 cd•s/m^2. The timing of the data acquisition and stimulus delivery was under computer control. Changes in light reflectance from the ocular fundus following the stimulus, such as a darkening (a decrease in light reflectance) or a brightening (an increase in light reflectance), were measured. The optical signal was calculated as follows: *(1)* The gray-scale values of the image obtained after the stimulus were divided, pixel by pixel, by those obtained during a 0.5-s period before the stimulus, and *(2)* This ratio was rescaled to 256 levels of gray-scale resolution to show the stimulus-induced reflectance changes.

8. Appendix 2: Technical Details of fOCT

System

We used a fiber-based OCT imaging system (**Fig. 6.7B**). It consists of a Mach-Zehnder-type heterodyne interferometer constructed from single-mode fibers for flexibility. Light from a broadband source (AFC Technologies, Canada) operating at an output power of 30 mW and having a central mean wavelength (λ_0) of 1.31 μm and a spectral width ($\Delta\lambda$) of 50 nm is split into sample and reference beams in the respective ratio of 19:1 by coupler 1. Optical frequencies of the sample and reference beams are shifted by acousto-optic modulators (AOM, Asahi glass, Japan). We use AOMs to introduce a constant and stable phase delay between the interfering beams. Both the reference and sample beams, after passing through the circulators, illuminate the reference mirror and cortex, respectively. The reflected lights were recombined at coupler 2. An interference beat signal that has a beat frequency of 250 kHz is detected only when the path lengths of the interferometer arms are matched to within the coherence length of the source that is calculated to be 34 μm in free space. Heterodyne detection was done with a lock-in amplifier (EG&G, USA) and the amplitude of the demodulated components was fed into a computer via a 16-bit A/D converter. The reference mirror M was mounted on a motorized stage and scanned at a speed of 2 mm/sec.

The sample arm viewing the animal side consisted of an objective lens of numerical aperture 0.08 and was also fitted with a CCD camera. This allowed simultaneous viewing of the cortical surface with the introduction of visible light from an auxiliary laser source (wavelength 680 nm). The whole unit was mounted on a manipulator unit (**Fig. 6.7C**) that has five degrees of freedom of translation along three axes and rotation and tilt (the flexibility is needed for making the probing light beam normal to the cortical surface). Galvano scanners were installed so as to perform surface scans. The animal-related fluctuation in the signal was reduced by conducting measurements in synchronization with heartbeat and respiration and by keeping the brain surface immobile using agarose.

Animals and Surgery

Each cat was anesthetized with a mixture of 70% N_2O and 30% O_2 supplemented with 1–2% isoflurane, paralyzed with Pancuronium bromide (0.1 mg/Kg/hr), and artificially ventilated by a respirator unit. Contact lenses were fitted to the eyes to protect the cornea from drying. The pupils of the eyes were dilated with 0.5% tropicamide and 0.5% phenylephrine hydrochloride. The head of the animal was held tightly by attaching it to a metal rod. A stainless steel chamber (18 mm inner diameter) was fixed onto the skull with dental acrylic cement by aseptic surgery and was placed above area 18 (coordinates A10P5 of Horsheley). After removal of the dura mater, the inside of the chamber was filled with 1.2% agarose (Agarose-HGS, Nacalai tesque, Japan, gel strength 1.5%) containing dexamethasone and antibiotics and was sealed with a round glass cover slip and a silicon gasket. Rectal temperature, ECG and expired CO_2 were continuously monitored during both OCT experiments and surgery.

OCT Scan

The stimuli were the same as that used for OISI and consisted of square-wave gratings (white = 8 cd/m2, black=0 cd/m2) having a spatial frequency and moving at a velocity of 4 degrees/sec. The stimulus set consisted of five patterns with control or blank (mean luminance 4 cd/m^2), horizontal (0°), vertical (90°) and oblique gratings (45°, 135°) and were presented in a random order. All stimuli were generated with a VSG2/3 graphics video board (Cambridge Research Systems, UK). The center of thc visual field was roughly estimated by projecting images of optic discs onto a screen in front of the animal. The distance of the CRT screen (200–300 mm) was adjusted to have the best focus of optic discs and surrounding vessel patterns. A total of 40 trials were obtained for each stimulus. In a single trial, data acquisition was done for 8 s, during which time 16 *x-z* frames (128×100 pixels corresponding to 1 × 1 mm) were obtained. The inter-stimulus interval (ISI) was 5 s. A schematic is shown in **Fig. 6.7D**. Stimuli appeared with a delay of 2 s after the acquisition onset and persisted for 2 s.

OCT Data Correction and Analysis

First, the scans were corrected for any misalignment of the surface position by a correlation-based procedure. For a single stimulus, we obtained a total of 640 scans. To compensate for the small variations in the surface position over different scans, we used correlation analysis. Out of these 640 scans, we selected an arbitrary scan (i) with the reflectivity detected as a function of depth being $R_s(x,z,i)$ at a lateral position x for a stimulus s. Next, we calculated the correlation between the ith scan and rest of the scans (j) using the following equation:

EMBED Equation.DSMT4

$$\Omega(\Delta z; i, j) = \frac{\int_{-\infty}^{\infty} R_s(x, z, i) R_s(x, z + \Delta z, j) dz}{\sqrt{\int_{-\infty}^{\infty} R_s(x, z; i) dz \int_{\infty}^{\infty} R_s(x, z + \Delta z; j) dz}} \quad (6.1a)$$

This operation gives the position of the correlation peak that corresponds to the amount of shift necessary for the jth profile to be in alignment with the ith profile.

Pixel noise was removed with a smoothing filter of window size 27 ×21 μm.

Next, the ratio of the post-stimulus over pre-stimulus scans was calculated for all the grating stimuli and the control condition as follows:

$$\gamma_s(x,z,t) = \frac{R_s^{post}(x,z,t)}{\Sigma_{Prescans} R_s^{Pre}(x,z,t)} \tag{6.2}$$

Here R_s is the reflectivity at position (x, z) at time t. "post" and "pre" indicate post-stimulus and pre-stimulus scans. The division operation removes the unchanging common variation and extracts only changes due to visual stimulation. Next, the ratio was averaged for all the scans obtained for each stimulus $< \gamma_s(x,z) >$. Finally, the differential OCT signal $< \gamma_{diff}(x,z) >$ was calculated as,

$$< \gamma_{diff}(x,z) > = < \gamma_{grating}(x,z) > - < \gamma_{control}(x,z) >. \tag{6.3}$$

With the above equation, by subtracting the differential OCT signal of the control, we could remove noise fluctuations such as respiration artifacts that were locked to the recording but not to the grating stimulus . We have restricted this discussion mainly to the results obtained by calculating the difference between $<\gamma_s$ (x,z)> that were obtained for two orthogonal gratings. The spatial map has been smoothed with a moving average filter roughly 100×115 μm.

References

1. Blasdel, G.G. and Salama, G. (1986) Voltage-sensitive dyes reveal a modular organization in monkey striate cortex. *Nature* 321 (6070), 579–585
2. Grinvald, A. et al. (1986) Functional architecture of cortex revealed by optical imaging of intrinsic signals. *Nature* 324 (6095), 361–364.
3. Ts'o, D.Y. et al. (1990) Functional organization of primate visual cortex revealed by high resolution optical imaging. *Science* 249 (4967), 417–420.
4. Bonhoeffer, T. and Grinvald, A. (1991) Iso-orientation domains in cat visual cortex are arranged in pinwheel-like patterns. *Nature* 353 (6343), 429–431.
5. Malonek, D. et al. (1994) Optical imaging reveals the functional architecture of neurons processing shape and motion in owl monkey area MT. *Proc R Soc Lond B Biol Sci* 258 (1352), 109–119.
6. Roe, A.W. and Ts'o, D.Y. (1995) Visual topography in primate V2: Multiple representation across functional stripes. *J Neurosci* 15 (5 Pt 2), 3689–3715.
7. Wang, G. et al. (1998) Functional architecture in monkey inferotemporal cortex revealed by in vivo optical imaging. *Neurosci Res* 32 (1), 33–46
8. Uchida, N. et al. (2000) Odor maps in the mammalian olfactory bulb: Domain organization and odorant structural features. *Nat Neurosci* 3 (10), 1035–1043.
9. Tsunoda, K. et al. (2001) Complex objects are represented in macaque inferotemporal cortex by the combination of feature columns. *Nat Neurosci* 4 (8), 832–838.
10. Vanzetta, I. and Grinvald, A. (1999) Increased cortical oxidative metabolism due to sensory stimulation: Implications for functional brain imaging. *Science* 286 (5444), 1555–1558
11. Vanzetta, I. et al. (2004) Columnar resolution of blood volume and oximetry functional maps in the behaving monkey; implications for FMRI. *Neuron* 42 (5), 843–854

12. Fukuda, M. et al. (2005) Localization of activity-dependent changes in blood volume to submillimeter-scale functional domains in cat visual cortex. *Cereb Cortex* 15 (6), 823–833
13. Tsunoda, K. et al. (2004) Mapping cone- and rod-induced retinal responsiveness in macaque retina by optical imaging. *Invest Ophthalmol Vis Sci* 45 (10), 3820–3826
14. Hanazono, G. et al. (2007) Intrinsic signal imaging in macaque retina reveals different types of flash-induced light reflectance changes of different origins. *Invest Ophthalmol Vis Sci* 48, 2903–2912
15. Bowmaker, J.K. et al. (1980) Microspectrophotometric demonstration of four classes of photoreceptor in an old world primate, Macaca fascicularis. *J Physiol* 298, 131–143
16. Kilbride, P.E. et al. (1983) Determination of human cone pigment density difference spectra in spatially resolved regions of the fovea. *Vision Res* 23 (12), 1341–1350
17. Kilbride, P.E. et al. (1989) Human macular pigment assessed by imaging fundus reflectometry. *Vision Res* 29 (6), 663–674
18. Elsner, A.E. et al. (1993) Mapping cone photopigment optical density. *J Opt Soc Am A* 10 (1), 52–58
19. Mandelbaum, J. and Sloan, L.L. (1947) Peripheral visual acuity. *Am J Ophthalmol* 30, 581–588
20. Birch, D.G. et al. (1987) The relationship between rod perimetric thresholds and full-field rod ERGs in retinitis pigmentosa. *Invest Ophthalmol Vis Sci* 28 (6), 954–965
21. Pulos, E. (1989) Changes in rod sensitivity through adulthood. *Invest Ophthalmol Vis Sci* 30 (8), 1738–1742
22. Osterberg, G. (1935) Topography of the layer of rods and cones in the human retina. *Acta ophthalmol* 13 (Suppl 6), 6–97
23. Curcio, C.A. et al. (1987) Distribution of cones in human and monkey retina: Individual variability and radial asymmetry. *Science* 236 (4801), 579–582
24. Packer, O. et al. (1989) Photoreceptor topography of the retina in the adult pigtail macaque (Macaca nemestrina). *J Comp Neurol* 288 (1), 165–183
25. Sutter, E.E. and Tran, D. (1992) The field topography of ERG components in man – I. The photopic luminance response. *Vision Res* 32 (3), 433–446
26. Weinhaus, R.S. et al. (1995) Comparison of fluorescein angiography with microvascular anatomy of macaque retinas. *Exp Eye Res* 61 (1), 1–16
27. Bonhoeffer, T. and Grinvald, A. (1996) Optical Imaging Based on Intrinsic Signals: The Methodology. In *Brain Mapping* (Toga, A.W. and Mazziotta, J.C., eds.), pp. 55–97, Academic Press, New York
28. Yao, X.C. et al. (2005) Rapid optical coherence tomography and recording functional scattering changes from activated frog retina. *Appl Opt* 44 (11), 2019–2023
29. Bizheva, K. et al. (2006) Optophysiology: depth-resolved probing of retinal physiology with functional ultrahigh-resolution optical coherence tomography. *Proc Natl Acad Sci U S A* 103 (13), 5066–5071
30. Srinivasan, V.J. et al. (2006) In vivo measurement of retinal physiology with high-speed ultrahigh-resolution optical coherence tomography. *Opt Lett* 31 (15), 2308—2310
31. Huang, D. et al. (1991) Optical coherence tomography. *Science* 254 (5035), 1178–1181
32. Bouma, B.E. and Tearney, G.J. (2002) *Handbook of Optical Coherence Tomography*, Marcel Dekker Inc.
33. Malonek, D. et al. (1997) Vascular imprints of neuronal activity: relationships between the dynamics of cortical blood flow, oxygenation, and volume changes following sensory stimulation. *Proc Natl Acad Sci U S A* 94 (26), 14826–14831
34. Tomita, M. et al. (1983) Effects of hemolysis, hematocrit, RBC swelling, and flow rate on light scattering by blood in a 0.26 cm ID transparent tube. *Biorheology* 20 (5), 485–494
35. Holthoff, K. and Witte, O.W. (1998) Intrinsic optical signals in vitro: A tool to measure alterations in extracellular space with two-dimensional resolution. *Brain Res Bull* 47 (6), 649–655
36. Rajagopalan, U.M. et al. (2003) Functional optical coherence tomography to reveal functional architecture of cat visual cortex in vivo. In *Proceedings of SPIE* (Vol. 5140), pp. 77–83
37. Maheswari, R.U. et al. (2003) Novel functional imaging technique from brain surface with optical coherence tomography enabling visualization of depth resolved functional structure in vivo. *J Neurosci Methods* 124 (1), 83–92
38. Lazebnik, M. et al. (2003) Functional optical coherence tomography for detecting neural activity through scattering changes. *Opt Lett* 28 (14), 1218–1220
39. Hubel, D.H. and Wiesel, T.N. (1977) Functional architecture of macaque monkey visual cortex. *Proc R Soc Lond B Biol Sci* (198), 1–59
40. Rajagopalan, U.M. et al. (1999) An optical coherence tomographic imaging system for investigating cortical functional

organization of brain: A simulation study. In *The International Society for Optical Engineering* (Vol. 3749), pp. 400–401

41. Bonhoeffer, T. and Grinvald, A. (1993) Optical imaging of the functional architecture in cat visual cortex: the layout of direction and orientation domains. *Adv Exp Med Biol* 333, 57–69
42. Hanazona, G. et al. (2008) Evaluating neural activity of retinal ganglion cells by flash-evoked intrinsic signal imaging in macaque retina. *Invest ophthalmol vis sci* in press

Part III

Electrophysiology

Chapter 7

Methods for Studying Functional Interactions Among Neuronal Populations

Nandakumar S. Narayanan and Mark Laubach

Abstract

How do populations of neurons work together to control behavior? To study this issue, our group simultaneously records from populations of neurons across multiple electrodes in multiple brain regions during operant behavior. Here, we describe methods for quantifying the relationship between neuronal population activity and performance of operant behavioral tasks. We describe statistical techniques, based on time- and trial-shuffling, that can establish the significance of correlations between multiple and simultaneously recorded spike trains. Then, we describe several approaches to studying functional interactions between neurons, including principal component analysis, cross-correlation analysis, analyses of rate correlations, and analyses of shared predictive information. Finally, we compare these techniques using a sample data set and discuss how the combined use of these techniques can lead to novel insights regarding neuronal interactions during behavior.

Key words: Neural coding, neural ensemble, population activity, correlated variability, JPSTH, multi-electrode recordings.

1. Introduction

Recent developments in electrode recording technology have facilitated recordings of neuronal population activity *(1, 2)*. Such techniques can generate data from hundreds of simultaneously recorded neurons in multiple brain areas *(3, 4)*. To analyze such data, a multivariate approach is applied to firing rates of simultaneously collected neurons *(5)*. This approach can generate stable predictions from neuronal ensemble data about a behavioral variable or can be used as control signals for prosthetic devices *(3)*.

Analyses of neuronal population data do not automatically offer insight about how neurons interact during behavior. This is

Fahmeed Hyder (ed.), *Dynamic Brain Imaging: Multi-Modal Methods and In Vivo Applications, vol. 489*

DOI 10.1007/978-1-59745-543-5_7 Springerprotocols.com

in part due to the complications induced by neuroanatomy. In the cerebral cortex, neurons form thousands of synapses with other cortical neurons in neighboring and distant brain areas and have descending and recurrent connections to subcortical systems *(6)*. Moreover, cortical neurons are typically embedded in a processing unit such as a column or local microcircuit *(7)*. As such, cortical neurons can have distinct non-synaptic functional relationships with neurons within and outside of their local processing units.

There have been mixed opinions in the literature about whether functional interactions between neurons underlie the encoding of behaviorally relevant information. Some studies have argued that interactions have little impact on information processing by neurons *(8–12)*. Other studies have presented strong evidence that interactions between neurons may be a central feature of neural coding *(3, 13–19)*. To establish the relevance of functional interactions to understanding how populations of neurons control behavior, neuronal population data must be studied with a variety of techniques and must be studied in awake, behaving animals using behavioral tasks that challenge the brain areas of interest (see Refs. *(20, 21)* for an example of this approach).

Here we describe seven distinct techniques for investigating functional interactions between neurons: Functional grouping by principal component analysis, cross-correlation, joint-peristimulus time histograms, rate correlations, trial-by-trial rate correlations, predictive interactions, and network interactions. We also establish the significance of interaction based on time-shuffling or trial-shuffling spike trains. We then apply each of these techniques to a population of simultaneously recorded neurons from the rodent dorsomedial prefrontal cortex (dmPFC) and motor cortex. We use each of these techniques to generate insights into how neurons interact; and then we combine these techniques to derive a portrait of neuronal interactions across each cortical area.

2. Methods

2.1. Neurophysiological Recordings

The multi-electrode data used here were collected from the rodent dorsomedial prefrontal cortex (dmPFC; comprising prelimbic and anterior cingulate cortex) and motor cortex of one rat during performance of a delayed-response task with a fixed 1.0 s delay *(20)*. In this task, animals had to initiate and maintain a lever press for 1.0 s, and then release the lever promptly to get a liquid reward (**Fig. 7.1A**). Twenty-one neurons (11 in motor cortex, 10 in dmPFC) were isolated from one animal during delayed-response performance. Methods used to acquire

A

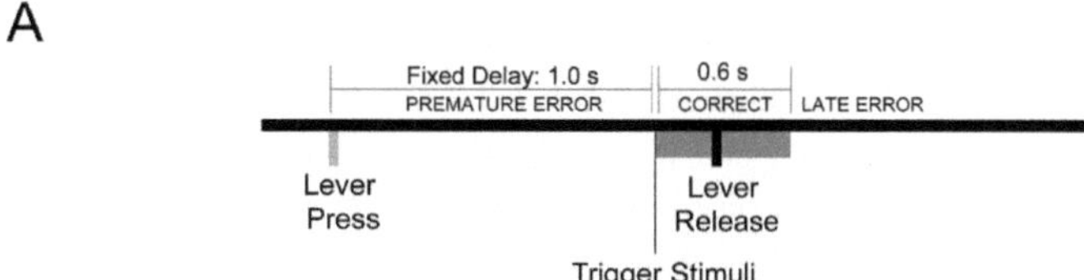

B

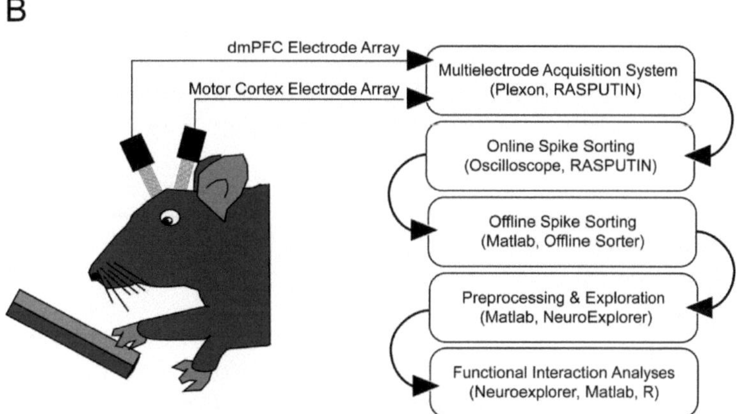

Fig. 7.1. Methods for multi-electrode recordings in freely behaving animals. (A) Sequence of events in delayed-response tasks. Rats were trained in a fixed-delay task, in which they had to initiate trials by pressing a lever until the end of a fixed (1.0 s) delay period. Rats had to release the lever within a designated response window to collect a liquid reward (*Correct* responses). If the lever was released before the trigger stimulus (*Premature error* responses) or after the response window (*Late error* responses), rats experienced a timeout period. After rats performed the fixed delay task, animals were re-trained on a delayed-response with two delays for three days. (B) Methods for collection of multi-electrode data. Briefly, neurophysiological data are collected from the awake, behaving rats implanted with arrays of electrodes (2 16-wire microwire electrode arrays in dmPFC and motor cortex) during performance of a delayed response task. Data are streamed to a computer, sorted online using an oscilloscope, cleaned offline using principal component analysis, loaded into MATLAB and R for subsequent multivariate analyses.

such data as well as the response properties of single neurons in these areas are described elsewhere in detail *(14, 20, 21)*. Briefly, electrical signals from each electrode in dmPFC and motor cortex are amplified, analog filtered, and streamed to a computer using a Multi-channel Acquisition Processor (MAP; Plexon, Dallas, TX). Putative single neuronal units were identified online using an oscilloscope and audio monitor. Cable noise and behavioral artifact is then cleaned using principal component projections of potential single units in Plexon's Offline Sorter. Peristimulus Peristimulus arrays of activity around behavioral events of interest are then constructed and explored using NeuroExplorer (Nex Technologies; Littleton, MA). All further analyses are performed using custom-written scripts for MATLAB (The Mathworks, Natick, MA) (**Fig. 7.1B**).

2.2. Code and Sample Data

We assume familiarity with data exploration in MATLAB. In each section, where possible, we provide a 'snippet' of MATLAB code that can be rapidly run on random data; this code is meant to give a flavor of each analysis and should be used for exploratory analysis. The exact code and techniques we use for estimating interactions often rely on toolboxes for MATLAB, and span several hundred lines of code. For these reasons, the code included here is relatively simple but demonstrates the principles that are relevant in studying functional interactions; the exact code along with a sample data of 11 motor cortex neurons is on our website: http://spikelab.jbpierce.org/Resources/FunctionalInteractions.

3. Statistical Significance of Functional Interactions

In any analysis of functional interactions, one must establish clear criteria for statistical significance to determine if interactions observed are due to chance. Although published formulas exist for determining the significance of many statistical tests, we suggest a statistical approach in which the empirical probability value for any test statistic should be computed for each analysis of interest. This approach has power because assumptions and data distributions accounted for by published formulas may be violated by particular data sets of interest. The best way to establish empirical significance is to reapply the same statistical tests to the same data in which the dimension of interest is randomly permuted. Most commonly, this dimension is either in *time* or in *trials* (**Fig. 7.2**).

Using shuffling, one might test the significance of a functional interaction by:

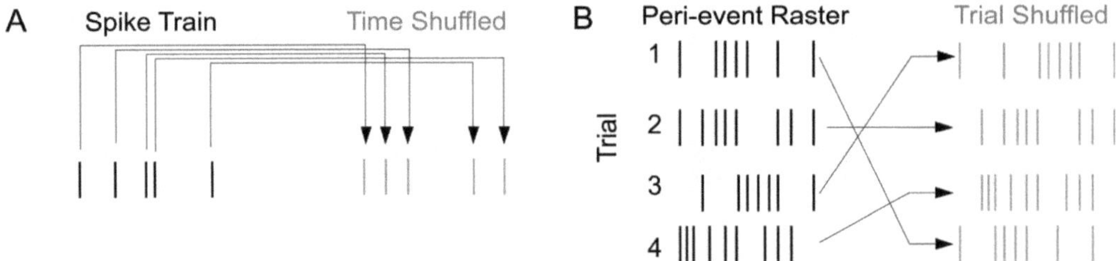

Fig. 7.2. Principles of shuffling can be used to assess empirical significance. (A) Time shuffling: In measures of correlation between spike trains, statistical significance can be assessed by comparing data of interest to test statistics generated from time-shuffled data in which spikes are shifted in time. This preserves basic statistics (i.e., firing rate), while destroying temporal information. (B) Trial shuffling: In measures of correlation between neurons over trials, statistical significance can be assessed by comparing data of interest to test-statistics generated from trial-shuffled data in which spikes from one trial are shifted to another trial. In a 'shift predictor', they are shifted by one trial; however, to account for non-stationarities, it is best to simply scramble the trial order. This preserves basic statistics (i.e., peristimulus time histograms, firing rate), while destroying trial-by-trial relationships with other neurons and with behavior.

1. Shuffling spike trains in time or in trials.
2. Applying functional interaction analysis of interest to shuffled data.
3. Repeat steps 1 and 2 as many times as computationally feasible, and obtain a distribution of test statistics from functional interaction analysis. Ten or one-thousand iterations are preferable for defining probability distribution; however, even one hundred iterations can establish significance at $p < 0.05$.
4. Establish a significance threshold by determining what values can be expected by chance at a probability of less than $p < 0.05$ (1 in 20; more stringent thresholds can be used as needed).

For instance, if one is interested in correlations in time, one should compare test statistics derived from correlations to test statistics derived from time-shuffled data (**Fig. 7.2A**). Time shuffling is appropriate to considering spike trains, which are a series of spikes in time recorded by a data acquisition system corresponding to the timing of action potentials. To time-shuffle a spike train, one can simply generate a random series of spikes matched to the length of the spike train of interest. For instance, to generate a spike train 10 s long at 10 Hz:

```
randSpiketrain = sort(rand(1,100) * 10); % random
timestamps,10s @ 10 Hz
```

More sophisticated temporal distributions (Poisson, bursting) can be generated by providing structure to the random data.

Experiments are commonly performed in 'trials', wherein a set of conditions is repeated many times while neuronal activity is tracked. In this scenario, trial-by-trial relationships become important in considering functional interaction between neurons; therefore, destroying this relationship for each neuron preserves neurons' task modulation but decorrelates neurons' trial-by-trial relationships. If one is interested in correlations over trials, one should compare test statistics derived from functional interaction analysis to test statistics derived from trial-shuffled data (**Fig. 7.2B**). To trial-shuffle data, spike trains associated with one trial are switched with a randomly selected trial. This process is also readily achieved in MATLAB with the command `randperm`:

```
myData = rand(149, 10); % random perievent
matrix: 149 trials 10 bins
trialShuffledData = myData(randperm(size(myData,1)),
:); %shuffled data
```

These methods illustrate that data can easily be shuffled in time or with respect to trials. As detailed above, one should repeat shuffles many times, and compare functional interactions of interest with functional interactions derived from trial- or time- shuffled data. When running the same functional interaction analysis over many pairs of neurons, one must consider how to approach multiple comparisons. At a $p < 0.05$, 1 in 20 interactions will be significant according to chance. One option is to use a simple correction for multiple comparisons (such as the Bonferonni correction). Alternatively, one can compare the number of functional interactions above a significance level to the number expected by chance via a X^2test.

4. Functional Grouping by Principal Component Analysis

In exploring functional interactions, the first step is to identify neurons with similar response properties. This can be achieved rapidly via techniques such as principal component analysis (PCA), a standard linear transform that uses singular value decomposition to project multivariate data on a series of axes (or components) to minimize their co-variance (**Fig. 7.3A**). To use PCA, one should:

1. Arrange perievent histograms from neurons of interest into a matrix where rows are neurons and columns are bins.
2. Apply SVD to generate components.

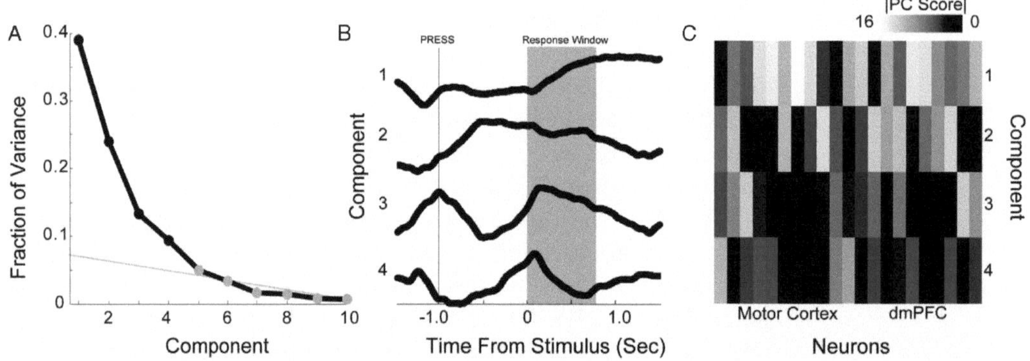

Fig. 7.3. Functional groups identified by principal component analysis (PCA). After peristimulus time histograms are extracted, PCA is run to generate a series of components that minimize variance. (A) Scree plot fraction of variance explained by each component. Statistical methods or line-fitting can be used to determine significant components in an automated way; here, we selected four components by eye. (B) Components as a function of time plotted over the trial in relation to stimulus onset on correct trials only. Note that lever press happens 1.0 s prior to the stimulus, lever release occurs 0–0.6 s after the stimulus (mean 0.28 s), and reward acquisition occurs 0.1 s after lever release. Component 1 seems to be response-related; component 2 seems to be related to holding the lever down; component 3 seems to be related to lever press and release movements, and component 4 seems to be involved in stimulus anticipation. (C) Components projected onto recorded neurons from motor cortex and dmPFC. Note that strong components (> 10) tend to occur on multiple neurons, suggesting common response properties across the population. Absolute value plotted – we are interested in only the strength of the scores.

3. Determine the variance accounted for by each component.
4. Visualize the components.

An excellent review on the mathematical details on PCA is Reyment and Jöreskog (1996). The method has advantages over other linear transformations in that it does not assume basis vectors and is not iterative; rather, its basis vectors are derived directly from the data set. Other linear transforms may be applied and have other advantages.

PCA is based on singular value decomposition, which can be run in MATLAB using the command svd:

```
randData = rand(49, 10); % random peri-event
matrix: 49 neurons 10 bins
[comps, s] = svd(randData', 0)  % SVD of the peri-event
matrix
% plot variance accounted for by components
figure; plot(l ./ sum(l)); xlabel('Components');
figure; hold on; plot(comps(:, 1:3));
xlabel('Time(bins)');
l = diag(s).^2/(size(randData,2)-1);  % percent
variance of components
scores = randData * comps;  % determine scores on each
components
figure; imagesc(scores); xlabel('Neurons'); colorbar; %
visualize scores
```

Note the above example is based on randomly generated data simulating a perievent perievent matrix of 49 neurons over 10 bins. The svd command outputs the components. A key portion of PCA is that the results are easily interpretable in terms of variance when the variance is uniform across the data; for these reasons, it is ideal to work with data normalized to unit variance (i.e., Z scores, where one unit equals one unit of standard deviation; not used above for simplicity).

A scree plot of the variance represented by each component (the variable 'l') above is instructive in guiding further analysis. Each number in this variable corresponds to a component; the raw value indicates how many neurons have variance explained by a particular component. Components that are interesting must account for significant variance. One rule of thumb is to extrapolate a line based on lower components; components that fall above this line are of interest; components near this line should be ignored. If desired, more stringent statistical criteria can be used for this determination. The larger the data, the more components may be of interest; however, rarely are more than 5–7 components of further interest. In random data, there is usually only one component (a flat line) that meets these criteria.

One can plot the components with the hope of interpreting them. Using this approach, we identified common response properties across our ensemble by performing PCA on the peri-stimulus peristimulus time histograms of our recorded neurons in dmPFC and motor cortex. Common forms of variance might correspond to functional groups of neurons within our ensemble.

Finally, the 'scores' of each principal component on particular neurons can be computed by matrix multiplying the components by the original data. By restricting components based on variance, dramatic dimension reduction can occur; for instance, in random data, where one component explains > 90% of variance, one can go from the original data of 490 values (10 bins X 49 simulated neurons) to 49 values (1 component X 49 neurons), a 90% reduction. The speed and performance of classification and clustering algorithms are improved by such dimension reduction.

We performed principal component analysis on neural data from prefrontal and motor cortices during delayed-response performance. When plotting the variance from each component, we noted that only 3–4 components were large, or greater than a straight line drawn from the smaller components (**Fig. 7.3A**).

The components isolated represented behavioral modulations. For instance, component 1 (inverted for clarity) had a large modulation related to responding, component 2 was related to holding the lever down, component 3 was modulated during pressing and releasing the lever, and component 4 was modulated in anticipation of the stimulus (**Fig. 7.3B**). The sign of a given PC carries information about the relative increases or decreases of firing rates by specific subsets of neurons. However, here we were not interested in this point. Instead, we focused on the issue of whether neurons had similar patterns of activity (i.e., similar changes in firing rates around the behavioral events). Therefore, in this example, we discarded the sign of the components.

To determine if there are functional classes within our data set, we examined the projection of PCA scores onto the neurons we recorded. For instance, neuron 1 in motor cortex has a large score for PC4 only (**Fig. 7.3C**) and therefore is likely involved in motor anticipation. To determine which scores were statistically significant, components were extracted from time-shuffled data. A principal component score of $|9.8|$ corresponded to a $p < 0.05$. Therefore, we interpret scores above this value as significant. We found that 11 neurons had significant projections of component 1 (6 in motor cortex, 5 in dmPFC), 8 neurons had significant projections of component 2 (3 in motor cortex, 5 in dmPFC) and 2 neurons had significant projections of component 3 and 4 (**Fig. 7.3C**).

PCA is a user-friendly and data-driven tool for exploration of multivariate data. However, there are several potential pitfalls regarding its use. First, the data set must be clean of artifacts (e.g., solenoids) and inconsistent spike sorting across channels can limit the interpretation of PCA. Second, the amount of variance accounted for by components must always be considered. With large data sets, it is tempting to consider smaller components (i.e., ~5). However, only a few neurons usually contribute to such components and we have found that such components are seldom informative about behavior. Third, caution must be applied in interpreting the components, in that the activity of a small subset of neurons at any given moment might drive a given component. To determine the robustness of the components, one should examine the scores for each component and check them against raw plots of neuronal activity, especially for neurons that have large weights on the components. In addition, we recommend that cross-validation techniques be used to assess the overall strength of components. Fourth, components themselves may not be truly orthogonal, and methods for rotating the components such as varimax or ICA, might be useful for interpretation *(22)*.

These concerns notwithstanding, this section illustrates how PCA can readily be used to identify common response properties among a group of simultaneously recorded neurons. Furthermore, these common response properties might correspond to functional groupings among neurons. While identifying patterns of responses, PCA cannot comment on specific interactions between neurons. To make such inferences, we move to other techniques.

5. Cross-Correlation

In neuroscience, a common and readily interpretable measure of functional interactions between neurons is cross-correlation *(23, 24)*. This method has been used in many neural systems to make powerful inferences about functional interactions as well as functional anatomy. Cross-correlation is designed to make inferences about synaptic connectivity between neurons (**Fig. 7.4A**). In investigating synaptic connectivity, one should observe narrow peaks (a few ms) in the latency of spikes of one neuron with respect to another. If such a peak is observed, then one can potentially make inferences about the number of synapses between each neuron. In certain, highly anatomically connected systems that are guided by topography (such as the auditory system), cross-correlation may provide detailed information about anatomical connectivity in lieu of intracellular recordings. Cross-correlation can also be applied to make inferences about local circuits *(25)*. Such inferences must be made with great care and require having well sorted single units. In extracellular recordings, waveforms recorded on the same wire may interact, diminishing the ability to spike sort. Furthermore, the recording system must sample at high rates to ensure waveforms do not collide. For these

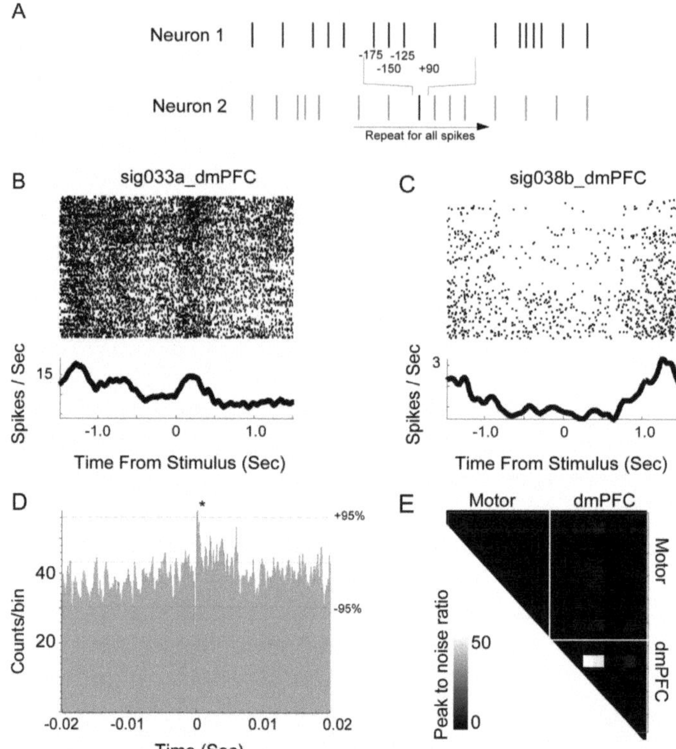

Fig. 7.4. Cross-correlation analysis between neurons in multi-electrode data. (A) In cross-correlation, for each spike train, the distance (in time) to every other spike of another spike train is recorded, and then a histogram constructed of distances in time of one spike in relation to another spike. (B) and (C) Peristimulus rasters for two dmPFC neurons; both neurons become inhibited prior to lever press, and recover firing rates after the lever is released. (D) These neurons have a strong, monosynaptic connection ~2–4 ms, suggesting they are connected or coactivated. These neurons are ~250 μm distant. (E) Cross-correlation across our ensemble of dmPFC and motor cortex neurons; the ratio of cross-correlation peaks to peaks in trial-shuffled data are plotted (this metric is used to rapidly look across a population of cells). Note there are few strong cross-correlations.

reasons, in our group we are hesitant to perform cross-correlative analysis between two neurons on the same wire.

To perform cross-correlations between spike trains from two neurons, one should:

1. Bin two spike trains of interest at 1 ms bins
2. Select one spike train as a reference spike train. For each spike in this spike train, determine the time latencies of spikes in the reference spike train to spikes of the other spike train. Typically, time latencies larger than ~200 ms are not of interest for synaptic interactions, greater than 200 ms may indicate slower interactions.
3. Once latencies are collected, the cross-correlation is simply the histogram of latencies.

4. To establish significance, peaks obtained in the cross-correlation of interest should be compared with peaks derived from time-shuffled data.

Cross-correlation measures the distance in time between the spikes of one neuron and the spikes of another neuron. If the spikes of one neuron tend to occur at a fixed time relative to the spikes of another neuron, a peak in the cross-correlogram should occur. Though the algorithm is simple, it is computationally intensive, and in MATLAB, relies on the compiled function `xcorr`:

```
t1 = sort(rand(1, 100) * 10); % random timestamps,10s @ 10 Hz
t2 = sort(rand(1, 100) * 10); % random timestamps,10s @
10 Hz
edges = 0:0.001:10;      % 1 ms bins
t1_binned = histc(t1, edges); % timestamps to binned
t2_binned = histc(t2, edges); % timestamps to binned
xc = xcorr(t1\_binned, t2_binned); % actual cross-
correlation
xc = xc(round(length(xc)/2)-200:round(length(xc)/2)+200);
```

In the above sample code, two random spike trains are generated lasting 10 s long, firing at approximately 10 Hz. In order for the cross-correlation to be performed, the spike trains must be binned; 1 ms bins is standard for visualizing short-latency synaptic interactions. In MATLAB, histc is a compiled routine that rapidly bins data. The xcorr command computes the cross-correlation. As we are interested only in short latency interactions, the cross-correlation is only kept at latencies ±200 ms.

For sessions from behaving animals (tens of minutes to hours), it is preferable to select epochs of interest on the order of tens of seconds for cross-correlation analysis. This selection helps in both computing the random cross-correlation, and in determining confidence intervals via time shuffling.

We performed cross-correlation analysis on neural data from prefrontal and motor cortices during delayed-response performance. In our example data set of 21 neurons, we find rare evidence for synaptic interactions in rodent frontal cortex. One pair of dmPFC neurons (4B and C) had a peak and a trough at 2–4 ms (**Fig. 7.4D**), hinting that this pair of neurons are monosynaptically connected or driven by common input. An examination of their response properties reveals that both cells become inhibited as animals initiate lever presses, and stay inhibited until after responses are complete and reward acquisition begins. These interactions provide a glimpse of how information might flow through the rodent dmPFC.

To determine which neurons had significant cross-correlation peaks, we calculated cross-correlation functions from time-shuffled data and compared the ratios of peak sizes for the observed and shuffled data sets. This procedure produced a single metric for the analyses of significant correlations. For shuffled data, the ratio of peak size of 1.1 corresponded to a $p < 0.05$. We interpreted ratios larger than this to be significant. A total of 35 pairs had significantly larger peaks than could be expected by chance. However, few of these pairs had strong correlations (mean peak to noise ratio: 1.23) (**Fig. 7.4E**).

When using the cross-correlations, several caveats must be considered. The first is that cross-correlation assumes stationarity of neurons. Under typical experimental conditions, many non-stationarities are present (anesthesia state, behavioral motivation, unit waveform drift, satiety states). Secondly, changes in rate can lead to spurious rate correlations – for this reason, a 'shift-predictor', i.e., the spike train from one neuron is shifted by the distance of one trial in time, is often subtracted from the cross-correlations to differentiate true interactions from simply rate covariations. Thirdly, in cross-correlative analyses, a sufficient number of spikes in both spike trains are required for statistical inference (Gerstein 1985). Fourthly, as discussed above, cross-correlation inferences rely on clear, artifact-free unit isolation. Fifth, the time-scale of neuronal interactions may influence the cross-correlation *(26)*.

Cross-correlation can also be used to look at long timescale interactions, i.e., on the order of several milliseconds or tens of seconds. Such long timescale interactions do not represent synaptic interactions; rather, they represent an increase between the firing rates of two neurons. While interesting, other techniques (such as rate correlations) may be better suited to such interferences.

These results illustrate how cross-correlation can be used to make inferences about connectivity in multi-electrode data. While this approach is perhaps the most common to studying functional interactions, the cross-correlation does not incorporate how correlations change as a function of behavior. To apprehend the dynamics of correlation, we use the joint-peristimulus time histogram.

6. Joint-Peristimulus Time Histogram

Since behavior provides a key window into neural responses (via perievent perievent or peristimulus peristimulus histograms), it is of great interest to know how relationships between neurons vary as a function of stimuli or events. Aertsen and colleagues *(27)* created the joint-peristimulus time histogram

(JPSTH) for this type of analysis. The JPSTH has been used to investigate relationships between neurons *(28)* and areas *(21)*. In many experiments, the same sequence of events is typically repeated for several 'trials'. It is interesting to know, then, how correlations between neurons relate to these events. Essentially, the JPSTH analysis looks at time-varying correlation between the neurons relative to stimuli or events (**Fig. 7.5**).

To perform the JPST analysis, one should:

1. Arrange the spike trains for two neurons of interest into *perievent perievent matrices*; i.e., a matrix in which rows are trials, and columns are bins (i.e., 25 ms) bins. Typically for JPSTH analysis, one is not interested in fast interactions; therefore, larger bin sizes (> 10 ms) are preferred.
2. For each trial, record the bin in which spikes occurred for each neuron. A Cartesian matrix can then be constructed in which

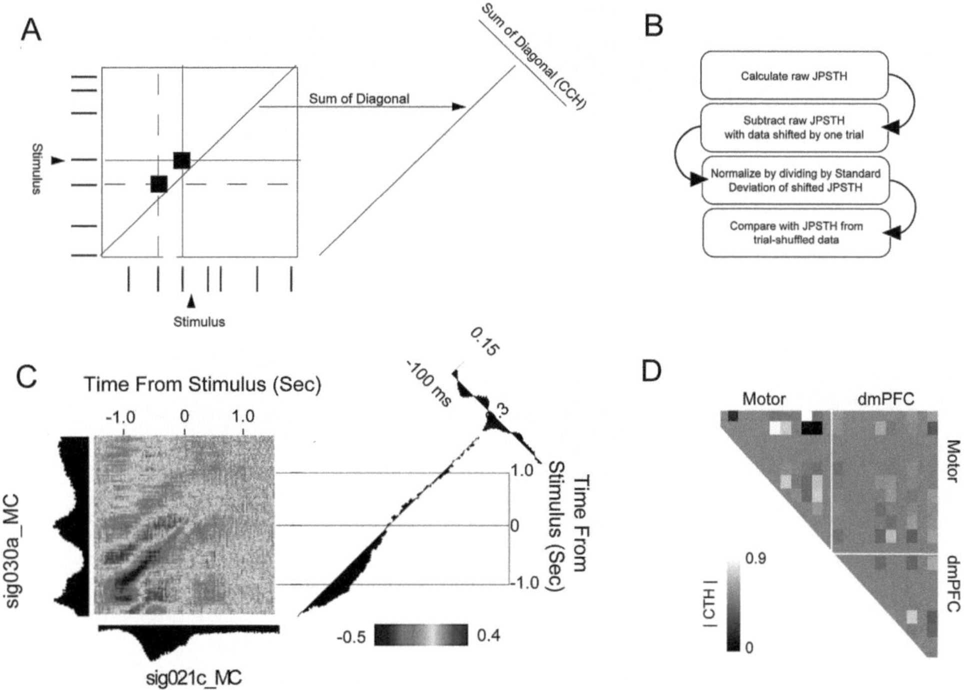

Fig. 7.5. Joint peristimulus time-histogram analysis. (**A**) For each trial, spike trains from one neuron are arranged on one axis, and spikes from another neuron are arranged on the orthogonal (i.e., vertical) axis. Coincident spikes over the course of the trial are recorded in the JPSTH matrix. This is repeated for every trial. (**B**) To correct for coincident spikes related to simple modulations in firing rate, the shift predictor is subtracted from the raw JPSTH, and the entire quantity is normalized to units of the shift predictor (by dividing by the standard deviation of the product of standard deviations of PSTHs). Finally, the JPSTH matrix is compared with an identical matrix computed from trial-shuffled data. (**C**) JPSTH for two motor cortex neurons reveals strong interactions prior to presentation of the stimulus. (**D**) JPSTH peaks across the ensemble.

the x axis are bins in which spikes occurred for neuron 1, and y axis are bins in which spikes for neuron y.

3. Over all the trials in the behavioral session, this Cartesian matrix can be summed to generate a *joint-peristimulus time matrix*. This matrix is referred to as the *raw matrix*.
4. It is important to correct for simple modulations in firing rate; to make this correction, perform steps 1–3 for perievent matrices in which trial orders are shuffled (preferred) or the trials are shifted by one trial. The matrix from these shuffled data is the *shift predicted matrix*.
5. Subtract the shift-predicted matrix from the raw matrix.
6. To compare between neurons, normalization is required. By dividing by the product of standard deviations of the peristimulus time histograms of each neuron, one can normalize the JPST matrix to units of correlation.
7. The two diagonals provide interesting data. The 45-degree diagonal provides a measure of time-varying correlation (also referred to as a correlation time histogram, or CTH), while the perpendicular diagonal (135 degrees) provides a measure of cross-correlation.

This analysis is a standard technique and is readily performed by existing software packages (NeuroExplorer; Nex Technologies, Littleton, MA; also available from http://mulab.physiol.upenn.edu/) and discussed in detail in the literature (**Fig. 7.5A and B**) *(27)*. Here, we provide a brief sample of code to calculate the raw matrix:

```
trials = 50; bins = 100;  % 50 trials, 100 bins
data1_binned = round(rand(trials, bins)); %sample
random data
data2_binned = round(rand(trials, bins)); %sample
random data
rawMatrix = zeros(bins, bins); %% preallocate matrices;
data1_hist = zeros(1, bins); data2_hist = data1_hist;
%% for loop to increment through trials
for i = 1:trials;
     x = find(data1_binned(i,:)>0); % find spikes
     y = find(data2_binned(i,:)>0); % find spikes
     % find matches and sum the spike matrix
     for j = 1:length(x); for k = 1:length(y);
     rawMatrix(x(j),y(k)) = rawMatrix(x(j),y(k))...
     + data1_binned(i,x(j))+data2_binned(i,y(k))-1 ;
     end; end;
     % increments the PSTH
     data1_hist(x) = data1_hist(x) + data1_binned(i,
x);
     data2_hist(y) = data2_hist(y) + data2_binned(i, y);
end
imagesc(rawMatrix); axis xy;
```

In above code, random data are generated. The loops then increments through the random data to find matches and to generate the raw JPSTH matrix. An optimized version of this approach would avoid for loops; this code is only used for clarity. Also, in order interpret the results, one must subtract a shift-predicted matrix, and divide by the product of standard deviations of PSTHs to normalize the matrix and compare between pairs.

A thorny problem is determination of significance of JPSTH arrays. Established approaches *(27)* assess significance either by determining if the modulation can be assessed by chance using a metric referred to as 'surprise'. We prefer a statistical approach with trial-shuffled data. In this approach, the CTH is calculated both for neurons of interest and for several iterations of trial-shuffled data. One can then investigate whether CTH values observed over a time epoch can be expected due to chance at a p value of 0.05.

A example of neurons with a significant JPST interaction is shown in **Fig. 7.5C**. This pair of motor cortex neurons reveals strong negative interactions both along the diagonal as well negative interactions before presentation of the stimulus (meaning that if the horizontal unit is firing, the vertical unit is not), and then a switch during the delay period when the vertical unit fires up to 2 s after delay activity in the horizontal unit. The cross-correlation reveals slow interactions on the order of 50–100 ms, suggesting that there are slow functional interactions between neurons.

We applied this analysis to the neurons in dmPFC and motor cortex in our example data set. In trial-shuffled data, JPSTH values of greater than $|0.2|$ corresponded to $p < 0.05$ *(20)*. Pairs of neurons in which one neuron did not have enough spikes ($<$ 10 times the number of trials) and from the same wires were excluded. Of 78 potential pairs, 15 (19%) had significant JPST interactions around the time of the stimulus (**Fig. 7.5D**), 9 in motor cortex, 2 in dmPFC, and 4 between motor and dmPFC.

Several potential pitfalls must be considered when using JPSTH. Foremost, all of the concerns of cross-correlation, including sufficient spikes and non-stationarity, apply when considering JPSTH. Non-stationarities become even more relevant because a confounding variable such as behavioral motivation may create spurious trial-by-trial relationships. The influence of such non-stationarities on JPSTH analysis is unknown. Our experience has been that the JPSTH should be normalized to facilitate comparisons with other neuronal pairs and that bin size is critical, and should be thoroughly explored for each data set of interest.

On the other hand, JPSTH is a flexible technique. Events need not be behavioral; they may be in reference to a third neuron or area *(29)*, and even continuous signals might be used in this analysis. Thus, JPSTH is a standard way for assessing functional interactions by examining the dynamics of correlations. Both cross-correlation and JPSTH lend themselves to pair-wise analysis of simultaneously recorded neurons.

7. Rate Correlations

Neurons may show correlations in firing rates in the absence of shared connections *(19, 21)*. Such interactions could occur over time (within trials) or over trials (within bins). Either of these types of functional interactions might correspond to functional units in the nervous system, such as cell assemblies *(30)*. Because spike trains are point processes, to compute correlations between them, one must convert them to a time series by binning (rate correlations / perievent analyses) (**Fig. 7.6A**). To compute rate correlations:

1. Bin two raw spike trains of interest into a series of bin counts.
2. Compute Pearson's correlation coefficient of binned firing rates.
3. Compare with what might be expected from random data.

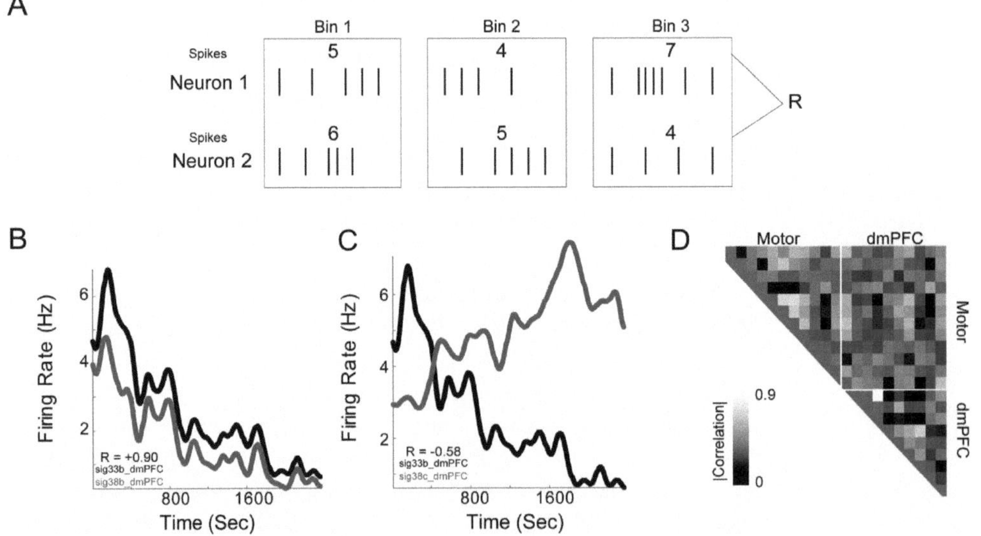

Fig. 7.6. Rate correlations. (A) This analysis simply bins neurons' firing rates over the session, and examines the correlation of neurons' binned firing rates over time. (B) Neurons could exhibit positive or (C) negative relationships. (D) Rate correlations over the ensemble (absolute value plotted – we are interested in only the strength of correlation) reveal several groups of neurons with strong rate correlations in both areas.

The above steps are simple and can be rapidly computed:

```
t1 = sort(rand(1, 200) * 10); % random
timestamps,10s @ 20 Hz
t2 = sort(rand(1, 50) * 10); % random timestamps,10s @ 5
Hz
edges = [0.25:0.25:10]; % edges for 50 ms bins
rate_t1 = histc(t1, edges); % bins for t1
rate_t2 = histc(t2, edges); % bins for t2
scatter(rate_t1, rate_t2); corrcoef(rate_t1,
rate_t2)
```

In the above code, two spike trains are generated, binned, and then correlated. This is a rapid way to look for relationships between the firing rates of two neurons and also can be useful in addressing non-stationarities that may threaten cross-correlation or JPSTH analyses.

We applied simple rate correlations to dmPFC and motor cortex recordings. We binned spike trains over 10 s over the 39 minutes animals were behaving, resulting in binned firing rates over a total of 235 bins. Note that over this epoch, considerable behavioral fluctuations could occur; however, for the present, we are only interested in relationships between neurons, and ignore all other parameters. In trial-shuffled data, correlations greater than $|0.14|$ corresponded to $p < 0.05$. Of 210 interactions, 109 pairs (52%) had significant rate correlations (67% of dmPFC pairs, 52% of motor cortex pairs, and 45% of dmPFC-motor pairs). Neurons could exhibit strong covariations over the session. For instance, two neurons could be strongly correlated (**Fig. 7.6B**) or anti-correlated (**Fig. 7.6C**) over the session. Most neurons did not have strong correlations (**Fig. 7.6D**). This result suggests that overall correlations were not governed by gross non-stationarities over behavioral sessions (i.e., degradation of unit isolation, artifacts in recording, decline in motivation); rather, the structure of correlations between specific pairs suggested that pairs of neurons exhibited strong functional interactions.

Perhaps the largest difficulty with rate correlations is that the animal behavior is uncontrolled. Correlations may be driven by behavioral variability, similar functional properties across neurons (compare with PCA), or even task-irrelevant behavior, such as grooming or movement artifacts. For these reasons, it is somewhat difficult to interpret rate correlations, especially in the absence of well-controlled operant behavior. In using rate correlations, all previous caveats apply (unit isolation, sufficient spikes). Furthermore, the number of bins should be carefully chosen. Too few bins lead to inflated correlation coefficients. Too many bins may destroy temporal structure. For this reason, any observed correlation must be explored at a variety of bin sizes.

These methods demonstrate how to use simple rate correlations to make inferences about functional relationships between neurons, and demonstrate how these techniques can be applied to simultaneous recordings from neural populations in rodent dmPFC and motor cortex.

8. Trial-by-Trial Rate Correlations

In essence, the JPSTH applies cross-correlation on a trial-by-trial basis. Gawne and Richmond *(31)* proposed the same analogy for rate correlations. This analysis is simple to calculate, as one can compare (via correlation) the mean or the variance of activity of two neurons over trials (**Fig. 7.7A**). To perform rate correlation, one should:

1. Construct perievent matrices for each neuron.
2. Sum the firing rates for each trial.
3. Compute Pearson's correlation coefficient of trial-by-trial firing rates.

These steps also can be easily computed with the following code, in which two perievent matrices are created from random data, and a correlation coefficient is calculated.

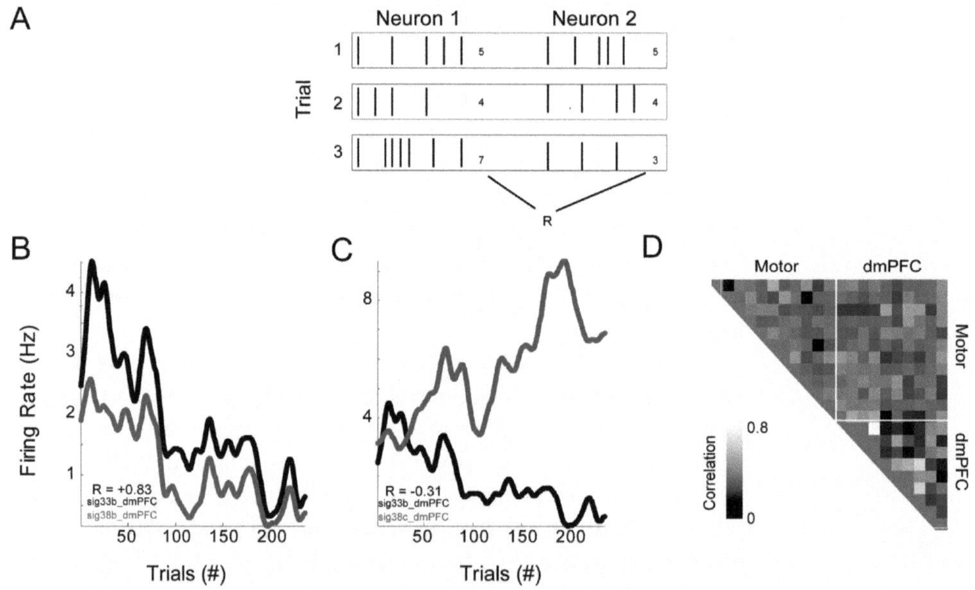

Fig. 7.7. Trial-by-trial rate correlations. (A) This analysis simply bins neurons' firing rates over the trials, and examines the correlation of neurons' firing rates over trials. (B) Neurons could exhibit positive or (C) negative relationships; same neurons as in Fig. 7.6. (D) Trial-by-trial rate correlations over the ensemble (absolute value plotted – we are interested in only the strength of correlation) reveal several groups of neurons with strong trial-by-trial correlations, particularly in dmPFC.

```
data1 = rand(70, 10); % random peri-event matrix: 70
trials, 10 bins
data2 = rand(70, 10); % random peri-event matrix: 70
trials, 10 bins
fr_data1 = mean(data1');  % sums along rows
fr_data2 = mean(data2');  % sums along rows
scatter(fr_data1, fr_data2); corrcoef(fr_data1,
fr_data2)
```

We applied trial-by-trial rate correlations to our sample dmPFC and motor cortex recordings over 235 correct trials; this number yielded similar power as our rate correlation analysis (above); over each trial, animals performed a stereotyped series of actions. In trial-shuffled data, we found that correlations greater than $|0.18|$ corresponded to $p < 0.05$. Of 210 interactions, 56 pairs (27%) had significant trial-by-trial rate correlations (69% of dmPFC pairs, 16% of motor cortex pairs, and 43% of dmPFC-motor pairs). Neurons exhibited strong trial-by-trial covariations. For instance, two neurons could be strongly correlated (**Fig. 7.7A**) or anti-correlated (**Fig. 7.7B**) over the session; these are the same neurons with rate correlations. Most neurons did not have strong correlations (**Fig. 7.7C**). Note that while motor cortex neurons had strong rate correlations, they tended not have strong trial-by-trial interactions. These methods, then, are able to separate resolve patterns of functional correlations.

9. Predictive Interactions

Thus far, we have discussed functional interactions in three domains – patterns of activity (PCA), timing of spikes (cross-correlation / JPSTH) and fluctuations in firing rate (rate and trial-by-trial covariations). Though these approaches are powerful in examining relationship between neurons, none of these approaches examines how neurons predict behavior. If neurons predict behavior in similar ways, then they may have functional interactions that are due to behavior.

To investigate relationships in how neurons predict behavior, we used methods from the field of *machine learning*, or *statistical pattern recognition*, to quantify information from spike trains about behavior on a trial-by-trial basis (see Refs. *(5)* for review of methods for spike train analysis and Refs. *(32–36)* for a review of the statistical issues). By comparing the predictions of behavior between two neurons, we can investigate whether neurons predict behavior in the same way (i.e., predict the same behavior on the same trials), or in different ways (i.e., predict behavior as would be expected by chance) (**Fig. 7.8A**). To perform statistical pattern recognition, one should:

1. Construct perievent matrices for each neuron.
2. Prepocess the perievent matrices (smoothing and decimation).
3. Reduce the dimensions of the data (via PCA or wavelet-based methods). This step is important for optimizing classifier performance; irrelevant dimensions will both slow and decrease classifier performance.

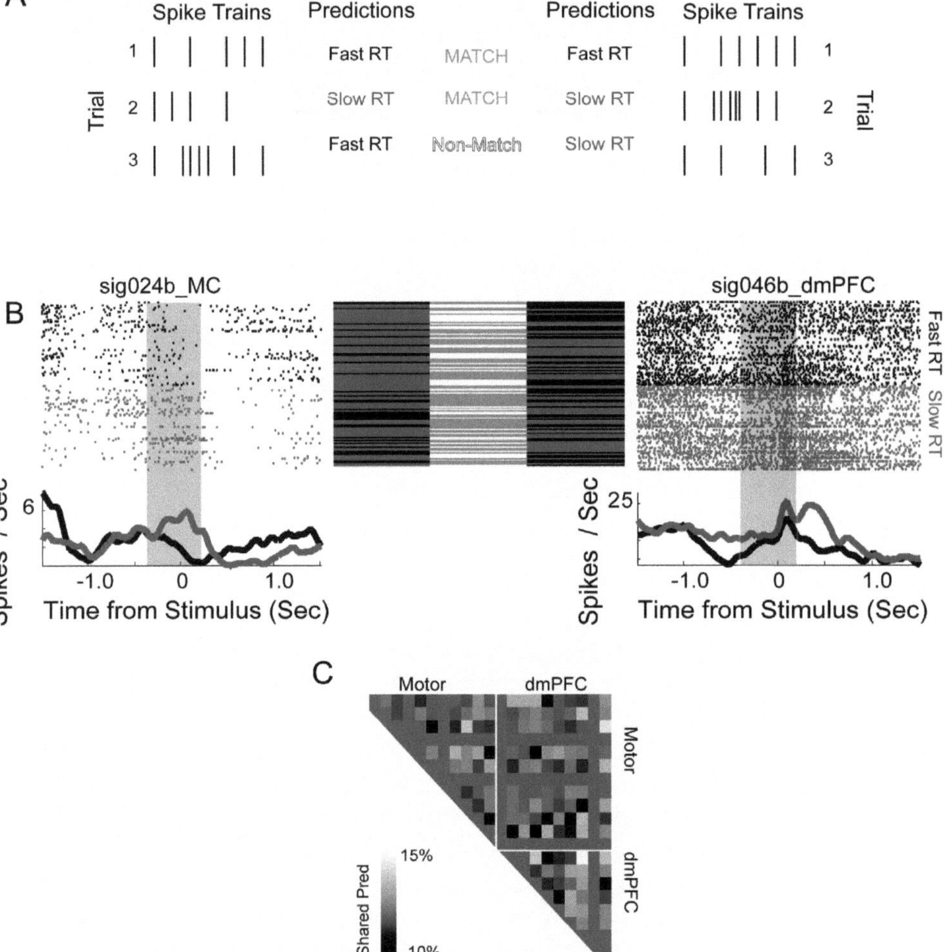

Fig. 7.8. Predictive interactions. (A) Statistical pattern recognition uses trial-by-trial spike train to generate predictions of behavior on a trial-by-trial basis. These predictions can be compared with predictions from another neuron. (B) Peristimulus rasters of neurons and their predictions of reaction time; black corresponds to trials with short RTs, red corresponds to trials with slow RTs. In between the peristimulus rasters, predictions are plotted: black means that a fast RT was predicted from spike trains; red means that a slow RT was predicted. Central column: Green indicates that the two neurons agreed in their predictive information; white indicates that the neurons disagreed. These neurons tended to agree more than could be expected by chance, and thus had an interaction in their predictive information. (C) Predictive interactions across the ensemble.

4. Segment the data into a training data and a testing set.
5. Training a classifier from training data to predict the behavioral outcome of testing data. Repeat at least 10 times (10 fold cross-validation) *(32)*.
6. Evaluate classifier performance with information theory or ROC analysis.
7. Repeat steps 1–5 for trial-shuffled data.

Implementation of these steps is highly involved and requires careful understanding of the methods used for preprocessing, dimension reduction, and classification. In essence, statistical pattern recognition uses neuronal data to predict a behavioral variable on each trial. Classifier performance can then be quantified by comparing classifier predictions of the behavioral variable (i.e., slow vs. fast RTs) with the actual behavioral variable (the RT that the animal actually made).

Below is a brief demonstration with random data using PCA and k-means clustering:

```
%make random data; 40 trials, 100 bins, 2 classes
a = rand(40, 100);  b = rand(40, 100) * 1.2; % 1.2 is the
separation coeff
data = ([a; b]);
groups = [ones(1, 40) ones(1, 40)*2];  % make a groups
vector
[u,s,v] = svd(data',0);  % pca for dimension reduction
U = u(:,1:3); %only first three components
scores = data * U; % obtain scores for PCA
% Use k means to cluster groups in high dimensional
space. This is being used % as a classifier WITHOUT
cross validation, and although it is not a strong % %
classifier, it illustrates the point; any classifier
will do
[pred1, C] = kmeans(scores, length(unique(groups)));
% Plot the PC scores in dimensional space; check for clusters
g1 = find(groups==1); g2 = find(groups==2); hold on;
scatter3(scores(g1, 1), scores(g1, 2), scores(g1, 3),
'k.');
scatter3(scores(g2, 1), scores(g2, 2), scores(g2, 3),
'r.');
scatter3(C(:, 1), C(:, 2), C(:, 3), 'b+');
confusionMatrix(1,1)=length((find(pred1(g1)==1)));%
true positive
confusionMatrix(1,2)=length((find(pred1(g1)==2)));%
false positive
confusionMatrix(2,1)=length((find(pred1(g2)==1)));%
false negative
confusionMatrix(2,2)=length((find(pred1(g2)==2)));%
true negative
```

In the above code, random data are created. In one class, a slight bias of 20% (represented by the factor 1.2 above) is generated. PCA is then performed for dimension reduction, and k-means clustering is used for classification. Classification is then assessed using a 'confusion matrix', or a 2 × 2 matrix comparing predicted (rows) vs. actual (columns) classification. Note that this approach, with 20% bias, generates relatively robust classification at minimal computational cost.

In our lab, we perform the above analysis somewhat differently *(5)*. We identify correct trials with reaction times in the lower quartile (selected to speed classification; equivalent results are possible for other splits of reaction time, or accuracy) *(14)* and construct perievent histograms (using 1 ms bins) for the epoch from 250 ms before lever release to 250 ms after the presentation of the trigger stimulus. The histograms were smoothed, using 10-fold decimation and low-pass filtering, and decomposed into a set of components that were ordered by relative variance. This latter step was carried out using wavelet packet analysis *(33)* based on the matching pursuit algorithm *(34)*. A 4-point 'symmlet' filter was used for the wavelet packet analysis.

Matching pursuit is a sequential decomposition algorithm that finds a series of signal components (up to 10 in this case) that account for decreasing levels of variance in the average perievent histograms of each neuron. This step is used in lieu of PCA above (as in Richmond and Optican 1987). This procedure is useful for reducing the complexity (i.e., dimensionality) of the data prior to analysis with statistical classifiers. By using matching pursuit, we reduced the dimensionality of the neuronal spike data, from 400 1-ms bins (the number of bins in a peristimulus time histogram) to typically 5 or fewer matching pursuit task-related features. Scores for matching pursuit components are defined from trials with the lower one-fourth of RTs. Scores were calculated as the dot product of the wavelet filter and the measured neuronal firing on each trial. The scores were then used to train statistical classifiers. Regularized discriminant analysis (RDA) *(35)* was used with leave-one-out cross-validation to predict trials with fast RTs versus trials with slow RTS (top quartile).

RDA is a non-parametric form of linear discriminant analysis that is appropriate for data with non-Gaussian distributions (i.e., the scores from matching pursuit) and that allows for non-linearity in the discriminant functions *(36)*. We normally use such an algorithm in lieu of k-means, which is used here for explanatory simplicity. Note that in the above sample code, cross-validation is not done; in general, it must be done to obtain measures of bias and variance. Results obtained from the classifiers were used to construct 'confusion matrices' – i.e., a list of trial outcomes versus the predictions of trial outcome made by the classifier in the form of a 2 × 2 table (for a data set with 2 possible outcomes there are 2 rows and columns in the confusion matrix). Whereas PCA and k-means are successful with a ~20% bias, using our approach, we can reliably predict biases of

~10%. However, this approach requires several libraries for the program R (http://cran.r-project.org/) for rapid classification. Examples are available on our website (spikelab.jbpierce.org/ Resources) and have been described in detail in previous publications *(5)*.

If neurons have a shared predictive relationship, then they should have shared predictions that are stronger than would be expected in trial-shuffled data. This is readily quantified by the following code:

```
for i = 1:10000
pred1 = zeros(1,100); pred2 = zeros(1,100);
pred1(find(rand(1,100)>0.7))=1;  % 70% prediction
pred2(find(rand(1,100)>0.6))=1;  % 60% prediction
match(i) = length(find(pred1==pred2));
end
```

The distribution of the variable match provides a probability distribution that is used to assess the significance of predictions.

We applied this approach to two neurons from dmPFC and motor cortex. Although these two neurons provided small amounts of predictive information (0.005 bits for the motor cortex neuron, 0.03 bits for the dmPFC neuron; **Fig. 7.8B**), we found that these neurons shared predictions 59% of the time, more than could be expected by chance (44±9%, $p < 0.05$). These data indicate that neurons that were weakly predictive of behavior could still share predictive relationships.

Across our population, we compared neuronal predictions with neuronal predictions from *within class* (i.e., within fast and with slow RTs) trial-shuffled data (**Fig. 7.8C**). We found that improvements in classification of 9% over random data corresponded to $p < 0.05$. We found that 12 (of 127; 10%) predictive interactions were greater than could be expected by chance. We would expect to find this number of significant predictive interactions at $p < 0.05$ by chance ($X^2 = 2.15$, $p < 0.14$).

We also compared predictive information on a trial-by-trial basis between dmPFC and motor cortex. The population of 10 dmPFC neurons provided 0.2 bits of information, whereas the population of 11 motor cortex neurons provided 0.5 bits of information. When predicting fast RTs, dmPFC and motor cortex shared predictions (76%) that could be explained by chance ($p < 0.05$ at 76%). On the contrary, when predicting slow RTs, dmPFC and motor cortex shared predictions (86%) were higher than could be explained by mere correlations with RT ($p < 0.05$ at 83%).

The use of statistical pattern recognition to explore trial-by-trial relationships in predictions between neurons should be approached carefully. This analysis is complex and reliant on understanding of procedures such as dimension reduction and classification. In pattern recognition, one must also worry about *bias* and *variance (32)*, which can readily influence trial-by-trial predictions.

However, these results indicate that dmPFC and motor cortex populations have functional interactions in their trial-by-trial predictive information (that is, their predictions about reaction times) only when predicting slow RTs. This type of analysis suggests that dmPFC neurons and motor cortex neurons functionally interact on slow but not fast RTs. This novel insight is an example of how predictive relationships between populations of neurons can be used to make inferences about how these populations interact.

10. Network Interactions: Synergy and Redundancy

As an extension of the preceding analyses of shared predictive relationships, one might ask how the predictive information of a two-neuron ensemble compared to the predictive information of each neuron individually. If two neurons provided more information individually than they do together, then they interact redundantly. On the other hand, if they provide more information together than they do individually, then they interact synergistically. This idea provides a framework *(14, 31, 37)* for interpreting network interactions (**Fig. 7.9A**)

To assess network interactions, one should:

1. Construct perievent matrices for one neuron.
2. Preprocess the perievent matrices (smoothing and decimation).
3. Reduce the dimensions of the data (via PCA or wavelet-based methods).
4. Segment the data into a training data and a testing set.
5. Training a classifier from training data to predict the behavioral outcome of testing data.
6. Repeat steps 1–5 for another neuron.
7. Repeat steps 1–5 for the two neurons together.
8. Compare information of the two neurons together with the information of each neuron individually.

If the predictive information of two neurons were greater than the sum of the predictive information individually, then we would call this a *synergistic* interaction. On the other hand, if the predictive information of two neurons were less than the sum of the predictive information of each neuron individually, then we would call this a *redundant* interaction. Finally if the predictive information of two neurons were equal to the sum of the

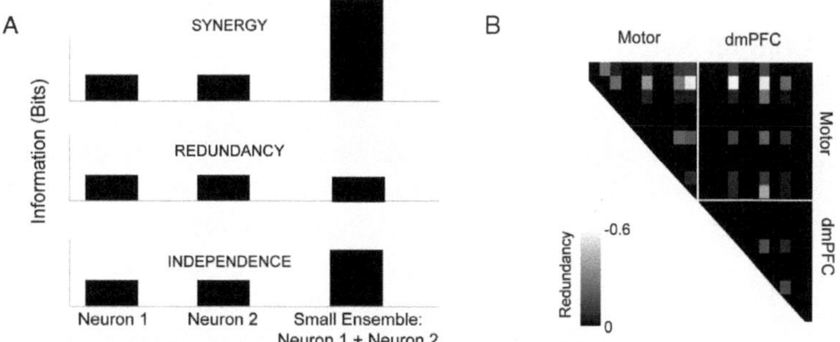

Fig. 7.9. Network interactions. (A) Two neurons predictive information can interact to produce more information together than individually (synergy), less information together than individually, or simply the linear sum together of their information individually (independence). (B) Redundancy interactions across the ensemble; we observed almost exclusively redundant interactions across the ensemble; note that this analysis is limited only to neurons with significant predictive information.

predictive information of each neuron individually, then we would call this an *independent* interaction.

Again, as with the previous section, the code for the pattern recognition is complex; on the other hand, the computation of whether neurons are synergistic, independent, or redundant is simple *(14)*:

$$P_{Ensemble} = I_{Ensemble} - \sum_{1}^{N} I_{Neuron}$$

Where $I_{Ensemble}$ is the predictive information of the ensemble, or two neurons together, and I_{Neuron}, is the predictive information of individual neurons in the ensemble.

We applied this analysis to our population of neurons recorded in dmPFC and motor cortex (**Fig. 7.9B**). First, we found for 36 pairs with predictive information, the mean $P_{ensemble}$ was –0.19 bits, suggesting overall redundant interactions. These results are convergent with previous studies from out lab *(14)*. We found that a $P_{ensemble}$ of –0.2 corresponded to $p < 0.05$. According to these metrics, 13 (of 36 pairs; 36%) pairs had significantly redundant interactions.

As an ensemble, we found that the population of neurons in both areas (dmPFC and motor cortex) provided 0.55 bits of predictive information. Motor cortex alone provided 0.5 bits of information, while dmPFC alone provided 0.2 bits of information. Thus combined, both areas had 0.15 bits of redundancy ($P_{ensemble}$= –0.15), below the significance level. These results suggest a lack of strong synergistic or redundant interactions *between* dmPFC and motor cortex.

11. Comparison of Functional Interactions with Multiple Techniques

We have presented an array of techniques for assessing functional interactions in this manuscript. Each technique we have presented describes a different aspect of how neurons interact, and provides a distinct perspective to neurons' functional interactions. To provide a multivariate portrait of functional interactions, we present a table describing the interactions between two pairs of neurons (**Table 7.1**).

For the first pair of neurons, (sig033b_dmPFC & sig038b_dmPFC), we find similar response properties (holding the lever down), strong cross-correlations (the ratio of peaks to peaks in random noise is shown), suggesting a synaptic relationship. We also find strong rate and trial-by-trial rate correlations; however, we find no instance of JPSTH interactions, predictive interactions, and hence, no network interactions. This type of pair, while interesting to many neuroscientists because it describes a monosynaptically connected pair of neurons at least 250 um distant from each other, has no correlations that change in relation to stimulus presentation and is poorly predictive of reaction time. This is a classic example of a *noise* correlation *(38)*, that is, despite the fact that these neurons themselves are having similar behavioral modulations, the communication between this pair of neurons is not directly related to behavior. This pair belies the importance of our approach of comparing correlative techniques, as no one measure of functional interactions provides a complete picture of neuronal interactions.

Table 7.1
Functional interactions

	Chance	Pair	Pair
	–	sig033b_dmPFC	sig040a_dmPFC
		sig038b_dmPFC	sig024a_MC
Principal component analysis	9.8	Hold component (PC 2)	No common component
Cross-correlation	1.1	12	1.0
JPSTH	0.2	0	0.35
Rate correlation	0.14	0.91	–0.03
Trial-by-trial correlation	0.18	0.83	0.17
Predictive information	~	0.72	0.83
Redundancy	–0.2	–0.002	–0.46

Another pair of neurons (sig040a_dmPFC and sig030a_MC) in different cortical areas has an opposite pattern. The neurons do not share components, and the cross-correlation, rate-correlation, and trial-by-trial correlation of these neurons is small; however, we find strong JPSTH correlations, predictive interactions, and redundant network interactions. This pair is easily ignored because of a lack of clear cross-correlations or trial-by-trial correlations; nonetheless, this pair has strong JPSTH and predictive relationships. These neurons likely have a strong *signal* (where signal is controlling response to the stimulus) correlation *(38)*, in that they seem to be working together to predict behavior. Their functional interactions may allude to top-down control of dmPFC on motor cortex *(21)*.

Finally, we integrate these analyses to provide a map of functional interactions (**Fig. 7.10**). In this case, we are not interested in significance; rather, we are interested in patterns of interactions including subthreshold values. For these reasons, we present a scaled correlation map across ensembles of neurons in frontal cortex.

We note that, in this data set, functional interactions in motor cortex common at the level of cross-correlation and JPSTH, but rare in rate or predictive interactions. On the contrary, in dmPFC, cross-correlation and JPSTH interactions are rare, but rate or predictive interactions are more common. Between dmPFC and motor cortex, particular combinations of neurons seemed to have robust functional correlations. These results demonstrate,

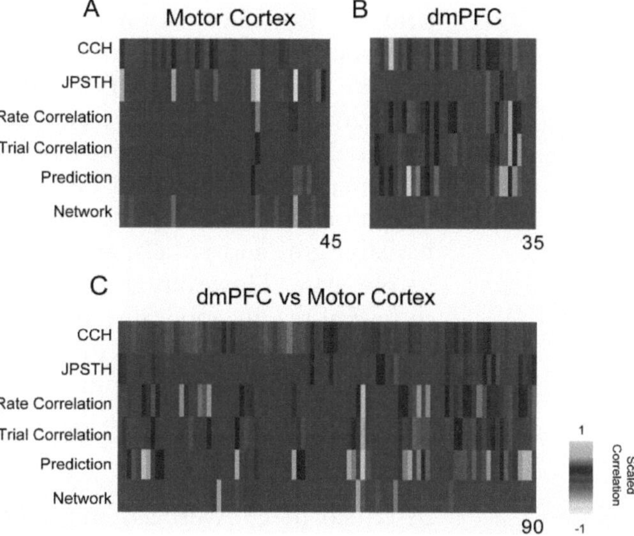

Fig. 7.10. Integrating functional interactions between measures. (A) Interactions within motor cortex; (B) dmPFC, and between dmPFC and motor cortex using techniques presented in this paper. For comparison, all interactions are normalized within a class of interactions, and presented as absolute value.

in principle, how functional interactions can be combined to elucidate patterns of interaction among cortical networks.

Thus, studying different kinds of functional interactions across an ensemble of simultaneously recorded neurons in multielectrode experiments affords novel insights into how neurons work together to encode information, and provides evidence that neurons can work together in diverse ways in service of behavior.

12. Notes

Here, we have described seven distinct approaches to investigating functional interactions between neurons: Functional grouping by PCA, cross-correlation, joint-peristimulus time histograms, rate correlations, trial-by-trial rate correlations, predictive interactions, and network interactions. We have described how to perform each analysis, limitations and applications of each analysis, and demonstrated how such analysis might work in an ensemble of 21 simultaneously recorded neurons from rodent dmPFC and motor cortex. By integrating and comparing functional interactions, we are able to gain insight on how relationships among populations of neurons contribute to the encoding of behavior.

While we have restricted our analysis of functional interactions to the methods discussed here, many investigators have developed an arsenal of tools with which to assail the problem of neuronal interactions. These include peer-prediction, in which pattern recognition is used to predict the spikes of another neuron *(39)*; information theory, in which a 'lookup table' is constructed to predict the information in one spike train from another *(37)*, detection of synchrony *(40)*, coherence (both in spike and field), in which frequency relationships between field potential and spikes from different channels are used to make inferences about local and distant processing *(41)*, as well as a host more under current development. Each of these techniques has strengths and pitfalls, and may be able to be readily applied to our data. However, though we have limited our discussion to several readily applicable and relatively simple techniques that demonstrate the principles of considering functional interactions between neurons, we recognize that future analyses using the techniques above or still being developed may further refine or expand our interpretations of how neurons interact.

What is the significance of the functional interactions described here? Averbeck and colleagues have examined correlated variability between neurons *(15–17, 38)*. They mention two ideas when considering neuronal correlations: *Signal* correlations, i.e., correlations that relate to an outside, measured variable; and *noise* correlations, i.e., correlations that do not relate to any measured variable. Some investigators have suggested that such noise correlations play a crucial role in neural processing *(40)*. Trial-shuffling plays a key role in discriminating between signal and noise correlations *(42)*, and is analogous to the within-class trial shuffling performed in our analysis of predictive information to determine if shared predictive relationships were related to high predictive information (signal correlations) or to shared predictive relationships (noise correlation). Furthermore, we report neurons with high noise and low signal correlations (**Table 7.1, 1st** column) as well as neurons with high signal and noise correlations (**Table 7.1, 2nd** column). However, such analyses typically require robust measurement of signal via a tuning curve; our task is too simple for such an analysis. Instead, we focus on using multiple approaches to develop a picture of neural interactions.

In exploring functional correlations between neurons, one must remember the anatomy of the brain areas of interest. For example, when recording between distinct cortical areas, one must understand the probability of direct synaptic connections and shared inputs. Furthermore, it is helpful to know in which other areas the neurons might interact (i.e., thalamus), and whether in other brain areas, projections to the neurons originate from the same neurons. While this information is difficult to get and requires painstaking anatomy, it is critical to interpreting data about how neurons interact. Within a cortical area, columnar organization, laminar information, receptive field properties, and recurrent connectivity all will help define and constrain interpretation of functional interactions between simultaneously recorded neurons *(6, 7)*.

Finally, the goal of studying functional interactions is to design clear experiments that test hypothesis about how neurons interact. For instance, we combined JPSTH analyses with techniques that inactivate dmPFC while recording from motor cortex to make insights about how the two areas might interact *(20)*. One could imagine future studies targeting microstimulation to groups with strong functional interactions, and investigate the role of stimulation of one neuron on some other (distant) neuron. Experiments of this nature will provide novel insights on how cortical neurons work together to control behavior.

Acknowledgments

We thank Eyal Kimchi and Nicole Horst for critical comments and helpful discussions. This work was supported by funds from the Tourette Syndrome Association, Kavli Institute at Yale, and the John B. Pierce Laboratory for ML and from an NIH training grant to the Yale Medical Scientist Training Program and Army Research Office for NSN.

References

1. Nicolelis, M.A., et al., *Simultaneous encoding of tactile information by three primate cortical areas.* Nat Neurosci, 1998. **1**(7): p. 621–30.
2. Nicolelis, M.A., ed. *Methods In Neuronal Ensemble Recording.* 1998, CRC Press: Boca Raton, FL.
3. Wessberg, J., et al., *Real-time prediction of hand trajectory by ensembles of cortical neurons in primates.* Nature, 2000. **408**(6810): p. 361–5.
4. Carmena, J.M., et al., *Learning to control a brain-machine interface for reaching and grasping by primates.* PLoS Biol, 2003. **1**(2): p. E42.
5. Laubach, M., N.S. Narayanan, and E.Y. Kimchi, *Single-neuron and ensemble contributions to decoding simultaneously recorded spike trains*, in *Neuronal population recordings*, C. Holscher, Editor. 2007.
6. Shepherd, G., *Synaptic Organization of The Brain.* 34d ed. 2003, Oxford: Oxford University Press.
7. Mountcastle, V.B., *Perceptual Neuroscience: The Cerebral Cortex.* 1998, Cambridge, MA: Harvard College.
8. Gochin, P.M., et al., *Neural ensemble coding in inferior temporal cortex.* J Neurophysiol, 1994. **71**(6): p. 2325–37.
9. Britten, K.H., et al., *A relationship between behavioral choice and the visual responses of neurons in macaque MT.* Vis Neurosci, 1996. **13**(1): p. 87–100.
10. Rolls, E.T., A. Treves, and M.J. Tovee, *The representational capacity of the distributed encoding of information provided by populations of neurons in primate temporal visual cortex.* Exp Brain Res, 1997. **114**(1): p. 149–62.
11. Rolls, E.T., et al., *Information encoding in the inferior temporal visual cortex: contributions of the firing rates and the correlations between the firing of neurons.* Biol Cybern, 2004. **90**(1): p. 19–32.
12. Reich, D.S., F. Mechler, and J.D. Victor, *Independent and redundant information in nearby cortical neurons.* Science, 2001. **294**(5551): p. 2566–8.
13. Zohary, E., M.N. Shadlen, and W.T. Newsome, *Correlated neuronal discharge rate and its implications for psychophysical performance.* Nature, 1994. **370**(6485): p. 140–3.
14. Narayanan, N.S., E.Y. Kimchi, and M. Laubach, *Redundancy and synergy of neuronal ensembles in motor cortex.* J Neurosci, 2005. **25**(17): p. 4207–16.
15. Averbeck, B.B. and D. Lee, *Neural noise and movement-related codes in the macaque supplementary motor area.* J Neurosci, 2003. **23**(20): p. 7630–41.
16. Averbeck, B.B., et al., *Neural activity in prefrontal cortex during copying geometrical shapes. II. Decoding shape segments from neural ensembles.* Exp Brain Res, 2003. **150**(2): p. 142–53.
17. Averbeck, B.B. and D. Lee, *Coding and transmission of information by neural ensembles.* Trends Neurosci, 2004. **27**(4): p. 225–30.
18. Dan, Y., et al., *Coding of visual information by precisely correlated spikes in the lateral geniculate nucleus.* Nat Neurosci, 1998. **1**(6): p. 501–7.
19. Vaadia, E., et al., *Dynamics of neuronal interactions in monkey cortex in relation to behavioural events.* Nature, 1995. **373**(6514): p. 515–8.
20. Narayanan, N.S., N.K. Horst, and M. Laubach, *Reversible inactivations of rat medial prefrontal cortex impair the ability to wait for a stimulus.* Neuroscience, 2006.
21. Narayanan, N.S. and M. Laubach, *Top-down control of motor cortex ensembles by dorsomedial prefrontal cortex.* Neuron, 2006. **52**(5): p. 921–31.
22. Laubach, M., M. Shuler, and M.A. Nicolelis, *Independent component analyses for quantifying neuronal ensemble interactions.* J Neurosci Methods, 1999. **94**(1): p. 141–54.
23. Aertsen, A.M. and G.L. Gerstein, *Evaluation of neuronal connectivity: sensitivity of cross-correlation.* Brain Res, 1985. **340**(2): p. 341–54.
24. Perkel, D.H., et al., *Nerve-impulse patterns: A quantitative display technique for three neurons.* Brain Res, 1975. **100**(2): p. 271–96.
25. Constantinidis, C., M.N. Franowicz, and P.S. Goldman-Rakic, *Coding specificity in cortical microcircuits: A multiple-electrode analysis of primate prefrontal cortex.* J Neurosci, 2001. **21**(10): p. 3646–55.
26. Brody, C.D., *Slow covariations in neuronal resting potentials can lead to artefactually fast cross-correlations in their spike trains.* J Neurophysiol, 1998. **80**(6): p. 3345–51.

27. Aertsen, A.M., et al., *Dynamics of neuronal firing correlation: modulation of "effective connectivity"*. J Neurophysiol, 1989. **61**(5): p. 900–17.
28. Vaadia, E., K. Kurata, and S.P. Wise, *Neuronal activity preceding directional and nondirectional cues in the premotor cortex of rhesus monkeys*. Somatosens Mot Res, 1988. **6**(2): p. 207–30.
29. Paz, R., et al., *Emotional enhancement of memory via amygdala-driven facilitation of rhinal interactions*. Nat Neurosci, 2006. **9**(10): p. 1321–9.
30. Tsukada, M., et al., *Dynamical Cell Assembly Hypothesis - Theoretical Possibility of Spatio-temporal Coding in the Cortex*. Neural Netw, 1996. **9**(8): p. 1303–1350.
31. Gawne, T.J. and B.J. Richmond, *How independent are the messages carried by adjacent inferior temporal cortical neurons?* J Neurosci, 1993. **13**(7): p. 2758–71.
32. Witten, I. and E. Frank, *Data Mining*. 2000, San Diego, CA: Academic Press.
33. Laubach, M., *Wavelet-based processing of neuronal spike trains prior to discriminant analysis*. J Neurosci Methods, 2004. **134**(2): p. 159–68.
34. Mallat, S. and Z. Zhang, *Matching pursuits with time-frequency dictionaries*. IEEE Transactions, 1993. **41**(12): p. 3397–3415.
35. Friedman, J., *Regularized discriminant analysis*. Journal of American Statisical Association, 1989. **84**(405): p. 166–175.
36. Hastie, T., R. Tibshirani, and J. Friedman, *The Elements of Statistical Learning*. Springer Series in Statistics. 2001, New York, NY: Springer-Verlag.
37. Schneidman, E., W. Bialek, and M.J. Berry, 2nd, *Synergy, redundancy, and independence in population codes*. J Neurosci, 2003. **23**(37): p. 11539–53.
38. Averbeck, B.B. and D. Lee, *Effects of noise correlations on information encoding and decoding*. J Neurophysiol, 2006. **95**(6): p. 3633–44.
39. Harris, K.D., et al., *Organization of cell assemblies in the hippocampus*. Nature, 2003. **424**(6948): p. 552–6.
40. Gray, C.M., et al., *Synchronization of oscillatory neuronal responses in cat striate cortex: temporal properties*. Vis Neurosci, 1992. **8**(4): p. 337–47.
41. Pesaran, B., et al., *Temporal structure in neuronal activity during working memory in macaque parietal cortex*. Nat Neurosci, 2002. **5**(8): p. 805–11.
42. Averbeck, B.B., P.E. Latham, and A. Pouget, *Neural correlations, population coding and computation*. Nat Rev Neurosci, 2006. 7(5): p. 358–66.

Chapter 8

Magnetoencephalography (MEG)

Andreas A. Ioannides

Abstract

Magnetoencephalography (MEG) encompasses a family of non-contact, non-invasive techniques for detecting the magnetic field generated by the electrical activity of the brain, for analyzing this MEG signal and for using the results to study brain function. The overall purpose of MEG is to extract estimates of the spatiotemporal patterns of electrical activity in the brain from the measured magnetic field outside the head. The electrical activity in the brain is a manifestation of collective neuronal activity and, to a large extent, the currency of brain function. The estimates of brain activity derived from MEG can therefore be used to study mechanisms and processes that support normal brain function in humans and help us understand why, when and how they fail.

Key words: Bioelectromagnetic forward and inverse problem, superconductivity, equivalent current dipole (ECD), magnetic field tomography (MFT).

1. Introduction

Brain function is founded on electrical activity within and between *neurons*. The activity of individual neurons is too small to be detected with sensors placed outside the head. The activity of neurons within circumscribed brain areas a few millimeters to a centimeter across is sufficiently differentiated from the activity of other neurons to justify thinking of these areas as segregated islands of processing. The processing within and interactions between these areas relies again on electrical activity, albeit, now it reflects not only neuron-to-neuron exchanges, but also collective action, coherent in time, from a large population of neurons that probably share similar properties (e.g. selectivity to some stimulus feature). This collective excitation of local network electrical activity in the brain is large enough to produce measurable changes

Fahmeed Hyder (ed.), *Dynamic Brain Imaging: Multi-Modal Methods and In Vivo Applications, vol. 489*

DOI 10.1007/978-1-59745-543-5_8 Springerprotocols.com

in both the electrical potential on the surface of the head, the electroencephalogram (EEG) and the magnetic field a few centimeters beyond, the magnetoencephalogram (MEG), which is the subject of this narrative.

MEG was introduced some 40 years ago *(1)* but stayed for much of the time within small academic groups mainly in physics departments, as a novelty area between physics and biology. Although what was needed for the technology to become clinically relevant was correctly identified early on *(2)*, the field reacted rather slowly because the requirements were beyond what academic departments or small companies could afford. Two other reasons delayed the emergence of MEG into modern neuroimaging. First, the signal of an MEG channel after the usual processing and averaging looks just like an EEG signal, so many saw MEG as an expensive EEG technology. Some still do so today. Second, it was known for well over a century that the mathematical problem of recovering the generators from the MEG and/or EEG signal has no unique solution *(3)*. The non-uniqueness of the bioelectromagnetic inverse problem is an undeniable theoretical fact, but a rather benign problem in practice. Evidence that real-time information about brain function was available from MEG at not only excellent temporal resolution but also at fine spatial detail became available in the 90s from novel analysis of multichannel MEG data *(4, 5)*. Eventually, helmet-like systems allowed the mapping of the instantaneous magnetic field all around the head in an instance. The analysis of the resulting signals provided, for the first time, a view of dynamics of brain function across the entire brain *(4)*.

From the numerous reviews of MEG, the 1993 work from the Helsinki group remains the most comprehensive and informative *(6)*. More recent reviews have emphasized how, despite the issues regarding the inverse problem, putative sources can be estimated *(7)* and, increasingly more often, how beam forming techniques can usefully scan the source space point by point *(8)*. Most of the techniques discussed in the literature use linear methods for extracting estimates for the generators. Heuristic analysis *(9)* and theory *(10)* suggest that a specific form of non-linearity is necessary for the solutions to possess expected properties for localized distributed sources. Dealing successfully with the computational penalty that goes with non-linearity, leads to reliable tomographic estimates of brain activity from instantaneous MEG signals. Magnetic Field Tomography (MFT) is the name given to the resulting method of extracting estimates of brain activity *(11)*. MFT solutions can scrutinize brain function at multiple spatiotemporal scales. In the spatial domain, the range covers details a few millimeters across (distinguishing activity within individual cytoarchitectonic areas) to mapping across almost the entire brain. In the time domain, events can be analyzed at timescales from a fraction of a millisecond to minutes and hours *(12)*. It thus becomes possible to extract from the same set of MEG data views of local and global activity and patterns of connectivity in this wide spatiotemporal scales of bewildering complexity *(13)*.

Here, we provide an overview of MEG, covering basic principles, measurements, source reconstruction, and finally examples to demonstrate the different types of output that can be extracted from the analysis of MEG recordings and by implication what is the relevance of such output to neuroscience. The next section, *Materials*, first describes in simple terms possible neural mechanism that may be responsible for coherent electrical activity from many neurons. It then outlines how the well-understood laws of electromagnetism and properties of the head explain the generation of, and differences between the EEG and MEG signals. Finally, this section outlines the instruments that allow the detection of the MEG signal outside the head. The *Forward and Inverse Problems* section outlines how one can compute the signal generated by a given configuration of generators (the forward problem) and how one can derive estimates for the generators from the measurements (inverse problem). The next section, *Output of MEG*, provides a flavor of the output of MEG analysis with emphasis on aspects that are unique to this technology. We finish with *Notes* where the advantages and disadvantages of MEG are discussed.

2. Materials

This section first outlines the likely neuronal mechanisms responsible for the generation of the MEG and EEG signal and then describes the basic elements of the hardware used for the detection of the MEG signal.

2.1. Neural Mechanisms

The smallest detectable MEG signal requires concerted action from many neurons numbering at a minimum many hundreds. These neurons must be arranged in a similar way in space and they must be activated in near synchrony. The very presence of a good size MEG and EEG signal is evidence for dual organization of neurons: a spatial organization in the way they are grouped together in space and large scale synchrony in the way their activity is organized in time. It is generally believed that relatively slow changes in electrical activity associated with post-synaptic potentials (PSP) at the apical dendrites of large pyramidal neurons are the main contributors to the MEG signal. Large pyramidal neurons are prime candidate generators of MEG signals because their elongated shape is ideal for producing strong primary currents. Furthermore they are arranged in parallel in the cortex so the net impressed current from nearby large pyramidal neurons will tend to sum up constructively. It is very likely that a large part of the MEG signal is indeed due to slow PSPs in the apical dendrites of pyramidal neurons, especially at frequencies well below 100 Hz. For this standard mechanism, typical estimates require about a million synapses to be simultaneously active to produce a measurable MEG signal *(6)*. MEG activity at frequencies well above 100 Hz is likely to be produced by synchronous action potentials *(14)*.

2.2. Recording the MEG Signal and Isolating the Contribution from the Brain

Although the MEG signal is generated by the collective activity of a large number of neurons, its strength is extremely weak compared to typical terrestrial magnetic fields. The earth's magnetic field is about a billion times as strong, while the usual urban environment at frequency ranges that overlap the ones of interest in MEG is still many orders of magnitude higher than the strongest MEG signal from a normal human brain. A pre-requisite for useful MEG measurements is therefore the availability of sensors that can detect the weak magnetic fields generated by the brain. Also required are methods that can exclude the large ambient fields and tools that can separate out the signal of interest from any remaining interfering signals from the environment and other signals generated by the body of the subject that are often considerably stronger than the signal of interest.

The basic MEG measurement relies on the detection of the electrical current in a small loop of wire, typically about one centimeter across, induced by the change in the magnetic field component perpendicular to the loop surface. Measurement of the induced current determines the value of the change in the magnetic field. Usually a set of coils is arranged as a gradiometer to emphasize nearby signals from the brain at the expense of distant sources. The detection of the minute magnetic field changes outside the head generated by electrical currents in the brain is measured by coupling the coil or gradiometer to an extremely sensitive "superconducting quantum interference device" (SQUID). SQUIDS, as the name implies, rely for their exquisite sensitivity on superconductivity and together with their sensing coils must be kept at extremely low temperatures, just a few degrees above absolute zero. To achieve this, sensing coils and SQUIDS are kept in a thermos-like container, the Dewar, which under normal operating conditions is filled with liquid helium. In modern systems, the bottom of the Dewar is shaped into a helmet with well over one hundred, nowadays a few hundred, sensing coils evenly distributed on its inner surface. Just a few millimeters away, on the other side of the insulating layer, at normal room temperature, a subject can safely place his/her head inside the helmet. Each sensing coil samples the local magnetic field and the full set of sensing coils can be "scanned" a few thousand times a second. Each scan delivers an independent measurement of the instantaneous magnetic field just outside the head. Although SQUID technology is now mature providing turn-key systems for whole-head MEG devices, its basic cryogenic requirements make it very expensive. In recent years, atomic magnetometry, a method of measuring magnetic fields based on the interaction of resonant light with atomic vapor, has emerged as a potentially cheaper alternative to SQUID based sensors *(15)*.

The second requirement, separating the signal of interest from the larger ambient background and other interfering signals, is achieved by a combination of passive shielding, use of gradiometer design either in hardware for the sensing coils coupled to the SQUIDS or in software using additional reference channels. Other signal processing techniques, e.g. Independent Component Analysis (ICA) *(16)* coupled to the use of information from auxiliary channels like the electrooculogram (EOG) and electrocardiogram (ECG) can effectively eliminate biological and other artifacts. The combination of the exquisite SQUID sensitivity with these hardware and software methods allows the measurement of the magnetic field generated by the brain with little contamination.

In a modern MEG system, typically a few hundred sensing coils, each coupled to its own SQUID, are housed at the bottom of the helmet-shaped Dewar, distributed so that they capture evenly the magnetic field just outside the head. The magnetic field for just one "timeslice" can be mapped by recording the signal from each sensor independently from, and for all practical purposes simultaneously with, the signal of each other sensor. In one second, a few thousand such timeslices can be recorded so that successive timeslices provide a movie of the instantaneous change in the magnetic field just outside the head. Since, as we will shortly describe, the speed of propagation from the generators to the sensors is the speed of light, the MEG signal change corresponds to the instantaneous change of the electrical current density in the brain generated by neuronal activity. The peaks of the signal generated by the brain are about two orders of magnitude higher than the device noise level, so the map of the magnetic field not only has exquisite time resolution (a fraction of a millisecond) but is also a very clean map of the topography of the magnetic field just outside the head.

3. Forward and Inverse Problem

The determination of the EEG and MEG signal from the knowledge of the sources, the electrical properties of their biological environment and the configuration of the measuring devices is known as the forward problem. The estimation of generator strength, location and timecourse from the EEG and MEG signal and the knowledge of electrical properties of their biological environment and the configuration of the measuring devices is known as the inverse problem. The laws of electromagnetism define what can be asked of the data and how the forward and inverse problems should be tackled, in particular what a priori assumptions can be made about the generators. The forward and inverse problem will be considered, describing in each case the theoretical framework established by the laws of electromagnetism and its implications for useful MEG (and sometimes EEG) applications

3.1. The Forward Problem

It is useful to separate the full current density into two terms. In general, we are interested in the first term known as impressed currents because they describe the active currents generated by energy-demanding neuronal activity. The remaining currents make up the second term; they describe the passive currents that flow as a result of the impressed currents in the biological medium. Impressed currents of an individual neuron cannot be directly detected by either MEG or EEG because they are too weak. Even under the most favorable conditions, a detectable signal can only be generated by the collective activity of at least many hundreds of neurons spread over 1 mm^2 or larger cortical patches. At this spatial scale, the appropriate terms that best separate the full current density into active and passive elements are referred to as primary current density and volume or return currents respectively. The primary current density depends on both intracellular currents and the local extracellular currents. The intracellular currents are closely related to the local impressed currents. Since these ionic flows are along axons and dendrites, the net contribution from a single neuron is a sum of vectors each pointing along the long axis of the corresponding active dendrite or axon. The overall primary current density generated by intracellular currents is the vector sum of contributions from active neurons, which is therefore strongly dependent on the overall arrangement of neurons. The flow of extracellular currents along the local conductivity gradients is determined mainly by cell membranes. For each focal neuronal activity, the local arrangement of cells determines the combined effect of both intracellular and extracellular currents and therefore shapes the resulting primary current density. The source space is a convenient label for the region of space where the primary current density can be non-zero, and it includes the entire brain. Primary currents can be thought of as the generators of the volume or return currents, i.e., the large-scale passive electrical current flowing in the "volume conductor", in the brain at large and bounded by the highly resistive skull. These large-scale passive electrical currents do not contribute to the magnetic field, except where they "twist" at boundaries with sharp changes in conductivity, especially the skull. In the special case that only concentric spherical boundaries of changes in conductivity are present, the magnetic field generated outside a conductor is given by an analytical expression *(17)*. Furthermore, the laws of electromagnetism and spherical symmetry define explicitly which generators can produce an external magnetic field and which are magnetically silent, i.e. they do not produce an external magnetic field no matter how strong they are. Specifically, radial components of the current density are magnetically silent sources. The magnetic field generated by tangential components of the current density can be written analytically in a form that depends on the center of the conducting sphere(s) and it does not depend on either the conductivities of the different compartments or the radii of the concentric shell(s), as long as the magnetic field is computed outside the conductor (last spherical shell). Finally, for a spherical conductor, the radial magnetic field, i.e. the component of the magnetic field (outside the head) pointing away from the center of the conducting sphere depends only on the primary currents. The magnetic field for realistic head shapes can be computed accurately and fast using a set of overlapping spheres appropriately chosen for each sensor *(18)*. The skull is smooth and nearly spherical; so, the convenient and relatively simple spherical model can provide an excellent estimate for the second term, except around openings like the eye sockets or parts of the skull that deviate substantially from the spherical model.

The EEG forward problem poses real difficulties in practice. The accurate computation of EEG signal is more demanding because it depends strongly on details of the conductivity profile. The differences in the forward problem for MEG and EEG signals have two main consequences. First, the relationship between neuronal activity is easier to model for MEG. On the one hand, the skull is transparent to magnetic fields and highly resistive to electrical currents (that must cross it to produce the scalp EEG) and, on the other, the effect of the conducting medium can be approximated by simple models for accurate computations of the magnetic field but have to be described in detail for the computation of the surface potential. Second, the EEG is influenced strongly by both radial and tangential electric currents while MEG is only sensitive to tangential ones.

The laws of electromagnetism endow both EEG and MEG signals with a direct relationship with the neuronal sources. Specifically, the electric and magnetic fields propagate from the (neuronal source) generator site with the speed of light. Since the sensors are just some centimeters away, for all practical purposes, the effect is immediate: a change in the source electrical activity in the brain produces an immediate change in the MEG and EEG signal. This is in sharp contrast with other neuroimaging methods like positron emission tomography (PET) and functional magnetic resonance imaging (fMRI) that rely on changes in blood flow or content (e.g. radioactive labeling or oxygenation) and therefore produce indirect correlates of neuronal activity with delays that are, at best, a good fraction of a second in the case of fMRI and minutes in the case of PET.

Finally, the forward problem is linear as a direct consequence of the linearity of the laws of electromagnetism. In other words, the electric and magnetic field generated by any combination of instantaneous current elements is simply the sum of individual contributions from each element. In the case of continuous primary current density, the instantaneous electric and magnetic field can be computed by integrating the contributions from each small volume element in the source space. In the case of a spherical model, the source space for MEG includes only regions where neurons and possibly white matter exists, any intervening regions and boundaries are not part of the source space as long as they do not generate primary currents.

3.2. Inverse Problem

In contrast to the forward problem, the inverse problem has no unique solution, a mathematical fact that was already demonstrated over 150 years ago *(3)*. Simply stated, it is impossible to reconstruct uniquely the electrical current density inside the head from MEG and/or EEG measurements. Even if we knew exactly the electrical potential on the surface of the head and the magnetic field everywhere outside the head, we would still be unable to determine the currents inside the head. In practice, non-uniqueness is much less of a problem than would appear from the dry mathematical statements. By definition, silent sources cannot be recovered and noise and sparse sensor coverage further limit what can be reliably extracted about the non-silent part of the current density vector. Nevertheless, what is often required of the data is to provide reliable estimates about which areas of the brain were preferentially activated by some stimuli or tasks and when. This limited objective is often satisfied with estimates of the timecourse of the non-silent part of the source configuration. The key question in practice is how accurately and reliably one can recover the non-silent part of the primary current density.

A unique solution of the biomagnetic inverse problem can be obtained by introducing constraints for the form of the generators. Two types of constraints are particularly popular *(6)*. The first assumes that the generators are one or more point-like sources, or current dipoles. Dipole source localization solutions are often interpreted as representatives for their neighborhood and are referred to as equivalent current dipoles (ECD). The second family of popular source localization methods assumes that the continuous current density can be written as a linear sum of (weighted) functions, each defining the sensitivity profile or lead fields for one of the sensors. These methods, known as minimum norm (MN) or weighted minimum norm (wMN) solutions, are popular because they lead to a linear system of equations which allows standard pseudoinverse techniques to define the inverse operator that can then be applied directly to the data. Theoretical scrutiny of the mathematical foundation of the inverse problem shows that neither current dipoles nor linear solutions are adequate. Minimum norm is not appropriate for tomographic localization for a rather subtle reason; although the forward problem is linear, the optimal algorithm for tackling the inverse problem cannot be linear *(10)*. The laws of electromagnetism provide no justification for expressing the full primary current density vector as a (weighted) sum of lead fields. Physics allows only the direction of the primary current density to be so represented, but this leads inevitably to a non-linear relationship between the measurements and the distribution of generators. This conclusion was reached first on the basis of simulation studies leading to the standard form of magnetic field tomography (MFT) *(9)*. In summary, the basic assumption of MFT is that a (weighted) linear expansion in terms of lead fields can represent only the *direction* of the current density. This is as much as can be deduced from the underlying physics for fixed detectors and conductivity profile. The strength of the current density must be determined more explicitly from the MEG signal itself. Specifically, the full current density must be obtained from a highly non-linear system of equations for each snapshot of data. It is precisely because linearity is lost, that a direct appeal to the data must be made on every timeslice of the data and a new non-linear system of equations must be solved each time. In this sense, MFT draws on all available information in the MEG signal. The advantage of the form of non-linearity introduced by MFT is the ability to recover activity that can be either spatially sharp or distributed, thus leading to tomographic description of the generators with practically no a priori assumptions. However, non-linearity comes with a heavy computational cost, but a rather affordable penalty today, thanks to modern computers.

To appreciate the subtle difference between various inverse problem approaches, it is necessary to consider in detail how the lead fields can be used to construct estimates of the unknown current density vector. The similar nature of different linear methods, e.g. of MN and wMN and LORETTA *(19)*, and how they differ from other non-linear methods like MFT and FOCUSS *(20)* can be best demonstrated by expressing the unknown current density as a series expansion of lead fields with different orders in the series modulated by the modulus of the current density raised to some power, as described in detail elsewhere *(10, 21)* and in outline in Appendix 1.

In the early 1990s, numerous comparisons between MFT and ECD models with computer generated data *(9)*, MEG signals evoked by stimuli *(22)* and epileptic activity *(23)*, all demonstrated the ability of MFT to identify distinct foci of activity as their strength changed while ECD models described them as wandering dipoles through the brain. In the last ten years, accurate MFT reconstructions have been demonstrated with many applications and with the emphasis recently placed on reconstructions of real-time data, i.e. for single timeslices of single trials *(4, 14)*.

4. Output of MEG

Neuroscience has, at its disposal, a plethora of exquisite techniques to study neural activity. In the vast majority of cases, the output of techniques is sufficiently constrained to limit the choices of the researcher to qualitatively one distinct category of output. The researcher has to perfect the technique to obtain the data of the highest possible quality, but she/he cannot change the qualitative nature of the neural events she/he is studying. For example, in a microstimulation experiment, one may worry about how large an area is excited but, one is certain that what is examined is the disruptive effect of injecting a current that perturbs the local neural interactions. Similarly, when one uses fMRI or PET, one can safely assume that one measures correlates of neural function mediated by blood supply and, hence, over delays of seconds. With MEG, the case is somewhat different because the method allows one to focus either at fine detail in space and time, and/or within a small area, or across the brain at timescales ranging from a fraction of a millisecond to hours. Quantitative changes in the choice of what spatiotemporal scales to consider imply sensitivity to qualitatively different neuronal events and organization. Using MFT, for example, to image in real time, one can follow changes in the brain at a fraction of a millisecond *(14)*, i.e. at a temporal resolution that is about one order of magnitude *higher* than the characteristic scale that it takes one brain area to influence another (this also demands working close to the noise level of the measurements). A safer but less ambitious approach, and still rather rare in the field, would study tomographically brain activity extracted from real time (un-averaged) MEG data filtered in the range say 3–200 Hz and after removing the interference from the mains and its harmonics and contributions from strong physiological sources like the eyes and heart. Such analysis would then map brain activity at about the timescale of transitions in the brain. The output of real-time tomographic analysis is an attempt to describe what is happening in the brain with minimal assumptions. Assuming that such a reconstruction is possible, statistics on the tomographic solutions can then be performed just as is done for fMRI data, but with far richer access to neuronal dynamics *(12, 13)*. The popular ECD method of analysis, using point-like models for the generators, will not work with real time MEG signals because rarely, if ever, the activity of the normal brain is dominated by a single focal source. Early MEG experiments with one or only a few sensors could only obtain signal topographies by repeating a task or delivering the same stimulus many times while recording the resulting MEG signal. Although this is no longer necessary with multi-channel systems, averaging remains a simple and powerful way of drastically reducing the complexity of the data. Averaging emphasizes activity that is precisely time-locked to an external stimulus which is more likely to be dominated by contributions that are fairly focal. As a result, ECD analysis often provides a very good fit for the average evoked response. Some of the apparently good fits obtained by ECD modeling of average data do reflect the true nature of the generators. However, the average makes up only a tiny part of the single trial MEG signal. Even the actual average MEG signal is often a collection of distinct responses that do not belong together *(5, 24)* and the apparent success of the ECD model could be a mirage with some of the real generators a fair distance away from the ECD loci *(25, 26)*. The likelihood always remains that a very large part of the activity related to the processing of the stimulus remains unexplained, lost in the process of averaging before any analysis is made, as recent studies have shown *(27, 28)*, but see reference *(29)* for a different view.

Despite the huge potential of MEG, its usefulness has been questioned. MFT has been for many years the only technique capable of real-time tomographic reconstructions, initially because its implementation demanded what, at the time, was super-computing resources *(30)*. What single trial MFT solutions described (tomographic estimates of real-time brain activity) was often confused with descriptive measures of power of the MEG signal and ECD fits to average data. Tapping into single trial data tomographically was giving a view of brain activity that was far more dynamic than the smooth evolution produced by fits to average data using ECDs that the MEG and EEG communities were familiar with for decades. Although this dynamic view was much more in step with invasive measures of activity, it was not adopted initially because the experience from ECD analysis and the non-uniqueness of the inverse problem had convinced many practitioners that tomographic analysis was impossible. In recent years, other techniques have emerged making better use of the information in the raw MEG signal than the averaging. Results obtained with these methods have vindicated many of the early MFT results. Beamformer techniques, in particular, have been increasingly used recently with good effect *(8)*. One of the most exciting results of MFT analysis was the identification of

V1 modulation well within 100 ms by activity in the amygdala some 40–50 ms earlier *(4, 31, 32)*. This early amygdala and V1 activations and their interaction were identified in the responses elicited by emotional faces, providing support for the existence of a low route to the amygdala as postulated by Le Doux *(33)*. In addition to the obvious theoretical importance of this observation for normal brain function, it was also found that early and late amygdala and V1 activations, and the interaction of these and other areas (fusiform gyrus and inferior frontal cortex) were different in normal subjects and schizophrenics *(31, 32, 34)*. These MFT findings were recently reproduced with remarkable consistency in the timing of amygdala and V1 activity as an event-related synchronization in response to fearful faces using beamformer techniques *(35)*.

We conclude this section with examples demonstrating how the nature of the output changes as we move from MEG signals to tomographic estimates of activity, and measures that describe the properties of groups of trials. We will show examples that progress systematically in complexity using either single trial data or the average of a small number of trials to mine the MEG data. The first example, **Fig. 8.1**, shows how strong features can be extracted directly from a signal signature, almost a direct reading of the MEG single trial signals. The data were collected from a median nerve stimulation that was strong enough to elicit a finger twitch. The first 11 responses are shown by the triggers collected simultaneously with the MEG signal (**Fig. 8.1a**). This unusually strong stimulus and the superficial location of the primary somatosensory cortex (S1) make the first cortical responses elicited by each stimulation almost visible in the raw signal. A virtual sensor can be constructed in this case (see below) which enhances the pattern of signal produced by S1 while reducing other signals. Despite this enhancement, the actual evoked response is barely discernible in the presence of other similar activations (**Fig. 8.1b**). The evoked response becomes easier to see in the zoomed version covering the third and fourth median nerve stimulations (**Fig. 8.1c**). In this figurine, two more properties of the MEG signal are evident. First, the evoked response is similar from trial-to-trial in its slow envelope but the fast transients riding on it are highly variable from trial to trial. Second, the virtual sensor (VS) output for the signal from the brain is always much higher than the system noise, a direct demonstration of the high dynamic range of MEG. The VS for the system noise is the barely distinguishable trace hugging the zero level horizontal axis; it is obtained by processing the MEG signal recorded when no subject is in the shielded room in exactly the same way as the signal from the subject. The VS is constructed by taking the difference of the means of the 7 strongest positive and 7 strongest negative MEG channels

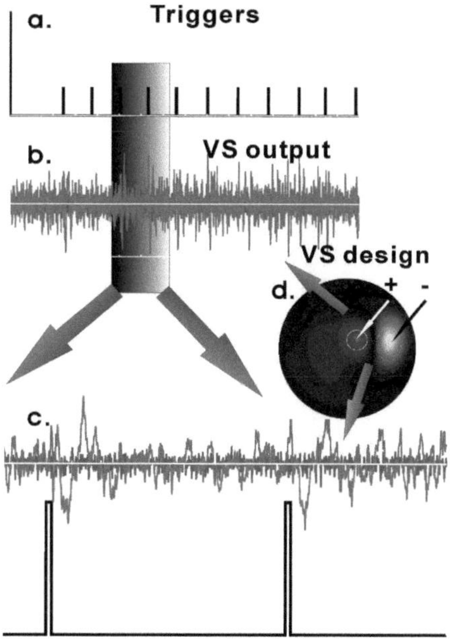

Fig. 8.1. Virtual sensor output for a strong signal. (**a**) The delivery of a median nerve stimulation is marked by a trigger and collected together with the MEG signal. (**b**) The MEG signal for the real measurement as marked by the triggers in (**a**) and for a recording of the noise level of the system (no subject in the MEG room). (**c**) A zoomed version of the VS output (*top*) and corresponding third and fourth triggers. (**d**) The VS is computed from the difference of means of the 7 most positive and 7 most negative MEG sensors. (*See* Color Plate)

(**Fig. 8.1d**). Weaker stimuli, e.g. typical visual stimuli, or median nerve stimulation below motor threshold, produce responses that are not easily discernible from the background activity.

The second example (**Fig. 8.2**) shows the reliable extraction of the expected (but relatively weak) activity through single trial tomographic (MFT) analysis using only a small number *(6)* of trials. **Figure 8.2a** shows single trial activations (purple and red curves) for left visual cortex and fusiform gyrus (FG) elicited by face and checkerboard stimuli presented in the lower right part of the visual field. The left and right columns show results for activation curves from the left dorsal V1 and FG respectively. The different rows show the responses when the presentation of the stimuli and the subject's attention is on one or other category of stimuli. It is critical to realize that these activation curves are extracted from tomographic solutions for each timeslice and single trial and they are much more specific than any signal or virtual sensor for that matter. Since the solutions are obtained independently for each timeslice and trial, simple but powerful estimates of the signal power (SP) and noise power (NP) and hence their ratio, signal to noise power (SNR) can be obtained from the activation curves of only six trials using 20 ms sliding windows across the

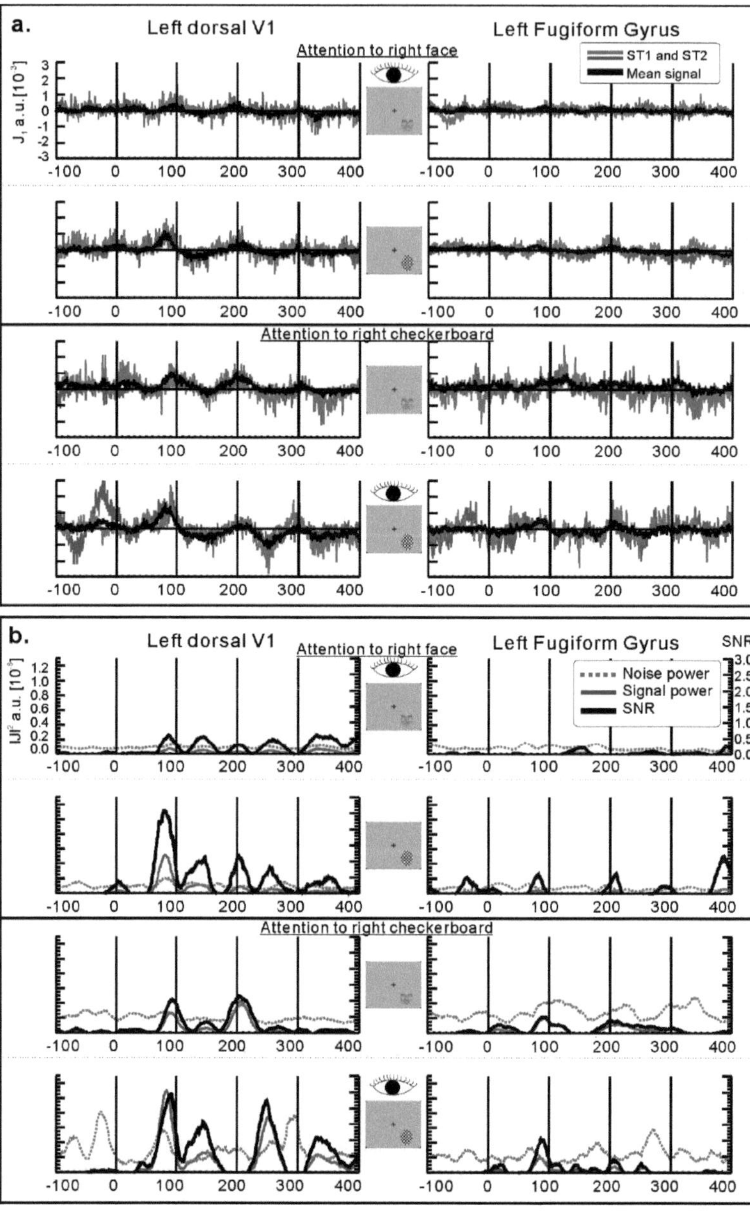

Fig. 8.2. (**a**) Two single trial activation curves extracted from tomographic MFT solutions with region of interest in the left dorsal V1 and the Left FG, for stimuli presented on the contralateral (*lower right*) part of the visual field. The stimuli in rows 1 and 3 were faces and in rows 2 and 4 checkerboard oval pattern. During a run, stimulus type and location varied randomly and the subject attended either to the faces on the right (*rows 1 and 2*) or to the checkerboards on the right (*rows 3 and 4*). For each case, two single trials and the average of the six trials used in the run are displayed. (**b**). The signal and noise power and their ratio (signal to noise ratio) are computed for the six trials and are displayed for each of the cases in (**a**). (*See* Color Plate)

latency axis. We compute SP, NP and SNR as described elsewhere *(36)* and give the relevant formulae in Appendix 2. **Figure 8.2b** shows the SP, NP and SNR for the six trials in the same order as part (A) of the same figure. It is evident that using only six trials in each set is enough for a faint response pattern to be seen in the average (heavy black line in **Fig. 8.2a**) and clear strong peaks in the SNR computed using the 20 ms sliding windows across the six single trials (**Fig. 8.2b**). The high SNR values are computed from SP and NP measures that show a much smoother behavior than the raw activation curves. It is evident in **Fig. 8.2b** that the peaks in SNR values are sometimes due to increase in SP; at others, due to decrease in NP and, at yet others, due to a combination of the two. The message from **Fig. 8.2** is that the use of measures that exploit the information in single trial tomographic analysis allows reliable extraction of information from a small number *(6)* of trials. Information of similar quality is available in the average signal only when a large number (typically hundreds) of trials are averaged. The average signal of large number of trials shows peaks that cluster around the SNR peaks of regional activations, often from more than one area. These results have significant implications for experimental designs. In general, it is highly desirable to have only a few trials of a given type within a session, for example to avoid habituation, or to include many different conditions in each run. Since this is incompatible with averaging hundreds of trials, the design of most experiments becomes a choice of lesser evils. The capability of single trial tomographic analysis of each timeslice therefore allows for more powerful experimental designs, as the next example demonstrates.

The last example summarizes the results of a recent study of illusory contour perception with stimuli presented at the center of the visual field and in each of the quadrants *(37)*. Using tomographic analysis of average data from relatively few trials in each run, we were able to study in the same experiment and for each part of the visual field, the effect of attention as reflected by different task demands. The MFT solutions were used to compute activation curves mainly within the primary visual cortex (V1/V2), the lateral occipital cortex (LOC) and FG for each subject, visual field presentation, condition and stimulus type. The location of the V1/V2, LOC and FG ROIs identified for each subject were transformed into the space of Talairach and Tournoux *(38)* and finally back-projected onto the MRI of one subject for display purposes. The statistical analysis of the activation curves from seven subjects demonstrated that a different mechanism operates for the processing of illusory figure processing when the stimuli are presented in the center and periphery of the visual field *(37)*. The results are displayed in **Figure 8.3** for stimuli presented in the center (**Fig. 8.3a**) and for stimuli presented in the four quadrants (**Fig. 8.3b**) of the visual field. In

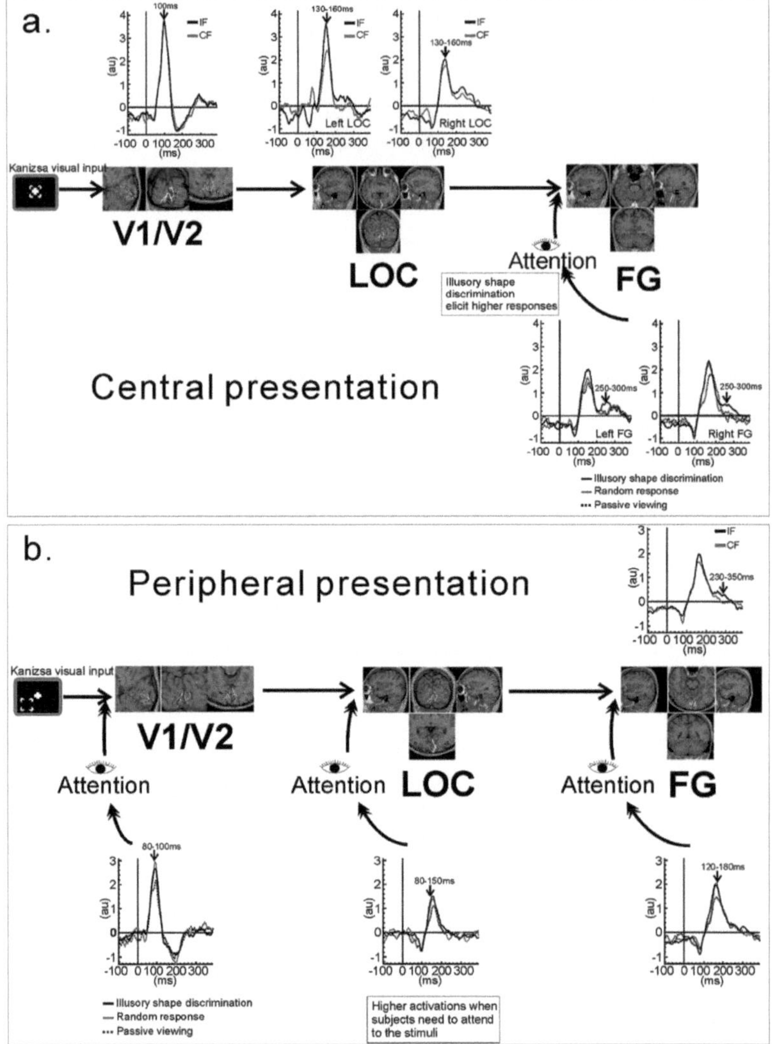

Fig. 8.3. Illusory figure processing with stimuli presented in the center (**a**) and periphery (**b**). In each case, the middle row shows cuts through each of the main areas identified by detailed MFT analysis. Dots on the MRI slices represent the center of regions of interest identified in each subject (after transformation to the space of Talairach and Tournoux *(38)*. The calcarine fissure and the V1/V2 borders (defined by an independent experiment for the subject whose MRI is used in the display) are also shown. The top row in each part shows the grand-average of the regional activations whenever a statistically significant difference between illusory contours present and absent was identified. The lower row shows similar grand-average activation curves when statistically significant differences were identified between tasks, requiring different attention levels.

parts (a) and (b) of **Fig. 8.3**, the middle row shows the location of the centers of V1/V2, LOC and FG ROIs (back-projected onto one MRI as described above) and superimposed on the nearest MRI slice used for common display. The top row (above the middle row showing the ROI locations) displays in separate curves the grand-averages for the illusory figures (IF) and control figures (CF); the control figures have the inducers rotated so no illusory contours are perceived. The activation curves are shown only for the cases when an IF effect is present, i.e. when ANOVA has shown that the response for IF and CF was significantly different. The activation curves below the middle row show grand-averages for the three different tasks: Classification task when subjects had to indicate the presence of an (illusory) shape (illusory shape discrimination); when the subjects had to respond randomly every time an image appeared, irrespective of the presence or absence of illusory figures (random response); and when subjects simply viewed the stimuli passively (passive viewing). The activation curves are shown only for the cases when a task effect is present, i.e. when ANOVA has shown that the response for different tasks was significantly different. These results showed for the first time a distinct difference between processing when stimuli were presented to the center and periphery of the visual field. In each and every task, stimuli presented at the center produce an IF effect first in V1/V2 (~100 ms) and then in the LOC and FG (130–160 ms); this IF effect thus appears to proceed independently of attention. For center presentation, attention effects show up late, after 200 ms in the FG activation. For presentation in the periphery, no IF effects are seen early, they are seen only after 200 ms mainly in the FG. Peripheral stimuli produce early attentional effects first in V1/V2 (80–100 ms), then in LOC (80–150 ms) and finally in the FG (120–180 ms).

In summary, the results presented show that very precise information can be extracted in both the temporal and spatial domain. Very simple techniques, almost like reading the raw signal are enough for strong and superficial generators (**Fig. 8.1**). Full tomographic techniques can be used to obtain real-time information across the brain in single trials or from a small set of trials as shown in **Fig. 8.2**. The ability to extract reliable information from a small number of trials is exploited in the last example. **Figure 8.3** shows when and where illusory figure processing takes place and how task demands modulate the allocation of attention in each area. These results were obtained from a detailed MFT analysis of average MEG data from seven subjects, using only a small number of trials in each average.

5. Notes

5.1. Disadvantages

The need for shielding and use of liquid helium makes MEG an expensive technology both in terms of the cost of hardware and the operating costs. Another disadvantage of MEG is the need for the subject to stay motionless while data are collected. MEG is insensitive to radial currents; so generators close to the center of the head (e.g. thalamus) and at the crest of gyri are close to silent sources. The patterns of activity identified with MEG are not very meaningful on their own because they lack anatomical context. The background anatomy must be provided by other methods, usually MRI, and the process of combining the background anatomy and the functional information requires considerable effort to ensure accurate coregistration between the two modalities for each subject and experiment.

5.2. Advantages

MEG is a completely non-invasive method with superior temporal resolution. With appropriate analysis methods, it can provide accurate localization of different brain regions activated simultaneously. The MEG signal depends weakly on the conductivity changes in the brain and simple models can provide accurate estimates of the magnetic field generated by a source in the brain. The insensitivity to radial sources adds to the discriminability of MEG, especially for sources in sulci.

6. Appendix 1: Lead Field Analysis

The measurements d_m (with m labeling sensors) depend on the primary current density vector $\mathbf{J}(\mathbf{r})$ via an integral over all regions with primary currents (the source space, Q),

$$\mathrm{d_m} = \int_{\mathrm{Q}} \varphi_m(\mathbf{r}) \cdot \mathbf{J}(\mathbf{r})\mathrm{d}\mathbf{r} \tag{8.1}$$

The vector function$\varphi_{\mathrm{m}}(\mathrm{r})$ is known as the lead field and it describes the sensitivity profile of the $\mathrm{m^{th}}$ sensor. The lead field is completely determined by the geometric properties of the coils making up each sensor and the conductivity details of the biological medium. Since only vectors with non-zero overlap with at least one lead field can contribute to the signal, we can express the unknown current density vector $\mathbf{J}(\mathbf{r})$ as a linear sum over the lead fields, modulated by some unknown function, ω, which, in its most general form, can be a function of location *and* of the unknown $\mathbf{J}(\mathbf{r})$ itself, i.e.,

$$J(r) = \sum_{\mathrm{m}} \mathrm{A_m} \varphi_{\mathrm{m}}(\mathrm{r})\omega(\mathrm{r},\mathrm{J}(\mathrm{r})) \tag{8.2}$$

The wMN solution follows from the tempting choice of setting $\omega(\mathrm{r},\mathrm{J}(\mathrm{r})) = \omega_{\mathrm{MN}}(\mathrm{r})$, i.e assuming that the function, ω, does not depend on any property of $\mathbf{J}(\mathbf{r})$,

$$J(r) = \sum_{\mathrm{m}} \mathrm{A_m} \varphi_{\mathrm{m}}(\mathrm{r})\omega_{\mathrm{MN}}(\mathrm{r}) \tag{8.3}$$

Substituting of (8.3) in (8.1) leads to a linear system of equations. The only choice to be made is for the spatial dependence of $\omega_{MN}(r)$. This choice is usually made to smooth over biases in the sensitivity of the sensors to different parts of the source space while "projecting out" source configurations that are already present in some baseline measurement. Linearity allows such tasks to be performed at the level of signal properties as described for example in the covariance matrices of active and control conditions. Although the (w)MN choice seems natural, it can not be justified a priori. It puts an enormous load on the weight factor, ω, demanding that just its spatial dependence can recover the strength and location of generators. Effectively, the simplicity and computational advantage of the linear models is bought at the expense of using only a small amount of the information in the data.

Generalized MFT admits a power expansion of $\omega(\mathbf{r},\mathbf{J}(\mathbf{r}))$ in the modulus of the current density,

$$\mathrm{J(r)} = |\mathrm{J(r)}|^{p+1} \sum_{\mathrm{m}} \mathrm{A_m} \varphi_{\mathrm{m}}(r) \omega_{\mathrm{p}}(\mathrm{r}) \tag{8.4}$$

Leading to a family of methods *(10, 21)*: MN, wMN and LORETTA *(19)* for $p = -1$, and a version of the FOCUSS *(20)* algorithm (corrected for gauge invariance) for $p = +1$. Standard MFT, as was initially selected via simulations corresponds to $p = 0$ *(9, 10)*.

The way standard MFT draws on the data makes the method computationally demanding but it also confers two contrasting and highly desirable properties that are necessary for accurate and unbiased localization. First, only for standard MFT (with $p = 0$) the right hand side of (8.4) depends on the modulus of the current density, just like the left hand side does, thus allowing sharp discontinuities in the current density vector with small values of the expansion coefficients. Second, standard MFT satisfies the principle of least sensitivity to both variations of the data and iterations of the non-linear norm constraints *(10)*. On the practical side, MFT allows only part of the function, ω, the a priori weight, $\omega_0(\mathbf{r})$, to be computed in advance from simulations with computer generated data *(9)* or in more general ways *(10)*. The full current density must be obtained from a highly non-linear system of equations for each snapshot of data. Specifically, the strength must be determined more explicitly from the MEG signal itself. A more detailed discussion about the theoretical basis and algorithmic implications of different choices of p can be found in *(10, 21)*, and a discussion of the pitfalls of choosing values other than $p = 0$ in *(32)*.

7. Appendix 2: Quantification of Signal and Noise in Timeseries

Consider the time-series, $x_i(t)$ representing either the signal of a sensor or the activation of a region of interest (ROI) for the ith single trial. For example, $x_i(t)$could be defined as$x_i(t) = \int_{ROI} J_i(r,t) \cdot \hat{u}_{ROI} d^3r$, with $J_i(r,t)$the instantaneous estimate for the current density vector at time t and trial (i) and $\hat{u}_{ROI}$ the direction of the current density vector at the maximum (modulus) of the MFT activation. A quantitative measure of the signal-to-noise ratio (SNR) can be derived from the ensemble of single trial timeseries using a conventional SNR estimator *(39)*. The spatial specificity of the MFT solutions allows such estimates to be made for relatively small segments of regional activations, and hence map their evolution across the latency axis. Around each timeslice t, we define aligned data segments $X_i(t,p) = \left[x_i(t-\frac{p-1}{2}),\ldots,x_i(t-1),x_i(t),x_i(t+1),\ldots,x_i(t+\frac{p-1}{2})\right]$ consisting of p samples. The noise power (NP) and signal power (SP) and SNR can then be defined using the following equations *(36, 40)*:

$$\bar{X} = \frac{\sum^N{}_{i=1} X_1(t,p)}{N}, NP = \frac{\sum^N{}_{i=1} \|\bar{X} - X_i(t,p)\|^2_{L_2}}{p(N-1)},$$

$$SP = \frac{1}{p}\|\bar{X}\|^2_{L_2} - \frac{1}{N}NP, SNR = \frac{SP}{NP}$$

For further discussion about these measures and their meaning see *(36, 41)*

References

1. Cohen, D. (1968) Magnetoencephalography: Evidence of magnetic fields produced by alpha-rhythm currents. *Science* 161, 784–786.
2. Fenwick, P. (1987) The inverse problem: A medical perspective. *Phys. Med. Biol.* 32, 5–9.
3. H.von Helmholtz (1853) ber einige Gesetze der Vertheilung elektrischer Ströme in körperlichen Leitern, mit Anwendung auf die thierisch-elektrischen Versuche. *Ann. Phys. Chem.* 89, 211, 353–233, 377.
4. Ioannides, A.A. (2001) Real Time Human Brain Function: Observations and Inferences from Single Trial Analysis of Magnetoencephalographic Signals. *Clinical EEG* 32(3), 98–111.
5. Liu, L., and Ioannides, A.A. (1996) A correlation study of averaged and single trial MEG signals: The average describes multiple histories each in a different set of single trials. *Brain Topogr.* 8, 385–396.
6. Hamalainen, M., Hari, R., Ilmoniemi, R.J., Knuutila, J., and Lounasmaa, O.V. (1993) Magnetoencephalography – Theory, Instrumentation, and Applications to Noninvasive Studies of the Working Human Brain. *Rev. Modern Phys.* 65, 413–497.
7. Darvas, F., Pantazis, D., Kucukaltun-Yildirim, E., and Leahy, R.M. (2004) Mapping human brain function with MEG and EEG: Methods and validation. *Neuroimage* 23 Suppl 1, S289–S299.
8. Hillebrand, A., Singh, K.D., Holliday, I.E., Furlong, P.L., and Barnes, G.R. (2005) A new approach to neuroimaging with magnetoencephalography. *Human Brain Map.* 25, 199–211.

9. Ioannides, A.A., Bolton, J.P.R., and Clarke, C.J.S. (1990) Continuous probabilistic solutions to the biomagnetic inverse problem. *Inv. Prob.* 6, 523–542.
10. Taylor, J.G., Ioannides, A.A., and Muller-Gartner, H.W. (1999) Mathematical analysis of lead field expansions. *IEEE Trans. Med. Imaging* 18, 151–163.
11. Ribary, U., Ioannides, A.A., Singh, K.D., Hasson, R., Bolton, J.P.R., Lado, F., Mogilner, A., and Llinas, R. (1991) Magnetic-Field Tomography of Coherent Thalamocortical 40-Hz Oscillations in Humans. *Proc. Natl. Acad. Sci. USA* 88, 11037–11041.
12. Ioannides, A.A. (2006) Magnetoencephalography as a research tool in neuroscience: State of the art. *Neuroscientist.* 12, 524–544.
13. Ioannides, A.A. (2007) Dynamic functional connectivity. *Curr. Opin. Neurobiol.* 17, 161–170.
14. Ioannides, A.A., Fenwick, P.B.C., and Liu, L.C. (2005) Widely distributed magnetoencephalography spikes related to the planning and execution of human Saccades. *J. Neurosci.* 25, 7950–7967.
15. Kominis, I.K., Kornack, T.W., Allred, J.C., and Romalis, M.V. (2003) A subfemtotesla multichannel atomic magnetometer. *Nature* 422, 596–599.
16. Bell, A.J., and Sejnowski, T.J. (1995) An information-maximization approach to blind separation and blind deconvolution. *Neural Comput.* 7, 1129–1159.
17. Sarvas, J. (1987) Basic mathematical and electromagnetic concepts of the biomagnetic inverse problem. *Phys. Med. Biol.* 32, 11–22.
18. Huang, M.X., Mosher, J.C., and Leahy, R.M. (1999) A sensor-weighted overlapping-sphere head model and exhaustive head model comparison for MEG. *Phys. Med. Biol.* 44, 423–440.
19. Pascualmarqui, R.D., Michel, C.M., and Lehmann, D. (1994) Low-Resolution Electromagnetic Tomography – A New Method for Localizing Electrical-Activity in the Brain. *Int J. Psychophysiol.* 18, 49–65.
20. Gorodnitsky, I., and Rao, B.D. (1992) Sparse signal reconstruction from limited data using FOCUSS: A re-weighted minimum norm algorithm. *IEEE Trans. Signal Process.* 45, 600–616.
21. Ioannides, A.A., and Taylor, J.G. (1999) Minimum norm, Magnetic Field Tomography and FOCUSS. Tohoku University Press, Sendai, pp. 228–231.
22. Ioannides, A.A., Singh, K.D., Hasson, R., Bauman, S.B., Rogers, R.L., Guinto, F.C., and Papanicolaou, A.C. (2007) Comparison of current dipole and magnetic field tomography analyses of cortical response to auditory stimuli. *Brain Topogr.* 6, 27–34.
23. Ioannides, A.A., Hellstrand, E., and Abrahamfuchs, K. (1993) Point and Distributed Current-Density Analysis of Interictal Epileptic Activity Recorded by Magnetoencephalography. *Physiol. Measur.* 14, 121–130.
24. Ioannides, A.A., Kostopoulos, G.K., Laskaris, N.A., Liu, L.H., Shibata, T., Schellens, M., PoghosyaN, V., and Khurshudyan, A. (2002) Timing and connectivity in the human somatosensory cortex from single trial mass electrical activity. *Human Brain Mapp.* 15, 231–246.
25. Moradi, F., Liu, L.C., Cheng, K., Waggoner, R.A., Tanaka, K., and Ioannides, A.A. (2003) Consistent and precise localization of brain activity in human primary visual cortex by MEG and fMRI. *Neuroimage.* 18, 595–609.
26. Tzelepi, A., Ioannides, A.A., and PoghosyaN, V. (2001) Early (N70m) neuromagnetic signal topography and striate and extrastriate generators following pattern onset quadrant stimulation. *Neuroimage* 13, 702–718.
27. Laskaris, N.A., Liu, L.C., and Ioannides, A.A. (2003) Single-trial variability in early visual neuromagnetic responses: an explorative study based on the regional activation contributing to the N70m peak. *Neuroimage* 20, 765–783.
28. Makeig, S., Westerfield, M., Jung, T.P., Enghoff, S., Townsend, J., Courchesne, E., and Sejnowski, T.J. (2002) Dynamic brain sources of visual evoked responses. *Science* 295, 690–694.
29. Furey, M.L., Tanskanen, T., Beauchamp, M.S., Avikainen, S., Uutela, K., Hari, R., and Haxby, J.V. (2006) Dissociation of face-selective cortical responses by attention. *Proc. Natl. Acad. Sci. U S A* 103, 1065–1070.
30. Liu, M.J., Hasson, R., and Ioannides, A.A. (1993) A transputer-based system for Magnetic Field Tomography. *Transputer Applications and Systems '93.* IOS Press, Amsterdam, pp. 1290–1297.
31. Ioannides, A.A., PoghosyaN, V., Liu, L.C., and Streit, M. (2002) Early amygdala activations in normal and schizophrenic subjects. *Soc. Neurosci. Abstr.*, p. 521.5.
32. Ioannides, A.A. (2007) MEG single-event Analysis: Networks for Normal Brain Function and Their Changes in Schizophrenia. *Complex Medical Engineering.* Springer, pp. 361–374.
33. LeDoux, J.E. (1996) *The Emotional Brain.* Simon and Schuster, New York.
34. Ioannides, A.A., PoghosyaN, V., Dammers, R., and Streit, M. (2004) Real-time neural activity and connectivity in healthy individuals and schizophrenia patients. *Neuroimage* 23, 473–482.

35. Luo, Q., Holroyd, T., Jones, M., Hendler, T., and Blair, J. (2007) Neural dynamics for facial threat processing as revealed by gamma band synchronization using MEG. *Neuroimage* 34, 839–847.
36. Laskaris, N.A., and Ioannides, A.E. (2001) Exploratory data analysis of evoked response single trials based on minimal spanning tree. *Clin. Neurophysiol.* 112, 698–712.
37. Abu Bakar, A., Liu, L.C., Conci, M., Elliot, M.A., and Ioannides, A.A. (2008, In press) Visual Field and Task Influence Illusory Figure Responses. *Human Brain Mapping.*
38. Talairach, J., and Tournoux, P. (1988) *Co-planar stereotaxic atlas of the human brain.* G. Thieme, Stuttgart ; New York.
39. Raz, J., Turetsky, B., and Fein, G. (1988) Confidence-Intervals for the Signal-To-Noise Ratio When A Signal Embedded in Noise Is Observed Over Repeated Trials. *Ieee Transactions on Biomedical Engineering* 35, 646–649.
40. Laskaris, N., Fotopoulos, S., Papathanasopoulos, P., and Bezerianos, A. (1997) Robust moving averages, with Hopfield neural network implementation, for monitoring evoked potential signals. *Evoked Potentials-Electroencephalography and Clinical Neurophysiology* 104, 151–156.
41. Liu, L.C., Fenwick, P.B.C., Laskaris, N.A., Schellens, M., PoghosyaN, V., Shibata, T., and Ioannides, A.A. (2003) The human primary somatosensory cortex response contains components related to stimulus frequency and perception in a frequency discrimination task. *Neuroscience* 121, 141–154.

Chapter 9

Functional Neuroimaging of Spike-Wave Seizures

Joshua E. Motelow and Hal Blumenfeld

Abstract

Generalized spike-wave seizures are typically brief events associated with dynamic changes in brain physiology, metabolism, and behavior. Functional magnetic resonance imaging (fMRI) provides a relatively high spatiotemporal resolution method for imaging cortical-subcortical network activity during spike-wave seizures. Patients with spike-wave seizures often have episodes of staring and unresponsiveness which interfere with normal behavior. Results from human fMRI studies suggest that spike-wave seizures disrupt specific networks in the thalamus and frontoparietal association cortex which are critical for normal attentive consciousness. However, the neuronal activity underlying imaging changes seen during fMRI is not well understood, particularly in abnormal conditions such as seizures. Animal models have begun to provide important fundamental insights into the neuronal basis for fMRI changes during spike-wave activity. Work from these models including both fMRI and direct neuronal recordings suggest that, in humans, specific cortical-subcortical networks are involved in spike-wave, while other regions are spared. Regions showing fMRI increases demonstrate correlated increases in neuronal activity in animal models. The mechanisms of fMRI decreases in spike-wave will require further investigation. A better understanding of the specific brain regions involved in generating spike-wave seizures may help guide efforts to develop targeted therapies aimed at preventing or reversing abnormal excitability in these brain regions, ultimately leading to a cure for this disorder.

Key words: Epilepsy, absence seizures, petit mal, fMRI, BOLD, EEG, rat, WAG/Rij, barrel cortex, thalamus, laser Doppler flowmetry, cerebral blood flow, CBF.

1. Introduction

The study of epilepsy can shed light on both normal and abnormal brain physiology. Generalized epileptic events such as absence seizures, with their characteristic spike-and-wave discharge (SWD), create transient and heterogeneous changes in neural activity throughout the brain. Dynamic functional imaging, with adequate temporal and spatial resolution, is an optimal

Fahmeed Hyder (ed.), *Dynamic Brain Imaging: Multi-Modal Methods and In Vivo Applications, vol. 489*

DOI 10.1007/978-1-59745-543-5_9 Springerprotocols.com

technique with which to comprehend these complex changes as they occur throughout the brain. The discovery of the blood oxygenation level dependent (BOLD) signal and the advent of functional magnetic resonance imaging (fMRI) have opened a new window into the neurological correlates of normal and abnormal brain activity. Absence seizures, because they are not usually associated with movement in either humans or animals, provide a rare opportunity to study ictal brain metabolism via a dynamic imaging modality such as fMRI, which is sensitive to patient movement. Animal models allow scientists to fully utilize the power of fMRI because researchers can simultaneously *(1)* control seizure onset and type, *(2)* map epileptic activity, and *(3)* invasively explore the neurological underpinnings and molecular mechanisms of fMRI and seizure genesis. The prevalence of fMRI in studying epilepsy reflects the excitement that this imaging modality has generated.

We will begin with a discussion of generalized epilepsy and SWD. Next, we will review possible relationships between neuroenergetics and fMRI signals in human and animal BOLD experiments. We will then move on to human and animal studies of SWD, review possible mechanisms for fMRI decreases in epilepsy, and finally, discuss future directions and practical applications of these findings.

2. Generalized Spike-Wave Seizures

Epileptic seizures are usually classified under two categories: Generalized seizures, which involve widespread regions in both hemispheres of the brain; and partial seizures, which involve focal brain regions *(1)*. In fact, recent evidence suggests that so-called "generalized" seizures affect focal brain regions more intensely while sparing others *(2–4)*, and "partial" seizures often involve widespread cortical-subcortical networks beyond the seizure focus *(5, 6)*. Nevertheless, the distinction between seizures that are predominantly unilateral ("focal") versus bilateral ("generalized") in origin is helpful in establishing broad syndromes for clinical diagnosis. A variety of patterns of electrical activity can be seen in both partial and generalized seizures, including rhythmic spike-wave discharges (SWD). Typical large-amplitude bilateral SWD are seen most commonly in absence epilepsy (**Fig. 9.1**), but are found in other forms of epilepsy as well *(2)*. Absence (*petit mal*) is considered a form of generalized epilepsy, and usually begins in childhood. Absence seizures consist of brief episodes of staring and unresponsiveness often accompanied by mild eyelid fluttering or myoclonic jerks. The duration of these seizures is usually less than 10 s, and post-ictal deficits are not common.

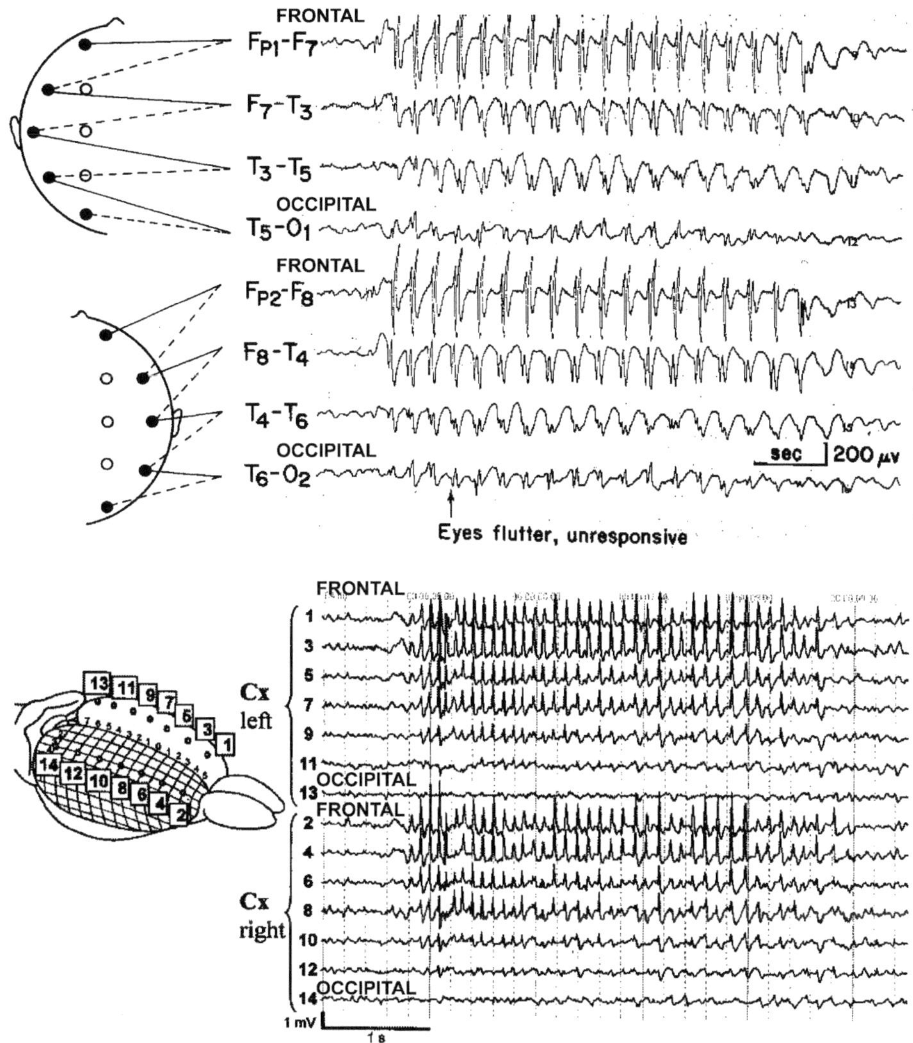

Fig. 9.1. Typical spike-wave EEG discharges during absence seizures show some regions intensely involved, and others relatively spared by seizures. (**A**) EEG recordings of a typical absence seizure from a 7-year old girl, reveals bilateral synchronous 3–4 Hz spike-wave discharges, with an anterior to posterior amplitude gradient. (Inset of electrode positions modified with permission from Fisch, B.J. (1991) Spehlmann's EEG Primer. Elsevier, Amsterdam. EEG recording modified with permission from Daly, D.D. and Pedley, T.A. (1990) Current Practice of Clinical Electroencephalography. Raven Press, New York.). (**B**) Electrocorticography from the surface of the WAG/Rij rat cortex during spike-and-wave seizures exhibits intense involvement of the anterior cortex and relative sparing of the occipital lobes. (Reprinted with permission from Meeren et al., 2002. Copyright 2002, Society for Neuroscience.)

Electroencephalogram (EEG) recordings during typical absence seizures reveal large-amplitude, bilateral 3–4 Hz spike-wave discharges (**Fig. 9.1**). SWDs are usually bilaterally symmetric, however, left or right amplitude preponderance occurs occasionally *(7)*. Although absence seizures are considered generalized epileptic events, SWD amplitude in humans is seen maximally in the frontal cortices and greatest towards the midline (**Fig. 9.1A**) *(8–11)*, and this distribution is also found in animal models of absence seizures *(12–15)* (**Fig. 9.1B**). Human and animal studies to date have implicated the cortex and thalamus in the generation of abnormal network oscillations involved in SWDs *(16–23)*. However, it has not been definitively determined whether it is an overall increase or decrease in neuronal activity in cortical and subcortical circuits that leads to spike-and-wave activity *(14, 15, 24–26)*. While the electrographic signature and distribution of SWDs have been characterized in previous studies, this review intends to discuss the question of metabolic activity and neuroenergetics during SWD. It is hoped that an extensive understanding of the neuronal activity changes during SWD might lead to improved treatment for absence seizures.

3. Neuroimaging in Generalized Epilepsy

Although EEG provides high temporal resolution, it is limited in its spatial sampling and cannot fully describe seizure activity throughout the brain. Neuroimaging techniques provide more comprehensive spatial sampling and can look deep into subcortical structures in which EEG recordings are not generally feasible in humans. The first goal of neuroimaging studies of SWD is the localization of seizure activity in specific brain regions as well as the identification of distributed networks during SWD. Because fMRI signals are only an indirect measure of neuronal activity, comparable animal models of SWD are necessary to *(1)* relate fMRI signal changes seen during SWD to underlying physiology and *(2)* aid the interpretation of fMRI changes seen in human SWD. Understanding the brain regions responsible for SWD generation yields numerous therapeutic implications including improved applications of deep brain stimulation, more effective medications, and the possibility of targeted gene therapy. A second goal of SWD fMRI research is to understand the brain's physiology during abnormal activity of spike-and-wave discharges. For example, the study of epileptogenesis might be furthered by identifying those areas which may be more susceptible to chronic activity-dependent changes *(27)*. Finally, a third goal is to relate fMRI changes throughout the brain to behavioral changes during seizures. Pairing behavioral analysis, including studies of impaired consciousness, with identification of the brain areas that demonstrate changes in fMRI signal will lead to a greater understanding of functional brain impairment during and between seizures *(3, 17, 28)*.

3.1. Non-fMRI Studies

Neuroimaging data of SWDs in the pre-fMRI era showed great variability. Global increases in cerebral metabolism or blood flow in human patients have been reported *(24, 29–32)* using single photon emission computed tomography (SPECT) studies or positron emission tomography (PET). At the same time, other PET, ^{133}Xe clearance, transcranial Doppler, and near-infrared spectroscopy (NIRS) studies of blood flow and metabolism have shown an absence of change, focal change, and generalized increases, decreases, and biphasic changes *(29, 33–43)*. One limitation of these studies is that the time resolution of Tc99m single photon emission computed tomography is approximately 30 s, the time resolution of fluoro-2-deoxy-D-glucose PET is approximately 30 min and the time resolution of ^{133}Xe clearance is a few minutes. These modalities may have trouble capturing the transient metabolic changes of absence seizures, which typically last less than 10 seconds. SPECT, PET, and ^{133}Xe are likely to integrate changes before, during, and after SWD episodes. Transcranial Doppler and NIRS have higher time resolution, but these methods lack sufficient spatial resolution. Animal studies have shown similarly confusing results (e.g. increased metabolism and decreased CBF during SWD in the same model) *(37, 44)*.

3.2. EEG-fMRI

In order to capture adequately the dynamic neuroenergetic changes during brief absence seizures (less than 10 s), an imaging modality must have two characteristics. First, simultaneous EEG must be taken so that it is possible to distinguish interictal (non-seizure) and ictal (seizure) images. Second, the modality must have sufficient temporal resolution to capture individual seizure or SWD events while also having sufficient spatial resolution to distinguish brain regions. Using the above criteria, fMRI is the most effective imaging modality currently in use to capture the complex dynamic changes in energy metabolisms and blood flow of SWDs. Experiments utilizing fMRI and simultaneous EEG-fMRI *(45–47)* have begun to explore the relationship between electrophysiology and neuroimaging changes associated with SWDs in human and animal models .

4. Relation Between fMRI and Neuronal Activity During Spike-Wave

Our interest in reviewing fMRI studies of SWD reflects *(1)* a belief that fMRI is the best modality to understand the neural activity underlying SWD and *(2)* a desire to understand the metabolic implications of the BOLD signal. "Neuronal activity" (ν), which includes presynaptic and postsynaptic membrane voltage changes associated with neural signaling, consumes energy (ATP). A large portion of this energy goes towards restoring equilibrium Na+ and K+ concentration gradients across the nerve membrane following the generation of action potentials and postsynaptic potentials *(48)*. Net oxygen delivery to the brain reflects the balance between cerebral blood flow (CBF) and metabolic rate of oxygen consumption (CMRO2) *(49, 50)*. Increased neural activity is linked to increased metabolism, which causes a CBF increase. Increased CBF causes an increase in oxygen delivery which normally exceeds metabolic demands, leading to an overall increase in blood and tissue oxygenation (pO_2). Since, as we will discuss shortly, deoxygenated hemoglobin reduces the BOLD fMRI signal intensity (S), increased oxygenation leads to an increase in S. Thus, increased neural activity and metabolism normally cause a paradoxical decrease in deoxygenated hemoglobin and an increase in the BOLD signal.

BOLD signal changes are related to baseline signal ($\Delta S/S$) and physiology by *(51–53)*:

$$\Delta S/S = A'[(\Delta CBF/CBF - \Delta CMRO_2/CMRO_2)/(1 + \Delta CBF/CBF) - \Delta CBV/CBV]$$

where A′ is a measurable physiologic and magnetic field dependent constant, and CBV is cerebral blood volume. Measurement of S depends on the transverse relaxation time of tissue water (T_2 for spin-echo, or $T_2{}^*$ for gradient echo image contrast), which, in turn, is related to red blood cell oxygenated hemoglobin *(49)*. This is because deoxyhemoglobin is an endogenous paramagnetic contrast agent and causes a decrease in T_2 (or $T_2{}^*$), leading to a decrease in the BOLD signal.

As a result of these complex interplays between CMRO2, CBF, and CBV, neural energetics indirectly couple neuronal activity with fMRI signal intensity (S). $CMRO_2$ is currently the most direct neuroimaging measure of neuronal activity because oxidative metabolism supplies the vast majority of neuronal energy. Even in periods of transient dramatic increases in neuronal activity, the majority of pyruvate and lactate molecules in the brain are eventually metabolized aerobically *(54)*. $CMRO_2$ can be measured by neuroimaging methods, but the BOLD fMRI signal is a more convenient, although indirect, method of mapping neuronal activity, and BOLD fMRI benefits from a higher spatiotemporal resolution than spectroscopic $CMRO_2$ measurements.

In summary, BOLD fMRI increases and decreases are usually interpreted, respectively, as increases and decreases in neuronal activity. However, as we will discuss next, it cannot always be assumed that this simple relationship holds, particularly under abnormal conditions such as epilepsy.

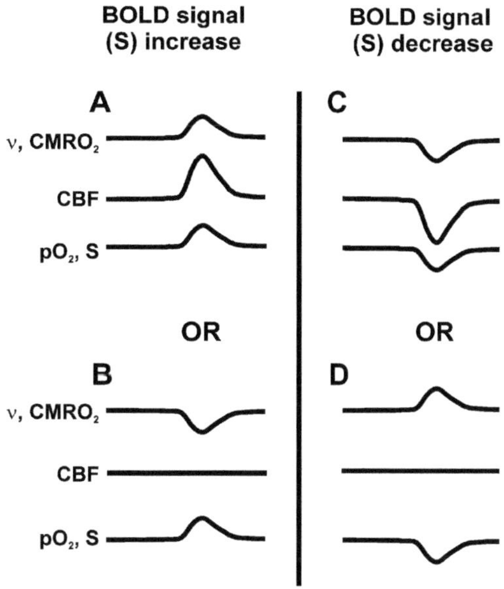

Fig. 9.2. Schematic representation of increases and decreases in the BOLD signal (S) as a result of possible changes in neuronal activity (ν), $CMRO_2$, cerebral blood flow (CBF), and tissue oxygenation (PO_2). (**A, B**) Examples of changes leading to increases in the BOLD signal. (**C, D**). Examples of changes leading to decreases in the BOLD signal. Note: Relative changes among ν, $CMRO_2$, CBF, and PO_2, S are not drawn to scale but are exaggerated for illustrative purposes. (Reprinted with Permission from Blumenfeld, 2007, *Epilepsia (93)*).

4.1. Increases

FMRI signal changes, resulting from changes in neuronal activity, must be interpreted cautiously due to the complexity of the underlying mechanisms contributing to the BOLD signal. Positive changes in the BOLD signal can result from two different mechanisms (**Figs. 9.2A, B**): (i) Increased neuronal activity (ν) and $CMRO_2$, accompanied by sufficiently increased CBF and oxygen delivery to exceed metabolic demands, which leads to increased pO_2 and S (**Fig. 9.2A**), or (ii) Decreased ν and $CMRO_2$, with no change (or minimal decreases) in CBF, so that oxygen delivery again exceeds metabolic demands, leading to an increase in pO_2 and in S (**Fig. 9.2B**). Although increased BOLD signal is usually thought to reflect *increases* in neuronal activity *(14, 55, 56)*, BOLD signal increase may also accompany *decreased* neuronal activity (if CBF does not change, or does not decrease sufficiently to reduce net oxygenation). Similarly, BOLD increases without changes in neuronal activity are possible if CBF is abnormally increased *(57)*.

4.2. Decreases

Like BOLD increases, BOLD decreases can reflect either decreases or increases (or no change) in neuronal activity (**Figs. 9.2C, D**), depending on the direction and magnitude of CBF changes. Usually, fMRI decreases are interpreted as *decreased* neuronal activity accompanied by an excessive decrease in CBF. This causes reduced net oxygen levels and therefore a decrease in pO_2 and in S (**Fig. 9.2C**) *(58, 59)*. However, as has recently been shown in some brain regions during intense seizure activity, it is also possible to observe BOLD fMRI decreases in the presence of increased ν and CMRO2 if CBF does not change, or does not increase sufficiently to meet metabolic demands, leading to a decrease in pO_2 and in S (**Fig. 9.2D**) *(60, 61)*. Similarly, it is possible for fMRI signals to decrease with little or *no change* in neuronal activity if CBF is abnormally reduced (e.g., by vascular steal or vasospasm).

Thus, fMRI signal change alone is not adequate to determine the underlying changes in neuronal activity in the imaged system. While the relationship between the BOLD signal and the underlying hemodynamics is being studied and characterized, it is generally true, under *normal* physiological conditions, that fMRI increases reflect increased neuronal activity and that fMRI decreases reflect decreased neuronal activity *(56, 59)*. However, additional study of these relationships is necessary to fully understand these relationships under abnormal circumstances such as those seen in epilepsy *(14, 15, 61, 62)*. Possible relationships between fMRI decreases and neuronal activity during SWD will be more fully discussed towards the end.

5. Human fMRI Studies of Spike-Wave

Attempts to identify BOLD changes in humans associated with SWD have yielded fascinating, though confusing, and sometimes contradictory results. Human studies to date have investigated patients with a broad spectrum of epileptic disorders, ages, and treatments. These studies have attempted to localize those brain regions associated with SWD while teasing apart the significance and meaning of the BOLD signal increases and decreases observed in their analyses. There remain many important unanswered questions regarding both human SWD as well as the physiological and neurological underpinnings of the BOLD signal. More extensive experimentation, in both human and animal models, is required. We will now discuss BOLD fMRI changes in specific anatomical regions during human SWD.

5.1. Cortical BOLD Changes

The importance of the cortex in SWD generation and maintenance has been widely shown in animal models *(2,18,23)*. Human fMRI experiments have reported cortical BOLD signal change to be predominately negative during SWD although positive changes have also been documented. Negative BOLD changes in the cortex during SWD have drawn the most intense scrutiny and speculation regarding their role in the clinical manifestations of absence seizures. Experiments have shown deactivations ranging from nearly the entire cortex *(25)* to smaller subsets of the cortex *(28, 63–67)* though the meaning of cortical BOLD decreases during SWD remains unknown. Experimenters have seen bilateral decreased BOLD signal change in the anterior and posterior interhemispheric regions, lateral frontal and parietal association cortices, and the posterior cingulate/retrosplenial/precuneus areas *(25,28,63–65,67,68)*. Several investigators *(63,66,69)* have noted the similarity between the BOLD deactivation pattern generated by SWD and those areas that characterize normal brain activity *(70)*. While much work has been done to pinpoint cortical decreases associated with SWD, variable results have been found and further study is necessary.

Reports of positive cortical changes in BOLD signal during SWD are less consistent and less pronounced than negative changes. Activations have been found in the bilateral precentral sulci *(64)*, mesial frontal cortex, bilateral insula *(69)*, bilateral motor cortex *(65)*, occipital cortex, and inferior parietal cortex *(66)*. Most of these studies have reported variable fMRI increases in the lateral frontal and parietal cortical regions *(28, 68)*. The variation and lack of consistency seen in labeling positive cortical BOLD changes suggests additional study is necessary.

5.2. Thalamic BOLD Changes

Like the cortex, the thalamus has long been implicated in SWD generation and maintenance *(2, 18, 23, 71)*. Human BOLD studies, especially those implementing continuous EEG-fMRI, have consistently found bilateral thalamic increases *(25, 26, 28, 63, 65, 66, 68)* though some studies have also found thalamic decreases *(26, 63)*. Some investigators have attempted to differentiate the BOLD signal changes in different thalamic nuclei, but such investigations are near the limit of the spatial resolution of fMRI. Thalamic activity is, at times, seen only at higher field strengths (3T) *(66)* and continued advances in fMRI technology may reveal more information regarding the thalamus' metabolic activity during SWD.

5.3. Basal Ganglia, Cerebellum, and Brainstem

BOLD changes outside of the cortex and thalamus have been reported, though these changes have received much less attention, and some studies have chosen to focus solely on the corticothalamic network. Increases in the cerebellum have been found *(28, 63, 69)* as have increases (probably artifactual) in the lateral ventricles *(69)* and white matter tracts associated with the thalamus *(65)*. Cerebellar activity during SWD has been previously noted *(72,73)*, but the cerebellum's role has not yet been defined. Decreases have also been found in the basal ganglia *(25, 28, 68)*. The importance of the basal ganglia and the brainstem in absence epilepsy has been demonstrated in animal models *(74–77)* and human patients *(78, 79)*. Currently, little emphasis has been placed on BOLD signal changes outside the thalamocortical network, but this is an area worthy of further study.

5.4. Discussion of Human fMRI Studies of Spike-Wave

SWD is a rhythm that arises out of normal brain circuitry and physiology but is a distortion of normal mechanisms *(2)*. It affects specific circuits in the brain while sparing others. Structures that are affected by spike-and-wave discharge are the same networks that are important for normal oscillations, and the involvement of these structures interrupts normal brain activity. Much attention has been paid to the possible connection between the behavioral deficits seen in absence seizures and the cortical deactivations seen in human BOLD studies *(69)*. This should not lead investigators to ignore prominent BOLD changes in subcortical structures or BOLD increases in the cortex.

fMRI increases in the thalamus have been observed repeatedly, but the mechanism that drives the BOLD changes is as yet unknown. Are the BOLD increases in the thalamus due to increased activity of excitatory or inhibitory neurons? Are neurons firing more or *less* in the thalamus during SWD? Thalamic decreases remain unexplained (as do the mechanism that underlies all BOLD decreases in SWD). It has also been suggested that SWD may be generated by more than one mechanism, which might explain the varied cortical BOLD signal changes *(25, 67)*. Finally, the relatively consistent BOLD changes seen in the thalamus, frontoparietal cortex, and posterior cingulate/retrosplenial/precuneus (**Fig. 9.3**) areas may represent an important functional network in SWD generation.

Cortical fMRI increases have not been given the same attention as cortical decreases in SWD. The possible role of these regions in SWD generation and behavioral deficits warrants further investigation. One author suggested that the variability in cortical signal might have less to do with SWD generation and more to do with the normal baseline activity of each individual *(66)*. If SWD leads to an interruption of normal activity, then baseline activity would dictate which cortical areas showed increases and decreases *(69)*.

5.5. Limitations of Human Studies to Date

Attempts to localize SWD using BOLD fMRI have produced inconsistent results. Studies have been limited by heterogeneous patient populations (e.g. age, medication, diagnosis, seizure activity) (reviewed in *(66)*). As has already been noted, stronger field strengths have allowed researchers to find signal changes that were not apparent at lower field strengths, and continued advances in fMRI research techniques may reveal a more consistent pattern of significant fMRI signal changes. Furthermore, while several researchers have connected their fMRI

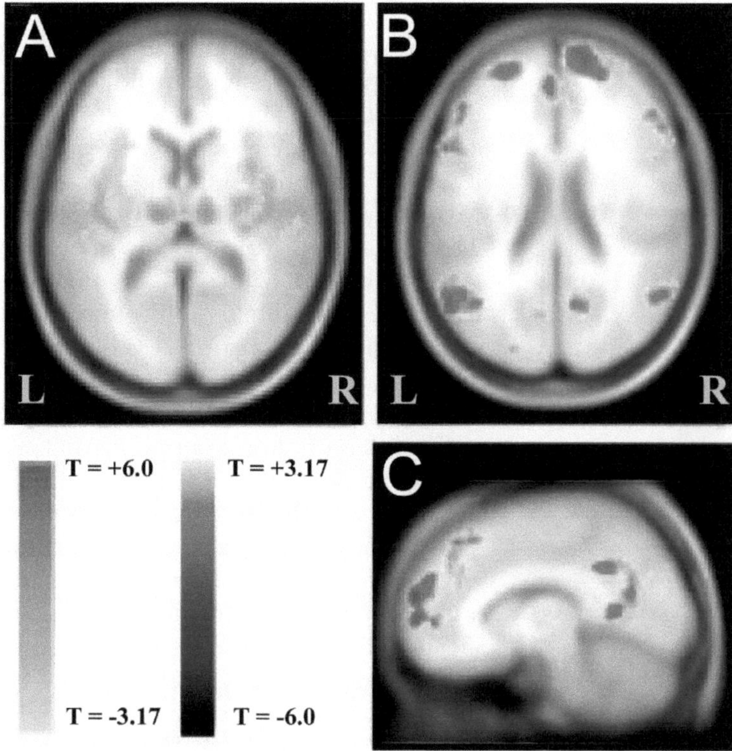

Fig. 9.3. fMRI changes in group analysis of generalized spike-wave seizures (15 patients with idiopathic generalized epilepsy). (**A**) Significant increases are shown in the bilateral thalami and in several cortical areas. (**B, C**). Decreases are shown in bilateral interhemispheric regions, lateral frontal cortex, and parietal association cortex. Modified with permission from Gotman et al, 2005 *PNAS (69)*. Color version of this figure can be viewed in the original publication *(69)* at http://www.pnas.org/.

data with the behavioral deficits often associated with absence seizures, simultaneous behavioral testing may allow researchers to understand the relationship between fMRI changes and behavioral changes *(28, 68)*. Thus, it will be critical in further studies to directly link fMRI signal changes with behavioral deficits. Finally, techniques to date have not necessarily differentiated BOLD activity based on seizure duration. CBF and brain metabolism may differ in brief SWD as opposed to in prolonged seizures or absence status epilepticus. Furthermore, fMRI signal changes and the spatial heterogeneity of SWD amplitude in different brain regions based on EEG (or MEG) have not yet been correlated.

6. Animal fMRI Studies of Spike-Wave

Animal studies are vital companions to human studies and provide advantages for studying epilepsy and conducting fMRI experiments. Animal models of epilepsy often allow investigators to control seizure type, frequency and onset, which enable the collection of the most consistent and reproducible data possible. Animals may be restrained and/or anesthetized to reduce movement artifact, and animal models make possible the study of fMRI and its neural correlates by allowing direct study of neural activity using combined imaging, electrophysiological measurements with microelectrodes and other techniques *(14, 15, 55, 56, 80)*. Tissue collected from specific brain regions identified by animal fMRI studies may uncover the molecular mechanisms giving rise to seizure susceptibility in those regions.

Of course, animal models of epilepsy are limited in that they are models. Results from animal models must be analyzed with the understanding that human and animal physiologies are crucially different. For example, SWD have a faster frequency in rodent models (7–8 Hz) than in humans (3–4 Hz), and SWD persist in adulthood in rodents, while they usually disappear in adolescence in humans. Other difficulties and technical challenges arise from the high magnetic field strength used to image small animals, including difficulty in obtaining simultaneous EEG during fMRI recordings *(15, 81)*, the sensitivity of imaging signals to movement artifact, and magnetic susceptibility artifact often found at air-tissue interfaces.

Anesthesia, if used, must be chosen with care. Anesthetic agents may suppress seizure activity or alter cerebral hemodynamics, introducing a confounding factor in fMRI analysis. While under anesthesia or otherwise, systemic physiology should be monitored in ventilated animals during fMRI studies. Changes in blood pressure and pCO_2 may confound imaging results. Ideally, SWD seizures should be imaged in animals without anesthesia *(81, 82)*. However, this is a technical challenge due to the extensive animal training required to habituate them to the recording procedures *(83, 84)*.

Human experiments have left many unanswered questions about the meaning of SWD-induced fMRI signal increases and decreases. It is hoped that accurate animal models may answer these questions by elucidating the relationships between the fMRI signal changes, underlying neuronal activity, and molecular mechanisms *(3)*. The main animal models of SWD studied with fMRI to date include spontaneous seizures in rat SWD models, and chemically induced seizures using gamma-butyrolactone (GBL).

6.1. WAG/Rij

Prominent animal models of SWD include spontaneous seizures seen in Wistar Albino Glaxo rats of Rijswijk (WAG/Rij). WAG/Rij rats have spontaneous SWD and are an established model of human absence epilepsy *(85)*. As in human data, there has been some inconsistency in reported BOLD signal changes during SWD. Some experimental data of fMRI recordings during spontaneous SWD activity in WAG/Rij rats have revealed no significantly negative BOLD changes associated with SWDs *(86)*. The same experiment showed widespread increases in the thalamus and in the cortex. No significant changes were found in the hippocampus *(86)*. A second group utilizing continuous EEG-fMRI found increased BOLD signal change in several regions of the cortex and subcortical structures, without major fMRI decreases (**Fig. 9.4**) *(15)*. Increases were seen in the somatosensory cortex, motor cortex, thalamus, basal ganglia, hippocampus, and brainstem (tectum and tegmental nuclei) and were mostly bilaterally and symmetrically distributed *(15)*. More recent studies with a higher field (9.4T) system detected both fMRI increases as well as decreases in specific brain regions in the same rodent model *(87)*. fMRI decreases were seen mainly in the basal ganglia, but were also occasionally present in small regions of the neocortex *(87)*.

Although absence epilepsy is considered a generalized seizure disorder, fMRI and electrical recordings of SWD have found that focal anterior regions of the brain were most intensively involved, while other brain regions remained relatively quiet (**Figs. 9.2** and **9.4**) *(13–15)*. Studies in rodent models have revealed focal abnormalities in voltage gated channel expression which may be related to epileptogenesis in this form of epilepsy *(27, 88)*. The ability to non-invasively image focal network involvement may ultimately lead to a better understanding of mechanisms in specific brain regions crucial for generation of SWD in both animal models and human patients.

6.2. GBL Rat Model

GBL is a precursor of γ-hydroxybutyrate and produces robust SWDs in rats, resembling petit mal status epilepticus *(82, 89)*. There is evidence that anesthesia limits the BOLD signal change in GBL-induced SWD *(82)* but awake animals have yielded interesting BOLD fMRI data. GBL-induced SWDs showed widespread negative changes in the cortex while also showing positive changes in the somatosensory and parietal cortices. The thalamus showed only significant positive BOLD changes. While these findings in some ways resemble reported changes during human SWD, important differences include the frequency of the discharges (6–7 Hz in GBL rat model vs. 3–4 Hz in human) and their duration (continuous status epilepticus in rat GBL model vs. brief episodes in human). The GBL model has some advantages compared to spontaneous rodent seizure models, since GBL-induced seizures are robust and long lasting , however, the spontaneous seizure models have the advantage of producing brief seizure episodes more similar to typical seizure durations in humans.

6.3. GBL Monkey Model

GBL seizures have also been studied in non-human primates. Advantages of primate models of SWD include SWD activity more closely resembling human SWD. Marmoset monkeys

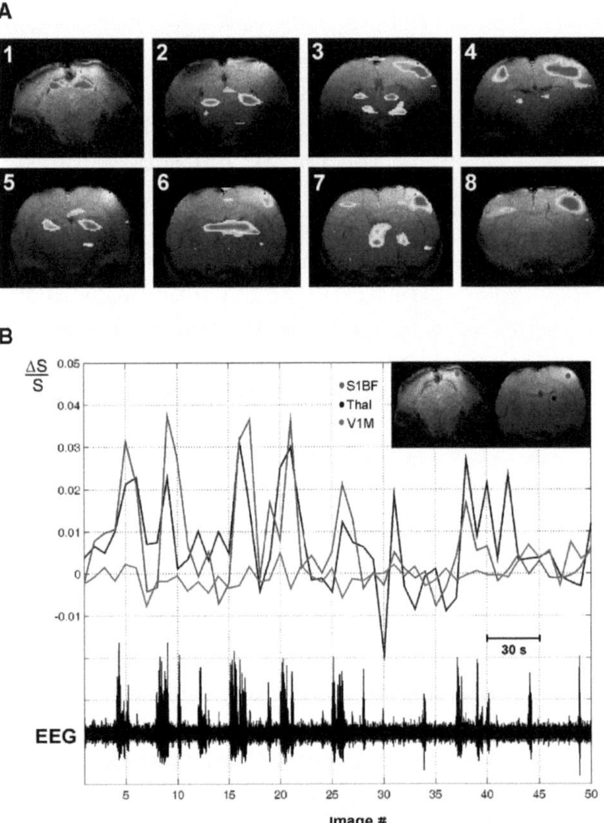

Fig. 9.4. Map of cortical and thalamic fMRI changes during spike-wave seizures in WAG/Rij rat. (**A**) Example of BOLD fMRI activations during SWD in a WAG/Rij rat under fentanyl-haloperidol anesthesia. t-maps were calculated from 23 pairs of images in which a pair constituted one image acquired at baseline and one image acquired during SWD. Results were overlaid onto corresponding high-resolution anatomic images. The eight coronal slices displayed were acquired from back to front (numbered 1–8) at 1-mm intervals. The first slice was at approximately –7.04 and the last slice was at +0.40 mm relative to bregma (Paxinos and Watson, 1998). The frontoparietal (somatosensory and motor) cortex, thalamus, and brainstem nuclei showed bilateral and relatively symmetrical BOLD signal increases, whereas no significant changes were seen in the temporal and occipital regions. t-map display threshold = 2. (**B**) Timecourse of BOLD fMRI signal changes shown with simultaneous EEG during spontaneous SWD. The onset of most SWD episodes, particularly those lasting more than 3 seconds, precedes increases in the BOLD fMRI signal ($\Delta S/S$) in barrel cortex (*red line*, S1BF) and thalamus (*blue line*, Thal). The primary visual cortex (*green line*, V1M) demonstrated no significant changes related to seizure activity. Same experimental run, and same animal as in A. Reproduced with permission from Nersesyan et al 2004B, *J Cerebral Blood Flow Metab (15)*. Color version of this figure can be viewed in the original publication *(15)* at http://www.nature.com/jcbfm/.

demonstrate 3 Hz SWD activity, which is more comparable to human SWD activity. Widespread increases were seen throughout cortical and subcortical structures including the thalamus and hippocampus. Unlike in GBL-induced seizures in rats, no significant negative BOLD changes were seen although changes in the hippocampus and the anterior cingulate were only seen during shorter time points. As in the rat model, GBL-induced seizures in monkeys most closely resembles status epilepticus *(90)*.

6.4. Discussion of Animal Studies of Spike-Wave

It is not yet clear why human fMRI studies of SWD show a mix of cortical increases and prominent decreases while animal models with brief episodes of SWD show mainly cortical increases. Similarly, it is unclear why prolonged SWD causes mainly BOLD increases in the monkey GBL model, but both increases and decreases in the rat GBL model. Some of this may reflect a lack of understanding of the fundamental mechanisms of fMRI increases and decreases during seizures. Direct recordings of neuronal

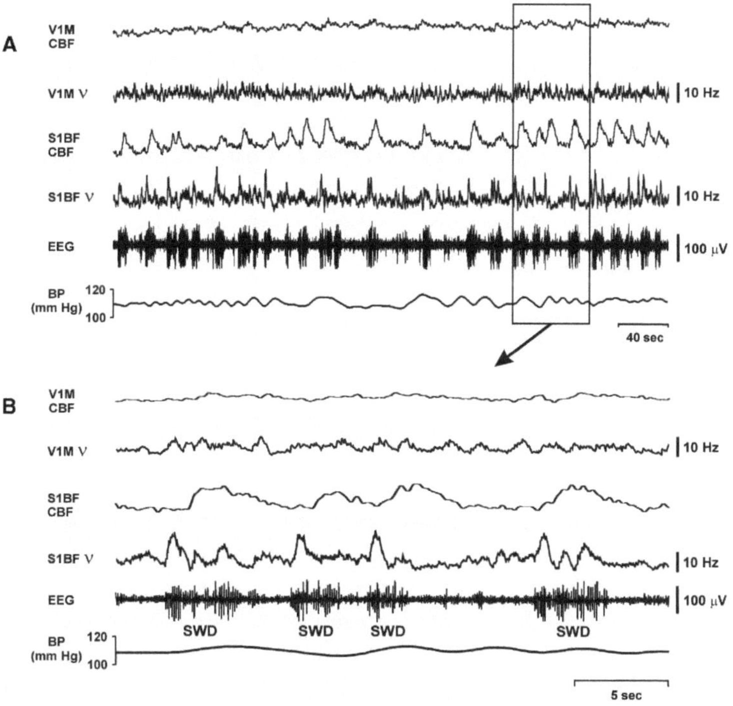

Fig. 9.5. CBF and neuronal activity recorded dynamically during SWD. (**A**) Combined laser Doppler flowmetry and extracellular multiunit data recorded changes in CBF and neuronal activity simultaneously during multiple episodes of SWD, along with EEG and arterial blood pressure (BP) monitoring in a WAG/Rij rat under fentanyl-haloperidol anesthesia. (**B**) Parallel increases in CBF and neuronal firing rate (ν) in barrel cortex (S1BF), and no or very small increases in primary visual cortex (V1M) induced by spontaneous SWD. Panel (**B**) shows the data from the boxed region of panel A on an expanded time scale. Reproduced with permission from Nersesyan et al 2004A, *J Cerebral Blood Flow Metab (14)*.

activity, which can be performed in animal models (**Fig. 9.5**), may provide a way to unravel the complex relationships between neuronal activity patterns and BOLD fMRI. Studies performed so far in WAG/Rij rats have demonstrated that regions with fMRI increases during SWD (e.g. barrel somatosensory cortex, S1BF, **Fig. 9.4**) exhibit increases in neuronal firing and CBF during SWD (**Fig. 9.5**). Meanwhile, regions with no fMRI signal changes during SWD (e.g. primary visual cortex, V1M, **Fig. 9.4**) show no changes in neuronal firing or CBF (**Fig. 9.5**). Further studies will be needed, particularly in regions showing fMRI decreases, to more fully understand these phenomena.

7. Possible Mechanisms of fMRI Decreases During Seizures

From the above discussion, it follows that BOLD fMRI increases during SWD most likely reflect focal increases in neuronal activity. However, the mechanisms of BOLD decreases during SWD and other forms of seizure activity are not well understood. Possible mechanisms of BOLD fMRI decreases include: *(1)* A *primarily vascular mechanism*, where blood supply decreases in specific brain regions during seizures. Primary causes of decreased blood flow could include (1A) *vasospasm* or (1B) *vascular steal.* Excessive decreases in local blood flow will decrease net oxygenation, leading to a decrease in BOLD. Significant decreases in oxygenation (hypoxia) could also secondarily cause some reduction in neuronal activity (**Fig. 9.2C**). *(2)* A *primarily neural mechanism* where cortical neuronal activity decreases in specific brain regions during seizures. If neurovascular coupling is normal, reduced neuronal activity usually leads to an excessive reduction in blood flow (**Fig.9.2C**), again causing a net decrease in oxygenation, and a decrease in the BOLD signal. A primary decrease in neuronal activity could occur in several ways: (2A) *Decreased neuronal activity in regions spared by seizures.* While seizures cause neuronal activity to increase in regions intensely involved in epileptic activity, network effect may cause neuronal activity to decrease in other regions of the brain which are *not* directly involved in the seizure. This could occur either through (2Ai) *active inhibition*, or (2Aii) *reduced excitation* in these regions. Examples of decreased neuronal activity and CBF have been observed while studying cortical brain regions spared by partial seizures in both animal models *(91)*, and in humans *(5, 6)*; (2B) *Decreased neuronal activity in regions involved by seizures.* A second primarily neural mechanism may occur in which some cortical regions which are *involved* in seizures could, nevertheless, show *reduced* neuronal activity. While seizure activity is generally thought to increase neuronal activity, SWD is associated with an alternating pattern, between increases in neuronal activity during the spike component, and decreases in neuronal activity during the wave *(92)*. It is possible that some brain regions may have subtle changes in intensity of firing during the spike or changes in duration of neuronal silence during the wave, which could lead to a mean *decrease* in neuronal activity during SWD, as discussed previously *(2, 93)*. *(3) Altered neuro-vascular coupling.* In this mechanism, unlike those above, the primary event is an *increase* in neuronal activity during seizures. However, this increase in neuronal activity is met by an inadequate increase in blood flow to match metabolic demands, leading to a BOLD fMRI decrease (**Fig. 9.2D**). Note that in this mechanism, blood flow can increase (unlike **Fig. 9.2D**); the main point is that the blood flow increase is inadequate. An example of this mechanism has recently been observed in a rodent model, where BOLD decreases in the hippocampus accompany intense seizure activity in the same region *(61)*. There are two main possibilities to explain abnormal neurovascular coupling during seizures: (3A) *Intense neuronal activity could overwhelm neurovascular coupling.* In this mechanism, sudden extreme increases in neuronal energy consumption could exceed the capacity of neurovascular coupling mechanisms to deliver adequate oxygen. (3B) *Dysregulation of the neurovascular coupling cascade.* In this mechanism, the inherent signaling cascade underlying neurovascular coupling could be altered acutely or chronically by seizures (or other causes), leading to an abnormal, blunted response to neuronal activity changes.

Which of the above mechanisms are involved in BOLD fMRI decreases during SWD in both humans and animal models remains to be determined by further investigations.

8. Discussion and Future Directions

We have seen that fMRI during SWD in both human and animal models can provide important information about abnormal network behavior during generalized seizures. Studies so far, have revealed focal bilateral increases in the frontoparietal cortex and thalamus, decreases in other specific cortical regions, and decreases in the basal ganglia. These investigations support the concept that "generalized" spike-wave seizures, in fact, arise from focal network dysfunction in specific regions of cortical-subcortical networks.

Many important questions remain which should be addressed by future studies. For example, do seizures in which consciousness is impaired differ from those in which consciousness is spared? Is the difference based on the brain regions involved? Are there different molecular mechanisms causing specific regions but not others to be involved during seizures? How do the circuit mechanisms differ for regions involved or spared by seizures? Is the impaired brain function and behavior during seizures caused by abnormally increased activity, abnormally decreased activity, or both? There remains much to learn about spike-and-wave discharges through the use of fMRI. Behavioral tests in both humans and animals must be added to simultaneous EEG-fMRI in order to fully correlate behavioral deficits with the anatomy of fMRI signal changes *(28, 68)*. More sophisticated analytic techniques must be developed in order to explore the fMRI time-course, as well as to better correlate the electrophysiology data with the hemodynamic response measured by fMRI data. Long-range network changes can be further investigated using resting functional connectivity studies based on the fMRI signals, and using technology such as diffusion tensor imaging (DTI). Numerous measurements, including CBF, CBV, and CMRO2, have yet to be undertaken in human patients or animal models to more fully understand the fundamental neuroenergetics of SWD. fMRI investigation of additional rat, mouse, and feline models of SWD *(23, 94, 95)* can also provide a more general understanding of SWD mechanisms. Finally, these techniques should be applied to understand developmental changes that occur during the development of epilepsy. Neuroimaging has the potential to provide a safe, non-invasive biomarker for epileptogenesis and its prevention, and could ultimately be used as a way to monitor the success of treatments aimed at suppressing spike-wave development *(27)*. Much exciting research remains to be done using fMRI to investigate the unknowns of spike-and-wave discharges and epilepsy in the future.

Acknowledgments

We thank Mi Hae Chung for assistance with the figures. This work was supported by NIH R01 NS055829, R01 NS049307, P30 NS052519, the Donaghue Foundation, and by the Betsy and Jonathan Blattmachr family.

References

1. ILAE, *Proposal for revised clinical and electroencephalographic classification of epileptic seizures. From the Commission on Classification and Terminology of the International League Against Epilepsy.* Epilepsia, 1981. **22**(4): p. 489–501.
2. Blumenfeld, H., *Cellular and network mechanisms of spike-wave seizures.* Epilepsia, 2005. **46** Suppl 9: p. 21–33.
3. Blumenfeld, H., *Consciousness and epilepsy: Why are patients with absence seizures absent?* Prog Brain Res, 2005. **150**: p. 271–86.
4. Blumenfeld, H., et al., *Selective frontal, parietal and temporal networks in generalized seizures.* Neuroimage, 2003. **19**: p. 1556–1566.
5. Blumenfeld, H., et al., *Positive and negative network correlations in temporal lobe epilepsy.* Cerebral Cortex, 2004. **14**(8): p. 892–902.
6. Blumenfeld, H., et al., *Ictal neocortical slowing in temporal lobe epilepsy.* Neurology, 2004. **63**: p. 1015–1021.
7. Ebersole, J.S. and T.A. Pedley, *Current Practice of Clinical Electroencephalography, 3rd*

Edition. 2003, Philadelphia, PA: Lippincott Williams & Wilkins.

8. Weir, B., *The morphology of the spike-wave complex*. Electroencephalography & Clinical Neurophysiology, 1965. **19**(3): p. 284–90.
9. Rodin, E. and O. Ancheta, *Cerebral electrical fields during petit mal absences*. Electroencephalography & Clinical Neurophysiology, 1987. **66**(6): p. 457–66.
10. Coppola, R., *Topographic display of spike-and-wave discharges*, in *Elements of Petit Mal Epilepsy*, M.S. Mysobodsky and A.F. Mirsky, Editors. 1988, Peter Lang: New York. p. 105–130.
11. Holmes, M.D., M. Brown, and D.M. Tucker, *Are "generalized" seizures truly generalized? Evidence of localized mesial frontal and frontopolar discharges in absence*. Epilepsia, 2004. **45**(12): p. 1568–79.
12. Vergnes, M., C. Marescaux, and A. Depaulis, *Mapping of spontaneous spike and wave discharges in Wistar rats with genetic generalized non-convulsive epilepsy*. Brain Research, 1990. **523**(1): p. 87–91.
13. Meeren, H.K., et al., *Cortical focus drives widespread corticothalamic networks during spontaneous absence seizures in rats*. Journal of Neuroscience, 2002. **22**(4): p. 1480–1495.
14. Nersesyan, H., et al., *Relative changes in cerebral blood flow and neuronal activity in local microdomains during generalized seizures*. J Cereb Blood Flow Metab, 2004. **24**(9): p. 1057–1068.
15. Nersesyan, H., et al., *Dynamic fMRI and EEG recordings during spike-wave seizures and generalized tonic-clonic seizures in WAG/Rij rats*. J Cereb Blood Flow Metab, 2004. **24**(6): p. 589–599.
16. Williams, D., *A study of thalamic and cortical rhythms in petit mal*. Brain, 1953. **76**: p. 50–69.
17. Blumenfeld, H. and J. Taylor, *Why do seizures cause loss of consciousness?* The Neuroscientist, 2003. **9**(5): p. 301–310.
18. Avoli, M., et al., eds. *Generalized Epilepsy*. 1990, Birkhauser: Boston.
19. Blumenfeld, H. and D.A. McCormick, *Corticothalamic inputs control the pattern of activity generated in thalamocortical networks*. Journal of Neuroscience, 2000. **20**(13): p. 5153–5162.
20. Kostopoulos, G., *Involvement of the thalamocortical system in epileptic loss of consciousness*. Epilepsia, 2001. **42**(3): p. 13–19.
21. McCormick, D.A. and D. Contreras, *On the cellular and network bases of epileptic seizures*. Annual Review of Physiology, 2001. **63**: p. 815–846.
22. Blumenfeld, H., *The thalamus and seizures*. Arch Neurol, 2002. **59**(1): p. 135–137.
23. Crunelli, V. and N. Leresche, *Childhood absence epilepsy: Genes, channels, neurons and networks*. Nature Reviews Neuroscience, 2002. **3**(5): p. 371–382.
24. Engel, J., Jr., et al., *Local cerebral metabolic rate for glucose during petit mal absences*. Annals of Neurology, 1985. **17**(2): p. 121–128.
25. Salek-Haddadi, A., et al., *Functional magnetic resonance imaging of human absence seizures*. Annals of Neurology, 2003. **53**(5): p. 663–667.
26. Aghakhani, Y., et al., *fMRI activation during spike and wave discharges in idiopathic generalized epilepsy*. Brain, 2004. **127**(Pt 5): p. 1127–1144.
27. Blumenfeld, H., et al., *Early treatment suppresses the development of spike-wave epilepsy in a rat model*. Epilepsia, 2008. **49**(3): p. 400–409.
28. Berman, R., et al., *Simultaneous EEG, fMRI, and behavioral testing in typical childhood absence seizures*. 2009, In press.
29. Theodore, W.H., et al., *Positron emission tomography in generalized seizures*. Neurology, 1985. **35**(5): p. 684–690.
30. Prevett, M.C., et al., *Demonstration of thalamic activation during typical absence seizures using H2(15)O and PET*. Neurology, 1995. **45**(7): p. 1396–1402.
31. Engel, J., Jr., D.E. Kuhl, and M.E. Phelps, *Patterns of Human Local Cerebral Glucose Metabolism During Epileptic Seizures*. Science, 1982. **218**: p. 64–66.
32. Yeni, S.N., et al., *Ictal and interictal SPECT findings in childhood absence epilepsy*. Seizure, 2000. **9**(4): p. 265–269.
33. Buchheim, K., et al., *Decrease in haemoglobin oxygenation during absence seizures in adult humans*. Neurosci Lett, 2004. **354**(2): p. 119–122.
34. Bode, H., *Intracranial blood flow velocities during seizures and generalized epileptic discharges*. European Journal of Pediatrics, 1992. **151**(9): p. 706–709.
35. De Simone, R., et al., *Changes in cerebral blood flow velocities during childhood absence seizures*. Pediatric Neurology, 1998. **18**(2): p. 132–135.
36. Klingelhofer, J., et al., *Do brief bursts of spike and wave activity cause a cerebral hyper- or hypoperfusion in man?* Neuroscience Letters, 1991. **127**(1): p. 77–81.
37. Nehlig, A., et al., *Absence seizures induce a decrease in cerebral blood flow: Human and animal data*. Journal of Cerebral Blood Flow & Metabolism, 1996. **16**(1): p. 147–155.

38. Sanada, S., N. Murakami, and S. Ohtahara, *Changes in blood flow of the middle cerebral artery during absence seizures.* Pediatric Neurology, 1988. **4**(3): p. 158–161.
39. Sperling, M.R. and B.E. Skolnick, *Cerebral blood flow during spike-wave discharges.* Epilepsia, 1995. **36**(2): p. 156–163.
40. Diehl, B., et al., *Cerebral hemodynamic response to generalized spike-wave discharges.* Epilepsia, 1998. **39**(12): p. 1284–1289.
41. Ochs, R.F., et al., *Effect of generalized spike-and-wave discharge on glucose metabolism measured by positron emission tomography.* Annals of Neurology, 1987. **21**(5): p. 458–464.
42. Park, Y.D., et al., *Focal cerebral metabolic abnormality in a patient with continuous spike waves during slow-wave sleep.* Journal of Child Neurology, 1994. **9**(2): p. 139–143.
43. Ferrie, C.D., et al., *Focal abnormalities detected by 18FDG PET in epileptic encephalopathies.[comment].* Archives of Disease in Childhood, 1996. **75**(2): p. 102–107.
44. Nehlig, A., et al., *Local cerebral glucose utilization in rats with petit mal-like seizures.* Annals of Neurology, 1991. **29**(1): p. 72–77.
45. Ives, J.R., et al., *Monitoring the patient's EEG during echo planar MRI.* Clin. Neurol., 1993. **87**: p. 417–420.
46. Goldman, R.I., et al., *Acquiring simultaneous EEG and functional MRI.* Clinical Neurophysiology, 2000. **111**(11): p. 1974–1980.
47. Gotman, J., et al., *Combining EEG and fMRI: A multimodal tool for epilepsy research.* Journal of Magnetic Resonance Imaging, 2006. **23**(6): p. 906–920.
48. Attwell, D. and S.B. Laughlin, *An energy budget for signaling in the grey matter of the brain.* Journal of Cerebral Blood Flow & Metabolism, 2001. **21**: p. 1133–1145.
49. Ogawa, S., et al., *On the characteristics of functional magnetic resonance imaging of the brain.* Annu Rev Biophys Biomol Struct, 1998. **27**: p. 447–474.
50. Hyder, F., et al., *Quantitative functional imaging of the brain: Towards mapping neuronal activity by BOLD fMRI.* NMR in Biomedicine, 2001. **14**(7–8): p. 413–431.
51. Ogawa, S., et al., *Functional Brain mapping by blood oxygenation level-dependent contrast magnetic resonance imaging.* Biophys J, 1993. **64**: p. 803–812.
52. Kennan, R.P., J. Zhong, and J.C. Gore, *Intravascular susceptibility contrast mechanisms in tissue.* Magn Reson Med, 1994. **31**: p. 9–21.
53. Weisskoff, R.M., et al., *Microscopic susceptibility variation and transverse relaxation: Theory and experiment.* Magn Reson Med, 1994. **31**: p. 601–610.
54. Shulman, R.G., F. Hyder, and D.L. Rothman, *Biophysical basis of brain activity: Implications for neuroimaging.* Quarterly Reviews of Biophysics, 2002. **35**(3): p. 287–325.
55. Logothetis, N.K., et al., *Neurophysiological investigation of the basis of the fMRI signal.* Nature, 2001. **412**(6843): p. 150–157.
56. Smith, A.J., et al., *Cerebral energetics and spiking frequency: The neurophysiological basis of fMRI.[see comment].* Proceedings of the National Academy of Sciences of the United States of America, 2002. **99**(16): p. 10765–10770.
57. Martin, C., et al., *Haemodynamic and neural responses to hypercapnia in the awake rat.* European Journal of Neuroscience, 2006. **24**(9): p. 2601–2610.
58. Stefanovic, B., J.M. Warnking, and G.B. Pike, *Hemodynamic and metabolic responses to neuronal inhibition.* Neuroimage, 2004. **22**(2): p. 771–778.
59. Shmuel, A., et al., *Negative functional MRI response correlates with decreases in neuronal activity in monkey visual area V1.* Nature Neuroscience, 2006. **9**(4): p. 569–577.
60. Suh, M., et al., *Neurovascular coupling and oximetry during epileptic events.* Molecular Neurobiology, 2006. **33**(3): p. 181–197.
61. Schridde, U., et al., *Negative BOLD with large increases in neuronal activity.* Cerebral Cortex, 2008. **18**: 1814–1827.
62. Stefanovic, B., et al., *Hemodynamic and metabolic responses to activation, deactivation and epileptic discharges.* Neuroimage, 2005. **28 (1)**: p. 205–215.
63. Hamandi, K., et al., *EEG-fMRI of idiopathic and secondarily generalized epilepsies.* Neuroimage, 2006. **31**(4): p. 1700–1710.
64. Archer, J.S., et al., *fMRI "deactivation" of the posterior cingulate during generalized spike and wave.* Neuroimage, 2003. **20**(4): p. 1915–1922.
65. Labate, A., et al., *Typical childhood absence seizures are associated with thalamic activation.* Epileptic Disorders, 2005. 7(4): p. 373–377.
66. Laufs, H., et al., *Linking generalized spike-and-wave discharges and resting state brain activity by using EEG/fMRI in a patient with absence seizures.* Epilepsia, 2006. **47**(2): p. 444–448.
67. Aghakhani, Y., et al., *Cortical and thalamic fMRI responses in partial epilepsy with focal and bilateral synchronous spikes.* Clin Neurophysiol, 2006. **117**(1): p. 177–1791.
68. Berman, R., et al., *Combined EEG and fMRI during typical childhood absence seizures at 3T.* Epilepsia. AES abstracts., 2005.

69. Gotman, J., et al., *Generalized epileptic discharges show thalamocortical activation and suspension of the default state of the brain*. Proceedings of the National Academy of Sciences of the United States of America, 2005. **102**(42): p. 15236–15240.
70. Raichle, M.E., et al., *A default mode of brain function*. Proc Natl Acad Sci U S A, 2001. **98**(2): p. 676–682.
71. Pinault, D., et al., *Intracellular recordings in thalamic neurones during spontaneous spike and wave discharges in rats with absence epilepsy*. Journal of Physiology, 1998. **509**(Pt 2): p. 449–456.
72. Kandel, A., Buzsaki, G., *Cerebellar neuronal activity correlates with spike and wave EEG patterns in the rat*. Epilepsy Reseach, 1993. **16**: p. 1–9.
73. Norden, A.D. and H. Blumenfeld, *The Role of Subcortical Structures in Human Epilepsy*. Epilepsy & Behavior, 2002. **3**(3): p. 219–231.
74. Paz, J.T., et al., *Activity of ventral medial thalamic neurons during absence seizures and modulation of cortical paroxysms by the nigrothalamic pathway*. J Neurosci, 2007. **27**(4): p. 929–941.
75. Tolmacheva, E.A., et al., *Cortical and limbic excitability in rats with absence epilepsy*. Epilepsy Res, 2004. **62**(2–3): p. 189–198.
76. Slaght, S.J., et al., *On the activity of the corticostriatal networks during spike-and-wave discharges in a genetic model of absence epilepsy*. J Neurosci, 2004. **24**(30): p. 6816–6825.
77. Deransart, C., et al., *Inhibition of the substantia nigra suppresses absences and clonic seizures in audiogenic rats, but not tonic seizures: Evidence for seizure specificity of the nigral control*. Neuroscience, 2001. **105**(1): p. 203–211.
78. Kohsaka, S., et al., *Brainstem activates paroxysmal discharge in human generalized epilepsy*. Brain Res, 2001. **903**: p. 53–61.
79. Kohsaka, S., et al., *Brainstem triggers absence seizures in human generalized epilepsy*. Brain Res, 1999. **837**: p. 277–288.
80. Schwartz, T.H. and T. Bonhoeffer, *In vivo optical mapping of epileptic foci and surround inhibition in ferret cerebral cortex*. Nature Medicine, 2001. 7(9): p. 1063–1067.
81. Van Camp, N., et al., *Simultaneous electroencephalographic recording and functional magnetic resonance imaging during pentylenetetrazol-induced seizures in rat*. Neuroimage, 2003. **19**: p. 627–636.
82. Tenney, J.R., et al., *Corticothalamic modulation during absence seizures in rats: A functional MRI assessment*. Epilepsia, 2003. **44**(9): p. 1133–1140.
83. Khubchandani, M., et al., *Stereotaxic assembly and procedures for simultaneous electrophysiological and MRI study of conscious rat*. Magnetic Resonance in Medicine, 2003. **49**: p. 962–967.
84. Sachdev, R.N., et al., *Experimental model for functional magnetic resonance imaging of somatic sensory cortex in the unanesthetized rat*. Neuroimage, 2003. **19**(3): p. 742–750.
85. Coenen, A.M. and E.L. Van Luijtelaar, *Genetic Animal Models for Absence Epilepsy: A Review of the WAG/Rij Strain of Rats*. Behav Genet, 2003. **33**: p. 635–655.
86. Tenney, J.R., et al., *FMRI of brain activation in a genetic rat model of absence seizures*. Epilepsia, 2004. **45**(6): p. 576–582.
87. Mishra, A.M., et al., *Physiology and imaging of increases and decreases in BOLD signals during spike-wave seizures in WAG/Rij rats*. Online at http://web.sfn.org/, 2007.
88. Klein, J.P., et al., *Dysregulation of sodium channel expression in cortical neurons in a rodent model of absence epilepsy*. Brain Research, 2004. **1000**: p. 102–109.
89. Snead, O.C., 3rd, et al., *Absence epilepsy: Advances in experimental animal models*. Advances in Neurology, 1999. **79**: p. 253–278.
90. Tenney, J.R., et al., *fMRI of generalized absence status epilepticus in conscious marmoset monkeys reveals corticothalamic activation*. Epilepsia, 2004. **45**(10): p. 1240–12407.
91. Englot D.J., et al., *Remote effects of focal hippocampal seizures on the rat neocortex*. 2009, In press.
92. Kostopoulos, G., et al., *A study of the transition from spindles to spike and wave discharge in feline generalized penicillin epilepsy: Microphysiological features*. Experimental Neurology, 1981. **73**(1): p. 55–77.
93. Blumenfeld, H., *Functional MRI studies of animal models in epilepsy*. Epilepsia, 2007. **48(Suppl. 4)**: p. 18–26.
94. Marescaux, C., M. Vergnes, and A. Depaulis, *Genetic absence epilepsy in rats from Strasbourg–a review*. Journal of Neural Transmission. Supplementum, 1992. **35**: p. 37–69.
95. Prince, D.A. and D. Farrell, "*Centrencephalic*" *spike and wave discharges following parenteral penicillin injection in the cat*. Neurology, 1969. **19**: p. 309–310.

Part IV

Functional Magnetic Resonance Imaging (fMRI)

Chapter 10

Tactile and Non-tactile Sensory Paradigms for fMRI and Neurophysiologic Studies in Rodents

Basavaraju G. Sanganahalli, Christopher J. Bailey, Peter Herman, and Fahmeed Hyder

Abstract

Functional magnetic resonance imaging (fMRI) has become a popular functional imaging tool for human studies. Future diagnostic use of fMRI depends, however, on a suitable neurophysiologic interpretation of the blood oxygenation level dependent (BOLD) signal change. This particular goal is best achieved in animal models primarily due to the invasive nature of other methods used and/or pharmacological agents applied to probe different nuances of neuronal (and glial) activity coupled to the BOLD signal change. In the last decade, we have directed our efforts towards the development of stimulation protocols for a variety of modalities in rodents with fMRI. Cortical perception of the natural world relies on the formation of multi-dimensional representation of stimuli impinging on the different sensory systems, leading to the hypothesis that a sensory stimulus may have very different neurophysiologic outcome(s) when paired with a near simultaneous event in another modality. Before approaching this level of complexity, reliable measures must be obtained of the relatively small changes in the BOLD signal and other neurophysiologic markers (electrical activity, blood flow) induced by different peripheral stimuli. Here we describe different tactile (i.e., forepaw, whisker) and non-tactile (i.e., olfactory, visual) sensory paradigms applied to the anesthetized rat. The main focus is on development and validation of methods for reproducible stimulation of each sensory modality applied independently or in conjunction with one another, both inside and outside the magnet. We discuss similarities and/or differences across the sensory systems as well as advantages they may have for studying essential neuroscientific questions. We envisage that the different sensory paradigms described here may be applied directly to studies of multi-sensory interactions in anesthetized rats, en route to a rudimentary understanding of the awake functioning brain where various sensory cues presumably interrelate.

Key words: Blood volume, glucose, glutamate, metabolism, neurovascular coupling, oxygen.

1. Introduction

Dynamic imaging of human brain function began approximately two decades ago with positron emission tomography (PET) (for

Fahmeed Hyder (ed.), *Dynamic Brain Imaging: Multi-Modal Methods and In Vivo Applications, vol. 489*

DOI 10.1007/978-1-59745-543-5_10 Springerprotocols.com

reviews see *(1, 2)*). Since the 1990s, however, nuclear magnetic resonance (NMR) – both imaging (MRI) and spectroscopy (MRS) – has played a major role in studies of in vivo neuroscience, both in animals and humans (for reviews see *(3, 4)*). The discovery of functional MRI (fMRI) by Seiji Ogawa in 1990 *(5)* further reinforced the role of functional imaging studies in neuroscience. The fMRI method, as originally proposed, depends on the paramagnetic effect of deoxyhemoglobin (in blood) upon the NMR transverse relaxation times of nearby water protons (in tissue) *(6)*. Since changes in the oxygen level in the blood determine the fraction of deoxygenated hemoglobin, the image contrast was fittingly termed blood oxygenation level dependent (BOLD).

After the initial demonstrations of the BOLD method in mapping dynamic brain function in humans during sensory stimulation *(7–10)*, the cognitive neuroscience community immediately embraced the method *(11)*. Today BOLD fMRI is arguably the most popular functional mapping tool for human studies, perhaps, in part, due to its non-invasive nature of application, relatively good spatiotemporal resolution, superior coverage of large parts of the brain, and the fact that experiments can be conducted on most clinical MRI scanners by slight adjustments. However, future utility of fMRI for diagnostic and treatment measures in humans is largely dependent on a better neurophysiologic interpretation of the BOLD signal change because the conventional fMRI map reflects changes in blood oxygenation, not the actual neuronal activity *(12–14)*. This particular goal seems to be best achieved in animal models *(15–17)* primarily due to the invasive nature of the non-NMR methods used *(18–20)* and/or pharmacological agents applied *(21–23)* to probe different features of cellular activity coupled to the BOLD signal change.

In the last decade, we have invested considerable research efforts towards developing various sensory stimulation protocols in rodents with fMRI *(24–26)*. Over that same period of time, coupled with state-of-the-art NMR advancements in high magnetic field scanners and improvements in other hardware components (e.g., gradient, shim, and radio frequency coils) *(27)*, we are now able to reproducibly *(28, 29)* measure relatively small BOLD signal changes (in rat or mouse brains) induced by peripheral stimuli with superior spatiotemporal resolution *(30, 31)*. Here we describe tactile (i.e., whisker, forepaw) and non-tactile (i.e., olfactory, visual) sensory paradigms applied to the anesthetized rat. We discuss features of peripheral stimulus delivery equipment needed to generate identical stimuli, both inside and outside the magnet. The focus is on development and validation of methods for stimulation of each sensory modality applied independently or in conjunction with one another. We demonstrate reproducibility of induced activations, as measured by changes in the BOLD signal and other non-NMR signals (e.g., electrical activity, laser Doppler blood flow), achieved with different sensory paradigms and discuss similarities and/or differences across the different sensory systems as well as the potential advantages they may have for the study of different neuroscientific questions.

The mammalian brain functions in a world full of sensory cues. The nature of stimuli perturbing the different sensory domains is quite diverse – from odorant molecules at the nose epithelium and photons bombarding the retina, to vibrations of the eardrum and mechanoreceptors on the skin. As a consequence, the brain structures dedicated to different senses are anatomically distinct, from the level of cell types to their vast synaptic interconnections *(32)*. Therefore, different peripheral stimuli of the rodent brain offer a rich platform on which to test effects of perturbations on functionally and anatomically heterogeneous populations of neural tissues. In addition, the vast literature on rodent brain anatomy, morphology, and physiology make possible viable extrapolation of results from such experiments.

Since the physical world around any creature impinges on all its sensory systems in parallel, a multi-dimensional representation of the surroundings is thereby needed for complex (e.g., hunting for prey) and even simple (e.g., avoidance of noxious conditions) behaviors. A central hypothesis in the design of controlled fMRI and neurophysiologic studies, as described here and executed similarly in many other laboratories around the world, is that sensory stimuli may have very different neurophysiologic outcome(s) when paired with a near simultaneous event in another modality. The different sensory paradigms described here may be applied directly to the study of multi-sensory interactions in anesthetized rats *(33)* and elucidation of such effects with high spatiotemporal precision is hypothesized to contribute to the understanding of the awake functioning brain *(34)*.

2. Materials and Methods

The design principle of our approach for reproducible stimulation of multiple senses can be summarized with the word *modularity*. The different sensory systems we aim to probe clearly necessitate specialized stimulus delivery systems, operating at appropriate timescales. However care was taken not to build mutually exclusive apparatus for the different senses (i.e., one stimulus modality should not prevent concurrent or simultaneous application of another). The basic requirement of accurate dynamic control over the stimuli called for computerized implementation. Therefore, all of our designs are based on a CED μ1401 analog-to-digital converter unit (or CED unit), in particular, its programmable analog and digital output capabilities, controlled by custom-written scripts running in the Spike2 environment (Cambridge Electronic Design, Cambridge, UK) on standard personal computers. All wires, pipes, and optical fibers needed to deliver the stimuli to the rodent were led into the magnetically shielded scanner room through small access holes in the sidewall. Exactly the same stimulus delivery mechanisms were applied to the rodents outside the magnet for bench neurophysiologic studies. All of the stimulation devices described below were home-built.

2.1. Stimulation Devices

2.1.1. Olfactory Stimulator

The design of the olfactometer is based on John Kauer's original idea *(35)*. We follow the design implemented in Lawrence Cohen's laboratory *(36)* but with specific modifications necessitated by the long delivery conduits needed between odorant source (outside the magnet room) and the subject (inside the magnet). Our olfactometer was designed and built to control the delivery of odorants mixed in air provided by an aquarium air pump. A schematic of the design is shown in **Fig. 10.1**. A charcoal filter was used to remove odorants present in room air before entering a system of solenoid valves (2 inlets and 1 outlet; Cole-Parmer Instrument, Vernon Hills, IL) and acrylic flow meters (maintained at 1 L/min; Cole-Parmer Instrument). The output of the flow meters was connected to long (~8 m) Teflon tubes that were led into the scanner room and then connected to a glass tube placed in front of the rat lying inside the magnet bore. The state of each solenoid valve (open/closed) was switched using a relay circuit controlled by the CED unit using Spike2 software. To provide a continuous steady-state environment to the nasal mucosa, the applied airflow was humidified and preheated to 28–30°C.

The olfactometer operates as follows: Filtered air enters the first solenoid valve (S1, always open except during the odor delivery) and the connected flow meter (FM) delivers that air to the freely breathing rat. The filtered air stream also enters a glass bottle with a specific odorant, from which the output is led through the second solenoid valve (S2). The input to a third solenoid valve (S3) is connected to a vacuum pump. During odorant stimulation, S2 is opened for a defined duration, causing a constant flow of the odorant, mixed in air, to be delivered into the glass tube in front of the rat. To rapidly end the stimulus, S3 is opened for a short period of time causing a transient vacuum in the delivery tube and thus sucking out all remaining molecules of the odorant. Each solenoid valve was switched in less than 50 ms. A straightforward extension to the system described above is one with multiple odorant containers, solenoid valves, and flow meters arranged in parallel to allow mixtures of odorants to be presented.

2.1.2. Visual Stimulator

The implemented design of the visual stimulus delivery system resulted from iterative modifications with the primary goal of

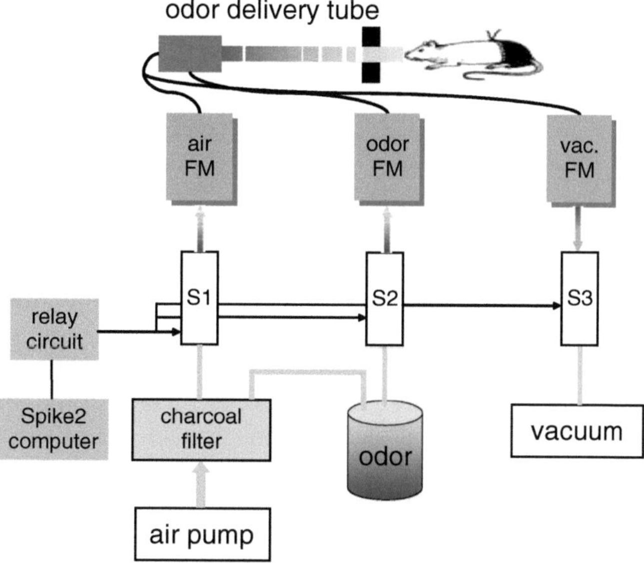

Fig. 10.1. Schematic representation of the odorant delivery system. Air, driven by a pump, was purified in charcoal filter to remove spurious odorants. Fast switching solenoid valves (S1, S2, S3) were used to control airflow through flow meters (FM). The solenoids were opened/closed by a relay circuit which in turn received input from a CED unit under the control of Spike2 software. S1 was closed only under stimulation. During odorant stimulation, S2 was opened, allowing filtered air to pass through the odor container and enter the delivery tube. At the end of a trial, S3 was opened to rapidly remove odor particles from the rat's vicinity.

being able to provide a variety of colored lights with wide ranging intensities and/or orientations, while the ambient lighting inside the room or the magnet bore remained constant. Therefore, our design differs from the stroboscopic-based *(37)* or dark-light adapted *(38)* systems which generate very robust intensity stimuli. The light was delivered to the rodent by polymer-based fiber optics (Fiber Optic Products, Clearlake Oaks, CA). Two thin (inner/outer diameters of 1.0/2.3 mm) polyethylene-coated optical fiber cables for independent stimulation of each eye were led into the scanner room. A custom-made fiberglass platform was secured in front of the animal (outside the magnet bore) that allowed the positioning of small cradles which in turn could be rotated to ensure optimal intensities and/or orientation of light source relative to each eye. Each semi-circular cradle, angled slightly down, was designed to securely hold a cylinder into which the optical cable was fixed. The light from the cables then illuminated the back of an acrylic lens (diameter of 6.4 mm) placed inside the cylinder. A piece of paper with alternating black-and-white stripes was glued to the back of the lens so that the dispersed light exiting the cable lead to a larger viewing angle and at the same time created better spatial contrast. The method has been shown to enhance responsiveness of primary visual cells in the rat *(39)*. Given the distance between the illuminated lens and the eyes (i.e., ~2.3 cm), we could then estimate the angle in the visual field subtended by each stimulus. With reference to an origin at the intersection of the sagittal midline, the coronal interocular line, and the horizontal plane through the eyes, the stimuli appear approximately at 35–50° azimuth and 10–40° elevation relative to the horizontal meridian.

In addition to a spatially specific and robust stimulus delivery method, our goal was to control accurately both the intensity (lux) and wavelength (color) of the stimuli. We opted for light emitting diodes (LEDs; Luxeon Star III, Lumileds Lighting, LLC, San Jose, CA), which come in a variety of colors (i.e., with well defined emission spectra) including "white." A power module (LuxDrive BuckPuck, LEDdynamics, Randolph, VT, USA), with a custom-made associated circuitry, was used to convert an applied voltage to a constant current fed to the LEDs because the output luminous intensity of LEDs is more linearly proportional to current. The controlling input voltage was applied by the CED unit using Spike2 software. We used acrylic collimator lenses to target the wide-angle LED light into the optical cables, which attached to cradles in front of the LEDs. The intensity of light entering each cable could thus be independently adjusted. Before each experiment, we calibrated the stimuli using a light meter (Extech Instruments, Waltham, MA). We measured, at four different pre-defined input voltages, the illuminance (in lux units) emitted from the stimuli.

2.1.3. Whisker Stimulator

The design of our whisker stimulator is based on accurate control of airflow through solenoid valves, similar to the case of the olfactory stimulator as discussed above. Aquarium air pumps were connected to two computer-controlled valves, the outputs of which were connected to long (~8 m) Teflon tubes that ran the length of the imaging bore to the rat's whisker pad. Whiskers on the chosen side(s) were trimmed to a length of ~20 mm. All non-stimulated whiskers, including the non-selected whiskers on the contralateral side, were trimmed away to avoid spurious activations. A lightweight masking tape (length and width of 20 and 6 mm, respectively) was fastened to the chosen whisker(s) to increase resistance to airflow while at the same time ensuring identical motion of the selected whisker(s). By alternative opening and closing of the solenoids, we generated short air puffs through the tubes near the whiskers. The distance between the tubes and the whiskers was typically about 2 cm. The default orientation of the tubes moved the whiskers in a rostrocaudal direction, but dorsal-ventral is also possible by changing the orientation of the tubes relative to the whisker pad. Alternating air puffs at rates of up to tens of Hz deflected the stimulated whiskers by ~2 mm.

2.1.4. Forepaw Stimulator

A pair of thin needle copper electrodes was inserted under the skin of the chosen forepaw(s). Square wave current pulses of 0.3 ms duration and variable amplitude (0.2–5 mA) were generated by use of an isolator unit (WPI, Sarasota, FL) connected to the electrodes. To control stimulus timing and frequency (typically 1–9 Hz), digital output signals were generated by the CED unit using Spike2 software.

2.2. Animal Preparation

All procedures were performed in accordance with protocols approved by the Yale Animal Care and Research Committee. Experiments were conducted on male rats (Charles River, Wilmington, MA; fed *ad libitum*) weighing between 180 and 350 g. In all rats, a femoral artery was cannulated with a polyethylene catheter (PE-50) to withdraw blood samples for blood gas analysis and to monitor arterial blood pressure, an intraperitoneal catheter (PE-10) was inserted for administration of soluble anesthetics, and a femoral vein was cannulated with a polyethylene catheter (PE-10) to administer other drugs. Blood pressure was continuously monitored. Ventilation parameters were adjusted to maintain arterial blood gas tensions (pCO_2, pO_2, pH) within normal range. A rectal temperature probe was inserted to monitor and maintain the core body temperature at ~37°C with a temperature-controlled recirculated warm water pad. All monitoring equipment, consisting of magnetic materials, were placed outside the scanner room and sampled by the CED unit for online recording. A standard block stimulation protocol was used in all experiments. We monitored analgesia depth by pain response to an automated electrical (5 mA, 0.3 ms, 1 Hz, 1 s) tail-pinch every 1/4 hour.

2.2.1. Forepaw and Whisker Studies

The rats (Sprague-Dawley) were initially anesthetized with 1.5–2% halothane in a mixture of 70% N_2O and 30% O_2. Tracheotomy was performed and the animal was artificially ventilated with 0.75–1% halothane in a mixture of 70% N_2O and 30% O_2 during all other surgical procedures. *D*-tubocurarine chloride (initial 0.5 mg/kg; supplemental 0.25 mg/kg/h; intravenous) was used to immobilize the rat. After all surgical procedures were completed, halothane was discontinued and anesthesia was maintained with α-chloralose (initial 80 mg/kg; supplemental 40 mg/kg/h; intraperitoneal).

2.2.2. Visual Studies

The commonly used albino strains of experimental rats, such as Sprague-Dawley, possess significantly poorer visual acuity than their wild counterparts *(40)*. The primary causes may be the lack of pigment in the retinal epithelium, leading to light scattering in the eye and abnormal development of retinothalamic *(41, 42)* and interhemispheric cortical connections *(42, 43)*. Therefore, we opted for the pigmented, "hooded", Long-Evans strain of rats for our visual experiments only. Preparation was similar to forepaw and whisker studies with the exception that isoflurane was used for induction (3–4%) and surgery (1.5–2%).

2.2.3. Olfactory Studies

Male Sprague-Dawley rats were anesthetized with urethane (1.3 g/kg, intraperitoneal) with additional doses (0.13 g/kg, intravenous) administered if necessary depending on the duration of the experiment. These rats were freely breathing (i.e., no tracheotomy).

2.3. Experimental Setup for fMRI Studies

We used a custom-designed cradle with a bite bar for head immobilization during fMRI scans. All stimulus delivery items were secured onto the holder with adhesive tape. In some cases, to record the electroencephalogram (EEG) simultaneously with fMRI, we used a pair of carbon fibers (WPI) placed bilaterally over the parietal cortex between the scalp and the skull *(44)*. All fMRI data were obtained on a modified 11.7 T Bruker horizontal-bore spectrometer (Bruker AVANCE, Billerica, MA) using a ^{1}H surface coil (diameter of 14 mm) radio frequency probe. The cradle was securely placed under the coil. The z axis position was dependent on the stimulus used and distance between coil and head (y axis) was minimized.

A voxel of ~100 and ~300 μL, respectively, for the bulb and brain was shimmed with fast, automatic shimming technique by mapping along projections (FASTMAP) *(45)* utilizing first and second order shim gradients. The static field inhomogeneity was optimized until the half-height line width of water in the shimming voxel was less than 15 and 20 Hz, respectively, for imaging the bulb and brain. The neuroanatomy was imaged with either the rapid acquisition relaxation enhanced (RARE) *(46)* or fast low-angle single shot (FLASH) *(47)* contrast sequence. For all brain studies, we used single-shot echo-planar imaging (EPI) *(48)* with sequential sampling *(49)* and the following parameters: Recycle time (TR) = 1000 ms; echo time (TE) = 15 ms; field of view (FOV) = 2.56×2.56 cm^2; image matrix = 64×64; number of slices = 3 or 6; slice thickness = 2 or 1 mm; voxel size = 320 or 160 nL; flip angle = 45–60°. For olfactory studies, we used FLASH contrast and the following parameters: TR = 500 ms; TE = 15 ms; FOV = 1.56×1.56 cm^2; image matrix = 64×64; number of slices = 20; slice thickness = 250 or 500 μm; voxel size = 15 or 30 nL; flip angle = 15–30°.

2.4. Experimental Setup for Electrophysiology and Laser Doppler Flowmetry

The rat was placed in a stereotaxic holder (David Kopf Instruments, Tujunga, CA) sitting on a vibration-free table in a Faraday cage adjacent to the scanner. For all somatosensory region recordings, we developed a combined dual-probe consisting of a high impedance microelectrode and a fiberoptic laser Doppler probe *(50, 51)*. The microelectrode was glued to the side of an 18 G spinal needle shaft (Terumo, Tokyo, Japan) and the bare laser Doppler probe was placed inside the needle shaft so that tips of each sensor protruded (approximately 1.5 mm) beyond the needle tip. The shaft of the needle was then placed in a microelectrode holder (Plastic One, Roanoke, VA) on the stereotaxic apparatus, and the tip of the dual-probe was advanced into the rat cortex (see below) to a depth corresponding to cortical layer 4, which could later be verified by histology *(52)*. For all olfactory bulb and visual recordings, only the microelectrode was used. The coordinates for all neurophysiologic measurements were guided by prior fMRI data *(16, 26)*.

Cerebral blood flow (CBF) was measured using a fiberoptic laser Doppler probe (830 nm; Oxford Optronix, Oxford, UK) sensitive to red blood cell flux. The bare fiber laser Doppler probe had a total diameter of less than 450 μm. CBF was dynamically recorded by the CED unit without any additional filtering. Electrical activity was measured with tungsten matrix microelectrodes consisting of two electrodes separated by 410 μm (FHC, Bowdoinham, ME) with high impedance (2–4 MΩ, tip size <1 μm). The pre-amplified electrical signal was digitized with the CED unit. The data were collected with a large bandwidth (10 Hz–20 kHz) and filtered into local field potential (LFP; 10–150 Hz) and multi unit activity (MUA; 300–1500 Hz) bands. The LFP data were represented in the raw arbitrary units of mV. From the MUA data, a template-matching algorithm (Spike2) was used to detect action potentials fired by individual neurons near the electrode tips to calculate spike rates (10 s bins) in units of Hz.

For recordings from somatosensory regions, tiny burr holes above contralateral and ipsilateral areas [forelimb area ($S1_{FL}$): 4.4 mm lateral and 1.0 mm anterior to bregma; whisker barrel field ($S1_{BF}$): 4.5–5.5 mm lateral 2.5–3 mm posterior to bregma] were thinned and the skull was carefully opened. Recordings from the visual areas were guided by locations from the Paxinos and Watson atlas *(53)*. For recordings from the olfactory bulb, tiny burr holes above the olfactory bulb were made (medial location: 0.5 mm lateral and 7.7 mm anterior from bregma; lateral location: 1.5 mm lateral and 8.7 mm anterior from bregma].

3. Results

3.1. Olfactory Stimulation

We examined odorant-induced activity patterns in the olfactory bulb with ester (isoamyl acetate) and aldehyde (hexanal) stimulations given to both nostrils in the same subject (**Fig. 10.2**). None of the odorants or their concentrations examined caused

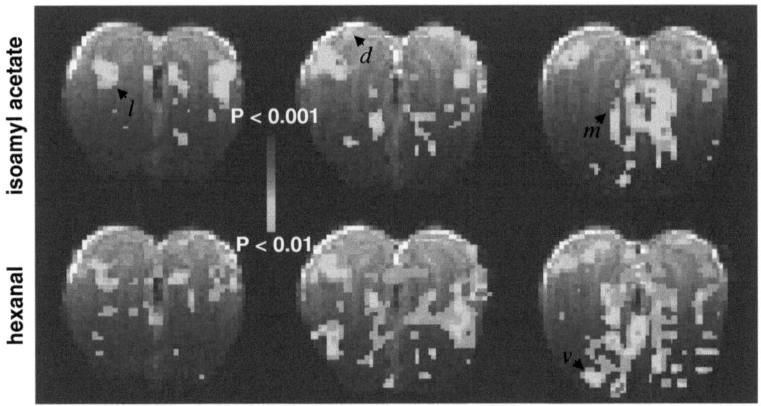

Fig. 10.2. BOLD responses from the olfactory bulb in Sprague-Dawley rats. The rat was exposed to isoamyl acetate (100%) and hexanal (100%) for a duration of 60 s each. The *t* maps were generated by comparing the mean signals in a 60-s baseline period before the odorant exposure. Odor-elicited activation patterns were imaged in 10 coronal slices with 500 μm thickness, of which three slices (4–6) are shown anterior-to-posterior (left to right). Both odors elicited distributed, yet non-identical responses throughout the bulb. Dorsal (*d*), lateral (*l*), and medial (*m*), and ventral (*v*) foci are identified by arrows. All data shown are from single trial runs. (*See* Color Plate)

any stimulation-induced variations in systemic physiology (data not shown). In agreement with prior fMRI findings using a variety of odorants *(26, 54, 55)*, the strongest BOLD activations were located mainly within the glomerular and olfactory nerve layers. Together, these layers comprise the outermost layers of the olfactory bulb and span about 100 μm in the thinnest region to about 500 μm in the widest region of the olfactory bulb. While both isoamyl acetate and hexanal elicited patterns with some degree of overlap in dorsal, lateral, and medial regions, hexanal generated much stronger ventral activations. In prior fMRI studies with esters and aldehydes *(54, 55)*, we found presence of ventral activations for aldehydes of different carbon lengths. For a given odorant, the activation patterns were not exactly symmetrical across the two bulbs, perhaps due to individual variations of air intake capabilities of each nostril. In all cases examined ($n = 5$), inter-bulbar asymmetries within one subject were much larger than intra-animal experimental variations. We refer to our previous studies *(26, 55)* for extensive discussion of activation patterns in the olfactory bulb.

Due to the strong medial and lateral BOLD activations observed with hexanal, we compared the BOLD signal changes with LFP recordings in the same locations (**Fig. 10.3**). The BOLD activation maps were used to identify the lateral and medial foci (**Fig. 10.3A,B**) for the LFP recordings (**Fig. 10.3C**). The strong BOLD response during hexanal stimulation (6–10%) is in good agreement with prior fMRI results for

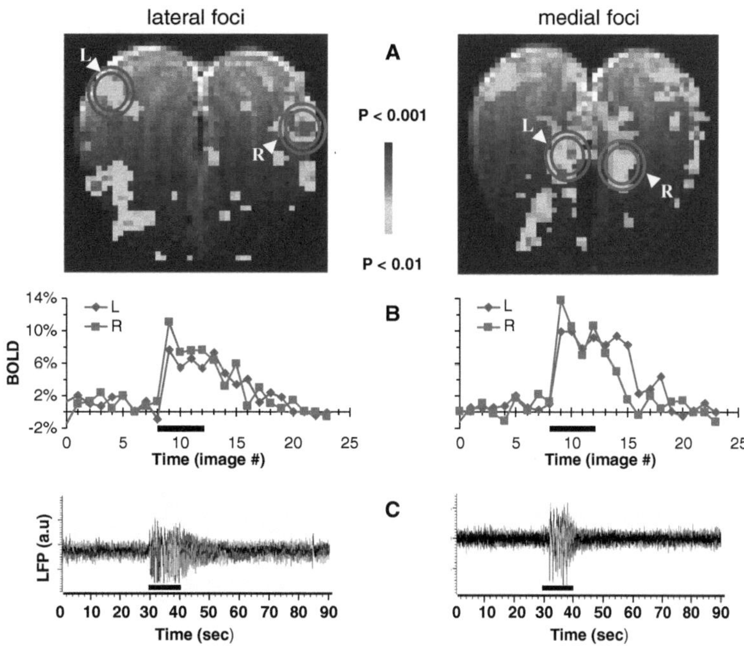

Fig. 10.3. Multi-modal responses from the olfactory bulb in Sprague-Dawley rats. Lateral (*left*) and medial (*right*) responses as measured by BOLD and LFP during hexanal (100%) exposure. (**A**) The BOLD activation maps from slices 8 (**left**) and 13 (**right**) situated from anterior-to-posterior in one subject where the entire olfactory bulb was imaged using 20 coronal slices with 250 μm thickness. The *t* maps were generated by comparing the mean signals in a 60-s baseline epoch before the odorant exposure. The circles shown in the BOLD maps identify the lateral (*left*) and medial (*right*) foci that were probed by electrophysiology (see below). (**B**) The BOLD responses from the circled (L,R) regions in each slice. Comparison of the two BOLD traces in the lateral (left) and medial (right) regions depicts the inter-bulbar differences (L,R). (**C**) The LFP responses measured from another animal but for the same odorant and the lateral/medial areas shown above. The lateral (*left*) and medial (*right*) LFP signals were made from the left and right olfactory bulbs, respectively. The exposure periods are indicated by the black bars in the middle and bottom rows. All data shown are from single trial runs. (*See* Color Plate)

isoamyl acetate *(56)*. The LFP recordings demonstrated complicated dynamic evoked patterns that suggest multiple frequency components, ranging between 5 and 25 Hz, where latencies of both medial and lateral responses were about 1 s. While these fMRI-guided LFP results partially agree with classical LFP recordings from the olfactory bulb *(57)*, further characterization of the LFP response is needed to distinguish between the latencies and the different frequency components *(58)*, for different regions and/or odorants.

3.2. Visual Stimulation

We investigated light-induced activity patterns in the visual areas of the brain (**Fig. 10.4**). There were no stimulation-induced variations in systemic physiology for the intensities (20–40 lux) or colors (white, green) of light examined (data not shown; *n* =

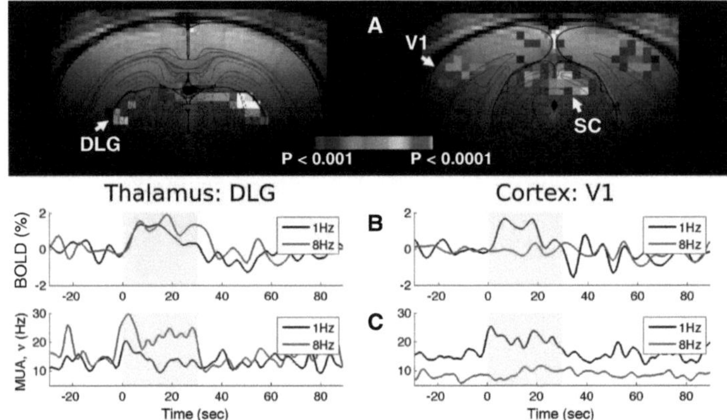

Fig. 10.4. Multi-modal responses from the visual cortex during bilateral light stimulation (white; 50 ms pulses; 30 lux) in Long Evans rats. (**A**) Cortical and subcortical BOLD responses with 1 Hz stimulation applied for 30-s duration. Data from five trials in one animal were averaged and the *t* maps were generated by comparing the mean signals in a 30-s baseline period before the stimulation block. The MRI data were overlaid on the structural outlines from the atlas of Paxinos and Watson *(53)*. Bilateral responses were observed in the dorsal lateral geniculate nuclei of the thalamus (DLG; left), the primary visual cortex (V1; *right*), and the dorsal layers of the superior colliculus (SC; *right*). At higher frequencies, there were weaker BOLD responses in the cortex. Results with green light were qualitatively similar to these results with white light. The time courses of (**B**) BOLD and (**C**) MUA responses to 1 Hz (*black line*) and 8 Hz (*red line*) stimuli in DLG (*left*) and V1 (*right*) from single trial runs. (**B**) The BOLD responses were calculated by averaging several activated voxels within a region-of-interest (in a single trial) from the maps shown above. (**C**) The MUA responses were obtained from another animal but under similar experimental conditions (*white*; 1 Hz, 50 ms pulses; 23 lux). The coordinates of DLG and V1 were guided by prior fMRI results: DLG: [–4.8 mm posterior, –3.8 mm lateral to bregma, –4.0 mm ventral from dura]; V1: [–5.6 posterior, –3.9 lateral to bregma, –0.9 mm ventral from dura at an angle of 10° from the vertical axis]. The 30-s stimulation period is indicated by the shaded region. (*See* Color Plate)

4). Bilateral, white light (50 ms pulses; 1 Hz; 30 lux) stimulation elicited reproducible bilateral BOLD activations (**Fig. 10.4A**) in the dorsal lateral geniculate nuclei of the thalamus (DLG), the primary visual cortex (V1), and the dorsal layers of the superior colliculus (SC). Both the amplitude (~2%) and locations (DLG, V1, SC) of the BOLD response are in good agreement with prior fMRI results from Long-Evans rats *(37)*. At higher stimulation frequencies, localized BOLD time courses from the DLG and V1 regions (**Fig. 10.4B**) revealed a weaker response in V1 and a slightly stronger response in DLG. The MUA signals from these regions were in general agreement with the fMRI findings (**Fig. 10.4C**). Results with green light (data not shown) were qualitatively similar to the results shown for white light.

3.3. Whisker Stimulation

We explored whisker stimulation induced activity patterns in the somatosensory area (**Figs. 10.5 and 10.6**). There were no stimulation-induced variations in systemic physiology for whisker

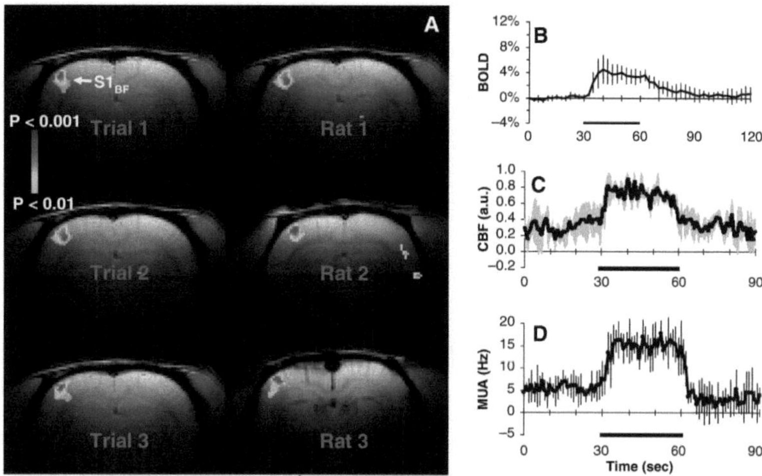

Fig. 10.5. Multi-modal responses from the contralateral whisker barrel field ($S1_{BF}$) during 8 Hz whisker stimulation in Sprague-Dawley rats. (**A**) Reproducibility of $S1_{BF}$ BOLD activation maps in the same subject (see Trial column; *left*) as well as other subjects (see Rat column; *right*). The *t* maps were generated by comparing the mean signals in a 30-s baseline epoch before the stimulation. All data shown are from single trial runs. Averaged time courses from the $S1_{BF}$ are shown for (**B**) BOLD ($n = 6$), (**C**) CBF ($n = 5$), and (**D**) MUA ($n = 5$). The 30-s stimulation period is indicated by the black bar. Vertical bars or gray shading represent standard deviations from the mean. (*See* Color Plate)

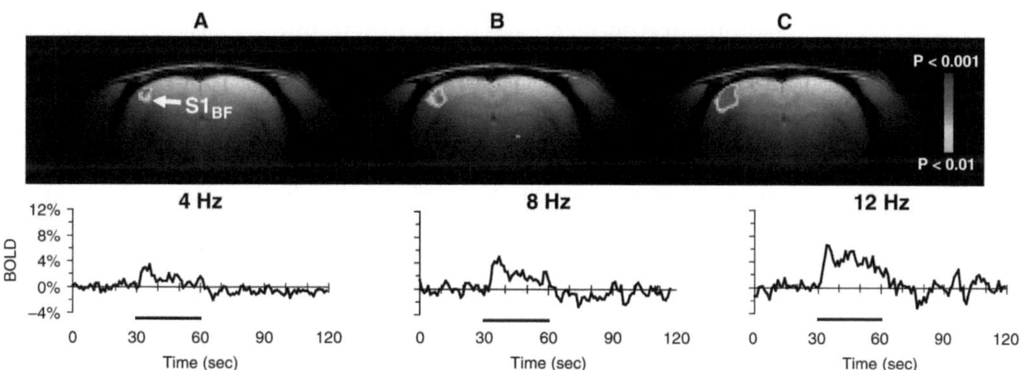

Fig. 10.6. Frequency-dependent BOLD responses from the contralateral whisker barrel field ($S1_{BF}$) in Sprague-Dawley rats. The *t* maps were generated by comparing the mean signals from 30-s epochs of baseline and stimulation. In the same subject, $S1_{BF}$ BOLD responses are shown for (**A**) 4 Hz, (**B**) 8 Hz, and (**C**) 12 Hz whisker stimulations. The top and bottom panels show the activation maps and time courses, respectively. All data shown are from single trial runs. The 30-s stimulation period is indicated by the black bar. (*See* Color Plate)

movement frequencies of less than 40 Hz (data not shown). Rostrocaudal movement of whiskers caused by air puffs over a wide range of frequencies reproducibly stimulated the contralateral, primary somatosensory area of the whisker barrel field ($S1_{BF}$). Repeated trials in the same subject and across subjects produced activations in approximately the same locations in the contralateral $S1_{BF}$ (**Fig. 10.5A**). We observed medial to lateral extent of contralateral $S1_{BF}$ activation, presumably because many whiskers were stimulated simultaneously. Given prior information about contralateral $S1_{BF}$ activation *(59–61)*, we hypothesize that the number of barrels corresponded approximately with the number of whiskers stimulated. The robust and reproducible contralateral $S1_{BF}$ activation was probably improved because whiskers on the ipsilateral side were clipped to remove spurious artifacts.

The averaged BOLD response with 8 Hz whisker stimulation (**Fig. 10.5B**) from the contralateral $S1_{BF}$ (~4%) was in good agreement with prior findings under similar experimental conditions *(25, 62)*. While the BOLD response was sustained for the duration of the stimulus, the signal usually decreased towards the pre-stimulus baseline as soon as the stimulus was terminated. In the same subject, we have used 30 s and 120 s duration stimuli and observed sustained BOLD response (data not shown). Localized $S1_{BF}$ measurements of CBF (**Fig. 10.5C**) and MUA (**Fig. 10.5D**) were well correlated with the fMRI results. Increasing the stimulation frequency augmented the BOLD response in the contralateral $S1_{BF}$ (**Fig. 10.6**). In the same subject, the BOLD response for 12 Hz stimulation was nearly twice the magnitude of the response for 4 Hz stimulation *(63)*.

3.4. Forepaw Stimulation

We evaluated forepaw stimulation induced activity patterns in the somatosensory area (**Figs. 10.7 and 10.8**). There were no stimulation-induced variations in systemic physiology for forepaw stimulation amplitudes of less than 5 mA and frequencies of less than 40 Hz (data not shown). Electrical stimulation (2 mA, 3 Hz) of the forepaw with 0.3 ms duration pulses for 30 s evoked a strong positive BOLD signal change in the contralateral, primary somatosensory area of the forelimb ($S1_{FL}$). No significant BOLD signal changes were observed in the ipsilateral $S1_{FL}$ region during unilateral forepaw stimulation. In a given subject, when either the left or the right forepaw was concurrently stimulated, there were no significant differences in the BOLD response between the two $S1_{FL}$ regions (**Fig. 10.7A**). The intra-subject reproducibility of BOLD activations with either forepaw stimulation (**Fig. 10.7A,B**) was nearly indistinguishable from the inter-subject reproducibility of BOLD activations (**Fig. 10.7C**), as assessed by the number of significantly activated pixels in the $S1_{FL}$ region and/or the averaged BOLD signal change in the $S1_{FL}$ region (data not shown). These results are in good agreement with prior fMRI observations *(16, 24)*.

The averaged BOLD response with 3 Hz forepaw stimulation (**Fig. 10.8A**) from the contralateral $S1_{FL}$ (~8%) was in good agreement with prior findings under similar experimental conditions *(16, 24)*. The BOLD response was sustained for the duration of the 30-s stimulus (**Fig. 10.8A**) and even when the stimulus duration was lengthened to several minutes (**Fig. 10.8B**). Increasing the stimulation frequency above 3 Hz decreased the

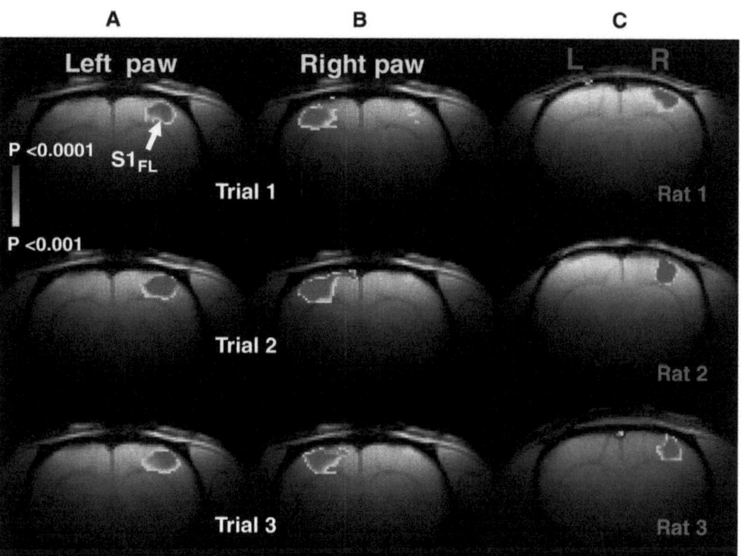

Fig. 10.7. BOLD responses from the contralateral forelimb area ($S1_{FL}$) during 3 Hz forepaw stimulation (0.3 ms pulses; 2 mA) in Sprague-Dawley rats. (**A**) Reproducibility of $S1_{FL}$ BOLD activation maps in the same subject during left paw and right paw stimulations (see Trial column; left and middle) as well as other subjects (see Rat column; *right*). The *t* maps were generated by comparison of the mean signals from 30-s baseline and stimulation periods. All data shown are from single trial runs. (*See* Color Plate)

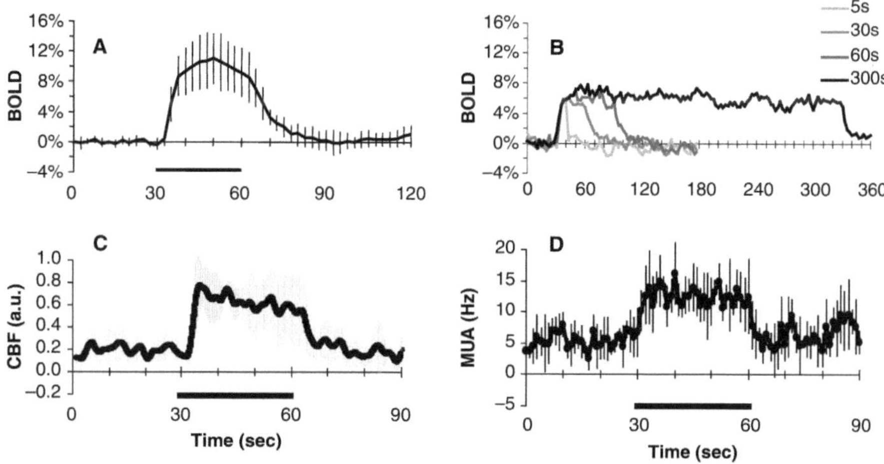

Fig. 10.8. Multi-modal responses from the contralateral forelimb area ($S1_{FL}$) in Sprague-Dawley rats (0.3 ms pulses; 2 mA; 3 Hz). BOLD time courses from the $S1_{FL}$ are shown for (**A**) averaged during 30-s stimulation ($n = 8$) and (**B**) a single subject where the stimulus duration was varied (5–300 s). The averaged time courses from the $S1_{FL}$ are shown for (**C**) CBF ($n = 5$), and (**D**) MUA ($n = 5$). The stimulation period is indicated by the black bar. Vertical bars or gray shading represent standard deviations from the mean.

BOLD response in the contralateral $S1_{FL}$ (data not shown), which is in good agreement with prior observations *(63–65)*. Localized $S1_{FL}$ measurements of CBF (**Fig. 10.8C**) and MUA (**Fig. 10.8D**) were well correlated with the fMRI results.

4. Discussion

4.1. Summary of fMRI and Neurophysiologic Results

We demonstrated the ability to deliver four independent sensory stimuli to anesthetized rats lying inside an MRI scanner and replicated identical conditions during neurophysiologic measurements outside the scanner. Noticeably, however, evoked BOLD responses with olfactory, visual, and whisker stimuli exhibited slight variability during the stimulation period (e.g., the BOLD response in the $S1_{BF}$ declines), whereas forepaw stimuli demonstrated relatively stable BOLD signal change. This is most likely a reflection of the nature of forepaw stimuli in which strong and highly synchronous barrages of identical efferent inputs invade the contralateral $S1_{FL}$, causing massive activity of cortical neurons, and which, in turn, may partially explain the more stable BOLD response during forepaw stimulation. It is unlikely that the forepaw stimulus is out of the physiological range, because the systemic physiology was unaffected by the stimulus as observed similarly with the other stimuli. Since we could tightly control the systemic physiology (i.e., pH $\approx$ 7.35; $pCO_2 \approx 37$ mmHg; $pO_2 > 100$ mmHg; blood pressure $\approx$ 95–110 mmHg; core temperature $\approx$ 37°C) for α-chloralose anesthetized rats which were given visual, whisker, and forepaw stimuli, we did not exclude a high number of animals from our studies. However, for the urethane-anesthetized rats receiving olfactory stimuli (i.e., no tracheotomy), temperature was the only parameter that could be efficiently controlled, and which unfortunately did not prove to be sufficient as we had to exclude nearly half of the experiments due to poor maintenance of systemic physiology. These issues are discussed further below (*see* **Section 4.2**).

The variable BOLD responses with different sensory inputs – which ranged from diffuse odorant maps (in the olfactory bulb) to highly localized somatosensory and visual foci (in the brain) – presumably reflected neurophysiologic variations across the sensory modalities as well as their neuroanatomic underpinnings. The localized neurophysiologic responses, measured here with the electrical (i.e., LFP or MUA) and/or the coupled CBF signal and which were in general agreement with the fMRI results, can be used to provide complementary insights to the multi-modal basis of the BOLD signal change *(12–14)*. Generally, the MUA responses from the brain (i.e., DLG, V1, $S1_{BF}$, and $S1_{FL}$) were quite comparable, with 5–7 Hz incremental change in ensemble firing rate for the respective stimuli. However, the CBF changes in the $S1_{BF}$ was nearly half that observed in the $S1_{FL}$. But the time lags to reach the CBF peak were nearly identical with whisker and forepaw stimuli, whereas the CBF responses reached peak 1–2 s earlier than the BOLD responses.

4.1.1. Non-tactile: Olfactory Stimulation

With the sole exception of the olfactory system, all peripheral sensory neurons synapse within domain-specific regions of the thalamus, before reaching the primary sensory cortical regions. When air containing odorant molecules is inhaled into the nasal cavities, a rich collection of receptor neurons in the olfactory epithelium (and which later comprise the outermost olfactory nerve layer in the olfactory bulb) converts the information into electrical form. As a molecule binds to a matching receptor, an impulse is sent to the olfactory bulb, which, in the rodent, is located just anterior to the frontal lobe and is separated into two "identical" hemispheres, each with a multi-layered organization *(66)*. Signals from receptor neurons arrive at the glomerular layer, where they are integrated by a large number of mitral/tufted cells that project their axons out of the olfactory bulb onto the olfactory cortex, specifically piriform and entorhinal cortices, amygdala, and the olfactory nucleus.

The olfactory bulb, more specifically the glomerular layer, has been shown "chemotopic maps" with exposure to specific odorants (i.e. responses to different odorants are processed in anatomically distinct locations in the glomerular space) using a variety of techniques *(67–71)* including fMRI *(26,28,54–56,72,73)*. Therefore, functional mapping at the level of individual glomeruli is important and this spatial resolution is quite commonly attainable with optical imaging *(58,68)*. However, this technical feat has also been achieved in the past with fMRI using voxel sizes less than 10 nL *(74, 75)*. For the current studies, however, we acquired data with high BOLD sensitivity in slightly larger voxels to cover the entire bulb at a temporal resolution superior to most other whole bulb mapping techniques. Furthermore, the minimally invasive approach of fMRI allowed repeated perturbations on the same subject to provide comparison across different experimental runs (from the same subject) without any averaging.

With our stimulation device, we were able to test response variability to different odorants (**Figs. 10.2 and 10.3**) as well as their concentrations (data not shown). We observed diffuse BOLD responses in the bulb, where a large proportion of the bulb was active for any given odorant and the activity patterns were specific to the odorant type (**Fig. 10.2**). While we correlated LFP signals to a few medial and lateral BOLD foci (**Fig. 10.3**), understanding the complex frequency components within the LFP responses and their relationship to the BOLD response remain a future area of research. Given that fMRI can inherently detect both hemispheres simultaneously, the hypothesized "identical" function of each bulb could also be tested in the future. Since we could not vary the incident direction of the odorants with the current delivery device, plans are underway to include spatial specific delivery tubes for directional delivery inside the magnet bore. However, all of these endeavors will require a next generation of fMRI hardware improvement (e.g., with parallel imaging *(76, 77)* using multiple smaller radio frequency coils) because of the smaller BOLD responses expected.

4.1.2. Non-tactile: Visual Stimulation

Though a predominantly nocturnal animal, the rodent makes use of its visual system when lighting conditions are favorable – e.g., for vision-based navigation *(78)*. Light enters the retina, which contains elongated photoreceptor cells that are full of light-sensitive photopigment molecules (or opsins). Upon interacting with incident photons, photoisomerization of rod and/or cone opsins initiates a cascade of processes that rapidly lead to a decrease in the intracellular concentration of cyclic GMP molecules, thereby changing the cell membrane's permeability to cations and thus altering its potential *(79)*. This graded signal is refined further by retinal networks before being transmitted, via the dorsal lateral geniculate nucleus of the thalamus, onto the primary visual cortex in the contralateral occipital lobe. A parallel pathway, bypassing the thalamus, transmits visual information to the superficial layers of the contralateral superior colliculus. The rat retina has two distinct types of cones – most sensitive to medium wavelengths (~510 nm; i.e., green) but some receptive to very short wavelengths (~360 nm; i.e., ultraviolet) – allowing discrimination between different colors *(80)*.

Our visual stimulation setup allowed us to vary the color and/or intensity of light pulses of different durations. At a constant pulse width of 50 ms and intensity of 20–30 lux, white (**Fig. 10.4**) and green (data not shown) light evoked well-localized cortical and sub-cortical responses which corresponded well with known retinotopic maps of both the V1 *(81, 82)* and the SC *(83)*. At higher stimulation frequencies, responses weakened in V1 and strengthened in DLG. These results confirmed a previous fMRI report that examined the effect of stimulus frequency on the cortical BOLD response *(37)* where there is an apparent inverse relationship between increasing flash frequency and BOLD response amplitude. A limitation of our method, however, is the inability to produce moving stimuli, such as drifting gratings typically employed in studies of the visual system to map response properties of single cells *(84)*. Since the optimal tuning properties of cells in the rat visual cortices appear to be randomly arranged *(85)*, the probability of recording signals from an optimally excited neuron within an ensemble is reduced. Regardless, we were able to record ensemble multi-unit discharges in the V1 and DLG and demonstrated their qualitative concurrence with the fMRI data. To relate these rat studies to results from other species where vision plays a more important role in their livelihood *(15, 18–20)*, studies are underway to examine the degree of temporal correlation between LFP, MUA, and BOLD signals in V1, SC, and DLG.

4.1.3. Tactile: Whisker Stimulation

The rat relies heavily not only on its exquisite sense of smell, but also on its vibrissa (i.e., whisker system). On either side of its face, the rat has approximately 35 whiskers organized in a grid-like pattern of rows and columns. The roots of these are long hairs wrapped in mechanoreceptors sensitive to minute vibrations. The signals generated by motion of the whiskers are transmitted via the trigeminal nerve to the ventral posterior lateral nucleus of the thalamus to whisker-specific regions in the primary somatosensory cortex. The conspicuous anatomical detail of densely packed clusters of cells, forming a topographic map of the contralateral whisker pad *(86)* led to the concept of "barrel" cortex *(87)*. During exploration of their environment, rats actively make contact with objects with their whiskers by voluntary head motions and facial muscle activity. The combined multi-parametric effects of the amplitude, frequency, and motion of whisker deflections upon impact provide the sensation of the object.

The whisker stimulation is a difficult model to apply reproducibly inside the magnet primarily because of space constraints. Nevertheless, our prior approach for whisker stimulation inside the magnet *(25, 62)* using Lorentz force to move a small conducting copper wire attached to the whiskers is unfortunately not easily reproduced identically outside the magnet. A good alternative is mechanical movement of whiskers *(88)*, but this works poorly inside a magnet. Since this method requires a non-magnetic manipulator which must span the entire length of the bore, dexterity becomes a problem and, therefore, there is limited control of the amplitude and/or frequency of the whisker stimulation. Thus, we developed a non-magnetic whisker stimulation device using air puffs *(63)* which can be easily used – either simultaneously or concurrently – with almost any other sensory stimuli (e.g., forepaw, olfactory, visual), both inside and outside the magnet, designed specifically for multi-sensory experimental paradigms. The frequency of whisker stimulation with this device could range from a few Hz to as much as 40 Hz, which seems to be the upper limit of naturalistic whisker stimulation *(89)*.

In the same rat or across a group of rats, whisker stimulation induced multi-modal responses (of BOLD, CBF, and MUA) in the contralateral $S1_{BF}$ (**Fig. 10.5**). The induced response seemed to peak at stimulation frequencies between 8 and 12 Hz (**Fig. 10.6**) which is within the natural whisking frequency in rodents *(89)*. Earlier studies *(90)* have shown that CBF in the rat $S1_{BF}$ increases linearly with higher whisker movement frequencies. However, future studies should investigate the neurophysiologic basis of the increased area of BOLD activation observed at the higher stimulation frequencies (**Fig. 10.6**). In summary, unilateral (and bilateral) whisker stimulation with air puffs resulted in fMRI and neurophysiologic signal increases in the contralateral $S1_{BF}$ for consistent and reproducible studies, both in the same subject and across a group of subjects. While this device provides easy control of whisking frequency, we are investigating more refined amplitude variations by controlling flow rates of compressed air.

4.1.4. Tactile: Forepaw Stimulation

While ambulatory functions fulfilled by the rodent's forelimbs and hind limbs may seem mundane compared to the perceptions of smell and whisking, somatosensory sensing by the limbs represents an important component of the multi-faceted environmental cues that a rodent must experience for survival. Upon stimulation of either limb, afferent signals arrive through the medial-lemniscal system at the contralateral ventral posterior lateral nucleus of the thalamus, prior to connecting with the limb-specific contralateral primary somatosensory cortex. Electrical stimulation of nerve fibers innervating the touch-sensitive limbs is the most widely used method of sensory stimulation of the rat, both in bench neurophysiologic *(91–94)* and fMRI *(95–101)* studies.

Forepaw stimulation in the rat is a very popular model in many fMRI laboratories *(95–98)* and can be easily applied identically inside and outside the magnet *(98–101)*. The pulse amplitude is typically in the range of a few mA and the pulse duration is usually a fraction of a ms. The electrodes are typically small enough to be placed subcutaneously in the forepaw's skin. It is easily applied and produces very well localized and strong electrical and hemodynamic cortical responses. During electrical forepaw stimulation, robust activations in the contralateral $S1_{FL}$ were reproducibly observed in different sessions as well as across many subjects under α-chloralose anesthesia (**Figs. 10.7 and 10.8**). The peak $S1_{FL}$ response was near 1–3 Hz forepaw stimulation frequency *(63–65)* which is quite different from the $S1_{BF}$ response during whisker stimulation. This model has been widely used for biophysical/physiologic characterization of the BOLD image contrast *(102–108)*, perhaps in part, because the averaged BOLD response localized in the contralateral $S1_{FL}$ is much larger than observed with other sensory modalities.

4.2. Challenges of In Vivo NMR and Bench Experiments

We have described some experiences with anesthetized rat preparations in our laboratory. But the awake experiment still has desirable features. However, the few fMRI studies that have reported on awake but restrained rats *(109–111)* do not demonstrate the same level of reproducibility in BOLD activations as anesthetized rats, perhaps in part, because the spontaneous BOLD oscillations can be as large as the induced BOLD responses *(111)* and furthermore, even the BOLD responses can be quite variable *(110)*. We, and many others, hypothesize that "the anesthetic state may be thought of as a tool for limiting the repertoire of physiological responses of cortical neurons so that the most anesthesia-resistant properties can be readily identified" *(112)* and studied under controlled settings. "Such maps should not lead to false conclusion, however, that they define only the functions carried out in a given cortical region, or that the sensory properties are reflected in exactly the same physiological manner as in the awake [state]" *(112)*. Successful in vivo fMRI and neurophysiologic studies depend on maintenance of normal physiologic state and sustaining the appropriate depth of anesthesia. Whether the subject is awake or anesthetized, optimal shimming conditions for fMRI scans are always desirable. Below, we discuss some insights into potential pitfalls and, where possible, how to avoid them.

4.2.1. Importance of Physiological State

The well being of experimental animals, whether under anesthesia or awake, should be monitored and, whenever possible, controlled. Usually, the overall physiologic state of the subject should be judged at three levels: Central nervous system (e.g., by EEG, MUA), cardiovascular system (e.g., by blood pressure, heart rate), and respiratory system (e.g., by pCO_2, respiratory rate). In our studies, all three components are controlled and measured to assess the anesthetized state. To provide external ventilation for the respiratory control, we used neuromuscular blocking agents which do not act on the central nervous system. Furthermore, we monitor analgesic level frequently. Blood pressure provides a continuous indicator of anesthesia/analgesia depth as mildly noxious stimuli should not cause increases in the instantaneous pressure of more than ~10% *(113)*. Additional doses of anesthetic agent(s) must be applied if there are indications of discomfort.

Arterial blood samples drawn periodically to monitor pCO_2 and pH provide very important measures of systemic physiology. In the rat, the pCO_2 is in the range 35–40 mmHg under optimal steady-state conditions. High pCO_2 leads to a reduced pH and dilatation of vessel walls in an attempt to increase flow and thereby remove more of the potentially toxic acidic substances from tissue. This situation will also affect the baseline CBF level and therefore affect the BOLD response and its temporal dynamics. Low pCO_2 has the opposite effect. Both situations interfere with the tissue's ability to adjust flow in response to activity changes, thereby constituting a serious *caveat* to the interpretation of hyperemic responses.

In freely breathing animals, blood gas tension is a crucial factor in determining the physiological state of the anesthetized state. Excluding olfactory studies, we mechanically ventilated through a tracheotomy and controlled both the rate and volume of breathing. Although buffering mechanisms in the blood exist to keep the pH within a narrow optimal range (~7.4), they are not able to do so indefinitely in a situation where breathing parameters are not suitably adjusted and pCO_2 levels span out of optimal range. The core temperature of the animal also influences the acid-base balance of the blood, and must be maintained stable (~37°C) despite the cool temperatures in the imaging bore (~18°C). In the case of a drop in pH of metabolic origin, such as acidosis caused by the anesthetic urethane *(114)*, periodic intravenous injections of sodium bicarbonate can be used to stabilize the animal and obtain reliable hyperemic responses.

4.2.2. Effect of Anesthesia

Anesthetics have been valuable in neuroscience to understand the effects of afferent inputs on localized regions of the brain *(115–117)* because most peripheral stimuli (i.e., auditory, visual, olfactory, limbs, whiskers) still reach the brain with anesthesia. Most anesthetics act on different levels of neuronal signaling by depressed axonal conductance of action potentials, modified dendritic and somatic integration, reduced presynaptic neurotransmitter release, and/or altered postsynaptic receptors. The $GABA_A$ receptor is commonly considered to be an important molecular target since it seems to be ubiquitously affected by most anesthetics. A notable exception is urethane, the effect of which appears to be mediated by the enhanced conductance of a specific potassium-channel subtype *(118)*. Anesthetics, inhaled or injected, reduce brain energy consumption rather uniformly across regions and cortical energy decline is typically monotonic with anesthetic depth *(119)*. Consistent with this observation, anesthetics also depress firing rates and impede slow field oscillations of excitatory pyramidal cortical neurons *(120)*.

Use of anesthesia during rodent fMRI and neurophysiology is needed to ensure immobilization and comfort during long experiments under physically constrained circumstances. Many reports on the "confounding effects" of anesthesia exist in the literature *(121, 122)*, pointing to the inherent limitation of studying neural systems under abnormal (i.e., non-awake) conditions *(123)*. Our belief is that although the full complexity of awake brain responses to sensory stimuli clearly cannot be reached under anesthesia, important insights into general functional properties of a given sensory system can be made *(29)*. Combined with the detailed knowledge of morphological, functional, vascular, and neuronal properties of rat brain tissue, results obtained in the anesthetized preparation can be extrapolated and generalized towards the awake brain. The common response features obtained under multiple different anesthetics to the same stimulation, in our laboratory and others, offers indirect support to the generality of the results.

4.2.3. Magnetic Susceptibility Effects and "Shimming"

Despite the high NMR sensitivity at 11.7 T, there are some challenges. Magnetic susceptibility is a property of matter indicating its propensity to become (para)magnetic upon introduction into a magnetic field. It is analogous to polarization of insulators in an electric field. The BOLD effect is a consequence of a dynamic process of "microscopic" magnetic susceptibility changes occurring as the concentration of oxygen-rich blood varies during a functional challenge *(5, 6)*. However, interfaces between substances with different magnetic susceptibilities lead to "macroscopic" signal loss near the boundary between media. In the rodent, regions with such interfaces leading to signal degeneration include the inferior temporal lobes adjacent to air-filled ear canals, the caudal occipital lobes in which the tissues separating the cortex from the cerebellum appear in the imaging plane, and the olfactory bulb embedded in bony and cartilaginous tissue surrounded by mucous membranes.

"Shimming" is a process of applying small corrective magnetic field gradients onto the main static magnetic field in order to optimize field homogeneity in and around the sample. It is a procedure to reduce structurally imposed signal loss, mainly at the media interfaces, by correcting for gross susceptibility-induced magnetic field inhomogeneities. Since the correction of "macroscopic" magnetic field inhomogeneities becomes increasingly important at higher fields, some automatic routines have been developed *(45, 124–126)*.

For high sensitivity fMRI studies, it is crucial that the surface coil be as close to the brain as possible. Therefore, it seems obvious that the removal of the scalp (skin, galea aponeurotica) and periosteum should improve the signal intensity from the brain since the whole brain moves closer to the coil. However, this approach will also move the air-tissue (bone) interface closer to the tissue of interest (brain), which in turn may lead to field inhomogeneities beyond the capabilities of the shim gradients. We avoid the scalp removal procedure whenever possible, focusing instead on careful positioning of the animal and shimming.

There is a special shimming problem in the case of olfactory bulb imaging. Since the nasal cavity is very close to the olfactory bulb, changes in the level of mucous secretion will affect the local susceptibility, and thereby the shimming. A stable physiological condition of the nasal epithelium during odor delivery experiments is required not only to prevent these adverse magnetic inhomogeneity effects on the measured signal, but also to ensure the odorant molecules' access to the epithelial odorant receptors. A common choice of anesthetic in rodent fMRI studies, α-chloralose, causes thick, mucoid oral-nasal discharge *(127)*. For odor stimulations, we used urethane – a long lasting anesthetic with excellent analgesic properties – the general use of which though must be discouraged for survival studies because of its carcinogenic and mutagenic effects. In addition to the choice of anesthetic, we carefully controlled the temperature of the incoming air to the animal by pre-heating it to 28–30°C, close to the core temperature maintained at ~37°C, to prevent mucous accumulation due to the low ambient temperature (~18°C) within the imaging bore.

4.3. Future Directions

In recent years, careful refinement of the BOLD effect to reveal its physiologic basis *(15–23, 29)* for future diagnostic use has been quite encouraging *(12–14)*. However, it should be acknowledged that these studies have primarily focused on neuronal functions where the role of their glial counterpart has been largely neglected *(128, 129)*. Therefore, another potential direction that animal models, such as the ones described here, may help with is the information about the glial activities that accompany neuronal and/or BOLD signal changes. But this endeavor would require development of glial-specific NMR and non-NMR methods to complement the plethora of neuron-specific measurements that currently exist and are routinely used in many laboratories.

Localized NMR data collection, for MRI as well as for MRS, is based on imposing linear field gradients which are rapidly alternated at high rates. The gradients are generated by currents passing through coils of various shape and orientation around the sample. The mechanical Lorentz force acting upon the gradient coils, particularly during EPI scans for fMRI, generates waves which are audible as loud "clicks" or "blips" in and around the scanner. Since this is a continuous noise source during our studies, its effect can be partly negated. We are currently investigating possibilities of applying noise-cancellation techniques for future studies. However, studying the auditory system is a considerable challenge with fMRI because of susceptibility-induced signal losses near the ear canals, in particular in the temporal auditory cortices. Nevertheless, some methods are being developed *(126)* which may alleviate the shimming troubles in such brain locations.

The developments presented here, and more plans underway, are readily applicable to studying effects of different tactile (i.e., forepaw, whisker) and non-tactile (i.e., olfactory, visual) modalities simultaneously. Meredith and Stein *(130, 131)* defined two classes of multi-sensory interactions within the receptive fields of cortical neurons: Response enhancement and response depression. Though it appears to be a general mechanism in the central nervous system, one region in which such multi-sensory convergence and synthesis takes place *en masse* is the SC, situated in the dorsal mid-brain, caudal to the thalamus. In the cat SC, Meredith and Stein *(130)* showed that a conservative estimate to the lower limit of the proportion of cells exhibiting multi-sensory interactions in the intermediate and deep laminae of the SC is 50%, the majority of which project efferent fibers to sensorimotor areas of the brain stem and spinal cord. These findings are consistent with the hypothesized role of the SC in the control of eye, head, and body movements underlying rapid orientation towards external objects. The results from our visual stimulation paradigm illustrate reproducible BOLD responses in the superficial laminae of the SC. Therefore, we are now in a position to begin fMRI and neurophysiologic studies of sensory integration with tactile and/or non-tactile sensory stimuli.

Acknowledgments

We thank Dr. Fuqiang Xu (Wuhan Institute of Physics and Mathematics, Wuhan, China) for help during preliminary stages of olfactory bulb imaging. We appreciate valuable inputs from Dr. Lawrence B. Cohen (Yale University) and his laboratory members (i.e., Drs. Ryota Homma and Efstratios Kosmidis) regarding the design of the olfactometer. We are grateful to Alyssa Siefert (a summer internship student from Carnegie Mellon University) and Scott McIntyre (an MRRC engineer) for the initial design of the visual stimulation device. We are also extremely appreciative of other MRRC/QNMR engineers and staff (i.e., Terry Nixon, Peter Brown, Mark Abildgaard, Bei Wang) for assistance in other parts of our experimental setup. This work was supported by personal grants to CJB (Dagmar Marshall's Fond, Frimodts Fond and Oticon Fonden), the Hungarian Research Foundation grant to PH (OTKA-T34122), and National Institutes of Health grants to FH (R01 MH-067528, R01 DC-003710, P30 NS-52519).

References

1. Raichle ME (1988) "Circulatory and metabolic correlates of brain function in normal humans" in Handbook of physiology – The nervous system V (Springer-Verlag, New York, USA) pp. 633–674
2. Raichle ME (1998) Behind the scenes of functional brain imaging: A historical and physiological perspective. Proc Natl Acad Sci USA. 95:765–772
3. Shulman RG, Blamire AM, Rothman DL, McCarthy G (1993) Nuclear magnetic resonance imaging and spectroscopy of human brain function. Proc Natl Acad Sci USA. 90:3127–3133
4. Shulman RG, Rothman DL, Behar KL, Hyder F (2004) Energetic basis of brain activity: Implications for neuroimaging. Trends Neurosci. 27:489–495
5. Ogawa S, Lee TM, Nayak AS, Glynn P (1990) Oxygenation-sensitive contrast in magnetic resonance image on rodent brain at high magnetic fields. Magn Reson Med. 14:68–78
6. Ogawa S, Menon RS, Tank DW, Kim SG, Merkle H, Ellermann JM, Ugurbil K (1993) Functional brain mapping by blood oxygenation level-dependent contrast magnetic resonance imaging. Biophys J. 64:803–812
7. Bandettini PA, Wong EC, Hinks RS, Tikofsky RS, Hyde JS (1992) Time course EPI of human brain function during task activation. Magn Reson Med. 25:390–397
8. Blamire AM, Ogawa S, Ugurbil K, Rothman D, Mccarthy G, Ellermann JM, Hyder F, Rattner Z, Shulman RG (1992) Dynamic mapping of the human visual cortex by high-speed magnetic resonance imaging. Proc Natl Acad Sci USA. 89:11069–11073
9. Kwong KW, Belliveau JW, Chesler DA, Goldberg IE, Weiskoff RM, Poncelet BP, Kennedy DN, Hoppel BE, Cohen MS, Turner R, Cheng HM, Brady TJ, Rosen BR (1992) Dynamic magnetic resonance imaging of human brain activity during primary sensory stimulation. Proc Natl Acad Sci USA. 89:5675–5679

10. Ogawa S, Tank DW, Menon R, Ellermann JM, Kim SG, Merkle H, Ugurbil K (1992) Intrinsic signal changes accompanying sensory stimulation: Functional brain mapping with magnetic resonance imaging. Proc Natl Acad Sci USA. 89:5951–5955
11. Posner MI, Raichle ME (1998) The neuroimaging of human brain function. Proc Natl Acad Sci USA. 95:763–764
12. Logothetis NK (2002) The neural basis of the blood-oxygen-level-dependent functional magnetic resonance imaging signal. Philos Trans R Soc Lond B Biol Sci. 357:1003–1037
13. Lauritzen M (2001) Relationship of spikes, synaptic activity, and local changes of cerebral blood flow. J Cereb Blood Flow Metab. 21:1367–1383
14. Hyder F (2004) Neuroimaging with calibrated fMRI. Stroke. 35 Suppl 1:2635–2641
15. Logothetis NK, Pauls J, Augath M, Trinath T, Oeltermann A (2001) Neurophysiological investigation of the basis of the fMRI signal. Nature. 412:150–157
16. Smith AJ, Blumenfeld H, Behar KL, Rothman DL, Shulman RG, Hyder F (2002) Cerebral energetics and spiking frequency: The neurophysiological basis of fMRI. Proc Natl Acad Sci USA. 99:10765–10770
17. Kim DS, Ronen I, Olman C, Kim SG, Ugurbil K, Toth LJ (2004) Spatial relationship between neuronal activity and BOLD functional MRI. NeuroImage. 21:876–885
18. Arthurs OJ, Williams EJ, Carpenter TA, Pickard JD, Boniface SJ (2000) Linear coupling between functional magnetic resonance imaging and evoked potential amplitude in human somatosensory cortex. Neuroscience. 101:803–806
19. Rees G, Friston K, Koch C (2000) A direct quantitative relationship between the functional properties of human and macaque V5. Nat Neurosci. 3:716–723
20. Niessing J, Ebisch B, Schmidt KE, Niessing M, Singer W, Galuske RA (2005) Hemodynamic signals correlate tightly with synchronized gamma oscillations. Science. 309:948–951
21. Kida I, Smith AJ, Blumenfeld H, Behar KL, Hyder F (2006) Lamotrigine suppresses neurophysiological responses to somatosensory stimulation in the rodent. NeuroImage. 29:216–224
22. Gsell W, Burke M, Wiedermann D, Bonvento G, Silva AC, Dauphin F, Buhrle C, Hoehn M, Schwindt W (2006) Differential effects of NMDA and AMPA glutamate receptors on functional magnetic resonance imaging signals and evoked neuronal activity during forepaw stimulation of the rat. J Neurosci. 26:8409–8416
23. Burke M, Buhrle C (2006) BOLD response during uncoupling of neuronal activity and CBF. NeuroImage. 32:1–8
24. Hyder F, Behar KL, Martin MA, Blamire AM, Shulman RG (1994) Dynamic magnetic resonance imaging of the rat brain during forepaw stimulation. J Cereb Blood Flow Metab. 14:649–655
25. Yang X, Hyder F, Shulman RG (1996) Single-whisker activation observed in rat cortex by functional magnetic resonance imaging. Proc Natl Acad Sci USA. 93:475–478
26. Xu F, Kida I, Hyder F, Shulman RG (2000) Assessment and discrimination of odor stimuli in rat olfactory bulb by dynamic fMRI. Proc Natl Acad Sci USA. 97:10601–10606
27. Ugurbil K, Adriany G, Andersen P, Chen W, Garwood M, Gruetter R, Henry PG, Kim SG, Lieu H, Tkac I, Vaughan T, Van De Moortele PF, Yacoub E, Zhu XH (2003) Ultrahigh field magnetic resonance imaging and spectroscopy. Magn Reson Imaging. 21:1263–1281
28. Schafer JR, Kida I, Rothman DL, Xu F, Hyder F (2006) Reproducibility of odor maps by fMRI in rodents. NeuroImage. 31:1238–1246
29. Maandag NJG, Coman D, Sanganahalli BG, Herman P, Blumenfeld H, Smith AJ, Shulman RG, Hyder F (2007) Energetics of neuronal signaling and fMRI activity. Proc Natl Acad Sci USA. 104:20546–20551
30. Kida I, Maciejewski PK, Hyder F (2004) Dynamic imaging of perfusion and oxygenation by fMRI. J Cereb Blood Flow Metab. 24:1369–1281
31. Kida I, Rothman DL, Hyder F (2007) Dynamics of changes in blood flow, volume, and oxygenation: Implications for dynamic fMRI calibration. J Cereb Blood Flow Metab. 27:690–696
32. Shepherd GM (2004) The synaptic organization of the brain (Oxford University Press, New York)
33. Wallace MT, Meredith MA, Stein BE (1993) Converging influences from visual, auditory, and somatosensory cortices onto output neurons of the superior colliculus. J Neurophysiol. 69:1797–1809
34. Wallace MT, Meredith MA, Stein BE (1998) Multisensory integration in the superior colliculus of the alert cat. J Neurophysiol. 80:1006–1010
35. Kauer JS (1974) Response patterns of amphibian olfactory bulb neurones to odour stimulation. J Physiol. 243:695–715

36. Zochowski M, Cohen LB, Fuhrmann G, Kleinfeld D (2000) Distributed and partially separate pools of neurons are correlated with two different components of the gill-withdrawal reflex in Aplysia. J Neurosci. 20:8485–8492
37. Van Camp N, Verhoye M, De Zeeuw CI, Van der Linden A (2006) Light stimulus frequency dependence of activity in the rat visual system as studied with high-resolution BOLD fMRI. J Neurophysiol. 95:3164–3170
38. Schulte ML, Pawela CP, Cho YR, Li R, Hudetz AG, Hyde JS (2007) Detecting responses to single light flashes in the rodent brain using laser Doppler and fMRI at 9.4T. Proc Int Soc Magn Reson Med. 1:619
39. Rooney BJ, Cooper RM (1988) Effects of square-wave gratings and diffuse light on metabolic activity in the rat visual system. Brain Res. 439:311–321
40. Prusky GT, West PW, Douglas RM (2002) Reduced visual acuity impairs place but not cued learning in the Morris water task. Behav Brain Res. 116:135–140
41. Lund RD, Lund JS, Wise RP (1974) The organization of the retinal projection to the dorsal lateral geniculate nucleus in pigmented and albino rats. J Comp Neurol. 158:383–403
42. Drager UC, Olsen JF (1980) Origins of crossed and uncrossed retinal projections in pigmented and albino mice. J Comp Neurol. 191:383–412
43. Abel PL, Olavarria JF (1996) The callosal pattern in striate cortex is more patchy in monocularly enucleated albino than pigmented rats. Neurosci Lett. 204:169–172
44. Nersesyan H, Hyder F, Rothman DL, Blumenfeld H (2004) Dynamic fMRI and EEG recordings during spike-wave seizures and generalized tonic-clonic seizures in WAG/Rij rats. J Cereb Blood Flow Metab. 24:589–599
45. Gruetter R (1993) Automatic, localized in vivo adjustment of all first- and second-order shim coils. Magn Reson Med. 29:804–811
46. Hennig J, Nauerth A, Friedburg H (1986) RARE imaging: a fast imaging method for clinical MR. Magn Reson Med. 3:823–833
47. Frahm J, Haase A, Matthaei D (1986) Rapid three-dimensional MR imaging using the FLASH technique. J Comput Assist Tomogr. 10:363–368
48. Mansfield P (1977) Multi-planar Image formation using NMR spin echoes. J Phys C. 10:L55–L58
49. Hyder F, Rothman DL, Blamire AM (1995) Image reconstruction of sequentially sampled echo-planar data. Magn Reson Imaging. 13:97–103
50. Nersesyan H, Herman P, Erdogan E, Hyder F, Blumenfeld H (2004) Relative changes in cerebral blood flow and neuronal activity in local microdomains during generalized seizures. J Cereb Blood Flow Metab. 24:1057–1068
51. Trubel HKF, Sacolick LI, Hyder F (2006) Regional temperature changes in the brain during somatosensory stimulation. J Cereb Blood Flow Metab. 26:68–78
52. Schridde U, Khubchandani M, Motelow JE, Sanganahalli BG, Hyder F, Blumenfeld H (2008) Negative BOLD with large increases in neuronal activity. Cereb Cortex. 18:1814–1827
53. Paxinos G, Watson C (1997) The Rat Brain in Stereotaxic Coordinates (Academic Press, New York, NY)
54. Yang X, Renken R, Hyder F, Siddeek M, Greer CA, Shepherd GM, Shulman RG (1998) Dynamic mapping at the laminar level of odor-elicited responses in rat olfactory bulb by functional MRI. Proc Natl Acad Sci USA. 95:7715–7720
55. Xu F, Liu N, Kida I, Rothman DL, Hyder F, Shepherd GM (2003) Odor maps of aldehydes and esters revealed by fMRI in the glomerular layer of the mouse olfactory bulb. Proc Natl Acad Sci USA. 100: 11029–11034
56. Schafer JR, Kida I, Rothman DL, Hyder F, Xu F (2005) Adaptation in the rodent olfactory bulb measured with fMRI. Magn Reson Med. 54:443–448
57. Beuerman RW (1975) Slow potentials of the turtle olfactory bulb in response to odor stimulation of the nose. Brain Res. 97:61–78
58. Lam YW, Cohen LB, Wachowiak M, Zochowski MR (2000) Odors elicit three different oscillations in the turtle olfactory bulb. J Neurosci. 20:749–762
59. Woolsey TA (1978) Some anatomical bases of cortical somatotopic organization. Brain Behav Evol. 15:325–371
60. Simons DJ (1983) Multi-whisker stimulation and its effects on vibrissa units in rat SmI barrel cortex. Brain Res. 276:178–182
61. Welker E (2000) Developmental plasticity: to preserve the individual or to create a new species? Novartis Found Symp. 228: 227–235
62. Yang X, Hyder F, Shulman RG (1997) Functional MRI BOLD signal coincides with electrical activity in rat whisker barrel. Magn Reson Med. 38:874–877
63. Sanganahalli BG, Herman P, Hyder F (2008) Frequency-dependent tactile responses in rat

brain measured by functional MRI. NMR Biomed. 21:410–416

64. Brinker G, Bock C, Busch E, Krep H, Hossmann KA, Hoehn-Berlage M (1999) Simultaneous recording of evoked potentials and T2*-weighted MR images during somatosensory stimulation of rat. Magn Reson Med. 41:469–473
65. Van Camp N, Verhoye M, Van der Linden A (2006) Stimulation of the rat somatosensory cortex at different frequencies and pulse widths. NMR Biomed. 19:10–17
66. Shepherd GM, Chen WR, Greer CA (2004) "Olfactory bulb" in The synaptic organization of the brain (Shepherd GM, Ed) (Oxford University Press, New York), pp. 165–216
67. Johnson, BA, Leon M (2000) Odorant molecular length: one aspect of the olfactory code. J Comp Neurol. 426: 330–338
68. Bozza T, McGann JP, Mombaerts P, Wachowiak M (2004) In vivo imaging of neuronal activity by targeted expression of a genetically encoded probe in the mouse. Neuron. 42:9–21
69. Fried HU, Fuss SH, Korsching SI (2002) Selective imaging of presynaptic activity in the mouse olfactory bulb shows concentration and structure dependence of odor responses in identified glomeruli. Proc Natl Acad Sci USA. 99:3222–3227
70. Meister M, Bonheoffer T (2001) Tuning and topography in an odor map on the rat olfactory bulb. J. Neurosci. 21:1351–1360
71. Mori K, Takahashi YK, Igarashi KM, Yamaguchi M (2006) Maps of odorant molecular features in the Mammalian olfactory bulb. Physiol Rev. 86:409–433
72. Xu F, Schaefer M, Kida I, Schafer JR, Liu N, Rothman DL, Hyder F, Restrepo D, Shepherd GM (2005) Simultaneous activation of mouse main and accessory olfactory bulbs by odors or pheromones. J Comp Neurol. 489:491–500
73. Martin C, Grenier D, Thevenet M, Vigouroux M, Bertrand B, Janier M, Ravel N, Litaudon P (2007) fMRI visualization of transient activations in the rat olfactory bulb using short odor stimulations. NeuroImage. 36:1288–1293
74. Kida I, Xu F, Shulman RG, Hyder F (2002) Mapping at glomerular resolution: fMRI of rat olfactory bulb. Magn. Reson. Med. 48:570–576
75. Liu N, Xu F, Marenco L, Hyder F, Miller P, Shepherd GM (2004) Informatics approaches to functional MRI odor mapping of the rodent olfactory bulb: OdorMapBuilder and OdorMapDB. Neuroinformatics 2:3–18
76. Jakob PM, Griswold MA, Edelman RR, Manning WJ, Sodickson DK (1999) Accelerated cardiac imaging using the SMASH technique. J Cardiovasc Magn Reson. 1:153–157
77. Pruessmann KP,Weiger M, Scheidegger MB, Boesiger P (1999) SENSE: Sensitivity encoding for fast MRI. Magn Reson Med. 42:952–962
78. Holscher C, Schnee A, Dahmen H, Setia L, Mallot HA (2005) Rats are able to navigate in virtual environments. J Exp Biol. 208:561–569
79. Sterling P, Demb JB (2004) "Retina" in The synaptic organization of the brain (Shepherd GM, Ed) (Oxford University Press, New York), pp. 217–269
80. Jacobs GH, Fenwick JA, Williams GA (2001) Cone-based vision of rats for ultraviolet and visible lights. J Exp Biol. 204:2439–2446
81. Adams AD, Forrester JM (1968) The projection of the rat's visual field on the cerebral cortex. Q J Exp Physiol Cogn Med Sci. 53:327–336
82. Gias C, Hewson-Stoate N, Jones M, Johnston D, Mayhew JE, Coffey PJ (2005) Retinotopy within rat primary visual cortex using optical imaging. NeuroImage. 24:200–206
83. Lund RD (1972) Anatomic studies on the superior colliculus. Invest Ophthalmol 11:434–441
84. Girman SV, Sauve Y, Lund RD (1999) Receptive field properties of single neurons in rat primary visual cortex. J Neurophysiol. 82:301–311
85. Ohki K, Chung S, Ch'ng YH, Kara P, Reid RC (2005) Functional imaging with cellular resolution reveals precise micro-architecture in visual cortex. Nature. 433:597–603
86. Woolsey TA, Van der Loos H (1970) The structural organization of layer IV in the somatosensory region (SI) of mouse cerebral cortex. The description of a cortical field composed of discrete cytoarchitectonic units. Brain Res. 17:205–242
87. Welker C, Woolsey TA (1974) Structure of layer IV in the somatosensory neocortex of the rat: description and comparison with the mouse. J Comp Neurol. 158:437–453
88. Lu H, Mazaheri Y, Zhang R, Jesmanowicz A, Hyde JS (2003) Multishot partial-k-space EPI for high-resolution fMRI demonstrated in a rat whisker barrel stimulation model at 3T. Magn Reson Med. 50:1215–1222
89. Simons DJ (1978) Response properties of vibrissa units in rat SI somatosensory neocortex. J Neurophys. 41:798–820

90. Gerrits RJ, Stein EA, Greene AS (1998) Blood flow increases linearly in rat somatosensory cortex with increased whisker movement frequency. Brain Res. 783:151–157
91. Ances BM, Zarahn E, Greenberg JH, Detre JA (2000) Coupling of neural activation to blood flow in the somatosensory cortex of rats is time-intensity separable, but not linear. J Cereb Blood Flow Metab. 20:921–930
92. Blood AJ, Pouratian N, Toga AW (2002) Temporally staggered forelimb stimulation modulates barrel cortex optical intrinsic signal responses to whisker stimulation. J Neurophysiol. 88:422–437
93. Buerk DG, Ances BM, Greenberg JH, Detre JA (2003) Temporal dynamics of brain tissue nitric oxide during functional forepaw stimulation in rats. NeuroImage. 18:1–9
94. Sheth SA, Nemoto M, Guiou M, Walker M, Pouratian N, Toga AW (2004) Linear and nonlinear relationships between neuronal activity, oxygen metabolism, and hemodynamic responses. Neuron. 42:347–355
95. Xu S, Yang J, Li CQ, Zhu W, Shen J (2005) Metabolic alterations in focally activated primary somatosensory cortex of alpha-chloralose-anesthetized rats measured by 1H MRS at 11.7 T. NeuroImage. 28:401–409
96. Morton DW, Keogh B, Lim K, Maravilla KR (2006) Functional brain imaging using a long intravenous half-life gadolinium-based contrast agent. Am J Neuroradiol. 27:1467–1471
97. Lowe AS, Beech JS, Williams SC (2007) Small animal, whole brain fMRI: Innocuous and nociceptive forepaw stimulation. NeuroImage. 35:719–728
98. Ogawa S, Lee TM, Stepnoski R, Chen W, Zhu XH, Ugurbil K (2000) An approach to probe some neural systems interaction by functional MRI at neural time scale down to milliseconds. Proc Natl Acad Sci USA. 97:11026–11031
99. Kuo CC, Chen JH, Tsai CY, Liang KC, Yen CT (2005) BOLD signals correlate with ensemble unit activities in rat's somatosensory cortex. Chin J Physiol. 48:200–209
100. Stefanovic B, Schwindt W, Hoehn M, Silva AC (2007) Functional uncoupling of hemodynamic from neuronal response by inhibition of neuronal nitric oxide synthase. J Cereb Blood Flow Metab. 27: 741–754
101. Huttunen JK, Grohn O, Penttonen M (2008) Coupling between simultaneously recorded BOLD response and neuronal activity in the rat somatosensory cortex. NeuroImage. 39:775–785
102. Lee SP, Silva AC, Ugurbil K, Kim SG (1999) Diffusion-weighted spin-echo fMRI at 9.4 T: microvascular/tissue contribution to BOLD signal changes. Magn Reson Med. 42:919–928
103. Silva AC, Lee SP, Yang G, Iadecola C, Kim SG (1999) Simultaneous blood oxygenation level-dependent and cerebral blood flow functional magnetic resonance imaging during forepaw stimulation in the rat. J Cereb Blood Flow Metab. 19:871–879
104. Mandeville JB, Marota JJ, Ayata C, Moskowitz MA, Weisskoff RM, Rosen BR (1999) MRI measurement of the temporal evolution of relative CMRO2 during rat forepaw stimulation. Magn Reson Med. 42:944–951
105. Kennan RP, Jacob RJ, Sherwin RS, Gore JC (2000) Effects of hypoglycemia on functional magnetic resonance imaging response to median nerve stimulation in the rat brain. J Cereb Blood Flow Metab. 20:1352–1359
106. Kida I, Hyder F, Behar KL (2001) Inhibition of voltage-dependent sodium channels suppresses the functional magnetic resonance imaging response to forepaw somatosensory activation in the rodent. J Cereb Blood Flow Metab. 21:585–591
107. Hyder F, Kida I, Behar KL, Kennan RP, Maciejewski PK, Rothman DL (2001) Quantitative functional imaging of the brain: Towards mapping neuronal activity by BOLD fMRI. NMR Biomed. 14:413–431
108. Liu ZM, Schmidt KF, Sicard KM, Duong TQ (2004) Imaging oxygen consumption in forepaw somatosensory stimulation in rats under isoflurane anesthesia. Magn Reson Med. 52:277–285
109. Lahti KM, Ferris CF, Li F, Sotak CH, King JA (1999) Comparison of evoked cortical activity in conscious and propofol-anesthetized rats using functional MRI. Magn Reson Med. 41:412–426
110. Peeters RR, Tindemans I, De Schutter E, Van der Linden A (2001) Comparing BOLD fMRI signal changes in the awake and anesthetized rat during electrical forepaw stimulation. Magn Reson Imaging. 19: 821–826
111. Sachdev RN, Champney GC, Lee H, Price RR, Pickens DR 3rd, Morgan VL, Stefansic JD, Melzer P, Ebner FF (2003) Experimental model for functional magnetic resonance imaging of somatic sensory cortex in the unanesthetized rat. NeuroImage. 19:742–750
112. Chapin JK, Lin CS (1984) Mapping the body representation in the SI cortex of anesthetized and awake rats. J Comp Neurol. 229:199–213

113. Kohn DF, Wixson SK, White WJ, Benson GJ (1997) Anesthesia and Analgesia in Laboratory Animals. (Academic Press, New York, USA)
114. Alfaro V, Palacios L (1997) Components of the blood acid-base disturbance that accompanies urethane anaesthesia in rats during normothermia and hypothermia. Clin Exp Pharmacol Physiol. 24:498–502
115. Adrian ED (1941) Afferent discharges to the cerebral cortex from peripheral sense organs. J Physiol. 100:159–191
116. Mountcastle VB (1957) Modality and topographic properties of single neurons of cat's somatic sensory cortex. J Neurophysiol. 20:408–434
117. John ER (1961) High nervous functions: Brain functions and learning. Annu Rev Physiol. 23:451–484
118. Sceniak MP, Maciver MB (2006) Cellular actions of urethane on rat visual cortical neurons in vitro. J Neurophysiol. 95:3865–3874
119. Siesjo BK (1978) Brain Energy Metabolism. (Wiley and Sons, New York, USA)
120. Antkowiak B (2001) How do general anaesthetics work? Naturwissenschaften. 88:201–213
121. Bonvento G, Charbonne R, Correze JL, Borredon J, Seylaz J, Lacombe P (1994) Is alpha-chloralose plus halothane induction a suitable anesthetic regimen for cerebrovascular research? Brain Res. 665: 213–221
122. Austin VC, Blamire AM, Allers KA, Sharp T, Styles P, Matthews PM, Sibson NR (2005) Confounding effects of anesthesia on functional activation in rodent brain: A study of halothane and alpha-chloralose anesthesia. NeuroImage. 24:92–100
123. Curtis JC, Kleinfeld D (2006) Seeing what the mouse sees with its vibrissae: A matter of behavioral state. Neuron. 50: 524–526
124. Shen J, Rothman DL, Hetherington HP, Pan JW (1999) Linear projection method for automatic slice shimming. Magn Reson Med. 42:1082–1088
125. Miyasaka N, Takahashi K, Hetherington HP (2006) Fully automated shim mapping method for spectroscopic imaging of the mouse brain at 9.4 T. Magn Reson Med. 55:198–202
126. Koch KM, Sacolick LI, Nixon TW, McIntyre S, Rothman DL, de Graaf RA (2007) Dynamically shimmed multivoxel 1H magnetic resonance spectroscopy and multislice magnetic resonance spectroscopic imaging of the human brain. Magn Reson Med. 57:587–591
127. Wixson SK, Smiler KL (1997) "Anesthesia and analgesia in rodents" in Anesthesia and Analgesia in Laboratory Animals (Kohn DF, Wixson SK, White WJ, Benson GJ, Eds) (Academic Press, New York, USA) pp. 165–200
128. Hyder F, Patel AB, Gjedde A, Rothman DL, Behar KL, Shulman RG (2006) Neuronal-glial glucose oxidation and glutamatergic-GABAergic function. J Cereb Blood Flow Metab. 26:865–877
129. Iadecola C, Nedergaard M (2007) Glial regulation of the cerebral microvasculature. Nat Neurosci. 10:1369–1376
130. Meredith MA, Stein BE (1986) Visual, auditory, and somatosensory convergence on cells in superior colliculus results in multisensory integration. J Neurophysiol. 56:640–662
131. Meredith MA, Stein BE (1996) Spatial determinants of multisensory integration in cat superior colliculus neurons. J Neurophysiol. 75:1843–1857

Chapter 11

Using fMRI for Elucidating Dynamic Interactions

Yul-Wan Sung and Seiji Ogawa

Abstract

For elucidation of information processing mechanism of the human brain, it is important to know dynamic characteristics of functional areas involved in stimulus processing and the functional network consists of those areas. Although it is very difficult to trace the dynamics of neuronal activities with slowly responding signals of functional MRI (fMRI), there is a possibility that some aspects of the dynamics can be probed up to a time scale of tens to hundreds milliseconds by fMRI. Here, we describe some evidences for such possibility in the visual information processing.

Key words: fMRI, Paired stimulus paradigm, interaction, inter stimulus interval (ISI), suppression.

1. Introduction

FMRI is the most useful method among non-invasive tools for mapping of brain function. The mapping study has been used to localize areas activated for processing the information generated in the brain by a stimulus. But to elucidate the functional roles of the areas involved in processing the stimulus, it is necessary to know the details of their functional characteristics and the temporal dynamics of the interaction among these areas in the network. Although fMRI is a highly effective method to spatially localize functional areas related to a given stimulus, it is very difficult to trace the dynamics of neuronal activities with the slow fMRI signal. However, there are cases where the dynamic effect can be detected up to a time scale of tens to hundreds ms by fMRI if neuronal activities interact with each other in a functional unit or between functional units. Here, we will show the potential of the conventional (or slow) fMRI by elucidating such dynamic interactions in the visual information processing of the human brain.

Fahmeed Hyder (ed.), *Dynamic Brain Imaging: Multi-Modal Methods and In Vivo Applications, vol. 489*

DOI 10.1007/978-1-59745-543-5_11 Springerprotocols.com

Visual information delivered to the primary visual cortex (V1) is passed onto the dorsal and/or the ventral pathways according to the content of the information. Contents related to the shape are processed on the ventral path and those related to motion or to location are on the dorsal path. The neuronal processing and the interaction among related areas proceed very fast. In the next section, we will show that such fast neuronal processing can be followed in some cases by a paired stimulus paradigm which allows an fMRI signal to respond in amplitude change rather than temporal change to these fast processes.

2. An fMRI Technique for Detecting Some Fast Neuronal Activity – A Paired Stimulus Paradigm

Information processing in the brain proceeds while many functional units deliver the information to each other or make feedback and modulate the activity of others. These processes occur at a fast temporal scale. It often happens that such activity-modulating interaction is suppressive. The interaction process can be probed by measuring the corresponding fMRI response if we can control the suppression by an external factor. This can be achieved by varying temporal relation between two stimulus inputs (paired stimulus paradigm). In this section, we will show, in experiments with the paired stimulus paradigm, the linkage of fMRI signal to fast neuronal interaction in the rat somatosensory cortex and also in the human visual cortex.

When one of the front paws of a rat was stimulated electrically, the response of SEP (somatosensory evoked potential) was strong in the contralateral area but was weak, if any, in the ipsilateral area (**Fig. 11.1(a)**). The corresponding responses of fMRI were also strong in the contralateral area and weak in the ipsilateral area with a different response shape (**Fig. 11.1(b)**). When one paw was stimulated and then the other side with a time delay, strong suppression of SEP at the contralateral area of the side of the second paw stimulation was observed. The suppression appeared at 40 ms inter stimulus interval (ISI) and the corresponding fMRI response was suppressed too (**Fig. 11.2**). This indicates that fast progressing of neuronal activities can be detected in a time scale of 40 ms as in this case. Furthermore, this suggests that the response to the second stimulation can be affected by the neuronal activation occurring earlier at a different site. Such event is an example of an interaction between functional units.

In an experiment to examine the suppressive response in the human visual area, two successive visual stimuli of the same kind were given with time delays (ISI) of about 100–1000 ms. The response to the second input at 200 ms ISI was almost completely suppressed and recovered at 1000 ms (**Fig. 11.3**). The state that

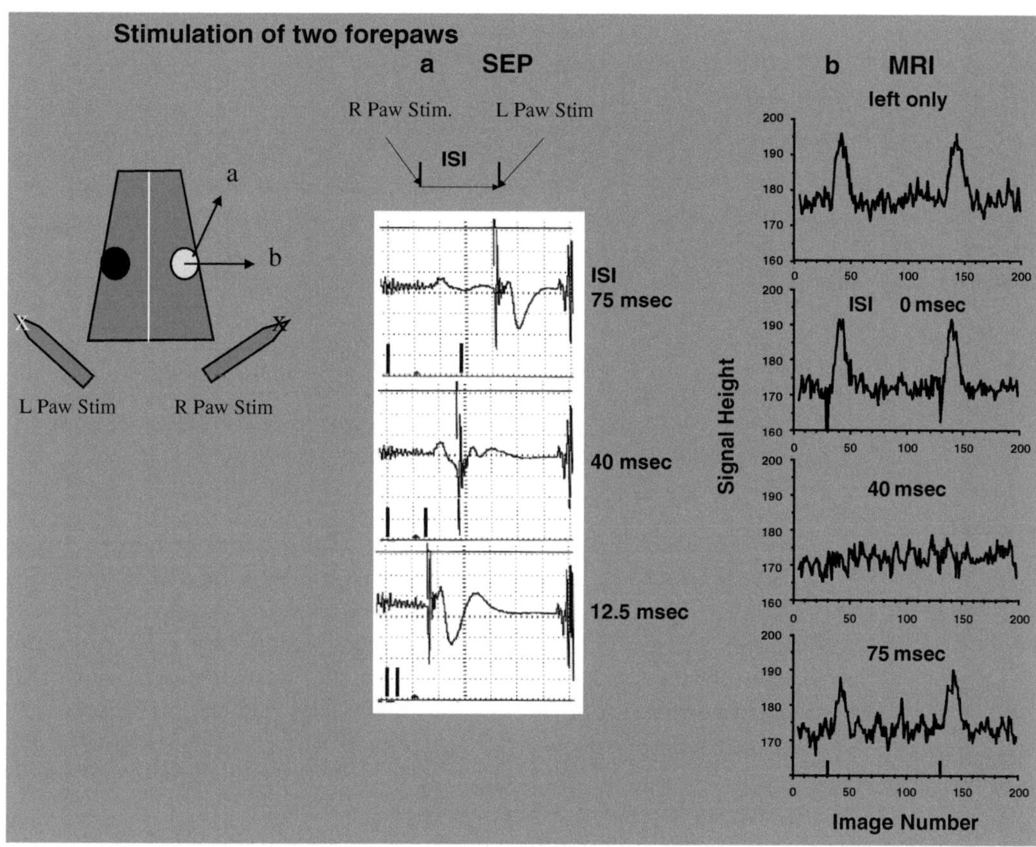

Fig. 11.1. (**a**) Somatosensory evoked potential (SEP) and (**b**) BOLD signal; two activation signals in a timecourse indicate separate trials for verifying reproducibility. Single forepaw electrical stimulation (1 mA, 300 usec pulse, 1 Hz 4 s) in Rat.

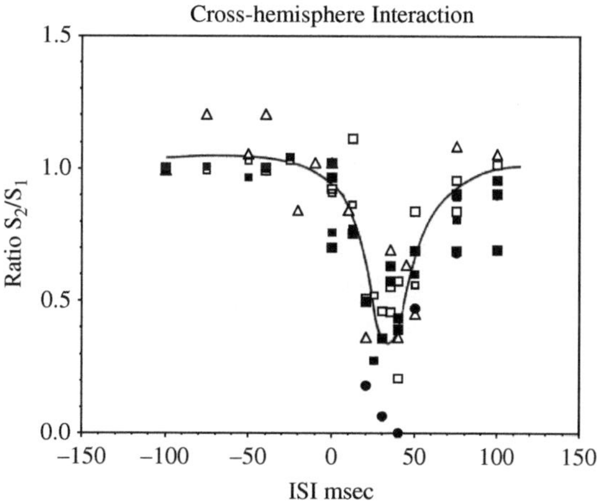

Fig. 11.2. Cross-Hemisphere interaction. SEP and BOLD responses with the two-forepaw stimulation. The symbols (Δ) are those of SEP measurements only, and filled circle symbols are those of MRI measurements without EEG electrodes. Open square symbols are SEP, and filled square symbols are BOLD respons. Paradigm I is flicker pattern, flashed once. Paradigm II is a moving pattern in which two bars were jumped with the speed of 4 degrees per 50 ms. Paradigm III is a repeated moving pattern of paradigm II with 200-ms ISI. One event consisted of 4 repetitions with 1.5 s lap-time.

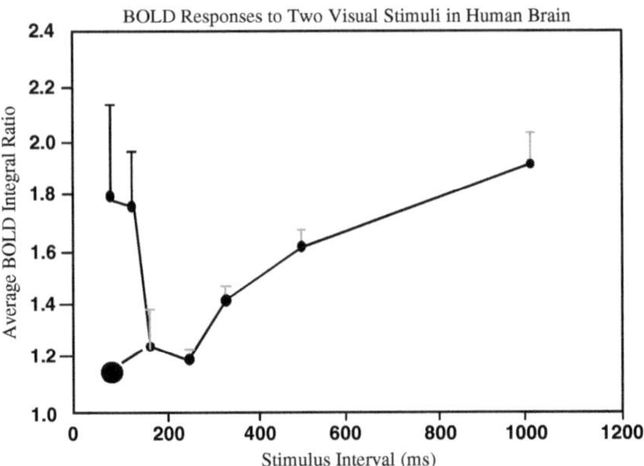

Fig. 11.3. ISI dependent BOLD responses to paired stimulation. Large filled circle indicates that the suppression can vary with stimulation contents.

is incapable of getting activated by a subsequent excitatory input is detected to be a suppressed state only when the excitatory input comes. The correlation between fMRI response and visual evoked potential seen in the human primary visual area with the same stimulation paradigm also supports that neuronal activity of this kind of interaction is reflected in fMRI signal *(1)*. The recovery from the suppressive state in terms of ISI can vary with stimulus contents as well as the areas that have different functional specificity as described in the later section *(2,-4)*.

In the following sections, we will show interaction between functional areas on the dorsal visual pathway with paired "moving" stimulation and on the ventral visual pathway with paired "shape" stimulation.

3. Functional Activation and Interaction at Motion Processing Areas

Evidences from electrophysiology of monkeys and non-invasive studies of the human brain indicate that visual area MT (or secondary area V5) is related to the primary motion processing. In fMRI studies on MT, the activation appeared more clearly when dots or bars are moved on the screen than when they are standing. Input characteristics of MT and interaction between functional units were probed here through an experiment using a paradigm with paired moving stimulations *(5)*.

In that experiment on MT, we used the following paradigms I, II and III shown in **Fig. 11.4**. In paradigm, a set of standing bars was flickered once. In paradigm II, the bars appeared moving by flickering one bar at a position and then shifted to another position with a time delay of 50 ms. In paradigm III, the moving pattern of paradigm II was repeated with 200 ms (50 ms from

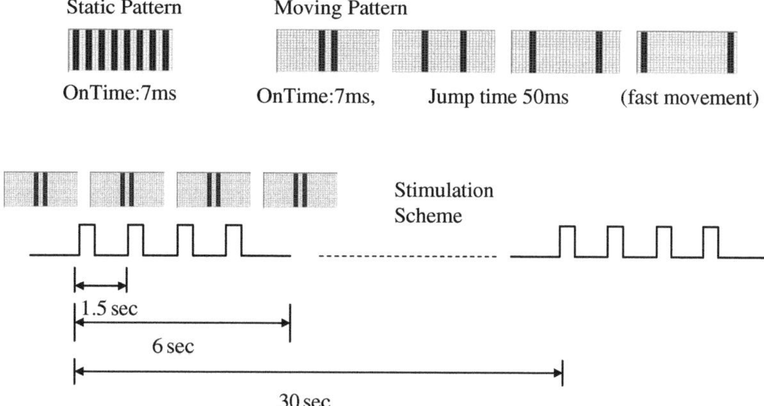

Fig. 11.4. Paradigms for MT activation. Bars are two degrees of visual angle. Paradigm I is flicker pattern, flashed once. Paradigm II is a moving pattern in which two bars were jumped with the speed of 4 degrees per 50 ms. Paradigm III is a repeated moving pattern of paradigm II with 200 ms ISI. One event consisted of 4 repetitions with 1.5 s lap-time.

the last of the bar set to the first of the second set) ISI, which was to induce a refractory response. Paradigms were repeated 4 times with a lap-time of 1.5 s making the period of total stimulation 6 s. Those paradigms were performed under a high contrast condition (90%) relative to the background and a low contrast condition (40%).

In V1 and MT, the difference between the responses in paradigm I and paradigm II was small under the high contrast condition (**Fig. 11.5**). Response trend in V1 with paradigm I, II and III were different from in MT. The responses in V1 increased

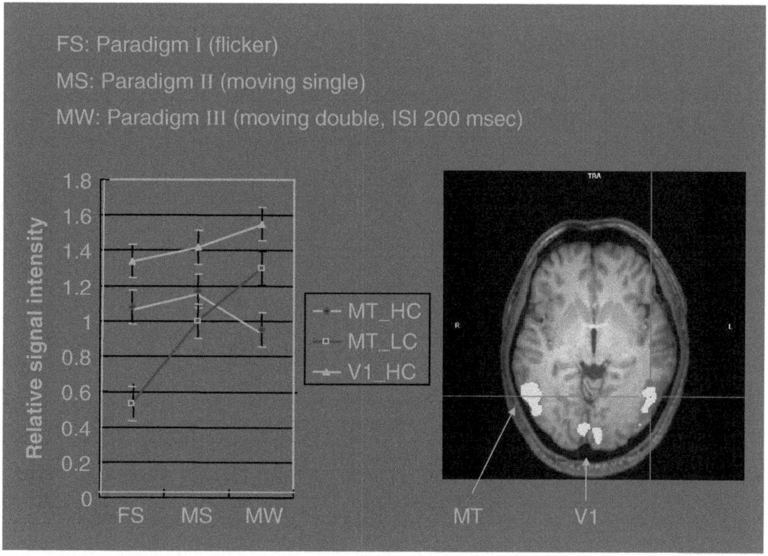

Fig. 11.5. Activation maps and paradigm-dependent BOLD signals at V1 and MT. HC: High contrast condition, LC: Low contrast condition.

in the order of paradigm I, II and III. The responses in MT increased from paradigm I to paradigm II and then decreased with paradigm III. The response with paradigm III was even smaller than with paradigm I and II. On the other hand, under the low contrast condition, there was no statistically significant activation in V1 with all the paradigms and the activation with paradigm I in MT decreased extensively in comparison with that under the high contrast condition. The response to the moving pattern of paradigm II with the low contrast was large and was not so different from with the high contrast case. The response to paradigm III under the low contrast condition was larger than that to paradigm II, i.e. the refractory suppression effect in the response to paradigm III was reduced at the ISI, showing some quicker recovery from the suppressed state. The observations of the strong response to non-moving flickering input pattern of paradigm I at high contrast and of the strong response to moving single of paradigm II at low contrast indicate that there is in MT an input non-specific to motion as well one specific to motion *(6)*.

The refractory response observed in V1 at the high contrast pattern presentation was similar to those reported earlier *(2)*. An interesting observation was that in MT, the paradigm III, paired two moving patterns, yielded the response even below the level with paradigm II where one moving pattern was shown once. Similar highly reduced activity below the single pattern input was also observed in various parietal areas such as BA7 (Brodmann area 7) and IPS (Intra parietal sulcus) (**Fig. 11.6**). There was apparent violation of causality in these observations. In the paired sequential presentation, the second event interfered with

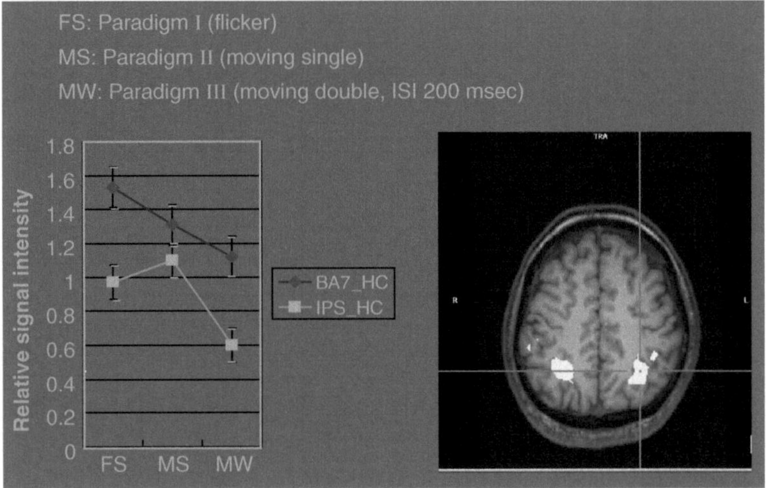

Fig. 11.6. Activation maps and paradigm-dependent BOLD signals at BA7 (Brodmann area 7) and IPS (Intra Parietal Sulcus). HC: High contrast condition, LC: Low contrast condition.

the first event that occurred 200 ms earlier. This suggested that the processing activity with paradigm III included not only those induced by the two sequential inputs but more by additional inputs beyond the two. It is known that MT receives inputs not only from V1 path but also directly from LGN. The latency of activation of neuronal population of MT is about 50–90 ms and is only just few milliseconds longer than V1 on an average *(7)*. There, one would expect the time delay between the inputs to MT from the two different channels, the input from LGN being sooner than the input from V1 channel. The reduction of signal in the paired presentation beyond the signal by the single presentation did not occur when the activation at V1 by low contrast presentation was very weak and the input to MT from V1 path was expected to be minimal. Therefore, the regular refractory suppression was observed for the straight MT input pair. When such extra suppression of signal as above is observed with the paired stimulus presentation, one can suspect that there are multi-inputs to the site beyond the sequential two inputs. Further experiments by varying ISI or changing the type of stimulus will elucidate the timing of these events.

4. Functional Activation and Interaction at Object Processing Areas

It is known that visual objects are processed along the ventral visual pathway in the human brain. With fMRI, functional areas involved in the processing of visual object information have been identified, but it is still difficult to assess how the neuronal processing progresses from lower to higher functional areas. It is widely accepted that object stimuli of the same or similar attribute are processed in a specialized functional area or in a network. Different stimuli of the same attribute are represented on different areas in the early visual cortices and the representation converges to specific sites for the processing of the common information in the stimuli. In progressing, the interaction between the inputs generated by the stimuli of a stimulus pair can be different from area to area, and the suppression is most pronounced when the inputs are identical. The identical input cases are well demonstrated in V1 area. The ISI for suppression and the strength of the suppression can be also different in those areas on the path. The extent of the suppression would be also dependent on contents of stimuli in a stimulation pair.

As a simple case, we will show an experiment in which two stimuli in a stimulation pair have spatial and temporal proximity. Two rectangular frames were presented in close spatial proximity with no partial overlap and with ISI (inter-frame) at 0, 50 and 240 ms (**Fig. 11.7**). Each trial consisted of five paired stimuli.

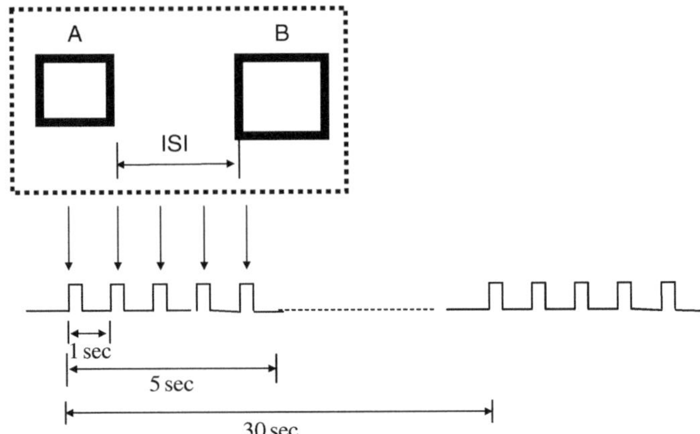

Fig. 11.7. Stimulation scheme. Two concentric rectangular frames were presented with inter-stimulus intervals (inter-frame) of 0, 50 and 240-ms. Smaller one was shown first, followed by larger one. The repetition time of paired stimuli, t1 is 1-s. The repetition time of trials, t2 is 30-s. The rectangular frames were not spatially overlapped in the display.

In each pair, the small frame was shown first, and after the ISI, the large frame followed. One would expected them to be represented by the activation at various sites in the early visual areas, and activation sites in the higher areas where shapes are recognized would converge to the specific sites for the shape information.

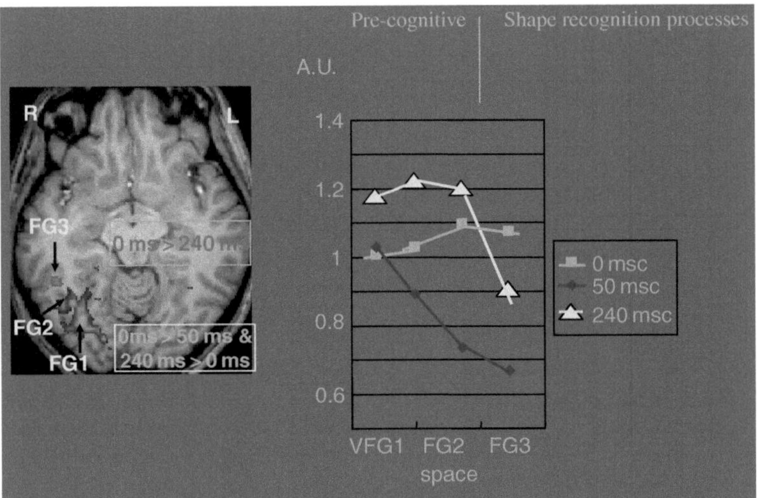

Fig. 11.8. Activation maps and trends of BOLD responses. Areas with positive responses for the difference of 0 and 50-ms (*orange*). Areas with positive response for the difference of 240 and 50-ms (*green*). FG1, FG2, and FG3 are areas in occipitotemporal region located at (30,-68,-9), (35,-57,-9), and (45,-46,-9). The responses were normalized by the response to the 0-ms ISI at V1. (*See* Color Plate)

Areas with greater fMRI response to the frame pair at the 240-ms ISI than to that at the 50-ms ISI were found only in the ventral occipitotemporal region (**Fig. 11.8**). Areas with greater response to the frame pair at the 0-ms ISI than to that at the 50-ms ISI were detected in the fusiform gyrus, the frontal cortex and the parietal cortex. FG1 is the local maximum area in the occipototemporal region that showed a greater response to the frame pair at the 240-ms ISI than to that at the 50-ms ISI. FG2 is the area that showed greater responses to the frame pairs at the 0-ms and 240-ms ISI's than to that at the 50-ms ISI. FG3 is the area that showed a greater response to the frame pair at the 0-ms ISI than to that at the 50-ms ISI (How about the comparison with 240-ms?). The area FG2, located at the posterior fusiform gyrus, was the only area that showed a significantly smaller response to the frame pair at the 50-ms ISI than to those at the 0-msand 240-ms ISI's. The reduction of the fMRI response in this area was paralleled by a reduction of visibility in our behavioral data where all participants reported that both frames were visible as separate frames at the 0-ms and 240-ms ISI's, but that the first frame was invisible at the 50-ms ISI. **Figure 11.4** shows the timecourses from this area. In the primary visual area V1, there was no significant difference among responses to the frame pairs at the ISI's of 0, 50- and 240-ms.

The responses in those occipitotemporal areas showed some trends that depended on ISI (**Fig. 11.8**). The response to the frame pair at the 0-ms ISI did not decrease from V1 to FG3, while the responses to the frame pairs at the 50-ms and 240-ms ISI's did decrease from V1 to FG3. In addition, there were differences between the pattern of response reduction at the 50-ms ISI and that at the 240-ms ISI. The response to the frame pair at the 50-ms ISI decreased progressively from V1 to FG3, while the response to the frame pair at the 240-ms ISI did not decrease from V1 to FG2, but abruptly decreased from FG2 to FG3. Although there was near loss of visibility of the small frame at 50-ms ISI, the relatively constant responses among 0-ms, 50-ms and 240-ms ISI's in the primary visual area indicate that there were neuronal activities in V1 corresponding to the first frame as well as to the second frame at 50-ms ISI. The two frames were presented in spatially separated positions on the display. In spite of the very low visibility of the small frame, the inputs related to the two frames did reach V1 at 50-ms ISI as much as at 0-ms and 240-ms ISI where the visibility was clear. It does indicate that the perception of the frame visibility does not occur at V1.

In areas of FG1 and FG2, the 240-ms ISI did not induce fMRI signal reduction for the paired stimuli but at 50-ms ISI, the fMRI responses showed reduction. The interaction between the two activities induced by the two stimuli was very short lived as judged from the ISI dependent signal reduction. The overlaps of activation between the local processing sites are not extensive and/or they are fairly independent. There is some similarity in left-eye and right-eye paired stimulation where the suppressive interaction between ocular dominance columns was relieved within short ISI *(4)*.

The following description on the locations and functional roles in the above areas has been reported: FG1 was located between visual area V4 and the lateral occipital area (LO), and FG2 was located in the posterior fusiform gyrus (pFG) *(8)*. The LO and pFG compose the lateral occipital complex (LOC) and are known to be involved in object processing *(9)*. It has been also reported that there exist populations of neurons that code complex shapes in primate V4 *(10)*. Reported human fMRI data from V4 showed larger responses for circular concentric patterns than for parallel grating patterns *(11)*, and human psychophysical data demonstrated the existence of context-dependent interactions as well *(12)*. Therefore, in the present experiment, the areas FG1 and FG2 are likely to be those sites that encoded the shape of frames with size/location distinction. The reduction of fMRI signal at 50-ms ISI could indicate the loss of effectiveness of the encoding. If the poor encoding process is related to the poor perception of visual information, the meta-contrast (backward suppression) phenomenon of the reduced visibility of the small frame at 50-ms ISI was associated with the signal reduction at FG2.

On the other hand, the fMRI response in FG3 showed reduction of signal at 240-ms ISI although the visibility of the two frames was not reduced at the ISI. FG3 is located in the dorsal extension of the fusiform face area (FFA) *(13)*. The FFA has been reported to be sensitive to circular concentric shapes as well as faces, which may account for the activation at FG3 to the concentric frames in this study *(11)*. It was also reported that FFA shows size invariance and suppressed responses to repeated stimuli *(14)*. Therefore, there is a good possibility that the paired inputs to FG3 were identical and that the reduction of the fMRI response at FG3 approached to the identical input cases of V1 (**Fig. 11.9**), where the activation by the second stimulus of an identical pair was severely suppressed when the ISI was 200 ms. These observations suggest that a site such as FG3, with suppression at the 240-ms ISI, is related to the processing for the size-independent shape recognition that follows shape encoding in area FG2 as discussed earlier.

This showed that there exist two stages in neuronal processing along the ventral visual pathway as indicated by the observed suppression phenomena. One is related to pre-recognition stage of an object shape and the other to the recognition stage and they were related to the suppression at 50 and 240 ms respectively.

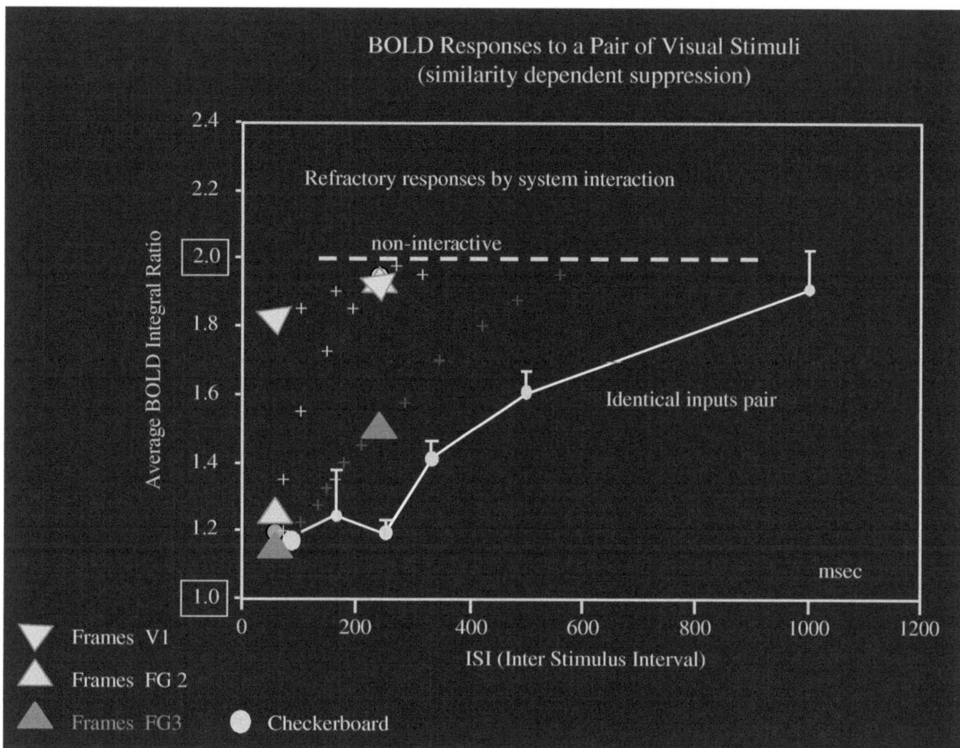

Fig. 11.9. Similarity dependent suppression of BOLD response. (*See* Color Plate)

Through this kind of dynamic interaction measured by fMRI suppression phenomena, we can probe the distinction of neuronal processing as well as the information flow in the object-processing pathway.

5. Summary

In applying fMRI to studies of brain science for functional activation, there are some known technical problems such as effects of large vessels, differentiation of global variation of brain circulation, local variations related to neuronal activation or deactivation and local blood circulation or modified supply. It is also challenging to chase fast neuronal processing through slow fMRI response and to know the interaction between functional units activated by a task or two. We have made efforts to tap the potential of fMRI. The paired stimulation paradigm provided potential in the dynamics of the interaction induced by neural activations in fMRI measurements. It is worth mentioning that, in such paradigms, the shortcomings of BOLD fMRI as described above, are likely minimized because the effect happens at the level of neuronal processes and the big vessel response, if it is local, will also show the same response characteristics as other smaller vessels. Selectivity of a site to inputs can be probed as well as the functional flow, and furthermore interactions in the functional network for processing a task could well be studied in fMRI measurements although it requires more work. The approach described here would be a promising fMRI technique.

References

1. Wan X, Riera J, Iwata K, Takahashi M, Wakabayashi T, Kawashima R (2006) The neural basis of the hemodynamic response nonlinarity in human primary visual cortex: Implications for nuerovascular coupling mechanism. Neuroimage 32: 616–625.
2. Ogawa S, Lee TM, Stepnoski R, Chen W, Zhu X-H, Ugurbil K (2000) An approach to probe some neural systems interaction by functional MRI at neural time scale down to milliseconds. Proc Natl Acad Sci USA 97: 11026–11031.
3. Sung Y, Kamba M, Ogawa S (2007) Progression of neuronal processing of visual objects. NeuroReport 18:411–414.
4. Zhu XH, Zhang XL, Tang S, Ogawa S, Ugurbil K, Chen W (2001) Probing fast neuronal interaction in the human ocular dominate columns based on fMRI BOLD response at 7 Tesla. Proc. Intl. Soc. Mag. Reson. Med 9, 287.
5. Ogawa S, Sung YW, Kamba M, Nagaoka T (2003) Probing the temporal dynamics of the functional system interaction in the brain by functional MRI. Shinkei Kenkyu no Shinpo 47: 902–908.
6. Tootell RB, Repps JB, Kwong KK, Malach R, Born RT, Brady TJ, Rosen BR, Belliveau JW (1995) Functional analysis of human MT and related visual cortical areas using magnetic resonance imaging. J Neurosci 15: 3215–3230.
7. Schmolesky MT, Wang Y, Hanes DP, Thompson KG, Leutgb S, Schall JD, Leventhal AG (1998) Signal timing across the macaque visual system. J Neurophysiol 79: 3272–3278.
8. Lerner Y, Hendler T, Ben-Bashat D, Harel M, Malach R (2001) A hierarchical axis of object processing stages in the human visual cortex. Cerebral Cortex 11: 287–297.
9. Grill-spector K, Kourtzi Z, Kanwisher N (2001) The lateral occipital complex and its role in object recognition. Vision Res 41: 1409–1411.
10. Pasupathy A, Connor CE (2002) Population coding of shape in area V4. Nat Neurosci 5: 1332–1338.
11. Wilkinson F, James TW, Wilson HR, Gati JS, Menon RS, Goodale MA (2000) An fMRI study of the selective activation of human extrastriate form vision areas by radial and concentric gratings. Curr Biol 10: R836–R838.
12. Habak C, Wilkinson F, Zakher B, Wilson HR (2004) Curvature population coding for complex shapes in human vision. Vision Res 44: 2815–2823.
13. Kanwisher N, McDermott J, Chun MM (1997) The fusiform face area: A module in human extrastriate cortex specialized for face perception. J Neurosci 17: 4302–4311.
14. Andrews TJ, Ewbank MP (2004) Distinct representations for facial identity and changeable aspects of faces in the human temporal lobe. Neuroimage 23: 905–913.

Chapter 12

Resting-State Functional Connectivity in Animal Models: Modulations by Exsanguination

Bharat B. Biswal and Sridhar S. Kannurpatti

Abstract

We studied the spatiotemporal characteristics of the resting state low frequency fluctuations in functional MRI (fMRI), blood oxygenation level dependent (BOLD) signal in isoflurane-anesthetized rats. fMRI-BOLD measurements at 9.4 Tesla were made during normal and exsanguinated condition previously known to alter cerebral blood flow (CBF) fluctuations in anesthetized rats. fMRI signal time series were low-pass filtered and studied by spectral analysis. During normal conditions, baseline mean arterial pressure (MAP) was 110 ± 10 mmHg and low-frequency fluctuations in BOLD signal were observed in the frequency range of $0.01-0.125$ Hz. Following blood withdrawal (exsanguination), MAP decreased to 68 ± 7 mmHg, resulting in an increase in the amplitude of the low-frequency fluctuations in BOLD signal time series and an increase in power at several frequencies between 0.01 and 0.125 Hz. Spatially, the BOLD fluctuations were confined to the cortex and thalamus spanning both hemispheres with sparse presence in the caudate putamen and hippocampus during both normal and exsanguinated states. Spatial distribution of the low frequency fluctuations in BOLD signal, from cross correlation analysis, indicate substantial inter-hemispheric synchrony similar to that observed in the conscious human brain. The behavior of the resting state BOLD signal fluctuations similar to CBF fluctuations during exsanguination indicates a myogenic dependence. Also, a high inter-hemispheric synchrony combined with different phase characteristics of the low frequency BOLD fluctuations particularly in the hippocampus relative to the cortex emphasizes distinct functional networks.

Key words: fMRI, BOLD fluctuation, CBF, resting state connectivity, brain, rat, isoflurane

1. Introduction

We first demonstrated a significant temporal correlation of resting-state low-frequency blood oxygenation level dependent (BOLD) signal fluctuations in functional MRI (fMRI) both

Fahmeed Hyder (ed.), *Dynamic Brain Imaging: Multi-Modal Methods and In Vivo Applications, vol. 489*

DOI 10.1007/978-1-59745-543-5_12 Springerprotocols.com

within and across hemispheres in primary sensorimotor cortex during rest. We defined resting-state connectivity (RSC) as significant correlation signal between functionally related brain regions in the absence of any stimulus or task. This result has been validated by a number of groups using different models *(1,2)* and is a subject which has been recently reviewed by Fox and Raichle *(3)*. This correlated signal arises from spontaneous low-frequency signal fluctuations (SLFs). These low-frequency spontaneous fluctuations have also been observed by several investigators using animal models and a variety of measurement techniques, including polarographic technique of tissue oxygenation *(4, 5)*, laser Doppler flowmetry (LDF) *(6)*, fluororeflectometry of NADH and cytochrome-aa3 *(7)*.

Testing hypotheses of the role of SLFs has involved attempts to determine their physiological origins. Cooper et al. *(8)* hypothesized that these fluctuations represent cellular maintenance of an optimum balance between cerebral blood flow (CBF) and oxidative metabolic rate (CMR_{O2}). Testing this hypothesis has involved manipulation of cerebral metabolism with anesthesia. These studies have involved comparison of activity during waking and anesthetized states in animals using various techniques including LDF *(6)* and fMRI in humans *(9, 10)*. Signal oscillations in the rodent brain vary with differing levels of halothane anesthesia, carbon dioxide level and nitric oxide synthase (NOS) blockade *(11)*. These results suggest support for the biophysical-origin hypothesis that affects the cerebral vasculature, either directly or indirectly. The neural mechanisms of slow rhythmic fluctuations have not yet been clearly defined, though with recent evidence indicating an active role for glia in neurovascular coupling *(12–15)*, the slow rhythmic fluctuations may have both neuronal and glial origins.

In spectrophotometric studies of the intramitochondrial redox state of enzyme cytochrome aa3 (CYTox) and cerebral blood volume (CBV), continuous slow oscillations and interhemispheric synchrony has been observed between these variables *(16)*. The relationship between CYTox and CBV oscillations seem to be independent of the physiological state as they have been observed during both awake state and sleep *(6, 16, 17)*, anesthesia *(5, 18, 19)* and during cerebral ischemia *(20, 21)*. Though there have been studies indicating the presence of metabolic oscillations in the absence of CBF oscillations – indicating that metabolic oscillations may be primary in origin – there is no concrete evidence to indicate so. On the contrary, flow oscillations can be linked to metabolic oscillations by the evidence indicating that NADH and/or cytochrome aa3 oscillations lagged behind CBV oscillations *(16, 17)*. These results indicate that spontaneous oscillations in the intra-mitochondrial redox state may, at least, in part be linked to rhythmic variations in CMR_{O2}.

It should be emphasized that the origin of the slow cerebral fluctuations of CBV and CYTox remains to be determined. It is unlikely to be entirely vascular ("vasomotion"), in view of the complex frequency/time and interhemispheric architecture of these fluctuations in cats and rabbits *(16, 17)*. There are a variety of neuronal, glial, and vascular phenomena that may offer their contributions to what finally appears as a measurable "fluctuation" *(22)*. For instance, glutamate-induced intracellular calcium waves within the glial syncytium may represent an energy-dependent indirect reflection of activity within focal neuronal fields. Such factors would need to be carefully dissected by future efforts. An interesting example of such multifactorial components of the slow fluctuations emerges from the study of CBV and CYTox during the transition from slow-wave sleep to REM sleep in the cat, as will be discussed in the following sections.

We first demonstrated a significant temporal correlation of SLFs, both within and across hemispheres, in primary sensorimotor cortex during rest *(9, 10)*. Nearly 74% of the time series from these voxels correlated significantly (after filtering the fundamental and harmonics of respiration and heart rates) while only a few voxel time courses ($< 3\%$) correlated with those in regions outside of motor cortex. Subsequently, Hampson et al. *(23)* demonstrated the presence of RSC in sensory cortices, specifically auditory and visual cortex. In their studies, signal from visual cortex voxels during rest (first scan) was used as a reference and correlated with every other voxel in the brain. Significant number of voxels from the visual cortex passed a threshold of 0.35, while only a few voxels from outside the visual cortex passed the threshold. They have demonstrated similar results in the auditory cortex *(24)*.

Lowe et al. *(25)* extended Biswal, Hyde and colleague's *(26)* results by showing such correlations over larger regions of sensorimotor cortex (i.e., across multiple slices). Xiong, Fox and colleagues *(27)* established relationships between motor and association cortex. Similar to earlier results, they observed RSC between sensorimotor cortex areas (primary, premotor, secondary somatosensory). Further, however, they observed RSC relationships between these motor areas and association areas, specifically anterior and posterior cingulate cortex, regions known to be involved in attention. Greicius et al. *(28)* observed RSC in anterior and posterior cingulate areas. Subsequent observation of activation during a visual attention task indicated similar cingulate activity.

These studies have established the foundation for "resting-state functional connectivity studies" using fMRI (e.g., *(23, 25,*

26, 28, 29)). Results from these studies form the basis for speculation regarding the functional role of RSC. Bressler et al. *(30)* have suggested that such correlated signal fluctuations may be a phenomenon representing the functional connection of cortical areas analogous to the phenomenon of "effective connectivity" defined by Friston et al. *(31)*. Thus, a family of cognitive-origin hypotheses (in contrast to biophysical-origin hypotheses) has emerged. Gusnard and Raichle *(29)*, for instance, have suggested that such coherence indicates the presence of a "default mode of brain function" in which a default network continuously monitors external (e.g., visual stimuli) and internal (e.g., body-functions, emotions) stimuli. Other cognitive-origin hypotheses suggest that low-frequency fluctuation is related to ongoing problem-solving and planning *(28)*. Biswal, Hyde and colleagues *(26)* and Xiong, Fox and colleagues *(27)* observed that analyses of resting-state physiological fluctuations reveal many more functional connections than those revealed by task-induced activation analysis. They hypothesized that task-induced activation maps underestimate the size and number of functionally connected areas and that functional networks are more fully revealed by RSC analysis.

Here, we discuss various animal studies that affect the CBF oscillations and hence the low frequency fluctuations in BOLD signal. The spatiotemporal characteristics of the low frequency fluctuations in brain oxygenation was mapped non-invasively using BOLD weighted fMRI imaging using physiological perturbations known to alter CBF oscillations. The resting state connectivity maps obtained from normocapnia and changes in mean arterial pressure (MAP) and differences between them are discussed. Low-frequency BOLD signal fluctuations were studied both in the time-domain and frequency-domain. These animal results were also consistent with earlier studies done in humans using fMRI. Functional maps obtained using the power spectrum of the frequencies and cross correlation analysis indicate a significant reduction in the resting state low-frequency BOLD physiological fluctuations in the cortical, sub-cortical and deeper brain structures during normal physiological states. Exsanguination led to an enhancement in the amplitude of the low frequency BOLD fluctuations and spatially expanded to most of the cortical, sub-cortical and deeper brain structures. The behavior of the resting state BOLD signal fluctuations similar to CBF fluctuations, during exsanguination, indicates a myogenic dependence. However, a high intra-hemispheric symmetry in the BOLD fluctuations with similar phase characteristics in other regions in this study except the hippocampus suggests that the fMRI-BOLD signal fluctuations in anesthetized rats may carry information from underlying neural activity.

2. Methods

2.1. Surgical Preparation

Seven male Sprague-Dawley rats (250–300 g; Harlan, Indianapolis, IN) were anesthetized with 1.2% isoflurane in oxygen. Body temperature was monitored with a rectal probe and maintained at 37.0 ± 0.5 using a homeothermic feedback heating system (Baxter K-MOD100, Gaymar Industries). Femoral arteries were cannulated with PE50 tubing for mean arterial blood pressure (MAP) measurements and blood withdrawal. Blood oxygen saturation was continuously monitored by a pulse oximeter positioned on the hind limb. The physiological parameters in control state were $S_aO_2 = 98 \pm 1\%$, pH$= 7.4 \pm 0.08$, $P_aCO_2 = 34 \pm 4$ mmHg and MAP$= 110 \pm 10$ mmHg. All procedures were approved by the research animal committee of Massachusetts General Hospital, Harvard Medical School.

2.2. FMRI Studies

fMRI experiments were performed using a 9.4 T/21 cm horizontal bore (Magnex Scientific) using a Bruker Advance console and custom-made surface-RF coil. In order to minimize motion artifacts, the rat was secured to the RF coil by a bite bar resting below the upper hard palate and over the snout along with a mask for the delivery of anesthetic gas. Coronal localization of slices was accomplished using an initially obtained mid-line sagittal slice and comparing it with the sagittal section from a stereotaxic rat brain atlas *(32)*. Five contiguous coronal slices were selected over the region –5 mm to 0 from the Bregma covering the somatosensory cortex, thalamus, caudate putamen and hippocampus. All catheters for mean arterial pressure (MAP) measurement and anesthesia delivery tubes were brought outside the magnet room of the MR scanner. Anatomical images were obtained before fMRI scanning using a RARE sequence with repetition time (TR) = 1 s, echo time (TE) = 19 ms, 256 × 256 matrix and field-of-view (FOV) = 3.0 cm. For fMRI-BOLD measurements, a single shot gradient EPI sequence was used to acquire multiple slices of images using a 128 × 128 matrix, TR/TE=2 s/15 ms, FOV=3.0 cm, slice thickness=1 mm. The resulting BOLD image had a spatial resolution of $0.24 \times 0.24 \times 1$ mm^3. Four hundred and fifty images were obtained in about 15 minutes in each scanning session. Studies were performed in sequential order starting with the rest scan, followed by exsanguination, where 8 ml/kg of blood was withdrawn from the femoral artery gradually in about 2–3 min. 8 ml/kg of blood was withdrawn since it led to a decrease in MAP almost near the autoregulatory limits in the present anesthetized rat preparation. MR images were acquired 5 min after withdrawal of the required volume. Finally the withdrawn blood was gradually replaced back into the arterial system at a rate slower than withdrawal. MR images were obtained 5 minutes after replacement was complete.

2.3. Data Analysis and Statistics

Low-pass filtered resting BOLD signal time courses from voxels in the sensorimotor cortex were cross-correlated with every voxel time course in the brain. Previously, it was observed that a significant temporal correlation was obtained with voxels from the sensorimotor and its associated cortex in humans *(25, 26)*. Very few voxels outside the sensorimotor cortex were reported to have significant correlation in humans. In the present study using the rat model, the sensorimotor cortex, caudate putamen, hippocampus and thalamus were defined as regions of interest according to the stereotaxic rat brain atlas *(32)*. A gaussian low-pass filter with a cutoff at 0.1 Hz was applied to all voxel time courses *(33)*. This reduced respiratory and cardiac signals and any corresponding aliased signals.

2.4. Frequency Estimation

The frequencies present in the voxel time-series data were calculated using Welch's averaged periodogram method *(34)*. Briefly, each data set was divided into eight sections, with 50% overlap between adjacent blocks. A Hamming window was then used for each section and the power spectrum computed. After the power spectrum was calculated from each segment, the power spectra were averaged for the eight segments. This was done on a voxel-wise basis for the entire brain. After the power spectrum was calculated, the frequency with the largest amplitude was identified as the dominant frequency noted for all the voxels.

2.5. Temporal Correlation

The temporal correlation between voxels was assessed by two different techniques. The first technique used cross-correlation analysis between six seed pixels from the center of region chosen from the function anatomy and was correlated with every voxel in that slice. Since we are only interested in the temporal correlation due to slow periodic spontaneous oscillations, a finite impulse response filter *(35)* was used to filter the high-frequency components from each of the data sets (as described above). Because of the short data size (110 time points), filter parameters were adjusted to minimize the generation of artifactual frequencies (sidelobes). The correlation coefficient was then calculated using the formula $r = X^{*}Y/(XY)$, where X and Y represent time course pixels from two voxels X and Y, and * represents the dot product. All voxels that passed a correlation of 0.3 were considered significant and locations were noted.

While resting state connectivity maps were generated to show the spatial location of regions that correlated significantly with a representative time course in the brain, they do not necessarily provide much information regarding the interaction between the different subregions. Connectivity patterns between different regions of interest (both for the visual and sensorimotor systems) were obtained using cross-correlation analysis. For each subregion chosen for the sensorimotor or visual systems, principal component analysis was computed *(36)*. Using PCA, a set of orthogonal data sets based on their energy content was determined. Briefly, PCA is a multivariate technique that replaces the measured variables by a new set of uncorrelated variables (principal components), arranged in the order of decreasing standard deviation (SD) or energy distribution. For this study, the first two components accounted for more than 75% of the standard deviation for each region of interest and were therefore used. The resulting time series (consisting of the first PCA) for each subregion was correlated with every other time component to obtain a pair-wise correlation matrix.

While significant temporal correlation typically represents similar time-series structure, it does not give any information regarding the frequency components that may give rise to high correlation coefficients. Therefore, in addition to cross-correlation, coherence maps were generated from which the contribution of specific frequencies could be estimated and compared with the temporal correlation map.

3. Results

BOLD weighted MR images at 9.4 T were obtained in isoflurane-anesthetized rats during various physiological conditions using a gradient echo EPI sequence. During normal physiological conditions, all animals imaged had MAP values ranging between 90 and 120 mmHg with baseline fluctuations in the fMRI-BOLD signal intensity in the frequency range 0.01–0.125 Hz. **Figure 12.1a,d** shows the spatially averaged power spectrum and the low-pass filtered BOLD signal time series from the whole brain during normal physiological condition for a typical rat.

After MR measurements during normal physiological condition, rats were exsanguinated where a small volume (8 ml/kg) of blood was gradually withdrawn from each animal resulting in hypovolemia and hypotension. MAP dropped from the normal control value of $110 \pm 10 - 68 \pm 7$ mmHg during exsanguination in all rats and led to an enhancement in the amplitude of the low frequency BOLD fluctuations. No significant change was observed in S_aO_2 during exsanguination which remained at

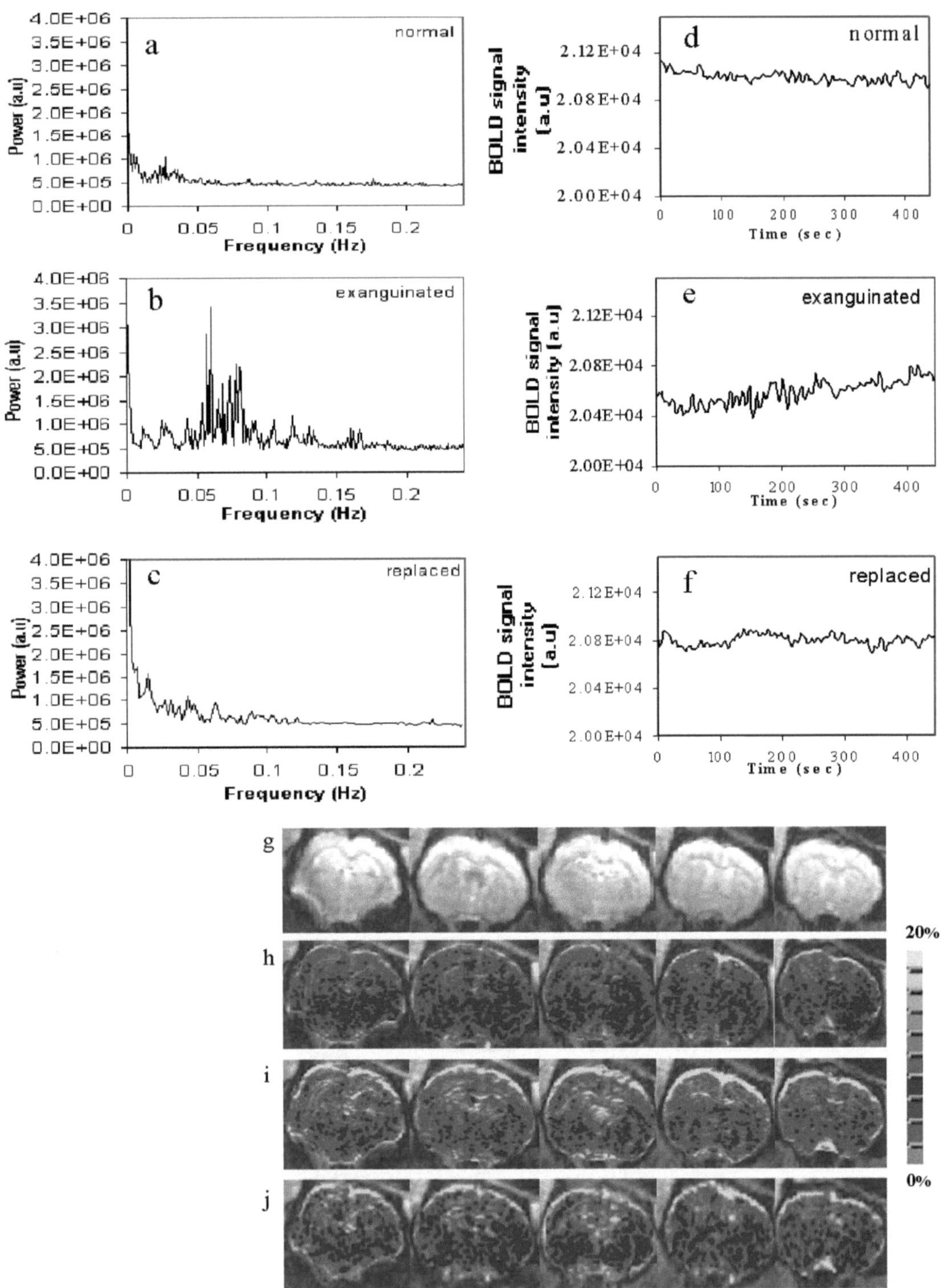

Fig. 12.1. Average power spectra of the BOLD signal, average filtered BOLD signal time series and standard deviation of the BOLD signal from the whole brain in a typical rat during normal, exsanguinated and blood replaced conditions. (**a–c**) average power spectra (**d–f**) average filtered BOLD signal time series, (**g**) anatomy (**h–j**) BOLD signal standard deviation maps during normal, exsanguinated and blood replaced conditions respectively. (*See* Color Plate)

98 ± 1%. This was accompanied by an enhancement in the magnitude of the power spectrum in the low frequency range. Two of the rats studied showed a very minimal (2%) decrease in the mean baseline BOLD signal intensity during exsanguination while five rats showed no significant change. Frequencies centered at 0.02, 0.03, 0.07, 0.10 and 0.125 Hz were significantly enhanced. **Figure 12.1b,e** shows the spatially averaged power spectrum of the BOLD signal and the low-pass filtered BOLD signal time series respectively from the whole brain in a typical rat during exsanguination.

Figure 12.1c,f shows the average power spectrum and the low-pass filtered BOLD signal time series from the whole brain in the same rat after blood replacement. The total time duration between blood withdrawal and replacement was around 30 minutes in all experiments. As indicated by the standard deviation maps from a typical rat, the enhancement in BOLD signal fluctuations was the maximum in cerebral cortex (**Fig. 12.1i**). Withdrawn blood volume, when replaced, led to a partial recovery of the BOLD signal fluctuations (**Fig. 12.1j**). An interesting observation was the enhancement in the fluctuations in the very low frequency range below 0.01 Hz after replacement of blood (**Fig. 12.1c**).

The regional distribution of the low frequency BOLD fluctuation was analyzed from specific anatomical regions namely cerebral cortex, caudate putamen, hippocampus and thalamus traced according to the rat stereotaxic atlas *(32)*. In any typical rat, the spatially averaged Fourier power of the low frequency BOLD fluctuations in the different anatomical regions were cerebral cortex>hippocampus>thalamus>caudate putamen (**Fig. 12.2a–d**). The same order in the Fourier power over anatomical regions was observed over all rats. Exsanguination also led to an increased power in most frequencies below 0.1 Hz over all rats.

Temporal characteristics and spatial distribution of the low frequency physiological fluctuations were analyzed using seed voxels within the previously described anatomical regions of interest (ROI). Randomly selected seed voxels within each ROI were correlated with voxel time courses from the whole brain after low-pass filtering (cutoff frequency of 0.1 Hz). **Figure 12.3a,c** shows typical activation maps obtained by correlating the time course of a seed voxel from the sensorimotor cortex, hippocampus and thalamus with all other regions of the brain. Though small in area, highly correlated clusters were observed across the sensorimotor cortex from either hemisphere with the seed voxel chosen from the sensorimotor cortex (**Fig. 12.3a**). However, correlated clusters were not detectable with seed voxels chosen from the hippocampus or thalamus (**Fig. 12.3b,c**). Exsanguination, which increased the amplitude of BOLD signal fluctuations

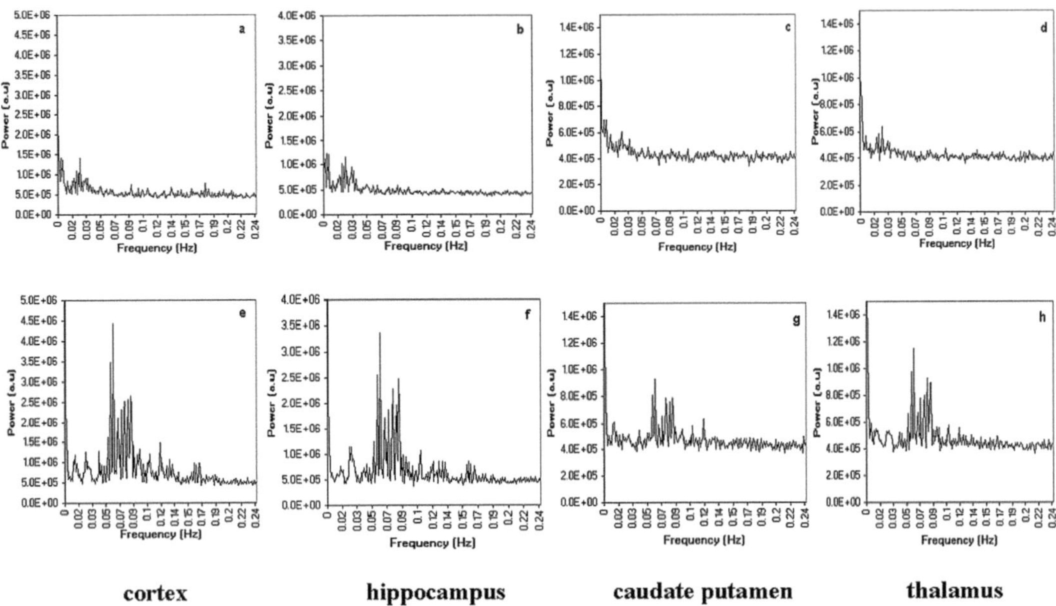

Fig. 12.2. Average power spectra of the BOLD signal from different regions of interest namely cortex, hippocampus, caudate putamen and thalamus in a typical rat during (**a–d**) normal and (**e–h**) exsanguinated conditions.

(**Fig. 12.1d**), led to an increase in the area of correlated voxels from both hemispheres after cross-correlation with a seed voxel time course chosen from different ROI's (**Fig. 12.3d–f**). Correlated voxels in the hippocampus and thalamus across both hemispheres emerged during exsanguination when correlated with a seed voxel obtained from the hippocampus (**Fig. 12.3e**) and thalamus (**Fig. 12.3f**). The correlation maps were distinct depending on the anatomical region of choice of the seed voxel in any single rat. The correlation maps were reproducible across all seven rats. **Figure 12.3g** shows the correlation map obtained after cross-correlating the time course of voxels from the whole brain with a seed voxel from the sensorimotor cortex during exsanguination across all seven rats. The pattern of activation was approximately similar and exhibited bilateral symmetry over each hemisphere over all rats.

Statistical analysis was performed across all rats to determine the extent of spatial correlation between regions. The average correlation coefficients of stereotaxically defined regions namely cerebral cortex, caudate putamen, hippocampus and thalamus (using seed pixels from the cerebral cortex) were determined over each rat and compared among all rats. The average correlation coefficient was the largest in the cerebral cortex, and the least in the hippocampus both during normal and exsanguinated conditions (**Table 12.1**). The correlation of cerebral cortex was relatively better with caudate putamen and thalamus when compared to the hippocampus. The average correlation coefficient for

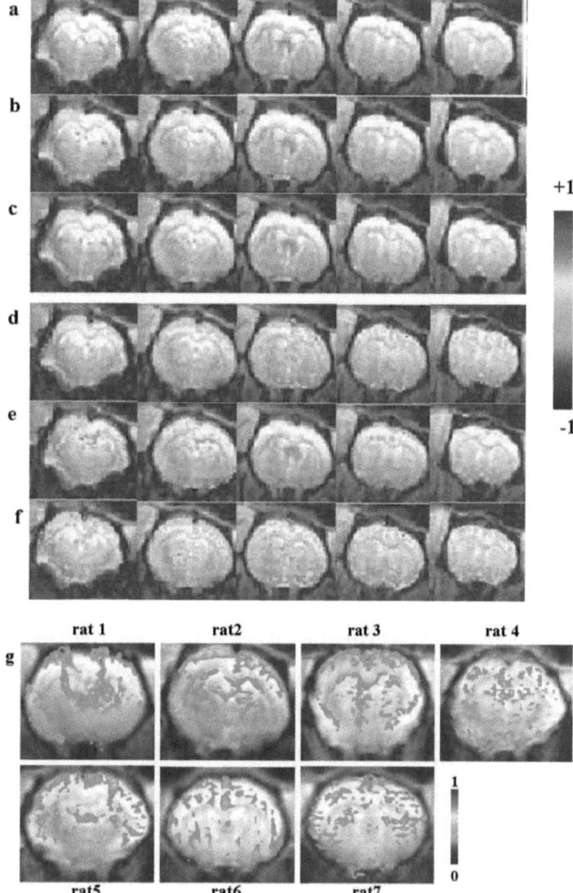

Fig. 12.3. Spatial correlation of low frequency BOLD signals from five contiguous slices from a typical rat. The anatomical underlay consists of a single EPI image of the brain and the functional overlay is the correlation coefficient in the absence of any stimulus during (**a–c**) normal and (**d–f**) exsanguinated conditions. Voxels from the whole brain were low-pass filtered (cut off 0.1 Hz) on a voxel wise basis and subsequently cross-correlated with the time course of a seed voxel obtained from (**a,d**) sensorimotor cortex (**b,e**) hippocampus and (**c,f**) thalamus. (**g**) Typical cross-correlation maps from a single coronal slice during exsanguinated conditions over all rats after cross-correlation with the time course of a seed voxel obtained from the sensorimotor cortex. A threshold ($\geq$0.3 for the correlation coefficient ($P < 10^{-6}$) was used to generate all correlation maps. The seed voxel location shows a high correlation coefficient value in the images. (*See* Color Plate)

each of the stereotaxically defined anatomical regions of interest increased during exsanguination when compared to normal physiological conditions. The average correlation coefficients from various anatomical regions of interest over all rats is shown in **Table 12.2**.

A sign and binomial test was performed under the hypothesis that an increase in mean correlation coefficient in the cortex with exsanguination was a "success" and a decrease in mean correlation

Table 12.1
Average correlation coefficients in stereotaxically defined anatomical regions of interest after cross correlation with 6 different seed voxels chosen from various spatial locations within the cerebral cortex region of interest from a typical rat (rat2)

Cerebral cortex		Caudate putamen		Hippocampus		Thalamus	
Normal	Exsanguinated	Normal	Exsanguinated	Normal	Exsanguinated	Normal	Exsanguinated
0.08	0.28	0.05	0.18	0.02	0.13	0.05	0.22
0.07	0.33	0.01	0.24	−0.03	0.14	0.02	0.28
0.09	0.19	0.03	0.17	−0.02	0.07	0.01	0.11
0.13	0.24	0.04	0.14	0.01	0.13	0.02	0.19
0.10	0.29	0.06	0.17	0.02	0.14	0.05	0.23
0.10	0.28	0.04	0.18	0.00	0.14	0.04	0.24

Table 12.2
Grand mean of the correlation coefficients in different anatomical regions of interest over all rats. The signal time series from six seed voxels in various spatial locations within the cerebral cortex region of interest was cross-correlated with all voxels from the cerebral cortex and other anatomical regions of interest namely caudate putamen, hippocampus and thalamus. Significance was tested considering 6 rats, which were "successful" under the hypothesis that exsanguination increases the average correlation coefficient in the cerebral cortex

Animal	Cerebral cortex		Caudate putamen		Hippocampus		Thalamus	
	Normal	Exsanguinated	Normal	Exsanguinated	Normal	Exsanguinated	Normal	Exsanguinated
Rat1	0.043	0.077	0.018	0.050	−0.013	−0.042	0.033	0.044
Rat2	0.098	0.266	0.038	0.136	0.001	0.128	0.027	0.212
Rat3	0.071	0.142	−0.035	0.038	0.043	0.031	0.095	0.055
Rat4	0.080	0.330	0.051	0.243	0.006	0.146	0.095	0.262
Rat5	0.056	0.049	−0.030	0.036	−0.021	−0.002	0.036	0.042
Rat6	0.075	0.352	0.038	0.236	0.026	0.131	0.035	0.278
Rat7	0.083	0.271	0.020	0.216	0.003	0.125	0.021	0.194
Mean	0.07	0.21	0.01	0.14	0.01	0.07	0.05	0.16
SD	0.13	0.12	0.04	0.10	0.02	0.08	0.03	0.11

Note: $P < 0.001$ cortex with respect to hippocampus during normal condition. $P < 0.0003$ cortex with respect to hippocampus during exsanguinated condition. $P < 0.008$ cortex with respect to caudate putamen during normal condition. $P < 0.001$ cortex with respect to caudate putamen during exsanguinated condition. $P < 0.19$ cortex with respect to thalamus during normal condition. $P < 0.0004$ cortex with respect to thalamus during exsanguinated condition.

coefficient was a "failure". In six out of a total of seven rats imaged, exsanguination led to an increase in the mean correlation coefficient in the cortex. Thus, with six "successful" trials out of a total of seven, and for a probability of success in each trial as 0.95, the chance of observing six or more successes in seven trials was 95.562%. This implied our underlying hypothesis to be valid. In six rats where our hypothesis was valid, the difference between the mean correlation in different anatomical regions was tested using a paired t-test during normal and exsanguinated conditions. As shown in **Table 12.2**, the error probability in the paired hypothesis test for significant difference between regions decreased during exsanguination when compared to normal condition.

Spatial distribution of the low frequency fluctuations in the whole brain was generated from specific frequency bands from the power spectra of the signal time series on a voxel wise basis. **Figure 12.4** shows color-coded maps of the Fourier power of specific frequency bands 0.01, 0.02, 0.03, 0.07 and 0.1 Hz from

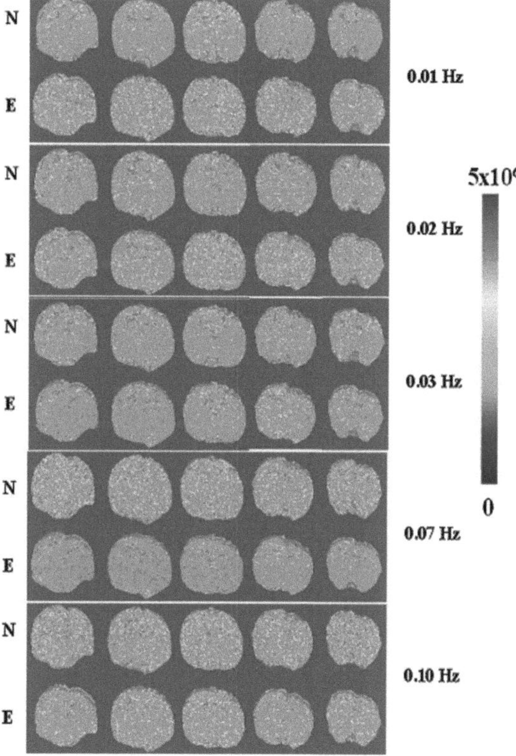

Fig. 12.4. Spatial maps of low frequency fluctuations in the BOLD signal at distinct frequencies in a typical rat during normal (N) and exsanguinated (E) conditions. Images were derived from specific frequencies namely 0.01, 0.02, 0.03, 0.07 and 0.10 Hz respectively from the power spectra of the BOLD signal time series on a voxel wise basis. Fluctuations in most of the low frequency bands increase spatially in the cerebral cortex during exsanguination. (*See* Color Plate)

a typical rat. Low-frequency fluctuations, as observed from the distinct frequency bands, originated predominantly from the cortical region spanning both hemispheres. The power of all frequencies below 0.1 Hz increased during exsanguination in the cerebral cortex.

4. Discussion

4.1. Baseline BOLD Signal Decrease with Hypotension

Exsanguination led to an enhancement in the magnitude of the low frequency fluctuations (**Fig. 12.1e**) despite variation in the baseline BOLD signal decrease. Two of the rats studied indicated a very minimal (2%) decrease in the mean baseline BOLD signal intensity during exsanguination while five rats showed no significant change. The small variation in the mean BOLD signal change observed in two rats may be due to inherent variability in autoregulatory limits in normal anesthetized rats *(37)*. A change in hematocrit as a result of hypovolemia can also cause a minimal decrease in the baseline BOLD signal. Variation in the increase in arteriolar CBV with autoregulated CBF during hypotension can influence the decrease in baseline BOLD signals at high fields *(38)*. However, the rate of MAP change determines the capacity of cerebral vasculature to maintain mean CBF levels. Barzo et al. *(39)* have observed that baseline CBF remains unchanged if hypotension is induced at a rate less than 24 mmHg/min in urethane anesthetized rats. The rate of decrease in MAP in our study was less than 24 mmHg/min as exsanguination was carried out by passive bleeding of the rat over a period of three minutes and is very unlikely to have contributed to the decrease in the baseline BOLD signal. As hypotension does not significantly affect the BOLD signal response to neural activation *(40)*, exsanguination-induced hypotension may not confound the hemodynamic fluctuations if influenced by underlying neural activity.

Amplitude of the BOLD signal fluctuations decreased substantially after replacement of withdrawn blood. While the MAP returned to the normal control levels, the amplitude of the BOLD signal fluctuations did not completely decrease to control levels. The hysteresis in the BOLD signal fluctuation amplitude cannot be attributed to hypotension alone since alternate methods that induce hypotension such as lower body negative pressure in anesthetized rats have shown to be relatively non-invasive where CBF fluctuation amplitude completely returned to normal control levels prior to hypotension *(41)*. Events such as plasma volume refill can occur during exsanguination by transfer of extravascular fluid into the circulation that can transport brain extracellular compounds into the blood stream *(42)*. It is possible that such circulating compounds may have influenced the flow fluctuations even after replacement of withdrawn blood. Further, it is also possible that the return to normal levels may be beyond the measurement window of the present study.

4.2. Dependence of BOLD Signal Fluctuations on CBF and CBV

LDF studies indicate that a decrease in MAP to threshold of autoregulatory limits does not decrease mean blood flow values but increases amplitude of CBF fluctuations, which fluctuates across the same mean value prior to drop in MAP *(5, 43)*. Furthermore, spontaneous fluctuations in BOLD and CBF signals and their dependence on MAP have been observed under different anesthesia *(43, 44)*. The hypocapnia dependent modulation in the amplitude of the low frequency BOLD signal fluctuations, suggests that they are strongly connected to spontaneous CBF fluctuations that have a similar dependence on MAP *(5)*.

BOLD signal response when acquired with a gradient-echo sequence is sensitive to vascular caliber and density. Differences in metabolic regulation and vascular density in different regions affect the BOLD contrast to noise ratio (**Fig. 12.1h**). The rat cortex has a larger blood volume than the thalamus and other deeper structures of the brain *(45)*. The Fourier power of the low frequency physiological fluctuations in the different anatomical regions and their enhancement in response to exsanguination were cerebral cortex>hippocampus>thalamus>caudate putamen (**Fig. 12.2**), following a vascular density or CBV weighted dependence of the BOLD signal fluctuations. The enhancement in the amplitude of the low frequency fluctuations in BOLD signal with a decrease in intravascular pressure and tone similar to LDF fluctuations during exsanguination *(5)* indicates a strong link between fluctuations in brain oxygenation and CBF fluctuations in the microvascular network. While this suggests a myogenic component in the generation of the observed low frequency fluctuations in BOLD signal, some of the results also support underlying neural activity. Though simultaneous measurements of neuronal activity were not carried out in the present study, direct measurement of neuronal signals (spike rate, LFP or EEG) from both hemispheres of the sensorimotor cortex during exsanguination would help clarify the myogenic and neuronal contribution to the observed BOLD signal fluctuations.

4.3. Implications for Resting State Connectivity

Connectivity maps in humans have been generated by cross-correlating the signal time course of every voxel in the brain with a seed voxel chosen from a region where brain activation from the respective sensory or motor stimuli is expected *(26, 46, 47)*. Using a similar analysis, during normal resting conditions in the anesthetized rat, cross-correlating a seed voxel from a region of interest encompassing the sensorimotor cortex, hippocampus and thalamus with the rest of the brain indicated sparse activation with coherent voxels in both hemispheres of the rat cortex (**Fig. 12.3a–c**). Exsanguination led to an increase in the number of correlated voxels across both hemispheres of the brain (**Fig. 12.3d–f**). The spatial extent of the correlation maps was dependent on the anatomical region of choice of the seed voxels that were used for the cross-correlation indicating distinct temporal characteristics of the fluctuations. The correlation maps indicate mostly similar phase characteristics of the low frequency BOLD signal fluctuations in the cortex and thalamus but a different phase in the hippocampus (compare **Fig. 12.3d,e,f**). Most of the cortical regions were negatively correlated with the hippocampal seed voxel (**Fig. 12.3e**). Within the sensitivity of the present experimental protocol, the distinct phase synchrony observed in the hippocampus may represent contribution from specific extravascular factors, which may be neurometabolic and/or neural signaling in origin. Thus, hypotension near the autoregulatory limits can sufficiently enhance the MR-sensitivity and, to a certain extent, may reflect the 'resting state' functional networks in the anesthetized rat model.

Aliasing of frequencies greater than the critically sampled ones can be a concern in the low frequency region. The frequencies of respiration and heart rate in the isoflurane-anesthetized rats were in the region of 1.2 Hz and over 5 Hz respectively and the TR has to be less than 100 ms to theoretically avoid aliasing *(48)*. Recently, using very low and high sampling rates (TR=125 ms and TR=3 s), DeLuca et al. *(49)* have studied the aliasing effects of physiological processes such as cardiac and respiratory fluctuations on the resting state fluctuations in the BOLD signal. The cardiac and respiration frequencies were spatially distinct from the resting state functional network and do not significantly affect the low frequency BOLD fluctuations. In order to minimize any aliasing effects, low-pass filtering (0.1 Hz cutoff) was applied to all data before the correlation analysis. Further, spatial distribution of prominent frequencies in the range between 0.01 and 0.1 Hz was ubiquitous across various anatomical regions with a strong presence in the cortex. No distinct presence of specific frequencies was observed in any one anatomical region either during normal or exsanguinated conditions (**Fig. 12.4**). As much of the fluctuation signal is located over multiple frequencies in the superficial part of the cortex, a strong fluctuation signal is also consistent on the base of the brain, midline and third ventricle (**Fig. 12.4**). Ventricular spaces are susceptible to artifacts from CSF pulsations from high frequency sources such as respiratory and cardiac cycles that may be aliased into the low frequency region in the power spectrum.

It is interesting to note, however, that a similar high correlation of the low frequency BOLD signal fluctuations in both hemispheres of the cerebral cortex existed for at least five different frequencies, which get augmented during exsanguination (**Fig. 12.4**). While this is unlikely due to aliasing effects in the low frequency range, it is possible that connected neuronal networks may be oscillating at multiple frequencies which is more consistent with neuronal than myogenic origin of the BOLD signal fluctuation.

4.4. Implication for Anesthesia Studies in Humans

The increase in the very low frequency below 0.01 Hz in the BOLD signal fluctuations during exsanguination, which remained after replacement of withdrawn blood (**Fig. 12.1b and c**), may indicate an alteration in autoregulatory equilibrium. Though the reasons behind the loss of CBF autoregulation are still controversial, anesthesia can have a major influence on autoregulation *(43, 50)*. The increase in the gain of the fluctuating feedback system that led to an increase in BOLD signal fluctuations during exsanguination may indeed be a result of an alteration of autoregulatory equilibrium *(43, 51, 52)*. These results are in concurrence with studies on anesthetized children undergoing fMRI which have shown significantly altered baseline BOLD signal fluctuations in the very low frequency range 0.02–0.04 Hz *(53)* when compared to BOLD signal fluctuations at 0.1 Hz in awake human subjects *(26)*. Prominent, very low frequency fluctuations (0.02–0.04 Hz) are suggested to occur in the absence of autoregulation *(18)*. A complete physiological monitoring and biophysical characterization of physiological fluctuations using the anesthetized rat model can be very useful in translation studies using anesthesia and fMRI in humans where invasive physiological monitoring can be a problem due to ethical reasons *(53)*.

5. Conclusions

Exsanguination significantly enhances the amplitude and spatial spread of the low frequency BOLD fluctuations to most of the cortical, sub-cortical and deeper brain structures in the isoflurane-anesthetized rat brain. The behavior of the resting state BOLD signal fluctuations similar to CBF fluctuations during exsanguination indicates a myogenic dependence. On the other hand, a high intra-hemispheric symmetry in the BOLD fluctuations with similar phase characteristics in most other regions except the hippocampus suggests that the fMRI-BOLD signal fluctuations in anesthetized rats may carry information from distinct functional networks. Hypotension near the autoregulatory limits in anesthetized rats can be used to improve the detection of distinct resting state neural networks despite the myogenic artefact.

Acknowledgments

This study was supported by NIH grant NS-39044 (BB).

References

1. Vincent JL, Patel GH, Fox MD, Snyder AZ, Baker JT, Van Essen DC, Zempel JM, Snyder LH, Corbetta M, Raichle ME. 2007. Intrinsic functional architecture in the anaesthetized monkey brain. *Nature*, 447:83–86.
2. Zhao F, Zhao T, Zhou L, Wu Q, Hu X. 2008. BOLD study of stimulation-induced neural activity and resting-state connectivity in medetomidine-sedated rat. *Neuroimage*, 39:248–260.
3. Fox MD, Raichle ME. 2007. Spontaneous fluctuations in brain activity observed with functional magnetic resonance imaging. *Nat Rev Neurosci*, 9:700–711.
4. Davis PW, Bronk DW. 1957. Oxygen tension in mammalian brain. *Fed. Proc*, 16: 689–692.
5. Manil J, Bourgain RH, VanWaeyenberge M, Colin F, Blockeel E, DeMey B, Coremans J, Paternoster R. 1984. Properties of the spontaneous fluctuations in cortical oxygen pressure. *Adv Exp Med Biol*, 169:231–239.
6. Hudetz, AG, Roman RJ, Harder DR. 1992. Spontaneous flow fluctuations in the cerebral cortex during acute changes in mean arterial pressure. *J Cereb Blood Flow Metab*, 12: 491–499.
7. Vern BA, Leheta BJ, Juel VC, LaGuardia J, Graupe P, Schuette WH. 1998. Slow fluctuations of cytochrome oxidase redox state and blood volume in unanesthetized cat and rabbit cortex. Interhemispheric synchrony. *Adv Exp Med Biol*, 454:561–570.
8. Cooper R, Crow HJ, Walter WG, Winter AL. 1966. Regional control of cerebral vascular reactivity and oxygen supply in man. *Brain Res*, 3:174–191.
9. Biswal BB, Van Kylen J, Hyde JS. 1997. Simultaneous assessment of flow and BOLD signals in resting-state functional connectivity maps. *NMR Biomed*, 10:165–170.
10. Biswal B, Hudetz AG, Yetkin FZ, Haughton VM, Hyde JS. 1997. Hypercapnia reversibly suppresses low-frequency fluctuations in the human motor cortex during rest using echo-planar MRI. *J Cereb Blood Flow Metab*, 17:301–308.
11. Hudetz AG, Smith JJ, Lee JG, Bosnjak ZJ, Kampine JP. 1995. Modification of cerebral laser Doppler flow fluctuations by halothane, PCO_2, and nitric oxide synthase blockade. *Am J Physiol*, 269:H114–H120.
12. Zonta M, Angulo MC, Gobbo S, Rosengarten B, Hossmann KA, Pozzan T, Carmignoto G. 2003. Neuron-to-astrocyte signaling is central to the dynamic control of brain microcirculation. *Nat Neurosci*, 6:43–50.
13. Filosa JA, Bonev AD, Nelson MT. 2004. Calcium dynamics in cortical astrocytes and arterioles during neurovascular coupling. *Circ Res*, 95:73–81.
14. Mulligan SJ, MacVicar BA. 2004. Calcium transients in astrocyte endfeet cause cerebrovascular constrictions. *Nature*, 431: 195–199.
15. Takano T, Tian GF, Peng W, Lou N, Libionka W, Han X, Nedergaard M. 2006. Astrocyte-mediated control of cerebral blood flow. *Nat Neurosci*, 9:260–267.
16. Vern BA, Schuette WH, Leheta B, Juel VC, Radulovacki M. 1988. Low-frequency fluctuations of cortical oxidative metabolism in waking and sleep. *J Cereb Blood Flow Metab*, 8:215–226.
17. Vern BA, Leheta BJ, Juel VC, LaGuardia J, Graupe P, Schuette WH. 1997. Interhemispheric synchrony of slow fluctuations of cortical blood volume and cytochrome aa3 redox state in unanesthetized rabbits. *Brain Res*, 775:233–239.
18. Dora E, Kovach AG. 1981. Metabolic and vascular volume fluctuations in the cat brain cortex. *Acta Physiol Acad Sci Hung*, 57:261–275.
19. Peltier SJ, Kerssens C, Hamann SB, Sebel PS, Byas-Smith M, Hu X. 2005. Functional connectivity changes with concentration of sevoflurane anesthesia. *Neuroreport*, 16: 285–288.
20. Golanov EV, Yamamoto S, Reis DJ. 1994. Spontaneous waves of cerebral blood flow associated with a pattern of electrocortical activity. *Am J Physio*, 266:R204–R214.
21. Mayevsky A, Ziv I. 1991. Fluctuations of cortical oxidative metabolism and microcirculation in the ischaemic brain. *Neurol Res*, 13:39–47.
22. Berridge MJ, Rapp PE. 1979. A comparative survey of the function, mechanism and control of cellular oscillators. *J Exp Biol*, 81:217–279.

23. Hampson M, Peterson BS, Skudlarski P, Gatenby JC, Gore JC. 2002. Detection of functional connectivity using temporal correlations in MR images. *Hum Brain Mapp*, 15:247–262.
24. Hampson M, Olson IR, Leung HC, Skudlarski P, Gore JC. 2004. Changes in functional connectivity of human MT/V5 with visual motion input. *Neuroreport*, 15:1315–1319.
25. Lowe MJ, Mock BJ, Sorenson JA. 1998. Functional connectivity in single and multislice echoplanar imaging using resting-state fluctuations. *Neuroimage*, 7:119–132.
26. Biswal B, Yetkin FZ, Haughton VM, Hyde JS. 1995. Functional connectivity in the motor cortex of resting human brain using echo-planar MRI. *Magn Reson Med*, 34: 537–541.
27. Xiong J, Parsons LM, Gao JH, Fox PT. 1999. Interregional connectivity to primary motor cortex revealed using MRI resting state images. *Hum Brain Mapp*, 8:151–156.
28. Greicius MD, Krasnow B, Reiss AL, Menon V. 2003. Functional connectivity in the resting brain: A network analysis of the default mode hypothesis. *Proc Natl Acad Sci U S A*, 100:253–258.
29. Gusnard DA, Raichle ME. 2001. Searching for a baseline: Functional imaging and the resting human brain. *Nat Rev Neurosci*, 10:685–694.
30. Bressler SL. Large-scale cortical networks and cognition, 1996. *Brain Res Brain Res Rev*, 20:288–304.
31. Friston KJ, Buechel C, Fink GR, Morris J, Rolls E, Dolan RJ. 1997. Psychophysiological and modulatory interactions in neuroimaging. *Neuroimage*, 6:218–229.
32. Paxinos G, Watson C. 1996. The Rat Brain in Stereotaxic Coordinate, Compact 3rd ed. Academic Press, New York.
33. Hamming RW. 1983. Digital filters, 2nd ed., *Prentice Hall*, Englewood Cliffs, New Jersey.
34. Welch PD. 1967. "The Use of Fast Fourier Transform for the Estimation of Power Spectra: A Method Based on Time Averaging Over Short, Modified Periodograms," *IEEE Trans. Audio Electroacoustics*, Vol. AU-15 (June 1967), pp. 70–73.
35. Biswal B, DeYoe, AE, Hyde JS. 1996. Reduction of physiological fluctuations in fMRI using digital filters. *Magn Reson Med*, 35:107–113
36. Joliffe IT. 1986. Principal Component Analysis, *Springer-Verlag*, New York.
37. Jones SC, Radinsky CR, Furlan AJ, Chyatte D, Qu Y, Easley KA, Perez-Trepichio AD. 2002. Variability in the magnitude of the cerebral blood flow response and the shape of the cerebral blood flow-pressure autoregulation curve during hypotension in normal rats. *Anesthesiology*, 97:488–496.
38. Kalisch R, Elbel GK, Gossl C, Czisch M, Auer DP. 2001. Blood pressure changes induced by arterial blood withdrawal influence bold signal in anesthetized rats at 7 Tesla: implications for pharmacologic MRI. *Neuroimage*, 14:891–898.
39. Barzo P, Bari F, Doczi T, Jancso G, Bodosi M. 1993. Significance of the rate of systemic change in blood pressure on the short-term autoregulatory response in normotensive and spontaneously hypertensive rats. *Neurosurgery*, 32:611–618.
40. Wang R, Foniok T, Wamsteeker JI, Qiao M, Tomanek B, Vivanco RA, Tuor UI. 2006. Transient blood pressure changes affect the functional magnetic resonance imaging detection of cerebral activation. *Neuroimage*, 31:1–11.
41. Herman P, Eke A. 2006. Non-linear analysis of blood cell flux fluctuations in the rat brain cortex during stepwise hypotension challenge. *J Cereb Blood Flow Metab*, 26: 1189–1197.
42. Kugelberg FC, Holmgren P, Druid H. 2003. Codeine and morphine blood concentrations increase during blood loss. *J Forensic Sci*, 48:664–647.
43. Jones SC, Williams JL, Shea M, Easley KA, Wei D. 1995. Cortical cerebral blood flow cycling: anesthesia and arterial blood pressure. *Am J Physiol*, 268:H569–H575.
44. Kannurpatti SS, Biswal BB, Hudetz AG. 2003. MAP reversibly modulates resting state fMRI-low frequency fluctuations in anesthetized rats. *Proc Intl Soc Magn Reson Med*, 11th Annu Meet, Toronto, Canada, pp. 1856.
45. Mandeville JB, Jenkins BG, Kosofsky BE, Moskowitz MA, Rosen BR, Marota JJ. 2001. Regional sensitivity and coupling of BOLD and CBV changes during stimulation of rat brain. *Magn Reson Med*, 45:443–447.
46. Stein T, Moritz C, Quigley M, Cordes D, Haughton V, Meyerand ME. 2000. Functional connectivity in the thalamus and hippocampus studied with functional MR imaging. *AJNR*, 21:1397–1401.
47. Cordes D, Haughton VM, Arfanakis K, Wendt GJ, Turski PA, Moritz CH, Quigley MA, Meyerand ME. 2000. Mapping functionally related regions of brain with functional connectivity MR imaging. *AJNR*, 21:1636–1644.
48. Kiviniemi V, Ruohonen J, Tervonen O. 2005. Separation of physiological very low frequency fluctuation from aliasing by switched sampling interval fMRI scans. *Magn Reson Imaging*, 23:41–46.

49. De Luca M, Beckmann CF, De Stefano N, Matthews PM, Smith SM. 2006. fMRI resting state networks define distinct modes of long-distance interactions in the human brain. *Neuroimage*, 29:1359–1367.
50. Halsey JH Jr, McFarland S. 1974. Oxygen cycles and metabolic autoregulation. *Stroke*, 5:219–225.
51. Vainionpaa V, Timisjarvi J. 1987. Spontaneous oscillation of the systemic arterial blood pressure during cardiopulmonary bypass in man. The effects of some drugs used during the operation. *Basic Res Cardiol*, 82:178–185.
52. Hudetz AG, Biswal BB, Shen H, Lauer KK, Kampine JP. 1998. Spontaneous fluctuations in cerebral oxygen supply. An introduction. *Adv Exp Med Biol*, 454: 551–559.
53. Kiviniemi V, Jauhiainen J, Tervonen O, Paakko E, Oikarinen J, Vainionpaa V, Rantala H, Biswal B. 2000. Slow vasomotor fluctuation in fMRI of anesthetized child brain. *Magn Reson Med*, 44:373–378.

Part V

Alternate Magnetic Resonance Methods

Chapter 13

Dynamic Magnetic Resonance Imaging of Cerebral Blood Flow Using Arterial Spin Labeling

Afonso C. Silva and Fernando F. Paiva

Abstract

Modern functional neuroimaging techniques, including positron emission tomography, optical imaging of intrinsic signals, and magnetic resonance imaging (MRI) rely on a tight coupling between neural activity and cerebral blood flow (CBF) to visualize brain activity using CBF as a surrogate marker. Because the spatial and temporal resolution of neuroimaging modalities is ultimately determined by the spatial and temporal specificity of the underlying hemodynamic signals, characterization of the spatial and temporal profiles of the hemodynamic response to focal brain stimulation is of paramount importance for the correct interpretation and quantification of functional data. The ability to properly measure and quantify CBF with MRI is a major determinant of progress in our understanding of brain function. We review the dynamic arterial spin labeling (DASL) method to measure CBF and the CBF functional response with high temporal resolution.

Key words: Functional magnetic resonance imaging; arterial spin labeling; animal models; cerebral blood flow; spatial resolution; temporal resolution.

1. Introduction

A major goal of neuroscience and neurophysiology is to seek a deeper understanding of brain organization in terms of its anatomical and functional units *(1)*. Modern neuroimaging methods, such as functional magnetic resonance imaging (fMRI) *(2)*, positron emission tomography (PET) *(3)* and optical imaging of intrinsic signals (OIS) *(4)*, have an increasing role in mapping elemental functional units in the cerebral cortex. These functional imaging modalities rely on the coupling of cellular activity to the hemodynamic regulation of energy supply and waste removal – called "cerebrovascular coupling" – to detect the changes in

Fahmeed Hyder (ed.), *Dynamic Brain Imaging: Multi-Modal Methods and In Vivo Applications, vol. 489*

DOI 10.1007/978-1-59745-543-5_13 Springerprotocols.com

cerebral blood flow (CBF), blood volume (CBV) or blood oxygenation, quantities that act as indirect surrogate markers of focal neuronal activity *(5)*. Through the "cerebrovascular coupling", homeostasis of the brain parenchyma is maintained by regulation of CBF on a precise spatial and temporal domain *(6–8)*. Because the spatial and temporal resolution of neuroimaging modalities is ultimately determined by the spatial and temporal specificity of the underlying hemodynamic signals, characterization of the spatial and temporal profiles of the hemodynamic response to focal brain stimulation is of paramount importance for the correct interpretation and quantification of functional data *(5)* and significant effort has been placed on understanding the nature of the cerebrovascular coupling *(9, 10)*.

In the spatial domain, the CBF response to neural activity is well localized within the active cortex. In rat primary somatosensory cortex (S1), for example, elemental vascular units supply individual whisker barrels *(11–13)*. The units are functionally linked for precise focal regulation of CBF, with the highest resting values and the highest changes being localized in layer IV *(14)*. The physiological basis of the differences in CBF increments between different layers may be pericytes placed strategically to regulate capillary blood flow to meet local demands *(15, 16)*, or differences in capillary density accompanying differences in synaptic function *(12, 17, 18)*. The match between microvascular structures and neural columns *(13)* ensures the CBF response is spatially specific to the activated column *(19)*. Indeed, since the mid 1990's, several fMRI and OIS studies successfully mapped elemental functional units in the cortex, such as individual whisker barrels in rat S1 *(20–22)*, individual digits in human sensorimotor cortex *(23, 24)*, cortical laminae in olfactory bulb *(25, 26)* and S1 *(27–29)* of rodents, and ocular dominance or orientation columns in the cat and human visual cortex *(19, 30–39)*.

In the temporal domain, the hemodynamic response consists of several processes with different time scales. Through cerebrovascular coupling, local decreases in vascular resistance lead to an increase in local blood volume and flow. The increase in blood flow results in hyperoxygenation of the capillary bed that drains into the veins, causing a signal that forms the basis of BOLD contrast *(2)*. The increased hyperoxygenation becomes prominent in medium to large draining veins, and thus this response is often considered to be spatially nonspecific. Upon cessation of the increased electrical activity, the neurovascular coupling causes restoration of the capillary and arteriolar volume, resulting in restoration of local CBF and of the imaging signal, once the transit of oxyhemoglobin across the local cerebral vasculature is completed *(40)*. Thus, the temporal resolution of neuroimaging techniques is limited by vascular transit times. Recently, much effort has been placed on determining the "hemodynamic impulse response" (HIR) function as a way to understand the minimal evolution of vascular events so that responses to complex neural activity can be modeled and predicted *(40–45)*. For example, in rats, the OIS response to a 2-s-long whisker or forepaw stimulus begins 0.5–1 s after stimulus onset, peaks at 2.5–3 s, and returns to baseline by 4–5 s *(40, 42–46)*. In humans, the full-width-at-half-maximum (FWHM) is 5–7 s *(47–49)*, and can be improved to 3–4 s if the contribution from large vessels is removed *(50)*.

There is growing evidence indicating that the fundamental spatial and temporal characteristics of the hemodynamic response are fine enough to resolve subcortical activity and suggesting that hemodynamic regulation occurs at a spatial scale that is much finer than the resolution of typical current human or animal fMRI, and implies that higher information content can be obtained with further technological improvements. However, it is not clear to what extent hemodynamic signals will be able to map elemental neuronal populations, and thus continued research on understanding the spatial and temporal evolution of the hemodynamic response will be essential to increase the applicability of neuroimaging to the study of functional brain organization. The ability to properly measure and quantify CBF with MRI is a major determinant of progress into our understanding of brain function. In the present work, we review the dynamic arterial spin labeling (DASL) method to measure CBF and the CBF hemodynamic response with high temporal resolution.

2. Dynamic Arterial Spin Labeling

Cerebral blood flow can be measured with MRI by using endogenous arterial water as a perfusion tracer according to a number of approaches collectively known as arterial spin labeling (ASL) *(51–54)*. The major advantages of ASL are that non-invasive and quantitative measurements of CBF can be performed and repeated indefinitely. Unlike other MRI methods for measuring CBF that use exogenous contrast agents, in ASL, the tracer is imaged against the brain tissue water background. Therefore, the general principle of ASL is to differentiate the net magnetization of arterial water flowing proximally to the brain from the net magnetization of brain tissue water. The labeled arterial water flows through the brain, causing a net decrease in magnetization due to its mixing (with or without exchange) with the brain tissue water, which is proportional to the flow rate and therefore it may be used to calculate CBF in the conventional units of $\left[\frac{ml\ blood}{100\,g\ tissue\cdot min}\right]$. In addition, it is necessary to acquire two images, usually in an interleaved manner, to determine CBF: one with spin labeling, and another as a control. Thus, the ASL methods suffer from poor temporal resolution, which is typically on the order of a few seconds.

The formalism for ASL closely follows the one developed by Kety for monitoring the kinetics of freely diffusible tracers *(55, 56)*. In ASL, the brain tissue longitudinal magnetization can be described by the Bloch equations, modified to include the effects of CBF:

$$\frac{dM_b(t)}{dt} = \frac{M_b^0 - M_b(t)}{T_{1b}} + CBF \cdot [M_a(t) - M_v(t)] \qquad (13.1)$$

where M_b is the brain tissue magnetization per gram of tissue, M_b^0 is the equilibrium value of M_b, T_{1b} is the longitudinal brain tissue relaxation time constant, CBF is the cerebral blood flow expressed in units of $\left[\frac{ml\ blood}{g\ tissue \cdot s}\right]$, and M_a and M_v are the arterial and venous blood magnetization per ml of blood, respectively. The above equation describes brain tissue as a single-compartment that is constantly receiving blood from the arterial side and losing blood water on the venous side. Assuming water to be a freely-diffusible tracer, the venous magnetization equals the brain magnetization according to:

$$M_v(t) = \frac{M_b(t)}{\lambda} \qquad (13.2)$$

where λ is the brain-blood water partition coefficient, defined as the ratio of the amount of water per gram of tissue and the amount of water per ml of blood. In equilibrium, the amount of water delivered by the arterial vasculature to the tissue compartment must equal the amount of water leaving that compartment on the venous side:

$$M_a^0 = M_v^0 = \frac{M_b^0}{\lambda} \qquad (13.3)$$

Therefore, **Equation (13.1)** can be rewritten as:

$$\frac{dM_b(t)}{dt} = \frac{M_b^0 - M_b(t)}{T_{1app}} - 2 \cdot \alpha(t) \frac{CBF}{\lambda} M_b^0 \qquad (13.4)$$

with the apparent longitudinal relaxation time for tissue water in the presence of perfusion, T_{1app}, and the degree of labeling efficiency, $\alpha(t)$, defined as:

$$\frac{1}{T_{1app}} = \frac{1}{T_{1b}} + \frac{CBF}{\lambda}$$

$$\alpha(t) = \frac{M_a^0 - M_a(t)}{2M_a^0} \qquad (13.5)$$

Equation (13.4) tells us a number of important things about ASL. First, to cause a change in brain tissue magnetization related to perfusion, one needs to label blood (i.e., $\alpha(t)$ must be different than zero). Second, the perfusion rate does not instantly change brain tissue magnetization, but it does so with a time constant given by $T_{1\text{app}}$. Third, because CBF is only on the order of 60 ml blood/100 g tissue-minute, CBF/$\lambda \approx 0.01$ s^{-1}, the impact of CBF on $T_{1\text{app}}$ is much too small to allow CBF to be reliably measured from changes in relaxation rates. Traditionally, ASL techniques have been presented as belonging to one of two basic implementation categories *(52, 57)*. In the first approach, arterial water is continuously labeled proximally to the region of interest in the brain *(58, 59)*. This approach is referred to as continuous ASL (CASL). In the second approach, a single, yet large volume of arterial blood is dynamically labeled proximally to the region of interest and allowed to flow into the tissue prior to data collection *(60–63)*. This approach is generally referred to as pulsed ASL (PASL). A detailed comparison of CASL and PASL techniques can be found in *(64)*, and is beyond the scope here.

The CASL approach is attractive for providing better sensitivity than PASL. In CASL, arterial water is continuously saturated ($\alpha_0 = \alpha(0) \approx 0.5$) or inverted ($\alpha_0 \approx 1.0$) proximally to the brain for a period long enough to allow the establishment of a steady-state in brain tissue magnetization. For a constant degree of labeling efficiency $\alpha(t) = \alpha_0 e^{-\tau/T_{1a}}$, where α_0 is the labeling efficiency at the labeling site, τ is the transit-time from the labeling site to the detection site, and T_{1a} is the longitudinal relaxation of arterial water, a steady-state $M_b^{label} = M_b(t)|_{t>5T_{1app}}$ is reached in which:

$$\Delta M = M_b^0 - M_b^{label} = 2\alpha T_{1app} M_b^0 \frac{CBF}{\lambda} \tag{13.6}$$

Thus, the CBF can be obtained from two images obtained with ($\alpha \neq 0$) and without ($\alpha = 0$) labeling:

$$CBF = \frac{\lambda}{2\alpha T_{1app}} \cdot \frac{M_b^0 - M_b^{label}}{M_b^0} \tag{13.7}$$

Equation (13.7) shows that CBF depends on the partition coefficient λ and on three additional parameters: the longitudinal tissue relaxation time T_{1b}, the transit time τ from the site of labeling to the site of interest in the tissue, and the difference in tissue MR signal between the control and the labeled states of magnetization. Commonly, these parameters are obtained in separate experiments. Furthermore, the parameters are measured usually in single instances, precluding dynamic analysis of fast variations. In many applications, as for example in functional MRI experiments aimed at determining the hemodynamic response with high temporal resolution, or in measurement of vascular transit-time maps associated with diagnosis of cerebrovascular diseases, it would be essential to determine CBF dynamically. One efficient way to accomplish dynamic measurements of CBF with ASL is to impose a systematic and periodic variation of the degree of labeling $\alpha(t)$. According to **Equation (13.4)**, periodic variations in $\alpha(t)$ would cause periodic variations in the tissue magnetization, and a temporal analysis of the tissue response would yield dynamic quantification of T_{1b}, τ and CBF. This transient analysis has been named dynamic ASL (DASL) *(53, 65)*.

The dynamic analysis can be introduced by simply allowing the degree of labeling $\alpha(t)$ in **Equation (13.5)** to be a time-dependent, periodic function. In such case, **Equation (13.5)** can be integrated characterizing the time evolution after steady-state has been reached of a system subject to a specific time-varying labeling function, and is given by *(65)*:

$$M_b(t) = M_b^0 - 2\alpha(t)M_b^0 \frac{CBF}{\lambda} e^{-\tau/T_{1a}} \otimes e^{-(t-\tau)/T_{1app}} \qquad (13.8)$$

where $\otimes$ denotes the convolution product. From **Equation (13.8)**, it is important to note that independent of the labeling function, the tissue response has a time shift introduced by the transit time τ.

In theory, the labeling function may be arbitrarily defined as long as the arterial spins can be inverted accordingly, so that definition of the labeling function should be determined by the experimental goals. For example, the use of a periodic square labeling function of period 2Δ defined as follows:

$$\alpha(t) = \begin{cases} \alpha_0 \cdot \mathrm{e}^{-\tau/T_{1a}} & , \quad 0 < t \leq \Delta \\ 0 & , \quad \Delta < t \leq 2\Delta \end{cases} \qquad (13.9)$$

will produce a periodic tissue magnetization response, given by *(65)*:

$$M_b(\tau < t \leq \Delta + \tau) = M_b^0 \left[1 - 2T_{1app}\alpha_0 e^{-\tau/T_{1a}} \frac{CBF}{\lambda} \times \left(\frac{\left(1 - e^{-\Delta/T_{1app}}\right) e^{-(t-\tau)/T_{1app}}}{e^{\Delta/T_{1app}} - e^{-\Delta/T_{1app}}} + \left(1 - e^{-(t-\tau)/T_{1app}}\right) \right) \right]$$

$$M_b(\Delta + \tau < t \leq 2\Delta + \tau) = M_b^0 \left[1 - 2T_{1app}\alpha_0 e^{-\tau/T_{1a}} \frac{CBF}{\lambda} \times \left(\left(\frac{\left(1 - e^{-\Delta/T_{1app}}\right) e^{-\Delta/T_{1app}}}{e^{\Delta/T_{1app}} - e^{-\Delta/T_{1app}}} + \left(1 - e^{-\Delta/T_{1app}}\right) \right) e^{-(t-\Delta-\tau)/T_{1app}} \right) \right] \qquad (13.10)$$

Figure 13.1 shows the tissue response expressed by **Equation (13.10)** simulated for different frequencies of a square labeling function. As expected, the tissue response is shifted with respect to the labeling function by the transit time τ. Also, both amplitude and the phase of the tissue response are modulated by the labeling function. The phase modulation occurs due to the way the labeling function alters the magnetization of the arterial water flowing to the detection site, while the amplitude modulation occurs due to how fast the labeled and unlabeled states are switched when compared to the transit time and to the longitudinal relaxation rate. The DASL technique has the advantages of allowing continuous repetition of the ASL experiment in such a way that the tissue response can be analyzed to unravel in a dynamic manner the three parameters of interest: T_{1b}, τ and CBF. Determination of these parameters can be accomplished with high sensitivity, because the labeling period 2Δ can be made long enough so that the tissue response reaches steady-state (e.g., **Fig. 13.1c**) and, because the periodic nature of the

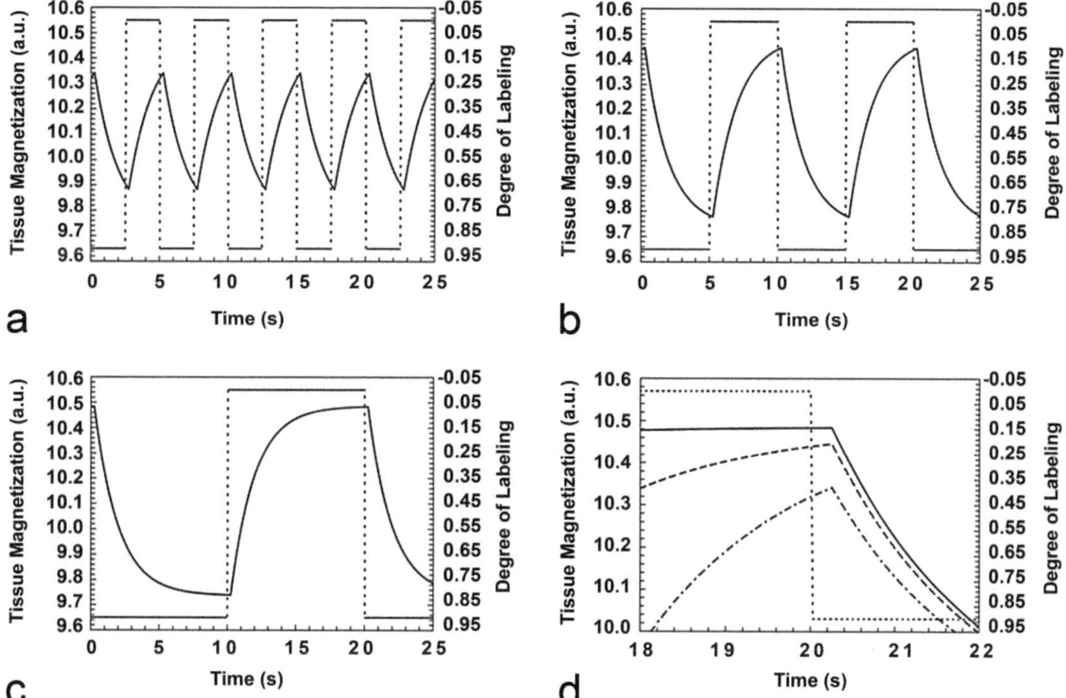

Fig. 13.1. Simulation of the tissue magnetization response to three different frequencies of the ASL function after steady state has been reached using the following parameters: $T_{1app} = 1.75$ s, $\tau = 0.25$ s, CBF = 150 ml / (100 g · min) and $\alpha_0 = 0.90$. Labeling frequencies of 0.20 Hz (**a**), 0.10 Hz (**b**) and 0.05 Hz (**c**) are shown. The dashed lines show the labeling function (*right-hand* scales), and the dark lines show the corresponding tissue responses in arbitrary units (*left-hand* scales). An expanded scale plot of the maximum amplitude reached in each situation is shown in (**d**). Note that different labeling frequencies modulate the amplitudes of the tissue response, while the delayed tissue response is independently governed only by the transit time τ from the labeling plane to the detection region.

response facilitates analysis of the data in the frequency domain, allowing for frequency filtering of noise and other components with frequencies outside the model. It is important to note that even though higher frequencies increase the number of cycles acquired in a given amount of time, thus improving the averaging process, their use may compromise the ability of accurately estimating T_{1app}, due to a decrease in the time span of evolution of the tissue magnetization and a decrease in the number of data points used for estimation of this parameter. On the other hand, the precision of estimation of the transit time τ is proportional to how accurately the temporal shift between the labeling function and the tissue response can be measured, and for this, a higher sampling rate is required. For accurate transit time estimation, the sampling rate of the DASL curve must be higher than the smallest expected transit time. This can be a significant limitation in studies of small animals where the transit times can be quite short, on the order of 100 – 350 ms *(53, 65, 66)*.

The DASL sampling scheme is particularly advantageous in the presence of periodic noise and other sources of signal fluctuation, such as the ones introduced by respiratory and cardiac cycles. In such cases, the periodic nature of the tissue response to a periodic labeling function greatly facilitates filtering of such noise sources. Efficient filtering can be accomplished by fitting the raw data to the model described by **Equation (13.10)**. A simple Fourier transform (FT) of the model determines the allowed spectrum of frequencies, while the FT of the data reveals the periodic perturbations present in the data. A complex filter comprised of real and imaginary components present in the allowed spectrum of frequencies can be applied to the data set to eliminate every frequency other than the ones predicted by the model.

Figure 13.2 illustrates the filtering process based on the DASL model for data obtained from an isoflurane-anesthetized rat at 7 T. The original DASL time course is shown in **Fig. 13.2a**. A labeling frequency of 0.05 Hz was used to generate the DASL time-course. The labeling time *TL* was 200 ms and the image repetition time *TR* was 250 ms. In spite of the good SNR, noise present in the data is quite visible. **Figure 13.2b** shows the real and imaginary components of the FT of the data, while **Fig. 13.2c** shows the frequency components allowed by the best fit of the data to the model in **Equation (13.10)**. **Figure 13.2c** shows the filtered data based on the frequencies allowed by the model. Negative amplitudes were removed by the filter, as well as the frequency component around 1 Hz corresponding to the respiratory rate. The filtered DASL time course is shown in **Fig. 13.2e**, demonstrating the effectiveness of the filter in removing the noise and other unwanted fluctuations. The advantage of this filtering approach is that even perturbations occurring at very low frequencies can be taken out of the data without

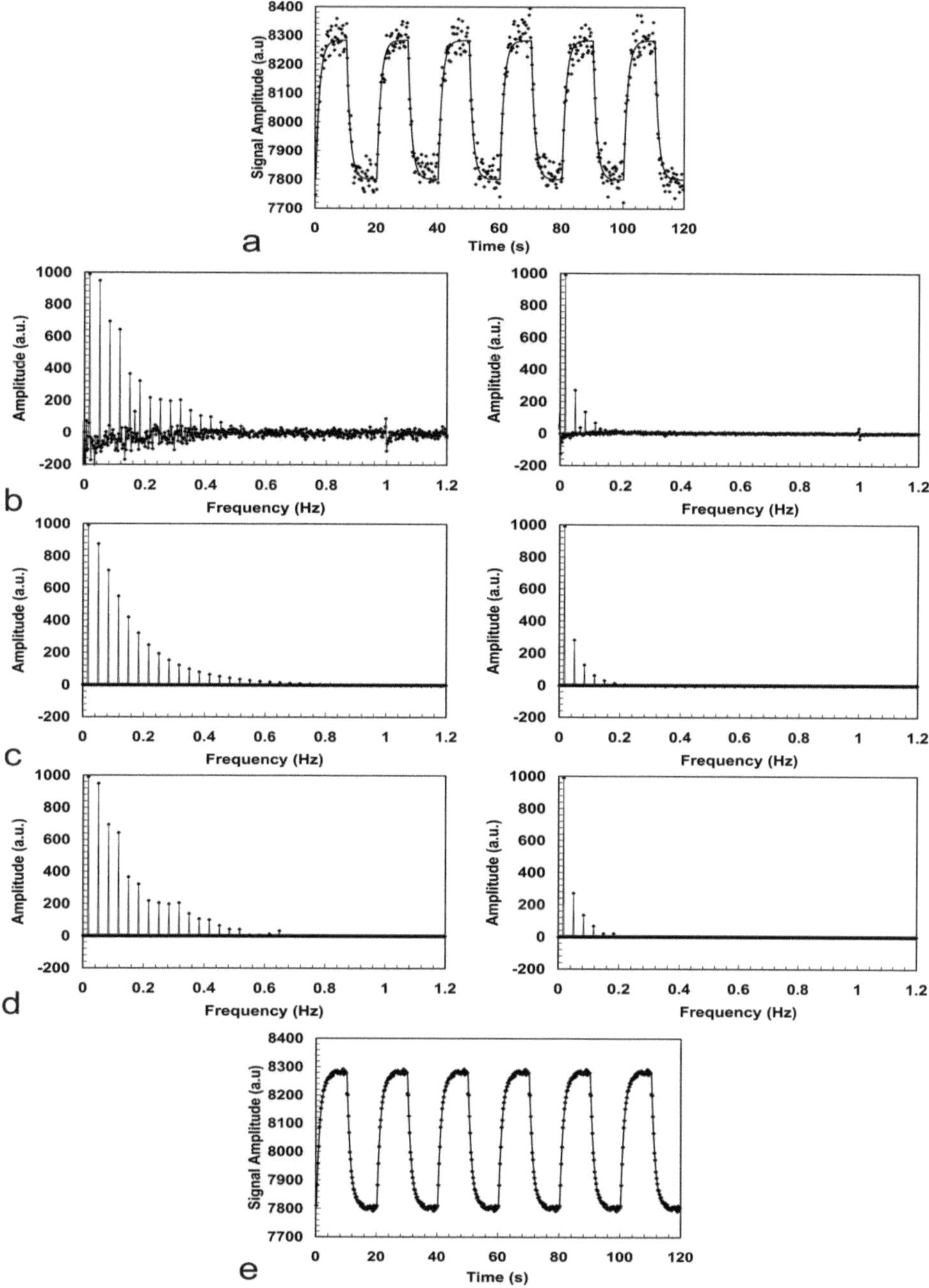

Fig. 13.2. Frequency filtering of DASL data obtained from an isoflurane-anesthetized rat at 7 T using a labeling frequency of 0.05 Hz. (**a**) DASL time-course. (**b**) The real (*left*) and imaginary (*right*) components of the FT of the data. (**c**) The real (*left*) and the imaginary (*right*) components of the FT of the best fit to the model expressed by **Equation (13.10)**. (**d**) The filtered data based only on the frequencies allowed by the model. In addition to removal of the noise, the peak around 1 Hz due to the respiratory frequency is also filtered out. (**e**) The filtered DASL time-course.

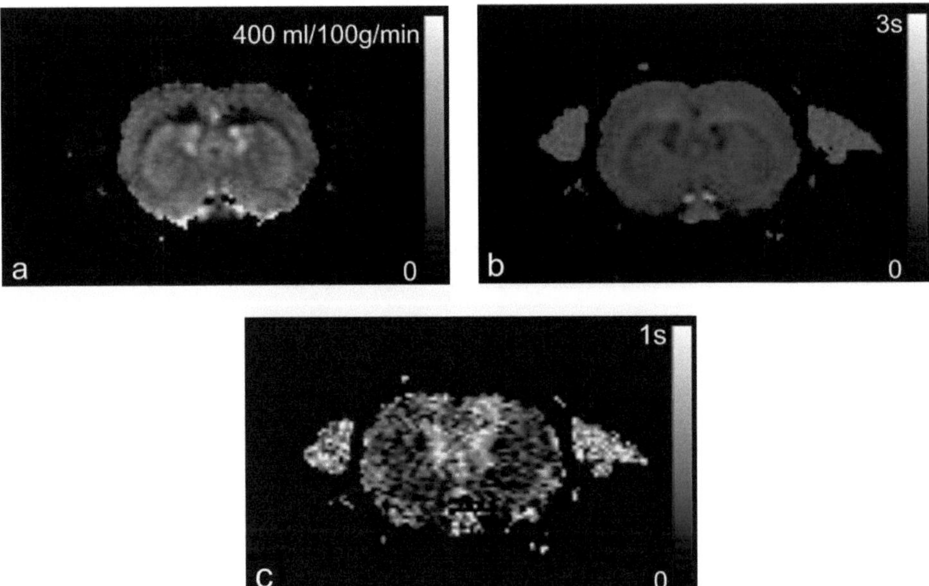

Fig. 13.3. Dynamic ASL parameter maps obtained from an isoflurane-anesthetized rat at 7 T. A 2 mm-thick coronal slice that included the forelimb area of primary somatosensory cortex was imaged using a GE-EPI sequence with the following parameters: $TR/TE = 250/20$ ms, $FOV = 25.6 \times 25.6\text{mm}^2$, matrix $= 128 \times 128$, image acquisition time 49 ms. (**a**) CBF map obtained with high SNR and showing excellent anatomical definition of gray and white mater. (**b**) $T_{1\text{app}}$ map. (**c**) Transit time map.

compromising the tissue response information. In theory, this should also enable to eliminate the contributions of large vessels to the CBF map. Assuming that the arterial protons flowing through the large vessels are flowing fast enough and do not exchange with tissue, they should not behave as predicted by the model in such a way that their contribution would be filtered out of the data set and, thus, not considered in the final fitting that would be used to generate a CBF map. After filtering, the CBF, $T_{1\text{b}}$ and transit time maps were obtained from the data and are shown in **Fig. 13.3a,c** respectively. The average whole-brain CBF value was 149 ml/(100 g ·min) (**Fig. 13.3a**). Note the high SNR of the CBF map. The transit-time map, shown in **Fig. 13.3c**, had an average value of 258 ms, which is in excellent agreement with previously published values for the transit-time in rats *(53, 65, 66)*.

3. Applications of DASL to Measurement of the Hemodynamic Response to Functional Activation

As mentioned above, the temporal resolution of ASL methods for quantifying CBF is inherently low, due to the need to acquire two distinct images and because proper perfusion contrast is achieved only when enough time is allowed for the labeled spins to travel into the region of interest and exchange with tissue spins. The

DASL technique is attractive in improving the temporal resolution of ASL, because data can be acquired with a high sampling rate while the magnetization state of the brain oscillates between the control and the steady-state conditions induced by the periodic labeling function.

In order to obtain high temporal resolution, short ASL RF pulses are used interleaved to an ultra-fast imaging sequence, such as echo-planar or spiral imaging *(67)*. A compromise must exist between using ASL RF pulses that are short enough to allow for high temporal resolution, but long compared to the imaging time so that high labeling duty-cycles (and thus labeling efficiency) can be maintained. Specifically, between the acquisition of two consecutive images with a short repetition time (*TR*), ASL is achieved by a labeling RF pulse of length *TL*. According to this scheme, labeling of the arterial spins can be done with a duty-cycle $k = TL/TR$. If *TL* is not too short compared to *TR*, the duty-cycle k is high enough to preserve the sensitivity of the CASL technique. The proper determination of quantitative CBF maps requires a correct measure of the degree of arterial spin labeling, $\alpha(t)$. The new definition of $\alpha(t)$ to account for the labeling duty-cycle k is given by:

$$\alpha(t) = \begin{cases} k \cdot \alpha_0 \cdot e^{-\tau/T_{1a}}, & 0 \le t \le \Delta \\ 0, & \Delta \le t \le 2\Delta \end{cases} \tag{13.11}$$

Equation (13.11) shows that the effect of using the DASL scheme as described above is to decrease the efficiency of labeling by the duty-cycle factor k. For example, ASL RF pulses of 78 ms duration were implemented in conjunction with a 30 ms readout EPI sequence *(27, 67)*. Under these conditions, CBF images could be formed every 108 ms, however, with a decreased labeling efficiency due to the lower duty-cycle $k = 0.72$. Similar DASL data, acquired using $TL = 200$ ms and $TR = 250$ ms, are shown in **Fig. 13.4.** In this case, $k = 0.8$.

To demonstrate the usefulness of DASL in probing the CBF response to functional brain activation, DASL experiments were performed during bilateral somatosensory stimulation in α-chloralose anesthetized rats at 7 T. A labeling frequency of 0.0167 Hz was used corresponding to a half-period $\Delta = 30$ s. The functional paradigm was defined so that 3 s-long stimulus was presented during the stationary period of the DASL cycles. The stimulus parameters were: current amplitude = 2 mA, pulse width = 333μs and repetition rate 3 Hz. **Fig. 13.4a** shows a t-score map of the BOLD functional response. Robust activation regions were present in both left and right primary somatosensory cortices. **Fig. 13.4b** shows the combined DASL-fMRI time-course. The functional hemodynamic response can be easily noticed on top of the DASL experiment. The control phase of the DASL

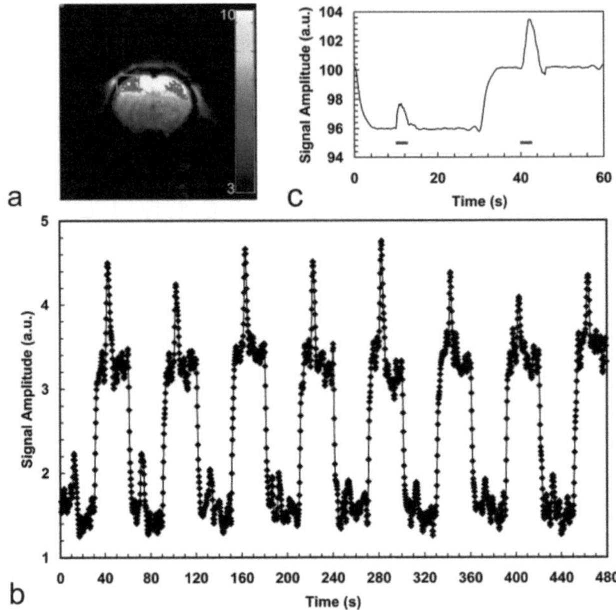

Fig. 13.4. Combined DASL-fMRI experiment. (**a**) t-score map of the BOLD functional response to bilateral stimulation of the rat forelimbs. Robust activation regions are present in both left and right primary somatosensory cortices. (**b**) Combined DASL-fMRI time-course. The functional hemodynamic is superimposed on top of the DASL experiment. (**c**) Data averaged into a single DASL cycle, obtained after fitting and filtering of the DASL curve. Stimulation periods are indicated by the horizontal bars underneath the hemodynamic response. The control phase of the DASL cycle displays a robust BOLD response, while the labeling phase shows the mixing of the functional BOLD and CBF contrast in anti-phase to each other.

cycle displays a robust BOLD response, while the labeling phase shows the mixing of the functional BOLD and CBF contrast in anti-phase to each other. Because functional increases in CBF lead to signal decreases in ASL, the resulting functional signal changes are smaller during the labeling half-cycles (when the BOLD contrast opposes the CBF contrast) than during the control half-cycles (when contrast is due to BOLD alone). After fitting and filtering of the DASL curve, the data was averaged into a single DASL cycle, shown in **Fig. 13.4c**. The hemodynamic response is superimposed on the DASL evolution. Stimulation periods are indicated by the horizontal bars underneath the hemodynamic response.

Once the CBF images are formed, analysis of the temporal characteristics of the CBF response to functional stimulation is desired. For this, a temporal deconvolution of the measured MRI time-course becomes necessary, because instantaneous changes in CBF cannot be quickly reflected in the MRI time-course. To show this, consider the longitudinal tissue magnetization obtained after

n RF pulses is *(68)*:

$$M_z\,(n \times TR) = \frac{M_0(1-e^{-TR/T_{1}app})}{1-\cos\theta \cdot e^{-TR/T_{1}app}} + \frac{M_0(1-\cos\theta)e^{-TR/T_{1}app}}{1-\cos\theta \cdot e^{-TR/T_{1}app}} \cdot \cos^n\theta \cdot e^{-nTR/T_{1}app} \qquad (13.12)$$

The first term in **Equation (13.12)** represents the longitudinal magnetization at the steady state condition, and the second term represents its dynamic evolution from the fully relaxed condition to the steady state condition. The basic principle of the ASL technique is the transfer of the longitudinal magnetization state of the endogenous labeled arterial water to the perfused tissue. According to **Equation (13.12)**, this transfer is limited by $T_{1\mathrm{app}}$, by TR, and by the excitation RF flip angle θ. Thus, it cannot occur instantly. As a consequence, quick changes in CBF such as the ones produced by focal brain stimulation are only reflected a few seconds later in the tissue magnetization, and thus the MRI time course measured in response to a quick change in perfusion is delayed. However, the transfer function that governs the delayed MRI response to step changes in CBF (and in $T_{1\mathrm{app}}$) is inherently present in **Equation (13.12)** in the form of the signal evolution from the equilibrium magnetization state M_0 to the new steady-state magnetization expressed by the first term in **Equation (13.12)**. Provided that it is measured during the MRI measurements of the functional CBF response, the transfer function can be used as a deconvolution kernel to remove the latency in the CBF response imposed by the MRI signal evolution. After the deconvolution process, the resulting CBF time course reflects accurately the dynamics of the actual CBF changes.

Using an acquisition scheme that consisted of employing short ASL RF pulses in conjunction with an ultra-fast imaging sequence, we were able to measure the CBF response to somatosensory stimulation in α-chloralose anesthetized rats with a temporal resolution of 108 ms *(27, 67)*. **Figure 13.5a** shows the MRI-estimated CBF time-course (gray), and the deconvolved CBF time course (black), in response to a 20 s-long stimulus. To obtain the true temporal dynamics of the CBF response, the MRI-estimated CBF curve was deconvolved with the initial 10 s of the control magnetization decay, generating the deconvolved CBF signal. It can be clearly seen how the CBF response measured with MRI is delayed with respect to the deconvolved curve. One drawback of the deconvolution process is that it adds oscillatory noise to the data. However, the CBF changes elicited by somatosensory stimulation in α-chloralose anesthetized rats are quite robust and immune to the small amount of noise introduced by deconvolution of the original functional ASL curve. One important advantage of the use of DASL acquisition strategies in functional MRI is that the control cycle of the ASL curve

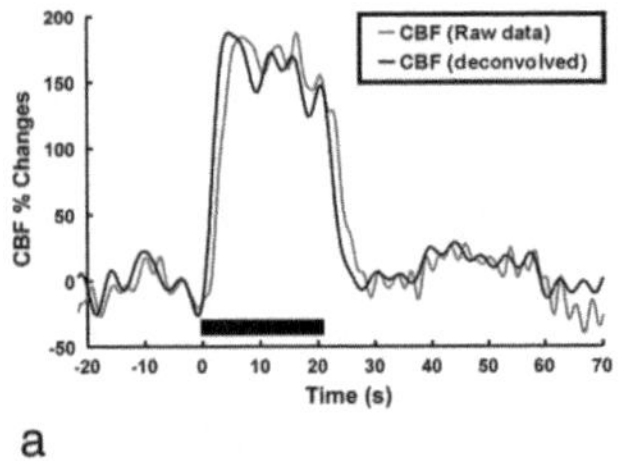

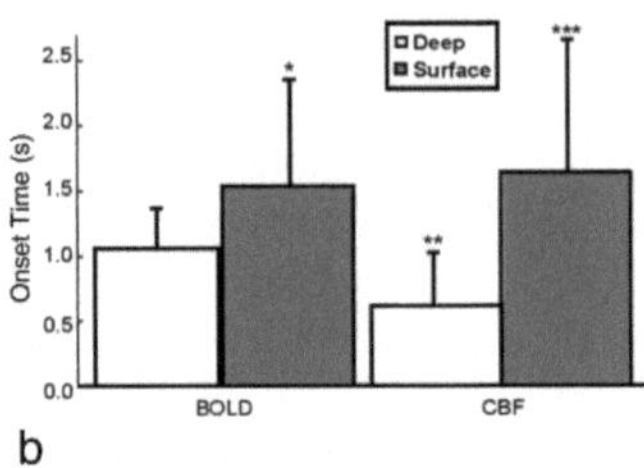

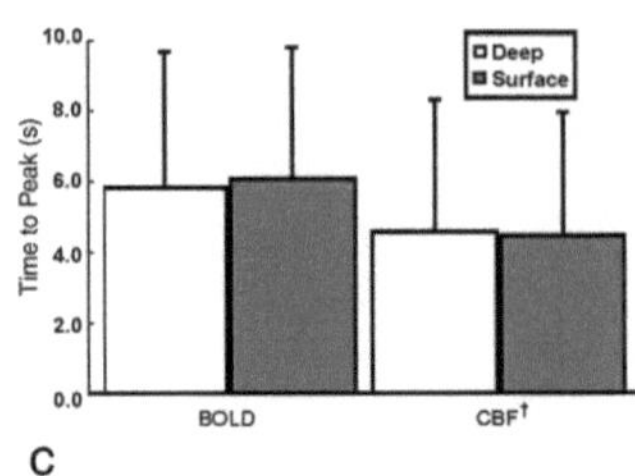

a b c

Fig. 13.5. Temporal response of CBF and BOLD fMRI signals. (**a**) Raw (*gray*) and deconvolved (*black*) CBF response curves obtained using ASL at 108 ms temporal resolution during electrical stimulation of the rat forepaw. The raw CBF curve was obtained at 9.4 T and was deconvolved with the tissue T1 decay curve to produce the true CBF response. (**b**) Averaged onset-times of CBF and GE-BOLD at 9.4 T in the surface (*gray bars*) and deep (*white bars*) regions of the somatosensory cortex. (*) The onset of the BOLD response in the cortical surface was significantly later than deeper in the cortex ($P < 0.03$). (**) CBF changes in the deep cortex occurred earlier than the corresponding BOLD changes ($P < 0.003$). (***) The onset of superficial CBF changes was significantly delayed compared to deep in cortex ($P < 0.004$). (**c**) Averaged time-to-peak of CBF and BOLD. There were no significant time-to-peak differences across regions within either BOLD ($P > 0.28$) or CBF ($P > 0.39$). (†) However, the CBF peak response occurred faster than the BOLD response in both regions ($P < 0.001$). Error bars = 1 std. dev. Adapted from ref. *(27)* with permission of Lippincott, William & Wilkins., and from ref. 67 with permission of Wiley-Liss Inc., Wiley Publishing Inc., a subsidiary of John Wiley & Sons.

can be used to obtain the BOLD functional changes so that this more popular fMRI contrast mechanism can be directly compared to the corresponding CBF changes.

Figure 13.5b shows the onset time of BOLD and CBF in the superficial and deep regions of the somatosensory cortex following the onset of stimulation. The onset of CBF changes in the deep layers of the somatosensory cortex occurred earlier than the corresponding BOLD changes ($P < 0.003$) *(27)*. However, in the superficial layers, the onset of the CBF response was delayed and it was similar to the latency of the superficial BOLD signal changes. **Figure 13.5c** shows the BOLD and CBF times-to-peak. The CBF peak response occurred faster than the BOLD response in both regions ($P < 0.001$) *(27)*. The heterogeneity of fMRI onset-times across the cortical depth layers is indicative of the effects of the vasculature on the fundamental temporal characteristics of the hemodynamic response, and makes a clear statement that the resolution of fMRI is fine enough to resolve subcortical activity and to study subcortical processes, such as columnar and laminar cortical communication.

Indeed, the shorter onset times found with CBF in the deeper layers of the cortex *(27)* were later confirmed with BOLD contrast obtained at higher spatial resolution *(28)* (Table 13.1). The BOLD onset times obtained at high spatial resolution ($200 \times 200\mu m^2$) from layers IV-V in the cortex *(28)* were as fast as the CBF onset-times obtained at a lower spatial resolution *(27, 67)*, suggesting that the onset-time of the hemodynamic response is faster than 600 ms and that improving the spatial resolution may further shorten the onset latency. In the latter study, layer-specific

Table 13.1
Onset time and time-to-peak for CBF and BOLD hemodynamic responses to somatosensory stimulation in rats

Contrast	CBF*		BOLD*		BOLD†		
Cortical region	Deep	Surface	Deep	Surface	Layers I–III	Layers IV–V	Layer VI
Onset Time (s)	0.6 ± 0.4	1.6 ± 1.1	1.1 ± 0.2	1.5 ± 0.8	1.27 ± 0.43	0.59 ± 0.17	1.11 ± 0.45
Time-to-Peak (s)	4.4 ± 3.7	4.4 ± 3.5	5.8 ± 3.9	6.0 ± 3.8	4.37 ± 0.48	4.13 ± 0.62	4.79 ± 0.51

* Data from Ref. *(27, 67)*. Spatial resolution: 470 × 470 × 2000μm^3. Temporal resolution 108 ms.
†Data from Ref. *(28)*. Spatial resolution: 200 × 200 × 2000μm^3. Temporal resolution 40 ms.

heterogeneity was maintained throughout the entire rising portion of the BOLD hemodynamic response, both in amplitude as well as in temporal aspect. Our data showed that layers IV–V in the rat somatosensory cortex have the strongest temporal resemblance to the stimulation paradigm, and that the onset times from those cortical laminae are shorter than the respective latencies from layers I–III and VI, suggesting that the hemodynamic response originates in the central region of the cortex and propagates up and down the cortical depth toward the supragranular and infragranular layers *(28)*. This finding is consistent with the hypothesis that changes in CBF start at the site of neural activity and propagate upstream toward the feeding arterioles and arteries *(69)*. Curiously, the shorter CBF and BOLD hemodynamic responses onset times in cortical layer IV is consistent with the expected order of neuronal current flow in the cortex *(70)*, thus suggesting that the hemodynamic response could be preserving the temporal order of cortical neuronal events. While CBF cannot respond as fast to increased neuronal activity as electrical events, still the cerebral vasculature could be constructed to follow electrical activity as much as possible *(12, 16)*, making fMRI signals useful for probing cortical activity with short latencies, such as laminar communication.

4. Future Applications of DASL

The property of DASL to measure the three basic parameters related to quantification of CBF make the technique attractive for the measurement of hemodynamic parameters both in scientific studies of human brain function as well as in clinical studies of cerebrovascular diseases. The periodic repetition of the labeling function gives DASL superior advantage in achieving CBF measurements with high SNR due to the multiple sampling of the

ASL data and due to the powerful noise reduction achieved by the fitting and filtering algorithms. Furthermore, DASL achieves unparalleled temporal resolution when compared to other ASL techniques, a fundamental feature to enable its application to the measurement of the hemodynamic response to functional brain stimulation and to the measurement of vascular transit times.

Recently, there has been renewed effort to map the perfusion territories of major cerebral arteries *(71)*. While other ASL methods have been developed to measure arterial territories in the brain, none of the proposed techniques to date have enough temporal resolution to allow detailed characterization of the changes in vascular perfusion patterns associated with cerebrovascular diseases such as focal ischemia or hemorrhage. DASL is likely to be particularly suited for these applications where the changes in vascular territories are accompanied by changes in transit-times due to alteration in the perfusion patterns.

In conclusion, DASL is destined to constitute a versatile experimental platform for studying the spatial and temporal characteristics of functional cerebral hemodynamics. Further optimization of the technique will allow analysis of the vascular nature of the BOLD and the CBF responses, as well as the changes in vascular transit-times and perfusion territories associated with functional hyperemia.

Acknowledgment

This research was supported by the Intramural Research Program of the NIH, National Institute for Neurological Disorders and Stroke.

References

1. Rakic,P. (2002) Evolving concepts of cortical radial and areal specification. *Prog Brain Res*, **136**, 265–280.
2. Ogawa,S., Tank,D.W., Menon,R., Ellermann,J.M., Kim,S.G., Merkle,H., Ugurbil,K. (1992) Intrinsic signal changes accompanying sensory stimulation: Functional brain mapping with magnetic resonance imaging. *Proc Natl Acad Sci U S A*, **89**, 5951–5955.
3. Phelps,M.E., Mazziotta,J.C., Huang,S.C. (1982) Study of cerebral function with positron computed tomography. *J Cereb Blood Flow Metab*, **2**, 113–162.
4. Lieke,E.E., Frostig,R.D., Arieli,A., Ts'o,D.Y., Hildesheim,R., Grinvald,A. (1989) Optical imaging of cortical activity: Real-time imaging using extrinsic dye-signals and high resolution imaging based on slow intrinsic-signals. *Annu Rev Physiol*, **51**, 543–559.
5. Villringer,A., Dirnagl,U. (1995) Coupling of brain activity and cerebral blood flow: Basis of functional neuroimaging. *Cerebrovasc Brain Metab Rev*, 7, 240–276.
6. Lauritzen,M. (2001) Relationship of spikes, synaptic activity, and local changes of cerebral blood flow. *J. Cereb. Blood Flow Metab*, **21**, 1367–1383.
7. Attwell,D., Iadecola,C. (2002) The neural basis of functional brain imaging signals. *Trends Neurosci*, **25**, 621–625.
8. Iadecola,C. (2004) Neurovascular regulation in the normal brain and in Alzheimer's disease. *Nat. Rev. Neurosci.*, **5**, 347–360.

9. Logothetis,N.K. (2003) The underpinnings of the BOLD functional magnetic resonance imaging signal. *J Neurosci*, **23**, 3963–3971.
10. Lauritzen,M. (2005) Reading vascular changes in brain imaging: Is dendritic calcium the key? *Nat Rev Neurosci*, **6**, 77–85.
11. Greenberg,J.H., Hand,P., Sylvestro,A., Reivich,M. (1979) Localized metabolic-flow coupling during functional activity. *Acta Neurol. Scand.*, **60**, 12–13.
12. Cox,S.B., Woolsey,T.A., Rovainen,C.M. (1993) Localized dynamic changes in cortical blood flow with whisker stimulation corresponds to matched vascular and neuronal architecture of rat barrels. *J. Cereb. Blood Flow Metab*, **13**, 899–913.
13. Woolsey,T.A., Rovainen,C.M., Cox,S.B., Henegar,M.H., Liang,G.E., Liu,D., Moskalenko,Y.E., Sui,J., Wei,L. (1996) Neuronal units linked to microvascular modules in cerebral cortex: Response elements for imaging the brain. *Cereb Cortex*, **6**, 647–660.
14. Gerrits,R.J., Raczynski,C., Greene,A.S., Stein,E.A. (2000) Regional cerebral blood flow responses to variable frequency whisker stimulation: An autoradiographic analysis. *Brain Res*, **864**, 205–212.
15. Ehler,E., Karlhuber,G., Bauer,H.C., Draeger,A. (1995) Heterogeneity of smooth muscle-associated proteins in mammalian brain microvasculature. *Cell Tissue Res.*, **279**, 393–403.
16. Harrison,R.V., Harel,N., Panesar,J., Mount,R.J. (2002) Blood capillary distribution correlates with hemodynamic-based functional imaging in cerebral cortex. *Cereb. Cortex*, **12**, 225–233.
17. Patel,U. (1983) Non-random distribution of blood vessels in the posterior region of the rat somatosensory cortex. *Brain Res*, **289**, 65–70.
18. Masamoto,K., Kurachi,T., Takizawa,N., Kobayashi,H., Tanishita,K. (2004) Successive depth variations in microvascular distribution of rat somatosensory cortex. *Brain Res*, **995**, 66–75.
19. Duong,T.Q., Kim,D.S., Ugurbil,K., Kim,S.G. (2001) Localized cerebral blood flow response at submillimeter columnar resolution. *Proc Natl Acad Sci U S A*, **98**, 10904–10909.
20. Narayan,S.M., Santori,E.M., Toga,A.W. (1994) Mapping functional activity in rodent cortex using optical intrinsic signals. *Cereb Cortex 1994 Mar-Apr;4(2):195-*, **4**, 195–204.
21. Yang,X., Hyder,F., Shulman,R.G. (1996) Activation of single whisker barrel in rat brain localized by functional magnetic resonance imaging. *Proc Natl Acad Sci U S A*, **93**, 475–478.
22. Yang,X., Hyder,F., Shulman,R.G. (1997) Functional MRI BOLD signal coincides with electrical activity in the rat whisker barrels. *Magn Reson Med*, **38**, 874–877.
23. Kurth,R., Villringer,K., Curio,G., Wolf,K.J., Krause,T., Repenthin,J., Schwiemann,J., Deuchert,M., Villringer,A. (2000) fMRI shows multiple somatotopic digit representations in human primary somatosensory cortex. *Neuroreport*, **11**, 1487–1491.
24. Overduin,S.A., Servos,P. (2004) Distributed digit somatotopy in primary somatosensory cortex. *Neuroimage*, **23**, 462–472.
25. Kida,I., Xu,F., Shulman,R.G., Hyder,F. (2002) Mapping at glomerular resolution: fMRI of rat olfactory bulb. *Magn Reson Med*, **48**, 570–576.
26. Schafer,J.R., Kida,I., Xu,F., Rothman,D.L., Hyder,F. (2006) Reproducibility of odor maps by fMRI in rodents. *Neuroimage*, **31**, 1238–1246.
27. Silva,A.C., Lee,S.P., Iadecola,C., Kim,S.G. (2000) Early temporal characteristics of cerebral blood flow and deoxyhemoglobin changes during somatosensory stimulation. *J. Cereb. Blood Flow Metab*, **20**, 201–206.
28. Silva,A.C., Koretsky,A.P. (2002) Laminar specificity of functional MRI onset times during somatosensory stimulation in rat. *Proc. Natl. Acad. Sci. U. S. A.*, **99**, 15182–15187.
29. Lu,H., Patel,S., Luo,F., Li,S.J., Hillard,C.J., Ward,B.D., Hyde,J.S. (2004) Spatial correlations of laminar BOLD and CBV responses to rat whisker stimulation with neuronal activity localized by Fos expression. *Magn Reson Med*, **52**, 1060–1068.
30. Bonhoeffer,T., Kim,D.S., Malonek,D., Shoham,D., Grinvald,A. (1995) Optical imaging of the layout of functional domains in area 17 and across the area 17/18 border in cat visual cortex. *Eur. J. Neurosci.*, 7, 1973–1988.
31. Malonek,D., Grinvald,A. (1996) Interactions between electrical activity and cortical microcirculation revealed by imaging spectroscopy: implications for functional brain mapping. *Science*, **272**, 551–554.
32. Malonek,D., Dirnagl,U., Lindauer,U., Yamada,K., Kanno,I., Grinvald,A. (1997) Vascular imprints of neuronal activity: Relationships between the dynamics of cortical blood flow, oxygenation, and volume changes following sensory stimulation. *Proc. Natl. Acad. Sci. U. S. A*, **94**, 14826–14831.
33. Menon,R.S., Ogawa,S., Strupp,J.P., Ugurbil,K. (1997) Ocular dominance in human V1 demonstrated by functional magnetic resonance imaging. *J Neurophysiol*, 77, 2780–2787.

34. Kim,D.S., Duong,T.Q., Kim,S.G. (2000) High-resolution mapping of iso-orientation columns by fMRI. *Nat. Neurosci.*, **3**, 164–169.
35. Cheng,K., Waggoner,R.A., Tanaka,K. (2001) Human ocular dominance columns as revealed by high-field functional magnetic resonance imaging. *Neuron*, **32**, 359–374.
36. Goodyear,B.G., Menon,R.S. (2001) Brief visual stimulation allows mapping of ocular dominance in visual cortex using fMRI. *Hum Brain Mapp*, **14**, 210–217.
37. Goodyear,B.G., Nicolle,D.A., Menon,R.S. (2002) High resolution fMRI of ocular dominance columns within the visual cortex of human amblyopes. *Strabismus*, **10**, 129–136.
38. Kim,S.G., Duong,T.Q. (2002) Mapping cortical columnar structures using fMRI. *Physiol Behav.*, 77, 641–644.
39. Fukuda,M., Moon,C.H., Wang,P., Kim,S.G. (2006) Mapping iso-orientation columns by contrast agent-enhanced functional magnetic resonance imaging: Reproducibility, specificity, and evaluation by optical imaging of intrinsic signal. *J Neurosci*, **26**, 11821–11832.
40. Sheth,S.A., Nemoto,M., Guiou,M.W., Walker,M.A., Toga,A.W. (2005) Spatiotemporal evolution of functional hemodynamic changes and their relationship to neuronal activity. *J Cereb Blood Flow Metab*, **25**, 830–841.
41. Narayan,S.M., Santori,E.M., Toga,A.W. (1994) Mapping functional activity in rodent cortex using optical intrinsic signals. *Cereb Cortex 1994 Mar-Apr;4(2):195-*, **4**, 195–204.
42. Narayan,S.M., Esfahani,P., Blood,A.J., Sikkens,L., Toga,A.W. (1995) Functional increases in cerebral blood volume over somatosensory cortex. *J Cereb Blood Flow Metab*, **15**, 754–765.
43. Berwick,J., Martin,C., Martindale,J., Jones,M., Johnston,D., Zheng,Y., Redgrave,P., Mayhew,J. (2002) Hemodynamic response in the unanesthetized rat: Intrinsic optical imaging and spectroscopy of the barrel cortex. *J. Cereb. Blood Flow Metab*, **22**, 670–679.
44. Martindale,J., Mayhew,J., Berwick,J., Jones,M., Martin,C., Johnston,D., Redgrave,P., Zheng,Y. (2003) The hemodynamic impulse response to a single neural event. *J. Cereb. Blood Flow Metab*, **23**, 546–555.
45. Berwick,J., Johnston,D., Jones,M., Martindale,J., Redgrave,P., McLoughlin,N., Schiessl,I., Mayhew,J.E. (2005) Neurovascular coupling investigated with two-dimensional optical imaging spectroscopy in rat whisker barrel cortex. *Eur J Neurosci.*, **22**, 1655–1666.
46. Narayan,S.M., Santori,E.M., Toga,A.W. (1994) Mapping functional activity in rodent cortex using optical intrinsic signals. *Cereb Cortex 1994 Mar-Apr;4(2):195-*, **4**, 195–204.
47. Friston,K.J., Holmes,A.P., Poline,J.B., Grasby,P.J., Williams,S.C., Frackowiak,R.S., Turner,R. (1995) Analysis of fMRI time-series revisited. *Neuroimage.*, **2**, 45–53.
48. Boynton,G.M., Engel,S.A., Glover,G.H., Heeger,D.J. (1996) Linear systems analysis of functional magnetic resonance imaging in human V1. *J. Neurosci.*, **16**, 4207–4221.
49. Aguirre,G.K., Zarahn,E., D'Esposito,M. (1998) The variability of human, BOLD hemodynamic responses. *Neuroimage*, **8**, 360–369.
50. de Zwart,J.A., Silva,A.C., van Gelderen,P., Kellman,P., Fukunaga,M., Chu,R., Koretsky,A.P., Frank,J.A., Duyn,J.H. (2005) Temporal dynamics of the BOLD fMRI impulse response. *Neuroimage.*, **24**, 667–677.
51. Detre,J.A., Zhang,W., Roberts,D.A., Silva,A.C., Williams,D.S., Grandis,D.J., Koretsky,A.P., Leigh,J.S. (1994) Tissue specific perfusion imaging using arterial spin labeling. *NMR Biomed*, **7**, 75–82.
52. Calamante,F., Thomas,D.L., Pell,G.S., Wiersma,J., Turner,R. (1999) Measuring cerebral blood flow using magnetic resonance imaging techniques. *J. Cereb. Blood Flow Metab*, **19**, 701–735.
53. Barbier,E.L., Silva,A.C., Kim,S.G., Koretsky,A.P. (2001) Perfusion imaging using dynamic arterial spin labeling (DASL). *Magn. Reson. Med.*, **45**, 1021–1029.
54. Golay,X., Hendrikse,J., Lim,T.C. (2004) Perfusion imaging using arterial spin labeling. *Top. Magn Reson. Imaging*, **15**, 10–27.
55. Kety,S.S. (1951) The theory and applications of inert gas exchange at the lungs and tissues. *Pharmacol. Rev.*, **3**, 1–41.
56. Kety,S.S. (1985) Regional cerebral blood flow: Estimation by means of nonmetabolized diffusible tracers — an overview. *Semin. Nucl. Med.*, **15**, 324–328.
57. Barbier,E.L., Lamalle,L., Decorps,M. (2001) Methodology of brain perfusion imaging. *J Magn Reson Imaging*, **13**, 496–520.
58. Detre,J.A., Leigh,J.S., Williams,D.S., Koretsky,A.P. (1992) Perfusion imaging. *Magn Reson Med*, **23**, 37–45.
59. Williams,D.S., Detre,J.A., Leigh,J.S., Koretsky,A.P. (1992) Magnetic resonance imaging of perfusion using spin inversion of arterial water. *Proc Natl Acad Sci U S A*, **89**, 212–216.
60. Edelman,R.R., Siewert,B., Darby,D.G., Thangaraj,V., Nobre,A.C., Mesulam,M.M., Warach,S. (1994) Qualitative mapping of cerebral blood flow and functional localization

with echo-planar MR imaging and signal targeting with alternating radio frequency. *Radiology 1994 Aug;192(2):513-*, **192**, 513–520.
61. Kwong,K.K., Chesler,D.A., Weisskoff,R.M., Donahue,K.M., Davis,T.L., Ostergaard,L., Campbell,T.A., Rosen,B.R. (1995) MR perfusion studies with T1-weighted echo planar imaging. *Magn. Reson. Med.*, **34**, 878–887.
62. Kim,S.G. (1995) Quantification of relative cerebral blood flow change by flow-sensitive alternating inversion recovery (FAIR) technique: Application to functional mapping. *Magn Reson. Med.*, **34**, 293–301.
63. Wong,E.C., Buxton,R.B., Frank,L.R. (1997) Implementation of quantitative perfusion imaging techniques for functional brain mapping using pulsed arterial spin labeling. *NMR Biomed.*, **10**, 237–249.
64. Wong,E.C., Buxton,R.B., Frank,L.R. (1998) A theoretical and experimental comparison of continuous and pulsed arterial spin labeling techniques for quantitative perfusion imaging. *Magn Reson. Med.*, **40**, 348–355.
65. Barbier,E.L., Silva,A.C., Kim,H.J., Williams,D.S., Koretsky,A.P. (1999) Perfusion analysis using dynamic arterial spin labeling (DASL). *Magn. Reson. Med.*, **41**, 299–308.
66. Zhang,W., Williams,D.S., Detre,J.A., Koretsky,A.P. (1992) Measurement of brain perfusion by volume-localized NMR spectroscopy using inversion of arterial water spins: Accounting for transit time and cross-relaxation. *Magn Reson Med*, **25**, 362–371.
67. Silva,A.C., Kim,S.G. (1999) Pseudo-continuous arterial spin labeling technique for measuring CBF dynamics with high temporal resolution. *Magn. Reson. Med.*, **42**, 425–429.
68. Gao,J.H., Holland,S.K., Gore,J.C. (1988) Nuclear magnetic resonance signal from flowing nuclei in rapid imaging using gradient echoes. *Med. Phys.*, **15**, 809–814.
69. Iadecola,C. (1993) Regulation of the cerebral microcirculation during neural activity: is nitric oxide the missing link? *Trends. Neurosci.*, **16**, 206–214.
70. Armstrong-James,M., Fox,K., Das-Gupta,A. (1992) Flow of excitation within rat barrel cortex on striking a single vibrissa. *J Neurophysiol*, **68**, 1345–1358.
71. Paiva,F.F., Tannus,A., Silva,A.C. (2007) Measurement of cerebral perfusion territories using arterial spin labelling. *NMR Biomed*, **20**, 633–642.

Chapter 14

Dynamic MRI of Small Electrical Activity

Allen W. Song, Trong-Kha Truong, Marty Woldorff

Abstract

Neuroscience methods entailing in vivo measurements of brain activity have greatly contributed to our understanding of brain function for the past decades, from the invasive early studies in animals using single-cell electrical recordings, to the noninvasive techniques in humans of scalp-recorded electroencephalography (EEG) and magnetoencephalography (MEG), positron emission tomography (PET), and, most recently, blood oxygenation level-dependent (BOLD) functional magnetic resonance imaging (fMRI). A central objective of these techniques is to measure neuronal activities with high spatial and temporal resolution. Each of these methods, however, has substantial limitations in this regard. Single-cell recording is invasive and only typically records cellular activity in a single location; EEG/MEG cannot generally provide accurate and unambiguous delineations of neuronal activation spatially; and the most sophisticated BOLD-based fMRI methods are still fundamentally limited by their dependence on the very slow hemodynamic responses upon which they are based. Even the latest neuroimaging methodology (e.g., multimodal EEG/fMRI) does not yet unambiguously provide accurate localization of neuronal activation spatially and temporally. There is hence a need to further develop noninvasive imaging methods that can directly image neuroelectric activity and thus truly achieve a high temporal resolution and spatial specificity in humans. Here, we discuss the theory, implementation, and potential utility of an MRI technique termed Lorentz effect imaging (LEI) that can detect spatially incoherent yet temporally synchronized, minute electrical activities in the neural amplitude range (microamperes) when they occur in a strong magnetic field. Moreover, we demonstrate with our preliminary results in phantoms and in vivo, the feasibility of imaging such activities with a temporal resolution on the order of milliseconds.

Key words: BOLD, fMRI, neuroimaging, noninvasive, Lorentz effect.

1. Introduction

Over the years, functional neuroimaging research has seen tremendous progress and greatly improved our understanding of brain function. The continuous pursuit to better measure neural activity has led to many exciting technical advances, including

Fahmeed Hyder (ed.), *Dynamic Brain Imaging: Multi-Modal Methods and In Vivo Applications, vol. 489*

DOI 10.1007/978-1-59745-543-5_14 Springerprotocols.com

single-cell electrical recordings, electroencephalogram (EEG), magnetoencephalogram (MEG), positron emission tomography (PET), and most recently, functional magnetic resonance imaging (fMRI). To date, however, these techniques are either invasive or limited in their ability to accurately localize neural activities in space or in time. Even when the latest multimodal neuroimaging methodology is used, there remain fundamental limitations that introduce sources of error and interpretative difficulties. For example, the simple combination of EEG (which has a high temporal resolution) and fMRI (which has a high spatial resolution) does not generally provide an unambiguous delineation of the spatiotemporal sequence of functional brain activity.

Among all the neuroimaging methods currently available, fMRI *(1–4)* has experienced a particularly explosive growth in recent years. The noninvasive nature of magnetic resonance imaging (MRI), along with its high apparent spatial resolution and moderate temporal resolution, rapidly engendered its emergence as one of the dominant techniques in functional brain research. Just about all of the fMRI contrast mechanisms developed thus far, however, have relied on indirect measures of neuronal activity. For example, the widely used blood oxygenation level dependent (BOLD) contrast relies on relative oxygenation changes, perfusion contrast relies on cerebral blood flow (CBF) changes, and cerebral blood volume (CBV) contrast relies on task-induced vessel expansion. Yet, none of the activities that these techniques measure are themselves thought to mediate information processing in the brain, but rather are hemodynamic ramifications of the neuronal activity itself. The hemodynamic modulations inevitably disperse the observed functional signal change both spatially and temporally, despite the high apparent spatiotemporal resolution of these techniques *(5, 6)*. Ideally, one would want to bypass these indirect markers and measure the activity of neurons directly. Motivated by the many advantages of MRI ideal for investigating brain function, researchers in the MR community have thus begun exploring and extending its use with the goal of being able to directly image transient neuronal activities.

Thus far, however, this type of direct MRI technique, albeit theoretically conceivable and probably the most intriguing, has still remained largely unattainable. The goal of noninvasive detection of neuroelectric activity with high spatiotemporal resolution has been extraordinarily challenging in the context of neuroimaging, as the electrical activity of neuronal tissue is extremely weak and the imaging voxels quite small. Indeed, even direct intracranial recordings of evoked local field potentials, which collectively acquire electrical signals from an area on the order of 1 cm^2, only measure potentials on the order of a few hundred microvolts and require hundreds of time-locked signal averages. Further, the electrical activities are also temporally transient and spatially inhomogeneous, exacerbating the difficulties in direct imaging of neural activity using MRI. Nonetheless, theoretical models and initial experiments have been proposed and carried out, which have helped shed light on the tantalizing possibility of direct and noninvasive MRI of neurons in action.

An early MRI study in 1989 by Joy et al. *(7)* provided intriguing results. The authors used spin echo phase imaging to assess the magnetic field perturbations induced by externally injected electrical currents, both in phantoms and in vivo. Their results showed that electrical currents in biological systems on the order of milliamperes could be detected by acquiring phase maps during stimulation. Since neuronal action potentials are essentially electrical depolarizations, these results suggested that it may be possible to image neuronal activity directly using MRI. It first appeared that the limitation was simply the low signal-to-noise ratio (SNR), which could be improved by using time-locked averaging of multiple trials, as is used in techniques such as event-related potential (ERP/EEG) and event-related field (ERF/MEG) recordings and with event-related BOLD fMRI. However, many technical challenges arose in addition to the SNR limitations, including the spatially incoherent and temporally transient nature of the neuroelectric activity, and the multiple confounding synchronized signals reflecting BOLD, CBV, and CBF changes or physiological noise. These proved to be extremely difficult to address, and as a result there were virtually no breakthroughs in the decade following the initial demonstration.

Nevertheless, interest was renewed in the late 1990's after a decade of explosive fMRI research based on the hemodynamic (e.g., BOLD) contrasts. While researchers across various disciplines continue to be deeply attracted to such fMRI methods, neuroscientists and physicists have relatively quickly reached their intrinsic spatiotemporal limitations to truly interpret the neural activities. Several groups have thus recently assessed the feasibility of using MRI for direct imaging of neuronal activation, more specifically by attempting to detect the minute magnetic field changes induced either by electrical currents in phantoms *(8–10)* or by neuronal currents during evoked or spontaneous brain activity in cell cultures or human subjects *(11–22)*. Despite some encouraging results, many issues remain controversial. For example, several simulations *(10, 15)* and experimental *(14, 17)* studies have shown that phase images are more sensitive to magnetic field changes induced by neuronal currents than are magnitude images, yet others claim the opposite *(12, 20, 21)*. Furthermore, several attempts at reproducing the positive results obtained in earlier studies have been unsuccessful *(13, 22)*. Whether conclusive or not, all of these studies were intrinsically limited by the small magnitude of the magnetic field changes induced by neuronal activation. Moreover, and probably more importantly, they did not take advantage of the strong main magnetic field and high-power gradient systems of modern MRI scanners.

To boost the signal detectability, we propose a new MRI technique termed Lorentz Effect Imaging (LEI) *(23, 24)*, which uses magnetic field gradients to significantly amplify and detect the Lorentz effect induced by spatially incoherent yet temporally synchronized neural-range electrical activity in a strong magnetic field. In the present work, we demonstrate its feasibility for imaging electrical currents on the order of microamperes with a temporal resolution on the order of milliseconds in gel phantoms and in vivo *(24, 25)*.

2. Theory

The LEI technique relies on the well known Lorentz effect, whereby a current-carrying conductor (or individual ions) exposed to a magnetic field experiences a Lorentz force equal to the cross product of the current vector (or electric charge) and the magnetic field. If the conductor (or individual ions) is surrounded by an elastic medium, this force induces a spatially incoherent displacement of the elastic medium in adjacent regions, resulting in a spatially incoherent displacement of the spins in these regions. In the presence of a magnetic field gradient, these spins experience a loss of phase coherence, which in turn results in a destructive signal summation within a voxel and, thus, a signal decay similar to that seen in the transverse relaxation effect. This contrast mechanism remains valid even for randomly oriented electrical current, or ionic flows, within a voxel, as the resultant spatially incoherent displacement of surrounding media (e.g. water molecules) still leads to a destructive signal summation, as shown in our recent manuscript *(26)*.

Since a magnetic field gradient also induces a loss of phase coherence of the static spins, resulting in an unwanted signal attenuation outside the regions of interest, balanced gradients (with positive and negative lobes of the same amplitude and duration) need to be applied, so that the phase shifts experienced by the static spins are rephased, as in diffusion-weighted imaging. These gradients must then be synchronized with the current such that it occurs only during either the positive or the negative lobe in order to preserve the phase shifts due to the Lorentz force-induced displacement. Furthermore, multiple cycles of such synchronized oscillating gradients can be used to greatly amplify the loss of phase coherence due to the Lorentz effect, and therefore increase the sensitivity of the technique. In this work, we use a gradient echo sequence with a series of oscillating gradients applied between excitation and data acquisition (**Fig. 14.1**).

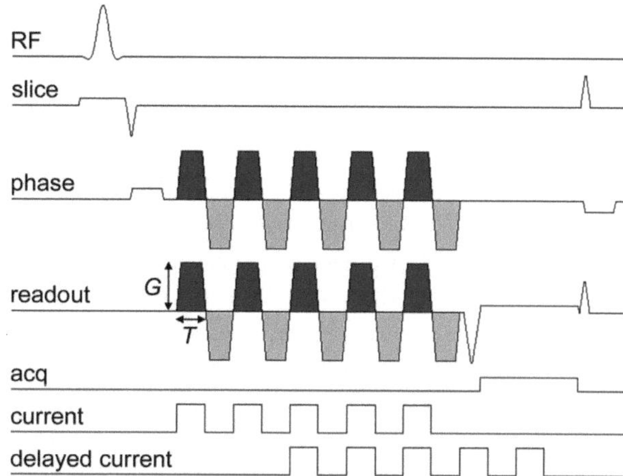

Fig. 14.1. Pulse sequence diagram for the phantom experiments. Gradient echo sequence with *n* (in this example, 5) cycles of oscillating magnetic field gradients (with positive and negative lobes of amplitude *G* and duration *T*, shown in *dark* and *light gray*) applied in the readout and phase encoding directions between excitation and data acquisition. The current is synchronized with the pulse sequence such that it is on only during the positive lobes of the oscillating gradients. In the third study (*bottom line*), the current is delayed with respect to the oscillating gradients (in this example, by two cycles).

Though LEI effect applies to ionic flows *(26)* and is not limited to a wire model, here we use the following simple model to help provide insight into the contrast mechanism of the LEI technique. Here, we consider a cylindrical current-carrying conductor oriented along the *y* axis, placed in a magnetic field oriented along the *z* axis, and surrounded by a homogeneous, isotropic, and linear elastic medium (**Fig. 14.2**). The resulting

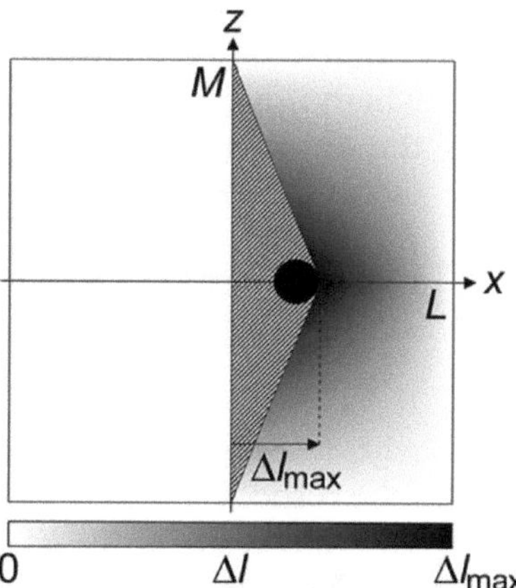

Fig. 14.2. A basic model of the signal loss. A current-carrying conductor (*black circle*) orthogonal to the plane of the figure, placed in a magnetic field oriented along the *z* axis, and surrounded by a linear elastic medium, experiences a Lorentz force resulting in a spatially incoherent displacement $\Delta l(x, z)$ of the elastic medium in the *x* direction ranging between 0 and Δl_{max}. It is assumed that there is no displacement in the $\{x < 0\}$ region and that there is an empty space in the hatched region (see text).

Lorentz force, oriented along the x axis and proportional to the current and the magnetic field, induces a displacement Δl_{max} of the conductor in the x direction, leading to a spatially incoherent displacement Δ l(x, z) of the surrounding elastic medium in the x direction. We assume that the deformation is elastic, i.e., the displacement is proportional to the applied force and inversely proportional to the Young's modulus of the elastic medium (Hooke's law). Furthermore, we assume that the conductor does not adhere to the elastic medium, so that it only induces a compression of the elastic medium on one side, but no dilation on the opposite side, thus leaving an empty space behind it. Because of the symmetry with respect to the z axis, we consider only the $\{z \geq 0\}$ region from now on. The maximum displacement experienced by the elastic medium is equal to the displacement of the conductor Δl_{max} and occurs at ($x = \Delta l_{max}$, $z = 0$). As a first order approximation, we assume that:

1. *(1)* There is no displacement in the $\{x \leq 0\}$, $\{x \geq L\}$, and $\{z \geq M\}$ regions (where L and M are defined in **Fig. 14.2**);
2. There is an empty space in the $\{0 < x < \Delta l_{max}(M{-}z)/M\}$ region (hatched in **Fig. 14.2**); (3) The displacement Δl at an arbitrary point (x, z) in the $\{\Delta l_{max}(M{-}z)/M \leq x \leq L\}$ region decreases linearly from Δl_{max} to zero as follows:

$$\Delta l(x, z) = \frac{L - x}{L - \Delta l_{\max}\frac{M-z}{M}} \frac{M - z}{M} \Delta l_{\max} \qquad (14.1)$$

In this region, the spin density is therefore equal to

$$\rho'(z) = \rho \frac{L}{L - \Delta l_{\max}\frac{M-z}{M}}, \qquad (14.2)$$

where ρ is the spin density in the absence of Lorentz force, and the phase shift obtained by applying n cycles of oscillating gradients (whose positive lobes are synchronized with the current) along the x axis is given by

$$\phi(x, z) = \gamma\, n \int_0^T G\, \Delta l(x, z)\, \mathrm{d}t \approx \gamma\, n\, G\, T\, \Delta l(x, z), \qquad (14.3)$$

where G and T are the amplitude and duration of one gradient lobe respectively, and γ is the gyromagnetic ratio ($2\pi \times 42.57 \times 10^6$ rad/T for protons). This phase shift is thus directly proportional to the displacement. To derive the right-hand side of **Equation (14.3)**, the displacement was assumed to occur over a time much shorter than T. If this were not the case, the resulting phase shift would be smaller.

The ratio of the signal intensity with and without Lorentz effect in a voxel of dimensions $L \times M$ can be computed by integrating the phase shift over the $\{0 \leq x \leq L; 0 \leq z \leq M\}$ region:

$$R = \frac{\sqrt{\left[\int_0^M \int_{\Delta l_{\max}(M-z)/M}^L \rho'(z)\cos\phi(x,z)\,\mathrm{d}x\mathrm{d}z\right]^2 + \left[\int_0^M \int_{\Delta l_{\max}(M-z)/M}^L \rho'(z)\sin\phi(x,z)\,\mathrm{d}x\mathrm{d}z\right]^2}}{\int_0^M \int_0^L \rho\,\mathrm{d}x\mathrm{d}z}$$

$$= \frac{1}{|\phi_{\max}|}\sqrt{\left(\int_0^{\phi_{\max}} \frac{\sin\alpha}{\alpha}\mathrm{d}\alpha\right)^2 + \left(\int_0^{\phi_{\max}} \frac{1-\cos\alpha}{\alpha}\mathrm{d}\alpha\right)^2}, \quad (14.4)$$

where $\phi_{max} = \phi(\Delta l_{max})$. Because of the complex nature of the compression of the elastic medium, the derivation of an analytical solution for a more general case is more difficult. Nevertheless, the simplified model developed here does give an insight into the signal loss mechanism of the LEI technique, while also providing a preliminary theoretical foundation for the quantitative evaluation of small electrical activity-induced MR signal changes as a function of the spatially incoherent displacement.

3. Methods

Building upon this initial theoretical analysis, phantom and in vivo experiments were designed and carried out to examine the contrast mechanism of the LEI technique, demonstrate its high spatial and temporal resolution, and assess its sensitivity for potential applications in biological systems.

3.1. Phantom Experiments

Two spherical gel phantoms (diameter 10 cm, 2.2% gelatin) were constructed. Phantom A contained a straight bundle of carbon wires (overall diameter 500 μm), whereas phantom B contained ten wires (diameter 100 μm) connected in parallel and oriented in random directions in three dimensions. The wires were connected via shielded cables to a square-wave pulse generator triggered by the positive lobes of the oscillating gradients, with a large resistor ($> 1K\Omega$) connected in series to minimize any current induced by the switching gradients that could contribute to the Lorentz effect.

All experiments were performed on a 4 T whole-body MRI scanner (General Electric Medical Systems, Milwaukee, WI, USA) equipped with a high power gradient system (40 mT/m maximum amplitude, 150 T/m/s slew rate), using a shielded quadrature birdcage head coil. The acquisition parameters were optimized based on the following considerations. **Equation (14.3)** shows that large values for n, G, and T should be used to amplify the loss of phase coherence and thus the resulting signal decay due to the Lorentz force-induced displacement. However, the increased diffusion weighting, quantified by the following b-factor: $b = (2/3)\,n\,\gamma^2\,G^2\,T^3$ (for one gradient axis) *(27)*, would result in a global signal attenuation. Since the phase shift is proportional to n, G, and T, whereas the b-factor is proportional to n, G^2, and T^3, it is preferable to use strong and short, rather than weak and long, gradient oscillations, as well as a large number of short gradient oscillations rather than fewer long ones. Consequently, we chose $G = 40$ mT/m and $T = 2$ ms. Furthermore, another trade-off for using a large number of gradient oscillations is the increased echo time (TE) needed to accommodate these gradients, resulting in a global signal attenuation due to T_2^* relaxation. We, therefore, experimentally determined that a value of $n = 15$, corresponding to a minimum TE of 71 ms, was optimal. With this choice of parameters, a maximum displacement Δl_{max} of 10 μm would result in a maximum phase shift $\phi_{max} \approx \pi$ and a signal loss $R \approx 20\%$, as estimated from Equations (14.3) and (14.4) respectively. The resulting b-factor is only 9 s/mm^2, thus causing a negligible signal attenuation due to diffusion weighting. Other imaging parameters were chosen as follows: Repetition time 1000 ms, flip angle 70°, field-of-view 12 cm, matrix size 256 × 128, and slice thickness 5 mm.

Three studies were carried out to assess the dependence of the LEI signal on the intensity of straight and randomly oriented electrical currents, as well as its dependence on the synchronization between the current and the oscillating gradients for a fixed current intensity.

The first study was conducted on phantom A to evaluate the sensitivity of the LEI technique using the simplest geometry. The phantom was positioned in the magnet with the wire orthogonal to the main magnetic field to maximize the Lorentz effect. Axial images were acquired with oscillating gradients applied in the direction orthogonal to both the wire and the main magnetic field, since the Lorentz force induced displacement occurs in that direction. Current pulses of 0, 5, 10, 20, 50, 100, 200, and 500 μA were applied in synchrony with the positive lobes of the oscillating gradients (in separate acquisitions), thus covering the range of values found in biological systems. Five averages were used for current intensities up to 20 μA to increase the SNR.

The second study was conducted on phantom B to demonstrate the feasibility of the LEI technique to detect currents flowing in multiple directions with a more complex geometry. All parameters were identical to those used in the first study, except that oscillating gradients were applied along both directions orthogonal to the main magnetic field. The LEI technique can detect displacements occurring in multiple directions whether oscillating gradients are applied along only one axis or both axes orthogonal to the main magnetic field. In either case, it is most sensitive to displacements occurring in the direction of the largest gradient. However, when oscillating gradients are applied along both axes rather than only one axis, this largest gradient is a factor $\sqrt{2}$ larger (assuming they have the same amplitude along both axes), and consequently the overall sensitivity is higher.

The third study was carried out for two purposes: First, to confirm that the observed signal loss is predominantly due to the intravoxel dephasing resulting from the spatially incoherent displacement of the gel rather than to the bulk displacement of the wire itself (as the wire does not generate any MR signal); and second, to demonstrate the high temporal resolution of the LEI technique. In this experiment, the current intensity was set at 500 μA and the current pulses were delayed with respect to the positive lobes of the oscillating gradients by 0–15 cycles, resulting in an overlap of 15–0 cycles between the two, respectively (**Fig. 14.1**, bottom line). As such, the Lorentz force induced displacement remained identical throughout the study, while the amount of loss of phase coherence and resulting signal decay due to the incoherent displacement was systematically varied. The study was conducted on phantom A with all other parameters identical to those used in the first study. The image acquired with no overlap between the current and the oscillating gradients (in which no signal change should occur) served as the reference, and, as such, was acquired using five averages to increase the SNR.

While the above experiments demonstrate LEI effect in a wire model, similar phenomenon is observed in ionic conductions in our recent work *(26)*, with LEI effect further amplified by the surrounding water molecules. This solution model of the LEI effect may be better suited to simulate the neural conductions in vivo.

4. In vivo *Experiments*

The concept of using synchronized oscillating gradients to increase the signal detectability from electrical currents was extended to in vivo experiments. To minimize potential confounds from hemodynamic modulations or physiological noise commonly seen in brain activation studies and ensure a precise timing control on the stimuli, these experiments were performed in the human median nerve by using electrical stimulation of the wrist to induce intrinsic sensory compound nerve action potentials.

Electrical stimulation of the median nerve was accomplished with a high-impedance electrical current stimulator (Grass S12; Grass-Telefactor, West Warwick, RI, USA) and two gold-plated disk electrodes secured on the ventral and dorsal sides of the right wrist directly over the median nerve (Note: Identical results were obtained when both electrodes were placed on the ventral side of the wrist directly over the median nerve). The current was delivered through the filtered penetration panel via shielded and twisted cables, with the shield grounded to the panel. All electrical switches were installed outside the magnet room, effectively isolating the electrical stimulation in the magnet room and thus removing any electrical interference with the MRI signal. The stimuli consisted of a series of biphasic rectangular current pulses with a duration of 1 ms and an amplitude ranging from 1.8 to 2.7 mA, which was set prior to each session to be just below the motor threshold for finger movement in order to avoid motion artifacts (electromyography measurements were performed for confirmation). The stimulator was triggered by the MRI scanner to ensure an accurate synchronization between the electrical stimulation and the pulse sequence.

The studies were performed on the same 4 T whole-body MRI scanner as used for the phantom experiments. All images were acquired by using a surface coil designed in-house and a gradient echo single-shot spiral imaging sequence with the following parameters: Repetition time 2,000 ms, TE 35.5 ms, flip angle 80°, field-of-view 20 cm, matrix size 64 × 64, and three contiguous axial slices (sagittal with respect to the forearm), which were each 15 mm thick and centered on the median nerve to ensure that the section of the nerve between the wrist and the elbow would be fully contained within one slice. A series of oscillating gradients with an amplitude of 36 mT/m and a duration of 5 ms for each lobe was applied along both axes orthogonal to the main magnetic field. Because the conduction time of the sensory activation in the human median nerve between the wrist and the elbow is about 4 ms *(28)*, this duration of 5 ms was chosen to ensure that sensory nerve action potentials could propagate in that section of the nerve within one lobe. High-resolution T_1-weighted images were also acquired at the same location for anatomical reference.

Four experiments were performed to examine the signal changes during stimulation. To ensure sufficient loss of phase coherence while limiting the T_2^* decay, a maximum of three cycles of gradient oscillations and three synchronized electrical pulses was used for the first experiment (Exp. 1, **Fig. 14.3a**). To establish a graded effect, a second experiment was carried out with only two cycles of gradient oscillations and two synchronized electrical pulses (Exp. 2, **Fig. 14.3b**). In addition, two control experiments were performed that were identical to Exp. 1 but with the electrical pulses delayed by 50 ms with respect to the oscillating gradients (Exp. 3, **Fig. 14.3c**) or without oscillating gradients (Exp. 4, **Fig. 14.3d**).

The activation paradigm was a block design consisting of seven alternating rest and stimulation periods, each lasting 20 seconds, during which ten image volumes were acquired. During the stimulation periods, electrical pulses were triggered to excite the median nerve, whereas during the rest periods, no stimulation was applied. Four runs were acquired for each experiment and averaged to increase the SNR.

After each run, subject bulk motion was assessed by computing the displacement of the image center-of-mass over time slice-by-slice, and runs with an in-plane displacement exceeding

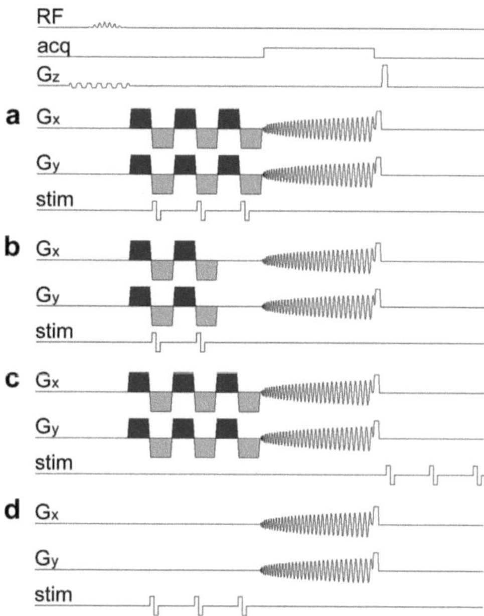

Fig. 14.3. Pulse sequence diagrams for the in vivo experiments. Timing of the radiofrequency excitation pulse (RF), the data acquisition window (acq), the magnetic field gradients on the slice selection (G_z), readout (G_x), and phase encoding (G_y) axes, and the current applied by the stimulator to the wrist (stim) for: (**a**) Three cycles of gradient oscillations (shown in *dark* and *light gray*) and three electrical pulses triggered at the onset of the negative gradient lobes (Exp. 1); (**b**) the same as (**a**) but with two cycles of gradient oscillations and two electrical pulses (Exp. 2); (**c**) the same as (**a**) but with the electrical pulses delayed by 50 ms (Exp. 3); (**d**) the same as (**a**) but without oscillating gradients (Exp. 4).

one pixel were discarded (~ 10% of the runs). Voxel-by-voxel linear detrending was applied to remove any linear drift in the MRI signal. A group student's *t*-test (one-tailed) was then carried out to detect significant differences between the images acquired during the rest and stimulation periods. The *t*-score maps were converted to *Z*-score maps and thresholded using a *Z*-score of $Z > 5$ (corresponding to a significance level of $P < 2.5 \times 10^{-7}$ uncorrected for multiple comparisons) and a cluster size of five voxels. Finally, the resulting activation maps were overlaid on the coregistered high-resolution anatomical images.

5. Results

5.1. Phantom Experiments

5.1.1. Current Intensity Dependence for a Straight Current

The results of the first phantom study are shown in **Fig. 14.4**. As expected, the central signal dip observed on the image acquired without current (**Fig. 14.4a**), which is due to the presence of the wire, becomes progressively larger and wider with increasing current intensities, as can be seen more clearly on the difference images (**Fig. 14.4b**). This signal loss, caused by the intravoxel

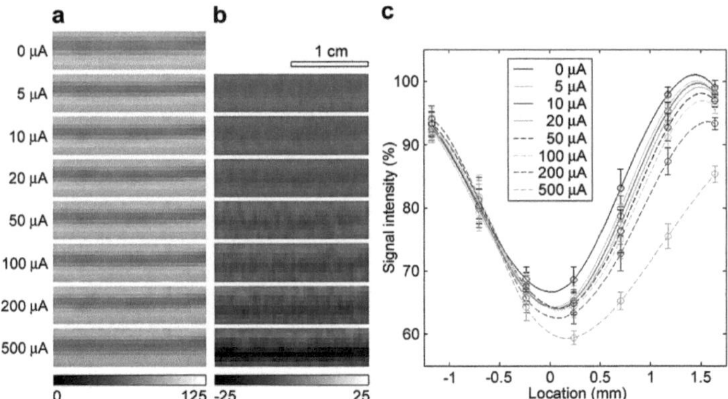

Fig. 14.4. Current intensity dependence for a straight current. (**a**) Images of a selected region of phantom A containing the wire acquired with different current intensities. (**b**) Corresponding difference images with the image acquired without current. (**c**) Signal intensity profiles across the wire averaged over the section shown in (**a**) and (**b**) for the different current intensities. The lines are computed using cubic spline interpolation. The signal intensity in (**a**), (**b**), and (**c**) is expressed in percentage of the maximum signal in the image acquired without current.

dephasing due to the Lorentz force induced incoherent displacement, occurs only on one side of the wire, as shown on the plot of the average signal intensity profiles across the wire (**Fig. 14.4c**). Similar results are obtained when the direction of the current is reversed, except that the signal loss occurs on the opposite side. The widening can reach up to 750 μm (full width at half maximum) for a current intensity of 500 μA, with a corresponding maximal signal loss of 25%. However, currents as low as 5 μA can still be detected.

Although the signal loss was only observed on one side of the wire, since the latter does not adhere to the gel, it is important to note that even if this were not the case (e.g., for a neuron surrounded by tissue), a dilation of the elastic medium on the opposite side would induce a spatially incoherent displacement as well, and thus, a loss of phase coherence similar to that induced by the compression of the elastic medium, resulting in a signal loss on both sides and making the technique more sensitive.

5.1.2. Current Intensity Dependence for Randomly Oriented Currents

The results of the second phantom study are shown in **Fig. 14.5** and demonstrate that the LEI technique does not require unidirectional currents. As for the straight wire in the first study, the signal losses correspond to the wire locations and become more pronounced with increasing current intensities, which can be clearly seen on both the original images (**Fig. 14.5a**) and the difference images (**Fig. 14.5b**). The overall signal loss dependence on the current intensity is highly comparable to that observed for the straight current. However, because of the directional dependence of the LEI signal, the segments of wire that were

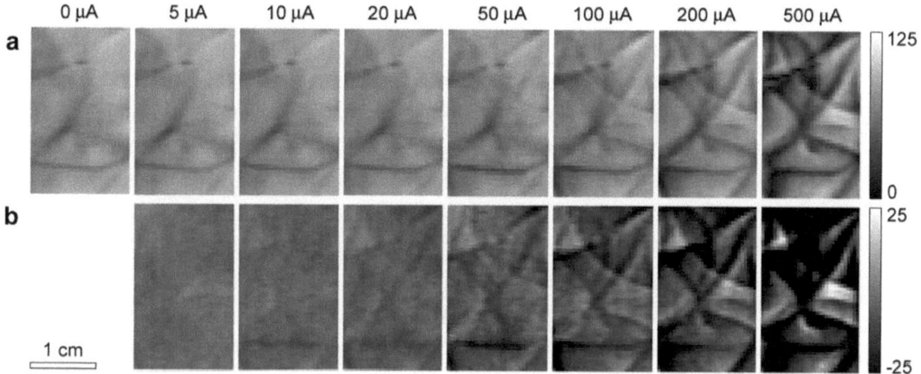

Fig. 14.5. Current intensity dependence for randomly oriented currents. (**a**) Images of a selected region of phantom B acquired with different current intensities. (b) Corresponding difference images with the image acquired without current. The signal intensity is expressed in percentage of the maximum signal in the image acquired without current.

not orthogonal to the main magnetic field experienced a smaller Lorentz force and therefore a smaller signal loss.

5.1.3. Dependence on the Synchronization between the Current and the Oscillating Gradients

The results of the third phantom study are shown in **Fig. 14.6** As expected, the signal loss progressively increases with the amount of overlap between the current pulses and the positive lobes of the oscillating gradients, which can be clearly seen on both the difference images (**Fig. 14.6b**) and the plot of the average signal loss within the entire region (**Fig. 14.6c**).

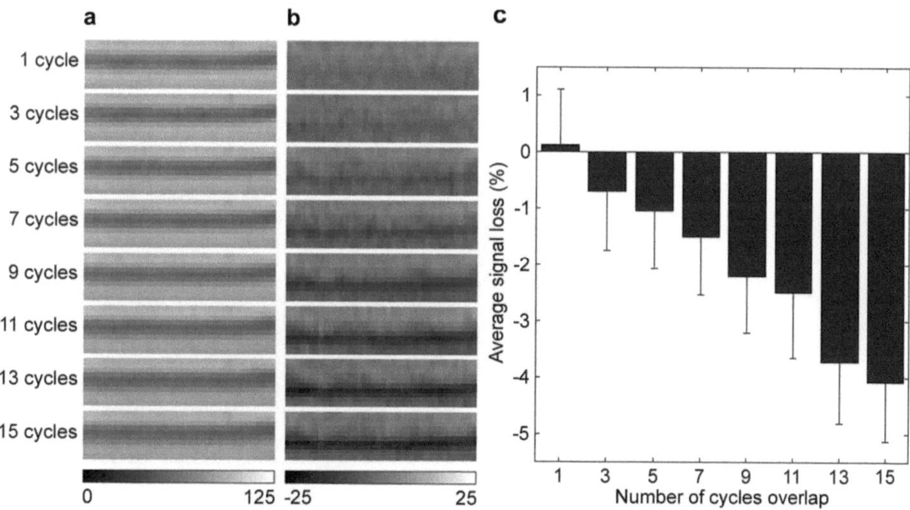

Fig. 14.6. Dependence on the synchronization between the current and the oscillating gradients. (**a**) Images of a selected region of phantom A containing the wire acquired with a 500 μA current intensity and different amounts of overlap between the current pulses and the positive lobes of the oscillating gradients. (**b**) Corresponding difference images with the image acquired without overlap, illustrating the signal loss as a function of the intravoxel dephasing, but not of the displacement itself. (**c**) Signal loss averaged over the region shown in (**a**) and (**b**) for the different amounts of overlap. The signal intensity in (**a**), (**b**), and (**c**) is expressed in percentage of the maximum signal in the image acquired without overlap.

In this study, the current intensity was fixed, so that the Lorentz force induced displacement remained identical for all conditions. On the other hand, the overlap between the current and the oscillating gradients was systematically varied, resulting in different amounts of loss of phase coherence. As such, these results demonstrate that the observed signal loss is predominantly due to the intravoxel dephasing resulting from the spatially incoherent displacement of the gel rather than the bulk displacement of the wire itself. Indeed, the signal loss from the bulk displacement itself is virtually undetectable as shown in the first image in **Fig. 14.6**, where only one cycle of overlap was used.

Furthermore, since the duration of one cycle of oscillating gradients is only 4 ms, these results also demonstrate that a temporal resolution on the order of milliseconds can be achieved with the LEI technique, which represents a dramatic improvement – at least two orders of magnitude – as compared to conventional BOLD fMRI.

5.2. In vivo Experiments

Figure 14.7 shows a representative functional activation map for Experiment 1, in which three cycles of gradient oscillations and three synchronized electrical pulses were used, overlaid on high-resolution T_1-weighted images of the forearm. Highly significant activation was found along the median nerve across subjects. The time course averaged over the activated region shows a systematic signal decrease of $(4.4 \pm 0.7)\%$ during the stimulation periods (**Fig. 14.8a**). The transitions between rest and stimulation periods exhibit no delay, in contrast to what is typically observed in conventional BOLD fMRI studies, which are limited by a hemodynamic delay of 3–6 seconds.

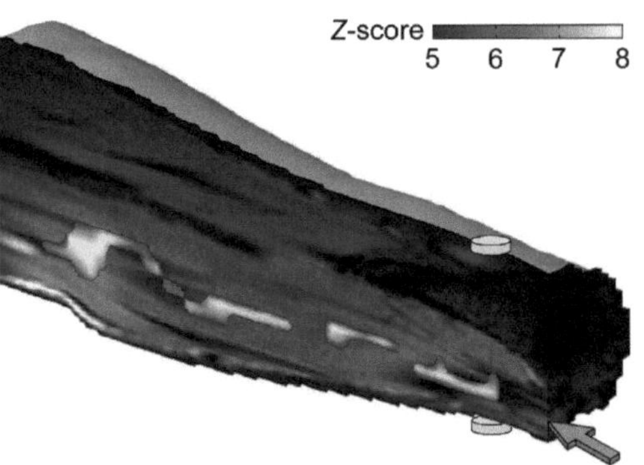

Fig. 14.7. Activation map showing the effect of neuroelectric activity in vivo in the human median nerve. The activation was obtained using three cycles of gradient oscillations and three electrical pulses synchronized with the negative gradient lobes (Exp. 1), and is overlaid on a stack of coregistered anatomical images. The discs represent the electrodes placed on the dorsal (*top*) and ventral (*bottom*) sides of the wrist. The arrow points to the median nerve.

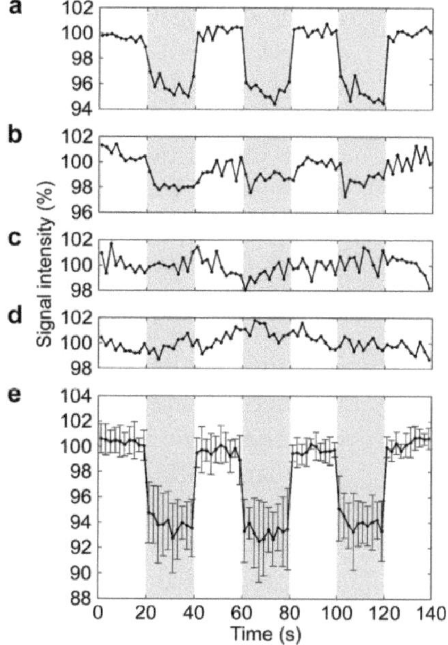

Fig. 14.8. Time courses during alternating periods of rest and electrical stimulation. These results were obtained using: (**a**) Three cycles of gradient oscillations and three electrical pulses synchronized with the negative gradient lobes (Exp. 1); (**b**) the same as (**a**) but with two cycles of gradient oscillations and two electrical pulses (Exp. 2); (**c**) the same as (**a**) but with the electrical pulses delayed by 50 ms (Exp. 3); (**d**) the same as (**a**) but without oscillating gradients (Exp. 4); (**e**) the same as (**a**) but averaged over seven different experimental sessions. The time courses in (**a–d**) were averaged over the activated region obtained in Experiment 1; the time course in (**e**) was averaged over the activated regions of all seven sessions. The error bars represent the standard deviation over the seven sessions. Each time course is normalized to the mean signal intensity during rest. The rest and stimulation periods are shown in white and gray, respectively.

Less significant activation was detected in Experiment 2, in which only two cycles of gradient oscillations and two synchronized electrical pulses were used. The time course averaged over the same activated region as in Experiment 1 shows a signal decrease of only $(1.5 \pm 0.6)\%$ during the stimulation periods (**Fig. 14.8b**). This finding illustrates that the sensitivity of our technique can be significantly decreased when there is insufficient loss of phase coherence, which is consistent with its contrast mechanism.

As expected, no activation was detected in either control experiment, in which the electrical pulses were delayed with respect to the oscillating gradients (Exp. 3) or in which no oscillating gradients were used (Exp. 4). The time courses averaged over the same activated region as in Experiment 1 show no systematic signal changes during the stimulation periods (**Figs. 14.8c and d**). These results further validate the contrast mechanism of our technique by demonstrating that the observed activation is indeed due to the loss of phase coherence generated by the oscillating gradients rather than the bulk displacement of the nerve itself, since this displacement is present in both control experiments, but is not synchronized with oscillating gradients. It should thus be emphasized that the magnitude of the displacement does not by itself determine the sensitivity of the technique, since, for a given displacement, the loss of phase coherence can be independently amplified by using more, stronger, and/or longer oscillating gradients, given sufficient SNR. Furthermore, the fact that the loss of phase coherence can be amplified by the oscillating gradients demonstrates that it is indeed caused by the spatially incoherent displacement rather than the magnetic field induced by the current. Finally, Experiments 3 and 4 also show that there are no artifacts due to eddy currents induced in the stimulation circuit by the oscillating gradients or electrical interference from the stimulation pulses, respectively, since no activation was detected when oscillating gradients or electrical pulses were applied alone.

It can therefore be derived that Experiment 2 is equivalent to Experiment 1 but with a one-cycle (i.e., 10 ms) temporal offset between the electrical stimulation and the oscillating gradients, since only electrical pulses that are synchronized with oscillating gradients contribute to the observed activation. As such, the significant difference between the results of Experiments 1 and 2 shows that our technique is sensitive to timing differences of the stimulation paradigm on the order of milliseconds, thus demonstrating its high temporal resolution.

Finally, to evaluate the test-retest reliability of our technique for directly imaging neuroelectric activity in vivo in a healthy median nerve, Experiment 1 was carried out in seven separate sessions on the same subject under identical experimental conditions to remove the dependence on subject and experimental variability. Highly significant activation was consistently detected along the median nerve in each session. The time course in the activated regions averaged over the seven sessions (**Fig. 14.8e**) highly resembles that obtained in a single session (**Fig. 14.8a**) and shows a systematic signal decrease of $(6.3 \pm 2.2)\%$ during the stimulation periods, confirming the consistency and robustness of our technique.

6. Discussion and Conclusions

The results presented here demonstrate the capability of the LEI technique for imaging spatially incoherent yet temporally synchronized electrical currents on the order of microamperes with a temporal resolution on the order of milliseconds. Accordingly, they provide a theoretical and experimental foundation for its potential application in imaging neural activities in the human brain.

Our studies have shown that the sensitivity of the LEI technique can be substantially improved by applying successive cycles of oscillating gradients and using optimized parameters. Moreover, this sensitivity can be further increased by using a higher field strength (since not only the SNR but also the magnitude of the Lorentz effect increases with field strength) and/or stronger oscillating gradients, as such advances in hardware become increasingly more available on modern MRI scanners. This is a clear advantage relative to methods that rely on the intrinsic magnetic field changes induced by the current, which are independent of the main magnetic field strength.

As shown here in the human median nerve, the LEI technique can potentially be applied to white matter tracts in the central nervous system to study the functional connectivity between various brain areas and to assess white matter integrity in diseases such as multiple sclerosis. Further, since our technique does not require the electrical current to be unidirectional, it can potentially be extended to image focal dendritic neuroelectric activity in gray matter, which, if successful, could have a tremendous impact on our ability to noninvasively study neuronal information processing in the brain. However, unlike for applications in the peripheral nervous system, the temporal delays between the stimuli and the neural activation in various cortical areas are often not known, making it difficult to synchronize the pulse sequence with the neuroelectric activity. Nevertheless, scalp ERP recordings could be used to help determine the proper delays, which could then be incorporated into the pulse sequence to allow time-locked detection of neural activation.

Despite the promising results presented here, direct noninvasive neuroimaging in the human brain using the LEI technique remains experimentally challenging. First, synchronized confounding factors, such as functional signals reflecting BOLD, CBV, and CBF changes, as well as physiological noise, can dominate the detected signal. Therefore, a careful design of the experimental paradigm is required to separate these slow effects from the rapid effects due to neuroelectric activity. In addition, as mentioned above, the criticality of the close timing synchrony between the neuroelectric activity and the oscillating gradients will require considerable further development for this technique to work successfully in gray matter. Moreover, even if all the timing information is known (e.g., using ERP or other neuroelectrical data), the need for extremely accurate synchronization between the stimulation and image acquisition will be highly demanding on present hardware capabilities. Nevertheless, such efforts to overcome these challenges and implement our technique for direct imaging of brain neuroelectric activity in vivo are currently underway.

With these challenges addressed, the potential of using LEI to directly and noninvasively image neural activations in real time can be realized to achieve both a high temporal resolution and spatial specificity as compared to conventional BOLD fMRI, and to help determine hierarchical organizations within activated neural networks. It is anticipated that, if even moderately successful, this technique could have a significant impact on neuroscience research.

Acknowledgments

This work was, in part, supported by the NIH (NS 50329, NS 41328) and NSF (BES 602529).

References

1. K.K. Kwong, J.W. Belliveau, D.A. Chesler, I.E. Goldberg, R.M. Weisskoff, B.P. Poncelet, D.N. Kennedy, B.E. Hoppel, M.S. Cohen, R. Turner, H.-M. Cheng, T.J. Brady, B.R. Rosen, Dynamic magnetic resonance imaging of human brain activity during primary sensory stimulation, Proc. Natl. Acad. Sci. USA 89 (1992) 5675–5679.
2. P.A. Bandettini, E.C. Wong, R.S. Hinks, R.S. Tikofski, J.S. Hyde, Time course EPI of human brain function during task activation, Magn. Reson. Med. 25 (1992) 390–397.
3. S. Ogawa, D.W. Tank, R. Menon, J.M. Ellermann, S.G. Kim, H. Merkle, K. Ugurbil, Intrinsic signal changes accompanying sensory stimulation: functional brain mapping with magnetic resonance imaging, Proc. Natl. Acad. Sci. USA 89 (1992) 5951–5955.
4. S. Ogawa, R.S. Menon, D.W. Tank, D.G. Kim, H. Merkle, J.M. Ellermann, K. Ugurbil, Functional brain mapping by blood oxygenation level-dependent contrast magnetic resonance imaging. A comparison of signal characteristics with a biophysical model, Biophys. J. 64 (1993) 803–812.
5. S.G. Kim, W. Richter, K. Ugurbil, Limitations of temporal resolution in functional MRI, Magn. Reson. Med. 37 (1997) 631–636.
6. R.L. Buckner, Event-related fMRI and the hemodynamic response, Hum. Brain Mapp. 6 (1998) 373–377.
7. M. Joy, G. Scott, M. Henkelman, In vivo detection of applied electric currents by magnetic resonance imaging, Magn. Reson. Imaging 7 (1989) 89–94.
8. J. Bodurka, A. Jesmanowicz, J.S. Hyde, H. Xu, L. Estkowski, S.J. Li, Current-induced magnetic resonance phase imaging, J. Magn. Reson. 137 (1999) 265–271.
9. J. Bodurka, P.A. Bandettini, Toward direct mapping of neuronal activity: MRI detection of ultraweak, transient magnetic field changes, Magn. Reson. Med. 47 (2002) 1052–1058.
10. D. Konn, P. Gowland, R. Bowtell, MRI detection of weak magnetic fields due to an extended current dipole in a conducting sphere: A model for direct detection of neuronal currents in the brain, Magn. Reson. Med. 50 (2003) 40–49.
11. H. Kamei, K. Iramina, K. Yoshikawa, S. Ueno, Neuronal current distribution imaging using magnetic resonance, IEEE Trans. Magn. 35 (1999) 4109–4111.
12. J. Xiong, P.T. Fox, J.H. Gao, Directly mapping magnetic field effects of neuronal activity by magnetic resonance imaging, Hum. Brain Mapp. 20 (2003) 41–49.
13. R. Chu, J.A. de Zwart, P. van Gelderen, M. Fukunaga, P. Kellman, T. Holroyd, J.H. Duyn, Hunting for neuronal currents: Absence of rapid MRI signal changes during visual-evoked response, Neuroimage 23 (2004) 1059–1067.
14. M. Bianciardi, F. Di Russo, T. Aprile, B. Maraviglia, G.E. Hagberg, Combination of BOLD-fMRI and VEP recordings for spin-echo MRI detection of primary magnetic effects caused by neuronal currents, Magn. Reson. Imaging 22 (2004) 1429–1440.

15. D. Konn, S. Leach, P. Gowland, R. Bowtell, Initial attempts at directly detecting alpha wave activity in the brain using MRI, Magn. Reson. Imaging 22 (2004) 1413–1427.
16. P.A. Bandettini, N. Petridou, J. Bodurka, Direct detection of neuronal activity with MRI: Fantasy, possibility, or reality?, Appl. Magn. Reson. 29 (2005) 65–88.
17. N. Petridou, D. Pleaz, A.C. Silva, M. Lowe, J. Bodurka, P.A. Bandettini, Direct magnetic resonance detection of neuronal electrical activity, Proc. Natl. Acad. Sci. USA 103 (2006) 16015–16020.
18. L.S. Chow, G.G. Cook, E. Whitby, M.N.J. Paley, Investigating direct detection of axon firing in the adult human optic nerve using MRI, Neuroimage 30 (2006) 835–846.
19. G.E. Hagberg, M. Bianciardi, B. Maraviglia, Challenges for detection of neuronal currents by MRI, Magn. Reson. Imaging 24 (2006) 483–493.
20. Y. Xue, J.-H. Gao, J. Xiong, Direct MRI detection of neuronal magnetic fields in the brain: Theoretical modeling, Neuroimage 31 (2006) 550–559.
21. L.S. Chow, G.G. Cook, E. Whitby, M.N.J. Paley, Investigation of MR signal modulation due to magnetic fields from neuronal currents in the adult human optic nerve and visual cortex, Magn. Reson. Imaging 24 (2006) 681–691.
22. L.M. Parkes, F.P. de Lange, P. Fries, I. Toni, D.G. Norris, Inability to directly detect magnetic field changes associated with neuronal activity, Magn. Reson. Med. 57 (2007) 411–416.
23. A.W. Song, A.M. Takahashi, Lorentz effect imaging, Magn. Reson. Imaging 19 (2001) 763–767.
24. T.-K. Truong, J.L. Wilbur, A.W. Song, Synchronized detection of minute electrical currents with MRI using Lorentz effect imaging, J. Magn. Reson. 179 (2006) 85–91.
25. T.-K. Truong, A.W. Song, Finding neuroelectric activity under magnetic-field oscillations (NAMO) with magnetic resonance imaging in vivo, Proc. Natl. Acad. Sci. USA 103 (2006) 12598–12601.
26. T.-K. Truong, A. Avram, A. W. Song, Lorentz effect imaging of ionic currents in solutions, J. Magn. Reson. 59 (2008) 221–227.
27. E.O. Stejskal, J.E. Tanner, Spin diffusion measurements: Spin echoes in the presence of a time-dependent field gradient, J. Chem. Phys. 42 (1965) 288–292.
28. J. Kimura, Electrodiagnosis in disease of nerve and muscle: Principles and practice, Oxford University Press, Oxford (2001).

Chapter 15

Advanced In Vivo Heteronuclear MRS Approaches for Studying Brain Bioenergetics Driven by Mitochondria

Xiao-Hong Zhu, Fei Du, Nanyin Zhang, Yi Zhang, Hao Lei, Xiaoliang Zhang, Hongyan Qiao, Kamil Ugurbil and Wei Chen

Abstract

The greatest merit of in vivo magnetic resonance spectroscopy (MRS) methodology used in biomedical research is its ability for noninvasively measuring a variety of metabolites inside a living organ. It, therefore, provides an invaluable tool for determining metabolites, chemical reaction rates and bioenergetics, as well as their dynamic changes in the human and animal. The capability of in vivo MRS is further enhanced at higher magnetic fields because of significant gain in detection sensitivity and improvement in the spectral resolution. Recent progress of in vivo MRS technology has further demonstrated its great potential in many biomedical research areas, particularly in brain research. Here, we provide a review of new developments for in vivo heteronuclear ^{31}P and ^{17}O MRS approaches and their applications in determining the cerebral metabolic rates of oxygen and ATP inside the mitochondria, in both animal and human brains.

Key words: In vivo ^{31}P MRS, in vivo ^{17}O MRS, in vivo heteronuclear MRS, brain, magnetic field, cerebral bioenergetics, brain metabolism, brain function, MRI.

1. Introduction

1.1. Cerebral Bioenergetics and Brain Function

The brain is an extraordinary organ; with its unique structure and complex functions, it distinguishes itself from other organs in many aspects. Unlike cardiac and skeletal muscles, the brain does not perform mechanical work. The major cellular functions in the brain include excitation and conduction resulting in unceasing electrophysiological activities in normal brain, even at rest. These integrated neuronal activities are essential for performing brain functions. Nevertheless, to sustain these activities

Fahmeed Hyder (ed.), *Dynamic Brain Imaging: Multi-Modal Methods and In Vivo Applications, vol. 489*

DOI 10.1007/978-1-59745-543-5_15 Springerprotocols.com

and functionalities, the brain has a high energy demand which is mainly fulfilled by the energy metabolism at the cellular level through adenosine triphosphate (ATP) production and utilization *(1–8)*.

The vast majority of ATP formation relies on the oxidative phosphorylation occurring inside the mitochondria. The activity of cytochrome oxidase, a key mitochondrial enzyme of oxidative energy metabolism, is regulated by the cell functional activities and their energy demand *(9, 10)*. At normal condition, the cerebral oxidative phosphorylation is expected to be tightly coupled and supported by the oxygen metabolism.

Figure 15.1 illustrates a simplified schema showing the major metabolic processes and ATP energy transportation occurring at various subcellular compartments and the associated hemodynamic processes for supporting normal brain function.

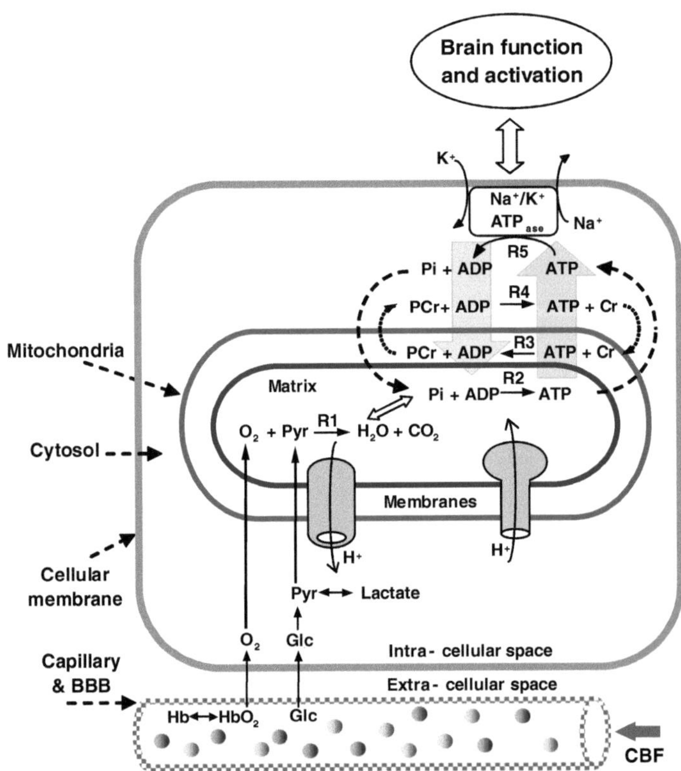

Fig. 15.1. Schematic diagram of simplified major brain network involving metabolisms and hemodynamics occurring in the capillary, subcellular compartments including the mitochondria and cytosol spaces. Oxygen dissociates from hemoglobin (Hb) and enters mitochondria and it is metabolized with pyruvate (Pyr) which is converted from glucose (Glc). This oxygen utilization via Reaction 1 (R1) is tightly coupled with four ATP reactions for generating ATP (R2), consuming ATP (R5), transporting and buffering ATP energy via the paired CK reactions (R3 vs. V4). These four ATP reactions constitute a chemical exchange network *PCr↔ATP↔Pi*. This neurovascular-metabolic coupled network is essential for brain function.

Oxygen and glucose are the major chemical substrates for brain metabolism and they are continuously supplied by the circulating blood in the capillary bed. Glucose is first converted to two pyruvate molecules in the cytosol, then the pyruvate enters the mitochondria and is metabolized oxidatively with diffused oxygen molecules through the mitochondrial respiratory chain and the cytochrome oxidase enzyme resulting in substantial oxygen utilization and final products of water and CO_2 (i.e., Reaction 1 or **R1** in **Fig. 15.1**). This oxygen metabolism is coupled with the oxidative phosphorylation for producing ATP from inorganic phosphate (Pi) and adenosine diphosphate (ADP) through the enzyme of F_1F_0-ATP_{ase} inside the mitochondria (i.e., Reaction 2 or **R2** in **Fig. 15.1**). In general, the mitochondrial oxidative phosphorylation dominates up to 90% of ATP production *(11)*. In contrast, ATP utilization mainly occurs in the cytosol space resulting in ADP and Pi products (i.e., Reaction 5 or **R5** in **Fig. 15.1**), ultimately, providing chemical energy for supporting various cellular activities and brain functions, in which a significant amount of ATP energy is used for neuronal signaling. This ATP utilization is particularly essential for maintaining and restoring the Na^+/K^+ ion gradients across the cellular membranes through the enzyme of Na^+/K^+-ATP_{ase} as well as for supporting the signaling process (e.g., neuronal transmission and cycling) at resting and activated brain states *(11–14)*.

The high energy demand in the brain causes extremely fast chemical cycling among ATP, ADP and Pi, which requires rapid transportation of these phosphate components between the cytosolic and mitochondrial compartments. This energy transportation could be partially accomplished by phosphocreatine (PCr) through the reversible creatine kinase (CK) reaction, thus, maintaining a stable ATP level in the brain cells *(15–17)*. There are at least two apparently coupled CK reactions: one occurring in the mitochondrial intermembrane space (i.e., Reaction 3 or **R3** in **Fig. 15.1**) and another one occurring in the cytosol space (i.e., Reaction 4 or **R4** in **Fig. 15.1**) although, in reality, these coupled CK reactions likely exist of these subcellular spaces. They play an important role in carrying the ATP molecules generated in the mitochondria into the cytosol for energy utilization, and then to bring the products of ADP and Pi back to mitochondria for sustaining ATP production. Therefore, PCr serves a vital role for energy transportation among subcellular compartments. Four ATP-related reactions (two reactions each for ATP_{ase} and CK) constitute a complex ATP metabolic process. They are tightly coupled and integrated with the oxygen metabolism and the hemodynamic process as depicted in **Figure 15.1** for controlling the dynamics of ATP production and utilization, all of which play a central role in the cerebral bioenergetics and function in normal and diseased brains. Development of in vivo tools being able to noninvasively detect and quantify the cerebral metabolic rate of oxygen utilization ($CMRO_2$) and ATP formation (CMR_{ATP}), therefore, become extremely important for studying and understanding the relation between ATP metabolism and cerebral bioenergetics and its impact on brain function.

1.2. In Vivo Heteronuclear MRS Approaches for Studying Brain Bioenergetics and Function

Although there are a number of techniques capable of studying brain metabolism, in vivo heteronuclear magnetic resonance spectroscopy (MRS) allows for noninvasively determining the physiological parameters and their changes that can be linked to brain metabolism, chemical kinetics and cerebral bioenergetics. The most commonly used in vivo heteronuclear MRS approaches for studying brain bioenergetics are in vivo ^{31}P, ^{13}C and ^{17}O MRS. Here, we will focus on ^{31}P and ^{17}O MRS in vivo.

In vivo ^{31}P MRS allows noninvasive assessment of numerous fundamental biochemical, physiological and metabolic events occurring inside a living brain (e.g., *(17–23)*). The primary information provided by in vivo ^{31}P MRS includes: intracellular pH, intracellular free magnesium concentration ($[Mg^{2+}]$) and high-energy phosphate (HEP) metabolites such as ATP, ADP and PCr. All these phosphate metabolites (except ADP) and Pi can be directly observed in an in vivo ^{31}P MR spectrum as demonstrated in **Fig. 15.2A** which was acquired from the human occipital-lobe at high field (7 tesla) *(24)*. The pH value can be calculated from the chemical shift difference (δ_i) between the Pi and PCr resonance peaks according to the following equation: *(25)*

$$pH = 6.77 + log\left[\frac{\delta_i - 3.29}{5.68 - \delta_i}\right] \tag{15.1}$$

and [ADP] can be calculated by the following equation:

$$[ADP] = \frac{[ATP]([Cr] - [PCr])}{K_{eq}[PCr][H^+]} \tag{15.2}$$

where K_{eq} is the equilibrium constant of the creatine kinase reaction and [Cr] is the total creatine concentration given by the summation of PCr and non-phosphorylated Cr.

Besides the aforementioned phosphate compounds and physiological parameters, other detectable phosphate compounds in the human brain ^{31}P MRS include: uridine diphospho (UDP) sugar (an important precursor for glycogen metabolism), diphospho nicotinamide adenine dinucleotides (NAD) involving oxidative respiratory chains, and four resolved phosphate compounds of glycerophosphoethanolamine (GPE), glycerophosphocholine (GPC), phosphoethanolamine (PE) and phosphocholine (PC), which are actively involved in the membrane phospholipid metabolism through phospholipid

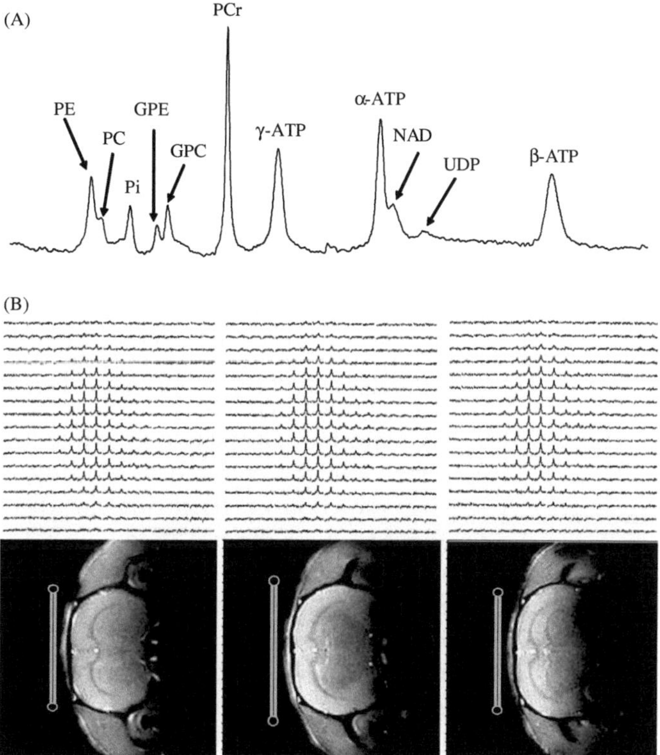

Fig. 15.2. (**A**) A typical ^{31}P MR spectra acquired from the human occipital lobe at 7T with a total sampling time of 6.4 minutes. The spectrum is characterized by excellent spectral resolution and NMR sensitivity, and a large number of well-resolved resonance peaks from phosphoethanolamine (PE); phosphocholine (PC); inorganic phosphate (Pi); glycerophosphoethanolamine (GPE); glycerophosphocholine (GPC); phosphocreatine (PCr); adenosine triphosphate (ATP); nicotinamide adenine dinucleotides (NAD) and uridine diphospho sugar (UDP). Adapted from Lei et al. of Reference 24. (**B**) 3D ^{17}O MRSI of natural abundance $H_2{}^{17}O$ (*top row*) and corresponding ^{1}H anatomical images (*bottom row*) of rat brain acquired at 9.4T. The ^{17}O surface RF coil positions and cross sections are indicated in the images.

biosynthetic enzymes *(26–29)*. In vivo ^{31}P MRS has been widely applied to study both normal and diseased brains and has led to a large number of publications (e.g., *(19, 21–24, 26–28, 30–42)*. The abnormality in the HEP metabolites has been frequently observed in the diseased brains using in vivo ^{31}P MRS, and these changes or the change of HEP concentration ratios have been commonly applied in many clinical studies.

Although the steady-state concentrations of cellular HEP metabolites are closely linked to the ATP metabolism, they are relatively stable under normal physiological conditions. This could be attributed to the rigorous regulation of the ATP metabolic reactions and their kinetics (see **Fig. 15.1**) for maintaining stable chemical energy supply of ATP in the brain. This is likely the case for the brain activation caused by brain stimulation and task performance. Thus, the measurements of the kinetics of ATP metabolism (i.e. reaction rate constants and fluxes) should logically be more sensitive and meaningful for quantifying the cerebral bioenergetics and its change under varied brain activity states as compared to the measurement of steady-state ATP concentrations. In vivo ^{31}P MRS, in combination with the magnetization transfer (MT) approach *(18, 43–48)*, is capable of noninvasively determining the chemical reaction fluxes related to the ATP metabolism and enzyme activity; and thus becomes an important tool for studying the brain ATP metabolisms and cerebral bioenergetics. One example is to apply this approach for measuring the metabolic flux of oxidative phosphorylation in the brain *(20, 49, 50)*. This useful in vivo approach and its new development will be described here.

Another alternative in vivo heteronuclear MRS approach which can directly and noninvasively assess the fluxes of oxidative metabolism occurring in the mitochondria (i.e., **R1** in **Fig. 15.1**) is the use of in vivo ^{17}O MRS approach for imaging $CMRO_2$ (e.g., *(51–54)* and a recent review article *(55)* and the references cited therein). The basic idea underlying this approach is to apply in vivo ^{17}O MRS for detecting the dynamic change of the ^{17}O-labeled metabolic water ($H_2{}^{17}O$) produced through the metabolism of inhaled $^{17}O_2$ gas, and ultimately for determining and imaging $CMRO_2$. Recent progress, especially at ultrahigh fields, has advanced the in vivo ^{17}O MRS methodology for imaging $CMRO_2$ and great promise has been demonstrated in the animal applications at high fields. This topic is also discussed here in some detail.

Although the metabolic fluxes measured by in vivo ^{31}P MRS and in vivo ^{17}O MRS reflect different metabolic reactions and pathways, these two MRS approaches are quite complementary. They are vital for investigating the coupling between the oxygen utilization and oxidative phosphorylation in the brain mitochondria under normal and pathological conditions.

2. Benefit of In Vivo ^{31}P and ^{17}O MRS at Ultrahigh Magnetic Field

2.1. Challenges of In Vivo Heteronuclear MRS

In practice, in vivo heteronuclear MRS faces many technical challenges due to its unfavorable sensitivity. For instance, the phosphate metabolites detected by in vivo ^{31}P MRS and the metabolic $H_2{}^{17}O$ detected by in vivo ^{17}O MRS are in a range of few millimolar concentrations, which are approximately 5000 times lower than the tissue water concentration, resulting in a much lower detection sensitivity. Thus, a high number of signal averaging is required to achieve reasonable spectral quality. This low sensitivity is even more manifest for the nuclear spins (e.g., ^{31}P and ^{17}O) with a relatively low magnetogyric ratio (γ) compared to in vivo ^{1}H MRS. This significantly limits the reliability, applicability, spatial and temporal resolutions of in vivo ^{31}P and ^{17}O MRS for general application, as well as clinical studies. One common trade off is to reduce the spatial and/or temporal resolution of in vivo MRS, thereby gaining detection sensitivity. However, in order to determine the nonuniform distribution of cerebral metabolites and chemical reaction fluxes in different brain regions, there is a high demand on improving the spatial resolution of in vivo MRS. Moreover, the dynamic changes caused by physiological perturbation (e.g., sensory stimulation) may occur in a relatively short time and the magnitudes of the changes are usually subtle compared to those observed under pathological states. Therefore, both reasonably high spatial and temporal resolutions are desired and thcy rely heavily on the achievable detection sensitivity, which poses the major challenge for in vivo heteronuclear MRS. One way to overcome this challenge is the use of high/ultrahigh field MRS. Besides the sensitivity gain at high fields, the spectral resolution of in vivo MRS is also significantly improved and many overlapped resonance peaks from different metabolites as observed at low fields become resolvable at ultrahigh fields as demonstrated in **Fig. 15.2A**.

2.2. Sensitivity Improvement of In Vivo Heteronuclear MRS at High Field

One of the most important advantages at ultrahigh fields is the potential gain in detection sensitivity. This is particularly crucial for the low γ nuclei MRS such as in vivo ^{17}O and ^{31}P MRS. However, it is not so straightforward to evaluate and compare detection sensitivity (or signal-to-noise ratio, SNR) at different magnetic field strengths (B_0). The apparent SNR achievable at a given field strength relies not only on the B_0 but also on other B_0-dependent parameters, such as the longitudinal relaxation time (T_1), the apparent transverse relaxation time (T_2*) and the repetition time (TR) for acquiring MRS signal. It is, in general, more useful to determine and evaluate the SNR of signal detected in a given unit sampling time under optimal acquisition condition, which means that the excitation flip angle (α) satisfies the Ernst equation of $\cos(\alpha_{opt}) = \exp(-TR/T_1)$ *(56)*. For a simple case with the single pulse and data acquisition scheme, such an averaged SNR can be quantified by **Eq. (15.3)** *(56–60)*,

$$SNR(\text{per unit sampling time}) \propto C\gamma^{2.5}B_0^{\beta}Q^{1/2}\left(\frac{T_2^*}{T_1}\right)^{1/2}(1-E_2^2)^{1/2}G(x) \quad (15.3)$$

where

$$G(x) = \left[2\frac{1-e^{-x}}{x(1+e^{-x})}\right]^{1/2} x = \frac{\mathrm{TR}}{\mathrm{T_1}} \text{ and } E_2^2 = \exp(-2\cdot a_t/T_2^*), \quad (15.4)$$

C is the concentration of the nuclear spin under observation; a_t is the spectrum acquisition time; Q is the radio frequency (RF) coil quality factor. C and γ are B_0-independent and can be treated as constants.

Equation (15.3) accounts for most B_0-dependent parameters that have influences in the apparent SNR, such as T_2* signal loss and the partial saturation effect on signal due to a relatively short TR of $\ll 5T_1$. For in vivo ^{31}P MRS applications, the term of $(1 - E_2^2)^{1/2}G(TR/T_1)$ approximates to 1.0 *(60)*, thus **Eq. (15.5)** is a simplification

$$SNR(\text{per unit sampling time}) \propto B_0^{\beta} Q^{1/2} \left(\frac{T_2^*}{T_1}\right)^{1/2}, \quad (15.5)$$

or

$$SNR(\text{per unit sampling time}) \propto B_0^{\beta *} \left(\frac{T_2^*}{T_1}\right)^{1/2}, \quad (15.6a)$$

if one applies the approximation of $Q \propto B_0^{1/2}$ resulting $\beta^* = \beta+1/4$ *(57, 60)*. The values for β^* and β based on the theoretical predictions are 1.75 and 1.5 respectively *(57)*.

For precisely quantifying and understanding the sensitivity of in vivo ^{31}P MRS as a function of magnetic field strength, it is necessary to quantify all B_0-dependent parameters used in **Eq. (15.6A)**.

2.2.1. Field Dependence of in vivo ^{31}P MRS Sensitivity

It has been demonstrated that both SNR and the quality of in vivo ^{31}P MRS acquired in human brain are improved at higher fields *(24, 61–63)*. Recently, we have quantitatively measured and compared the values of T_1, T_2* and SNR of PCr resonance peak acquired from the human occipitallobe at 4 tesla and 7 tesla *(60)*. We found that not only T_2* of PCr, but also T_1 of PCr (i.e. $T_{1,PCr}$) were shortened at 7 tesla compared to 4 tesla. This observation indicates that the field dependence of $T_{1,PCr}$ is likely determined by two competing relaxation mechanisms, *i.e.* the chemical shift anisotropy (CSA) and the dipolar interaction that simultaneously influence the ^{31}P longitudinal relaxation time *(60, 64, 65)*. In the case with dominant CSA contribution, the T_1 value decreases with increasing field strength following the relation of $1/T_1 \propto B_0^2$; in contrast, the T_1 value increases with increasing field strength when the dipolar interaction mechanism dominates. The opposite trends between these two relaxation mechanisms lead to a decrease in $T_{1,PCr}$ at higher field and offer an advantage for in vivo ^{31}P MRS as a shorter repetition time can be employed which allows more signal averaging within the same sampling time; as a result, improving SNR and leading to a total SNR gain of 56%

at 7T as compare to 4T. Moreover, the β value derived from the experimental measurements and **Eq. (15.5)** in this study was 1.4 *(60)*. This value is in an excellent agreement with the theoretically predicted β value of 1.5 as described in the literature *(57)*.

Another interesting observation of this study is that the linewidth broadening of the phosphate resonance peak with increased field strength is not linearly correlated with the field strength and, thus, at higher field strength, there is a small increase in the line width of phosphate signals *(60)*. Moreover, higher field strength increases the chemical shift dispersion (i.e., in Hz unit) of in vivo ^{31}P MRS linearly with field strength. The combination of these two factors (i.e., large increase in chemical shift dispersion and relatively small increase in linewidth broadening) improves the ^{31}P spectral resolution at 7T (see **Fig. 15.2A**). This improvement makes it possible to resolve many adjacent phosphate metabolite resonance peaks (e.g., α-ATP versus NAD) and it is especially helpful for resolving the Pi peak from other adjacent resonance peaks. The well-resolved Pi peak is crucial for measuring the ATP production flux and will be discussed later.

2.2.2. Field Dependence of in vivo ^{17}O MRS Sensitivity

^{17}O is a stable oxygen isotope with a spin quantum number of 5/2 and detectable by ^{17}O MRS. It possesses an electric quadrupolar moment and its natural abundance is only 0.037%, which is almost 30 times lower than that of ^{13}C and 2700 times lower than that of ^{1}H. Moreover, the γ value of ^{17}O is 7.4 times lower than that of ^{1}H. These factors attribute to the low inherent ^{17}O sensitivity that might be the major obstacle for the development and application of in vivo ^{17}O MRS approaches.

The relaxation mechanism of ^{17}O spin is distinct from ^{31}P spin. The ^{17}O quadrupolar moment can interact with local electric field gradients and the temporal fluctuations of this interaction induced by molecular motion dominate the ^{17}O relaxation processes and determine both T_1 and T_2 of $H_2{}^{17}O$ *(66)*, where T_2 is the transverse relaxation time. Theoretically, the T_1, T_2 and $T_2{}^*$ values of ^{17}O water spins are expected to be insensitive to B_0 *(55, 66)*. We have experimentally investigated and compared the ^{17}O relaxation times of the natural abundance $H_2{}^{17}O$ in rat brain at both 4.7 and 9.4 tesla. We confirmed that the ^{17}O relaxation times are B_0-independent *(67)*. Recent experimental evidence has indicated that the B_0-independence of ^{17}O relaxivity can further extend to a much higher magnetic field, such as 17.6 tesla *(68)*. The field independence of ^{17}O relaxivity reveals that ^{17}O sensitivity gain at higher fields is not compromised by the relaxation times, and this could lead to a substantial sensitivity gain when the field strength increases according to **Eq. (15.6B)** if the same acquisition parameters (e.g., TR and a_t) are applied at different field strengths,

$$SNR(\text{per unit sampling time}) \propto B_0^{\beta *}. \qquad (15.6b)$$

Our experimental results showed approximately a four-fold SNR gain in the in vivo ^{17}O MRS signal measured in the rat brain at 9.4 tesla compared to 4.7 tesla *(67)*. Moreover, the β* value deduced from these experimental measurements and **Eq. (15.6B)** was ~2 *(67)*, which is close to the theoretically predicted β* value of 7/4 *(57)*. The significant sensitivity gain provides adequate SNR for acquiring the 3 dimensional ^{17}O MRS imaging of the natural abundance $H_2{}^{17}O$ from the small brain of rat with a temporal resolution of ~10 seconds at 9.4T (see one example shown in **Fig. 15.2B**).

Clearly, the increase of B_0 should benefit all of the in vivo multinuclear MRS approaches. In vivo ^{17}O MRS probably benefits the most from the substantial sensitivity gain at ultrahigh fields. On the other hand, in vivo ^{31}P MRS can benefit from both the moderate sensitivity gain and the significant spectral resolution improvement, which is crucial for resolving many overlapping resonance peaks of different metabolites. All these benefits provided by high field strength are essential for improving the reliability of in vivo ^{17}O and ^{31}P MRS and for accurately determining cerebral metabolic fluxes.

3. How to Apply In Vivo ^{17}O MRS for Imaging $CMRO_2$

In vivo ^{17}O MRS methodology has two major applications for studying brain function and cerebral bioenergetics through imaging either cerebral blood flow (CBF) or $CMRO_2$ *(51–55, 67, 69–87)*. Both applications rely upon the measurement of brain $H_2{}^{17}O$ content and its dynamic change using in vivo ^{17}O MRS. The CBF measurement is based on monitoring the washout rates of inert $H_2{}^{17}O$ tracer in the brain tissue following an intravascular bolus injection of ^{17}O-labeled water. The $CMRO_2$ measurement is based on monitoring the dynamic changes of metabolically generated $H_2{}^{17}O$ in the brain tissue from inhaled ^{17}O-labeled oxygen gas *(55)*. There are two types of magnetic resonance approaches for monitoring brain $H_2{}^{17}O$ in vivo: a direct approach by using ^{17}O MRS detection, and an indirect approach by using ^1H magnetic resonance imaging (MRI) to measure the changes in T_2- or $T_{1\rho}$-weighted proton signals caused by the ^{17}O-^1H scalar coupling and proton chemical exchange (see recent review article of *(55)* and the references cited therein). Both direct and indirect approaches are suitable for CBF measurements. However, recent studies indicated that the in vivo ^{17}O MRS approach at ultrahigh fields seems to have more advantages for quantifying and imaging $CMRO_2$, which perhaps is the most important application of in vivo ^{17}O MRS. Here, the in vivo ^{17}O MRS approach for imaging $CMRO_2$ is described in detail.

3.1. Theory and Quantification of $CMRO_2$ Based on In Vivo ^{17}O MRS Approach

In general, there is a similarity between in vivo ^{17}O MRS approach and positron emission tomography (PET) approach *(88–91)* for measuring $CMRO_2$ through the use of isotropic labeled oxygen gas ($^{17}O_2$ for MRS and $^{15}O_2$ for PET). After gas exchange in the lung, the inhaled and labeled O_2 molecules quickly bind to hemoglobin (Hb) in the blood, forming O_2-hemoglobin complexes (HbO_2) (see **Fig. 15.1**). Through the feeding arteries and arterioles, the HbO_2 complex enters the brain capillaries. The O_2 molecules are dissociated from hemoglobin in the capillaries, then cross the brain blood barrier (BBB) in the form of free gas, diffuse into the brain tissue (intra- and extracellular space), and finally enter mitochondria where they are metabolized and produce the water with the isotope label following the chemical reaction *(3)*:

$$\text{Carbon substrate} + O_2 \rightarrow \text{Oxidized carbon substrate} + 2H_2O \tag{15.7}$$

According to this reaction (equivalent to **R1** in **Fig. 15.1**), one oxygen molecule produces two isotopic labeled water molecules ($H_2{}^{17}O$ for ^{17}O MRS and $H_2{}^{15}O$ for PET), which can move out of the mitochondria (traversing the opposite pathway as O_2 entry) and finally be washed out from the brain through venules and veins. The dynamic change of the isotope labeled water in the brain is tightly linked to the cerebral oxygen utilization rate. It provides the vital signal source for determining $CMRO_2$ for both ^{15}O PET and ^{17}O MRS.

Though the principle underlying the ^{17}O MRS approach for measuring $CMRO_2$ was historically adopted from the well-established ^{15}O PET approach, there is significant distinction between these two neuroimaging approaches. For instance, one complication in PET is that it is unable to distinguish the ^{15}O signal contribution from $^{15}O_2$ molecules versus the metabolically generated $H_2^{15}O$ molecules. This limits the robustness of PET technique for imaging $CMRO_2$ *(55, 90)*. In contrast, the ^{17}O MRS approach specifically detects the metabolically generated $H_2{}^{17}O$ without confounding signals from$^{17}O_2$ molecules because of the “invisibility” of $^{17}O_2$ in an in vivo ^{17}O spectrum *(55)*. This unique magnetic resonance specificity leads to a great advantage which significantly simplifies the in vivo ^{17}O MRS methodology for measuring and quantifying $CMRO_2$ *(55, 87)*. Nevertheless, the dynamic change of the metabolically generated $H_2{}^{17}O$ concentration in the brain during an $^{17}O_2$ inhalation is interplayed by three parallel processes (see **Fig. 15.3**): (i) Cerebral oxygen utilization for generating the metabolic $H_2{}^{17}O$ in the brain tissue, (ii) Cerebral blood perfusion resulting in $H_2{}^{17}O$

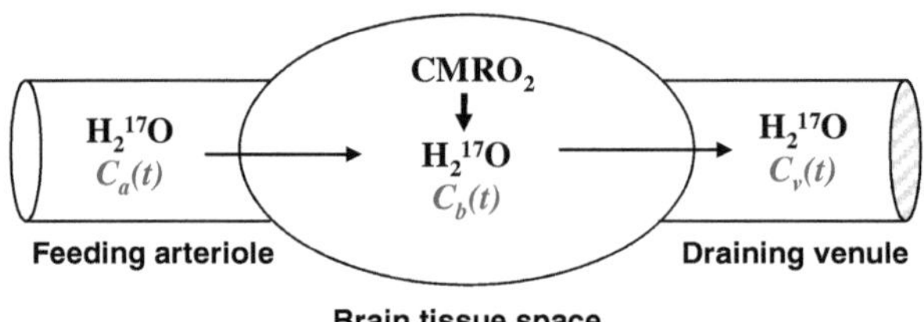

Fig. 15.3. Schematic illustration of a "complete model" describing three parallel processes of the ^{17}O -labeled metabolic water ($H_2{}^{17}O$) occurring in the brain when the ^{17}O-labeled oxygen gas molecules are introduced via an inhalation. In this model, only the metabolic $H_2{}^{17}O$ is considered because the ^{17}O-labeled O_2 is invisible by in vivo ^{17}O NMR. $Ca(t)$, $Cb(t)$ and $Cv(t)$ stand for the $H_2{}^{17}O$ concentration in arteriole, brain tissue and venule, respectively, as a function of the $^{17}O_2$ inhalation time.

washout from the brain, and (iii) Blood recirculation bringing the metabolically generated $H_2{}^{17}O$ in the entire body back to the brain. All contributions from these three processes have to be considered for quantifying $CMRO_2$. Based on the Kety - Schmidt theory *(92–94)*, the mass balance of the isotope labeled $H_2{}^{17}O$ in the brain tissue during an $^{17}O_2$ gas inhalation can be derived as *(52, 54, 55, 87)*:

$$\frac{dCb(t)}{dt} = 2\alpha f_1 CMRO_2 + mCBF\left(f_2 Ca(t) - \frac{nCb(t)}{\lambda}\right) \qquad (15.8)$$

where $C_a(t)$, $C_b(t)$ and $C_v(t)$ are the metabolic $H_2{}^{17}O$ concentrations in excess of the natural abundance of $H_2{}^{17}O$ concentration in the arterial blood, brain tissue and venous blood respectively, as a function of $^{17}O_2$ inhalation time (t, unit = minute); α is the ^{17}O enrichment fraction of the oxygen atoms in the inhaled $^{17}O_2$ gas; λ is the brain/blood partition coefficient *(95)*. The factor of two accounts for the fact that one O_2 converts to two H_2O molecules through oxidative metabolism according to **Eq. (15.7)**; f_1 and f_2 are two unit conversion factors *(54, 87)*. The correction factor m is used in **Eq. (15.8)** to account for the water permeability restriction across the brain blood barrier *(96)*, and n is another correction factor that accounts for the permeability restriction occurring when $H_2{}^{17}O$ molecules which are metabolically generated inside the mitochondria across the mitochondrial membranes *(54, 87)*. Both m and n depend on CBF *(54, 87)*. The function of $C_a(t)$ (or artery input function) is determined by the total metabolic $H_2{}^{17}O$ generated in all aerobic organs of a living body. It can approximate as a linear function of $^{17}O_2$ inhalation time (i.e., $C_a(t)=At$, A is a constant) *(52, 54, 78, 87)*. Under this approximation, the solution for solving the differential equation of **Eq. (15.8)** is:

$$CMRO_2(t) = \frac{\left\{\left[\dfrac{C_b(t) - \dfrac{Af_2\lambda^2}{mn^2 CBF}\left(\dfrac{mnCBF}{\lambda} te^{-\frac{mnCBF}{\lambda}t} + e^{-\frac{mnCBF}{\lambda}t} - 1\right)}{1 - e^{-\frac{mnCBF}{\lambda}t}}\right] - \dfrac{Af_2\lambda t}{n}\right\}}{\dfrac{2\alpha\lambda f_1}{mnCBF}}. \quad (15.9)$$

According to this equation, the $CMRO_2$ value at *each* data point measured at different inhalation time (t) can be calculated by using the experimentally measured CBF, A, n, $C_b(t)$ values and other known constants (f_1, f_2, m, $\alpha$and λ). The quantification approach based on **Eq. (15.9)** is a complete model which accounts for all required parameters for precisely determining $CMRO_2$ *(54, 55, 87)*.

3.2. Measurement and Imaging of $CMRO_2$ using In Vivo ^{17}O MRS Approach

The complete model as described by **Eq. (15.9)** requires multiple experimental measurements of four variables (A, n and $C_b(t)$ and CBF) in order to calculate $CMRO_2$. **Figure 15.4** illustrates the procedures for performing these in vivo ^{17}O MRS measurements and the results of a rat brain study with α-chloralose anesthesia at 9.4 T *(54, 55)*.

3.2.1. Imaging CBF

The CBF measurement was performed via bolus injection of a small amount of ^{17}O-enriched water into one internal carotid artery and monitoring the washout process of the $H_2{}^{17}O$ tracer in the brain using 3 dimensional ^{17}O chemical shift imaging (CSI) *(67)*. **Figure 15.4A** demonstrates the stacked plots of $H_2{}^{17}O$ spectra acquired from a single voxel (as indicated in the brain anatomy image) of 3 dimensional ^{17}O CSI data set in a representative rat before and after the $H_2{}^{17}O$ bolus injection. The peak height of the $H_2{}^{17}O$ spectra shows an exponential decay and its decay rate determines the CBF value in the CSI voxel *(54, 55, 67)*.

3.2.2. Imaging $C_b(t)$

The crucial step for $CMRO_2$ measurements is to monitor and image the dynamic changes of the metabolic $H_2{}^{17}O$ content in the brain (i.e., $C_b(t)$) during an inhalation of $^{17}O_2$ gas. **Figure 15.4B** illustrates the stacked plots of ^{17}O spectra of cerebral $H_2{}^{17}O$ from the CSI voxel acquired before, during and after a 2-minute inhalation of $^{17}O_2$ *(54)*. It indicates excellent ^{17}O sensitivity for detecting the cerebral $H_2{}^{17}O$ signal and its change during the inhalation; and the approximately linear increase of $H_2{}^{17}O$ during a short $^{17}O_2$ inhalation is evident, and the slope is tightly coupled to $CMRO_2$.

One practical challenge for most in vivo MRS approaches is the difficulty for measuring the absolute concentrations of metabolites of interest. Nevertheless, the natural abundance $H_2{}^{17}O$ signal which can be accurately measured in the

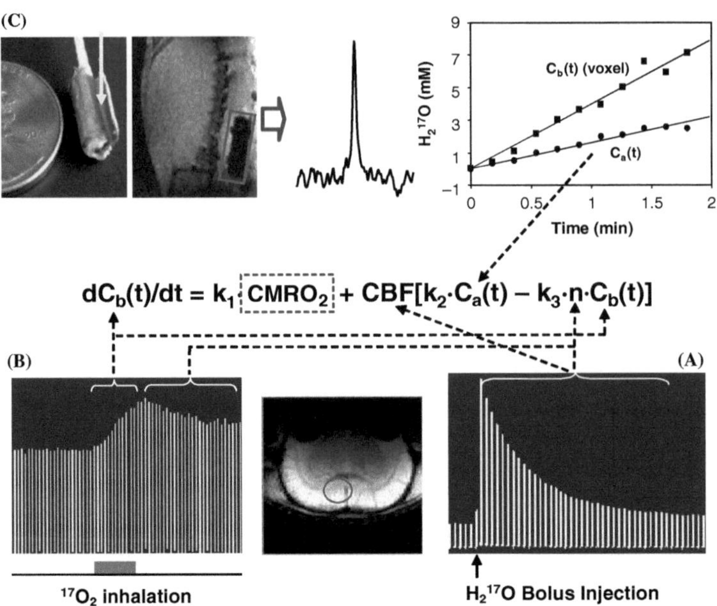

Fig. 15.4. Schematic diagram showing the multiple in vivo ^{17}O measurements for determining $CMRO_2$ using the complete model according to the mass balance equation of **Equation (15.8)** which links $Cb(t)$, $Ca(t)$, CBF and n with $CMRO_2$. To simplify the equation, three known constants of $2\alpha_1$, mf_2 and m/λ used in **Equation (15.8)** are replaced by k1, k2 and k3, respectively. (**A**) Stacked plot of the ^{17}O spectra of cerebral $H_2{}^{17}O$ tracer from one representative rat brain voxel as indicated by the circle in the anatomical image (*low center insert*). The spectra were acquired before and after a bolus injection of $H_2{}^{17}O$ for CBF measurements. (**B**) Stacked plot of the ^{17}O spectra of the metabolic $H_2{}^{17}O$ from the same representative voxel acquired before (natural abundance), during (as indicated by the *gray bar under the stacked plot*) and after a two-minute $^{17}O_2$ inhalation. (**C**) Measurement of $Ca(t)$ by using an implanted ^{17}O RF coil (*the left insert*). The middle insert illustrates a ^{17}O spectrum of natural abundance $H_2{}^{17}O$ obtained from the rat carotid artery blood by using the implanted RF coil before $H_2{}^{17}O$ inhalation. The right insert shows the time course of $Ca(t)$ (*circle symbol*) and $Cb(t)$ from a representative 3D ^{17}O CSI voxel (*square symbol*) in the same rat during a two-minute inhalation of $^{17}O_2$. Finally, the ratio between the ^{17}O signal decay detected after a bolus injection of $H_2{}^{17}O$ (see **Fig. 15.4A**) versus the ^{17}O signal decay detected after the cessation of $^{17}O_2$ inhalation (see **Fig. 15.4B**) gives the constant of n. Adapted from Zhu et al. of Ref *(54)*.

brain before the introduction of ^{17}O-labeled oxygen gas (see **Fig. 15.4B**) provides an excellent internal reference for quantifying the absolute concentration of $C_b(t)$ for each CSI voxel *(54, 87)*. This self-calibration method is independent of the ^{17}O detection sensitivity. This is particularly valuable when a surface RF coil with nonuniform spatial distribution of detection sensitivity is applied.

3.2.3. Other Measurements

The arterial input function of $C_a(t)$ was measured in vivo by an implanted ^{17}O RF coil *(97)* wrapped around a carotid artery (in the rat). **Figure 15.4C** illustrates the implanted ^{17}O RF coil, the natural abundance $H_2{}^{17}O$ signal detected from the rat carotid blood and $C_a(t)$ measured during a two-minute inhalation of $^{17}O_2$ *(54)*. The experimental result showed an approximately linear relation between the arterial $H_2{}^{17}O$ concentration and the $^{17}O_2$ inhalation time, and the linear fitting of $C_a(t)$ gave the value of constant A required by **Eq. (15.9)**. Note that the $C_a(t)$ (see **Fig. 15.4C**) and $C_b(t)$ (see **Fig. 15.4B**) measurements could be conducted simultaneously with the configuration of dual ^{17}O RF coils and receivers *(97)*.

Finally, the ratio between the decay rates of $H_2{}^{17}O$ signal measured after the cessation of $^{17}O_2$ inhalation (see **Fig. 15.4B**) versus that after a $H_2{}^{17}O$ bolus injection (see **Fig. 15.4A**) gives the constant of n, reflecting the $H_2{}^{17}O$ permeability restriction across the mitochondrial membranes *(54, 87)*.

3.2.4. Imaging CMRO$_2$ in Rat Brain

The values of $C_b(t)$, CBF and n measured from each ^{17}O MRS imaging (MRSI) voxel and the value of A measured from each ^{17}O inhalation measurement in the same animal as demonstrated in **Fig. 15.4** can be used to calculate the absolute $CMRO_2$ value as a function of inhalation time according to **Eq. 15.9. Figure 15.5A** shows one example of $CMRO_2$ time course from a representative ^{17}O MRSI voxel with a temporal resolution of 11 seconds *(54)*. It is evident that the $CMRO_2$ values are independent of $^{17}O_2$ inhalation time if the first two $CMRO_2$ values characterized with relatively large fluctuations are excluded. These $CMRO_2$ values were averaged for improving measurement accuracy. The same procedure and calculation can be applied to all ^{17}O MRSI voxels for generating 3D $CMRO_2$ images in the rat brain *(54, 77)*. **Figure 15.5B** demonstrates three adjacent $CMRO_2$ images in the coronal orientation from a representative rat brain. The averaged $CMRO_2$ and CBF values in the rat brains anesthetized with α-chloralose were 2.19 ± 0.14 μmol/g/min and 0.53 ± 0.07 ml/g/min (n=7), respectively *(54)*. The $CMRO_2$

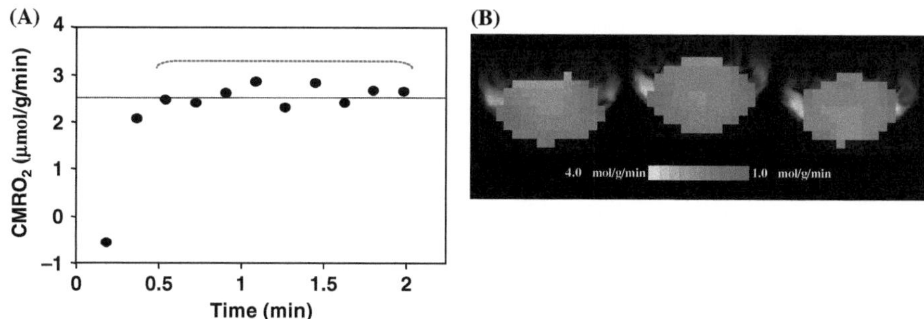

Fig. 15.5. (**A**) Plot of the calculated $CMRO_2$ values using the complete model as described by **Equation (15.9)** as a function of $^{17}O_2$ inhalation time. (**B**) Three-dimensional coronal $CMRO_2$ images of rat brain measured by in vivo ^{17}O MRS approach during a two-minute $^{17}O_2$ inhalation. Adapted from Zhu et al of Ref. *(54)*. (*See* Color Plate)

value is consistent with the reported values in the literature, which were measured by other independent techniques under similar physiological condition *(98–100)*.

3.3. Establishing a Robust and Noninvasive ^{17}O MRS Approach for Imaging $CMRO_2$

3.3.1. Noninvasive Approach

The major technical limitation of the complete model for determining $CMRO_2$ using in vivo ^{17}O MRS is the requirement of invasive measurements (e.g., CBF, $C_a(t)$, n). This could significantly limit the potential of this in vivo approach for broad biomedical applications, especially in humans. Thus, it is crucial to examine the feasibility of developing a completely noninvasive ^{17}O approach for imaging $CMRO_2$. Attempts have been made to simplify the experimental procedures and the models for determining $CMRO_2$ based on a number of approximations *(52, 53, 73, 79, 81, 87)*. We discuss one of these models, in which the invasive measurements could be eliminated by using the simplified model based on the Taylor's and Polynomial Theorems *(87)*. Briefly, the $C_b(t)$ time course can be expressed by a polynomial expansion

$$C_b(t) = a_1 t + a_2 t^2 + a_3 t^3 + \cdots . \tag{15.10}$$

In this expansion, the first-order (or linear) coefficient of a_1 is directly proportional to $CMRO_2$ according to the following equation *(87)*

$$CMRO_2 = \frac{a_1}{2\alpha f_1} \tag{15.11}$$

where α and f_1 are known constants (see above). Thus, using this *simplified model*, only the time course of $C_b(t)$ measured noninvasively by in vivo ^{17}O CSI approach is needed and it can be fitted to the polynomial function given by **Eq. (15.10)** to calculate the linear coefficient of a_1, and ultimately determining $CMRO_2$ according to **Eq. (15.11)**. For practical applications, a quadratic polynomial function provides a good approximation for fitting the time course of $C_b(t)$ with moderate measurement fluctuation *(101)*. We have demonstrated that the $CMRO_2$ value obtained based on the complete model with invasive procedures has no statistical difference from that based on the simplified model and quadratic function fitting where only a single noninvasive measurement of $C_b(t)$ is required *(87)*. Moreover, our results also validated that the linear fitting of $C_b(t)$ could provide a good approximation for determining $CMRO_2$ in the rat brain when the $^{17}O_2$ inhalation time is relatively short (e.g., 2 minutes) *(87)*. This is consistent with the prediction based on either **Eq. (15.8)** or **(15.10)**. During the initial $^{17}O_2$ inhalation period, both $C_a(t)$ and $C_b(t)$ have not built up significantly resulting in near zero value of the second term on the right side of **Eq. (15.8)**. Therefore, **Eq. (15.8)** can be approximated as a linear differential equation, and $C_b(t)$ becomes a linear function of the inhalation time. This is also evident in **Eq. (15.10)** in which the high-order terms become negligible when t is short. Therefore, it is possible to use the simplified model for determining and imaging $CMRO_2$ based on a single and completely noninvasive ^{17}O spectroscopic imaging measurement of $C_b(t)$ *(55, 86, 87)*.

Recently, we have further examined the simplified model for determining $CMRO_2$ under varied physiological conditions *(86)*. In this study , we compared the $CMRO_2$ results obtained with the complete model versus the simplified model using linear fitting of $C_b(t)$ under normothermia (37°C) condition and hypothermia (32°C) condition, which is a well known factor leading to significant suppression of both CBF and $CMRO_2$. **Figure 15.6** demonstrates an excellent consistency of the $CMRO_2$ results between the complete and simplified models for either the voxel-wised comparison (**Fig. 15.6A**) or the averaged $CMRO_2$ comparison (**Fig. 15.6B**) at both brain temperatures *(86)*. The comparison results reveal the validity of the simplified ^{17}O approach for imaging $CMRO_2$ applicable at a wide physiological range. Additional technical development which further advances the in vivo ^{17}O methodology could ultimately provide the simplest and completely noninvasively ^{17}O neuroimaging approach for imaging $CMRO_2$ (in humans).

3.3.2. Robustness and Reliability

Another merit of the ^{17}O $CMRO_2$ imaging approach is its ability for performing repeated $CMRO_2$ imaging measurements with a short interval between two measurements. This is due to the fact that the cerebral $H_2{}^{17}O$ concentration can quickly reach a new and steady level within a few minutes after the cessa-

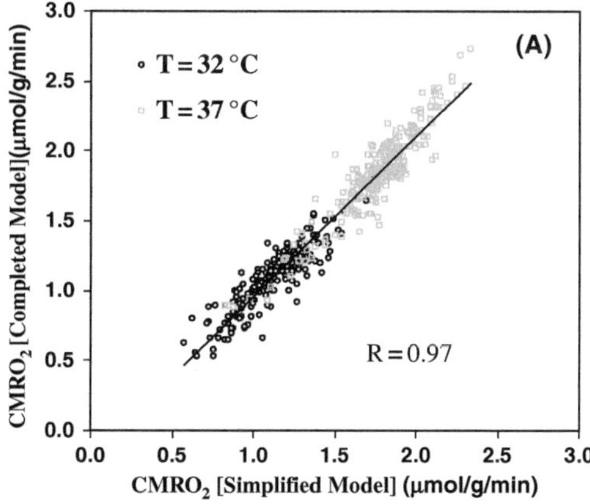

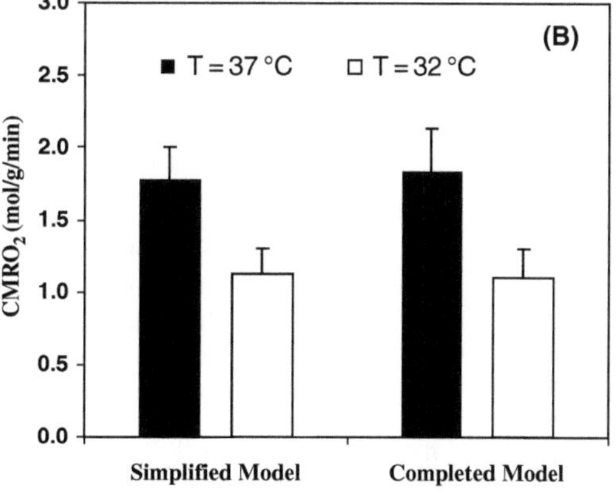

Fig. 15.6. (**A**) Voxel based $CMRO_2$ calculation and comparison using the completed and simplified models from a representative rat (total voxel number used was 224 for 32°C and 254 for 37°C, voxel size = 75μ). (**B**) Averaged $CMRO_2$ values in the same rat brain at normothermia (37°C) and hypothermia (32°C) conditions, calculated with simplified and completed model, respectively. Adapted from Zhu et al of Ref. *(86)*.

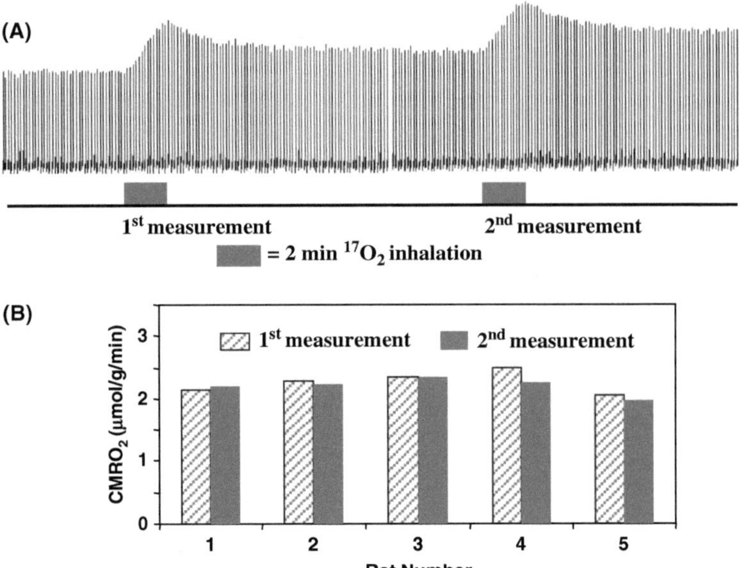

Fig. 15.7. (**A**) Stacked plots of $H_2{}^{17}O$ spectra from a representative voxel of 3D ^{17}O MRSI data acquired before, during and after two consecutive two-min $^{17}O_2$ inhalations in a rat brain. (**B**) The comparison results between two repeated $CMRO_2$ measurements in five rat brains. Adapted from Zhu et al of Ref. *(86)*.

tion of $^{17}O_2$ inhalation, which allows repeated $CMRO_2$ measurements in the same subject and experimental session (see **Fig. 15.7A**). **Figure 15.7B** shows the excellent reproducibility of repeated $CMRO_2$ measurements in five rats (1st and 2nd measured $CMRO_2$ values were 2.26 ± 0.18 and 2.20 ± 0.14 μmol/g/min giving a ratio of 1.03 ± 0.05 between the consecutive measurements) *(86)*. The results demonstrate the robustness and reliability of the simplified in vivo ^{17}O approach for noninvasively and rapidly imaging $CMRO_2$ repeatedly in a small brain of rat. This capability is particularly valuable for studies aiming at $CMRO_2$ changes induced by physiological or pathological perturbations in which multiple measurements are required under different conditions (e.g., control versus stimulation for brain function study). Therefore, the combination of the simplified model and ultrahigh field in vivo ^{17}O MRS may potentially provide an alternative neuroimaging modality for studying the central role of oxidative metabolism in brain function and neurological diseases *(55, 77)*.

3.3.3. Demonstration of in vivo ^{17}O MRS Application for Studying Brain Bioenergetics

It is well documented that the basal $CMRO_2$ is sensitive to the brain temperature (see *(3, 102)* and the references cited therein). However, most studies reported in the literature were based on the global $CMRO_2$ measurements of entire brain using the Kety-Schmidt method *(92, 93)*, and were limited by the lack of spatial information regarding regional $CMRO_2$. We have conducted a

$CMRO_2$ imaging study using 3D in vivo ^{17}O CSI combined with the simplified model at 9.4 tesla for quantifying absolute $CMRO_2$ values in the rat brain at normal brain temperature (37°C) (i.e., normothermia) and mild hypothermia (32°C) conditions *(86)*. **Figure 15.8A** illustrates an example showing three representative slices of 3D $CMRO_2$ maps from a rat brain under normothermic and hypothermic conditions. These images clearly show significant reduction of $CMRO_2$ crossing the entire brain induced by lowering brain temperature several degrees. This metabolic suppression occurring at hypothermia was consistently observed in all five rats studied (**Fig. 15.8B**), resulting in an average of 45% $CMRO_2$ reduction as compared to normothermic condition *(86)*. These results indicate that the established in vivo ^{17}O MRS approach is sensitive to determine the dynamic $CMRO_2$ change

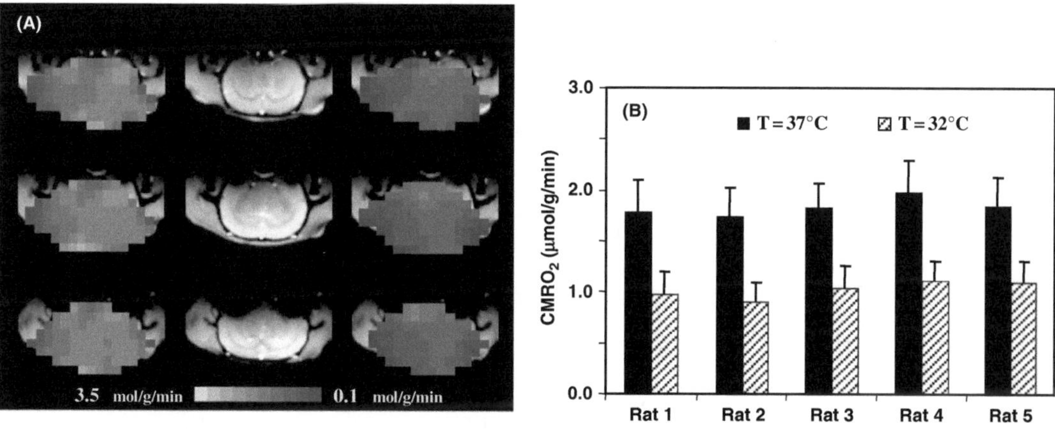

Fig. 15.8. (**A**) 3D $CMRO_2$ maps of a representative rat brain obtained at normothermia (*left column*) and hypothermia (*right column*), and their corresponding anatomic images (*middle column*). (**B**) Summary of $CMRO_2$ results measured at normothermia and hypothermia conditions ($n = 5$). Adapted from Zhu et al of Ref. *(86)*. (*See* Color Plate)

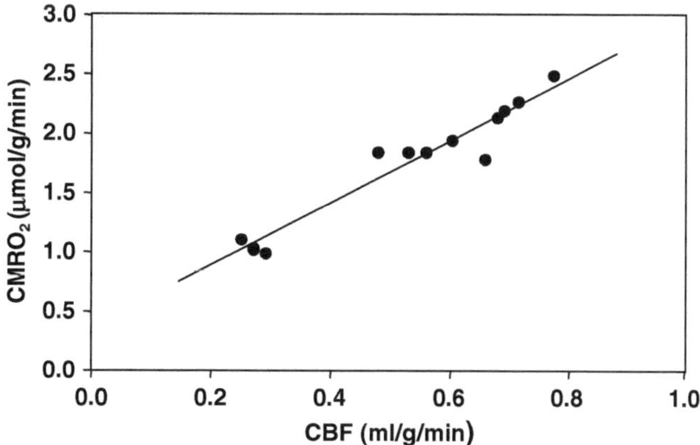

Fig. 15.9. Correlation of CBF and $CMRO_2$ values in the rat brains anesthetized with α-chloralose at brain temperature range of 32–37°C. The linear correlation coefficient (R) was 0.97. Adapted from Zhu et al. of Ref. *(86)*.

and its spatial distribution due to physiological perturbation. The measured $CMRO_2$ can be quantitatively correlated to other associated physiological parameter changes. **Figure 15.9** illustrates one example showing the quantitative relation between $CMRO_2$ and CBF: both of them were measured by in vivo ^{17}O MRS in the α-chloralose anesthetized rat under a wide range of physiological conditions from normothermia to hypothermia *(86)*. It shows a strong correlation between CBF and $CMRO_2$ with a linear correlation coefficient of $R = 0.97$ indicating a tight vascular-metabolic coupling in the rat brain.

4. How to Apply In Vivo ^{31}P MRS for Studying Cerebral ATP Metabolic Fluxes and Bioenergetics

The ATP metabolism for regulating both ATP production and utilization plays a fundamental role in cerebral bioenergetics, brain function and neurodegenerative diseases. Two important chemical reactions that contribute to the brain ATP metabolism are the ATP_{ase} and CK reactions. They are coupled together and constitute a three-^{31}P-spin chemical exchange system involving ATP, PCr and Pi (i.e., PCr↔ATP↔Pi). One vital function of this exchange system is to maintain a stable cellular ATP concentration ensuring continuous energy supply for sustaining electrophysiological activity in the brain. Logically, the measures of ATP metabolic fluxes should be more sensitive to the brain activity and energy state and their change than that of steady-state ATP and other HEP concentration. Therefore, they should provide a useful index reflecting cerebral bioenergetics under various brain states. It would be, thus, essential to find a noninvasive and reliable technique being able to assess the cellular exchange rates (or fluxes) of PCr↔ATP↔Pi in brain *in situ*. The sole approach for serving this purpose is the use of in vivo ^{31}P MRS combined with magnetization transfer method *(20, 43–47, 50, 103)*. However, to completely determine the kinetics and fluxes involved in the PCr↔ATP↔Pi exchange requires extensive measurements and information including three steady-state phosphate metabolite concentrations (i.e., [ATP], [PCr] and [Pi]) and their intrinsic T_1 values, and four pseudo first-order chemical reaction rate constants (forward and reverse rate constants for the CK reaction and the ATP_{ase} reaction, respectively) *(50)*. The products of the rate constants and their related phosphate concentrations can provide four ATP metabolic fluxes along both forward and reverse reaction directions in the *PCr↔ATP↔Pi* exchange system (see **Fig. 15.10**).

Several in vivo ^{31}P MT methods have been developed such as conventional two-spin magnetization saturation transfer (CST), inversion recovery transfer (IT) and two-dimensional chemical exchange spectroscopy (2D-EXSY) *(44, 45, 49, 104)*. They

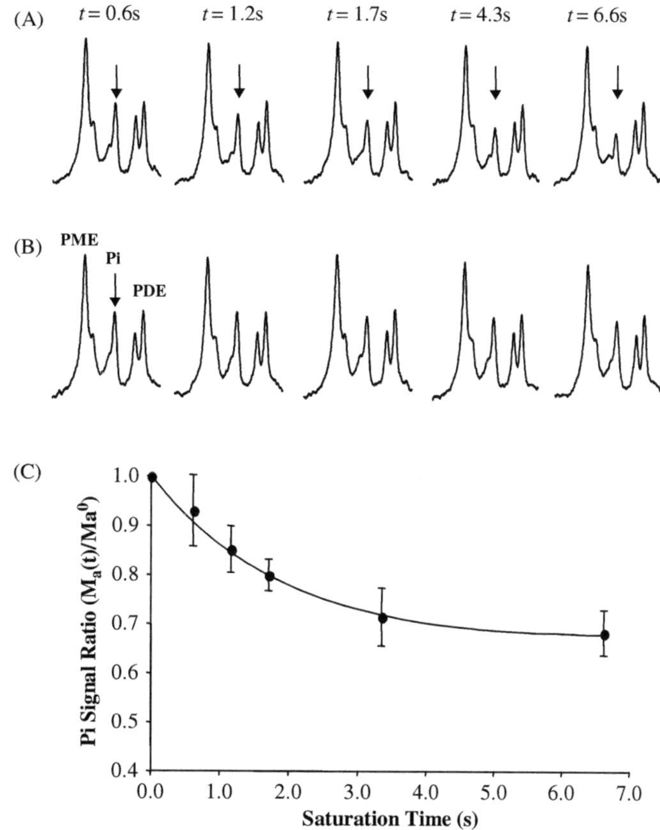

Fig. 15.10. (**A**) Progressive saturation measurements and averaged in vivo ^{31}P spectra (partially displayed) as a function of γ-ATP saturation time (t), and (**B**) their corresponding control ^{31}P spectra. The arrows point to the Pi resonance peak. (**C**) Normalized Pi signal change as a function of t and regression fitting (*solid line*) according to **Eq. (15.13a)**. Adapted from Lei et al. of Ref. *(20)*.

have been applied to physiological studies of ATP metabolism in a variety of organs, from hearts to brains *(20, 24, 45, 46, 49, 50, 103, 105–111)*. Among these methods, the CST method is most commonly used in biomedical research, perhaps due to its methodological simplicity and high efficiency. This method is particularly useful in measuring the forward rate constants and fluxes in the chemical exchange system of *PCr↔ATP↔Pi* when a frequency-selective RF saturation pulse train is applied to fully saturate the γ-ATP spin. Under this circumstance, the three-spin chemical exchange system of *PCr↔ATP↔Pi* can be treated as two independent two-spin chemical exchange systems (i.e., *PCr↔ATP* and *ATP↔Pi*); consequently, the forward rate constants and fluxes for both the CK reaction (i.e., **R4** in **Figs. 15.1 and 15.13**) and ATP_{ase} reaction (i.e., **R2** in **Figs. 15.1 and 15.13**) can be explicitly determined. These forward metabolic fluxes reflect the ATP synthesis or production rates catalyzed by ATP_{ase} and CK, respectively. However, the measurements of the reverse CK flux (*ATP→PCr*) using CST (i.e., by saturating PCr) resulted in an inequality between the forward and reverse CK fluxes *(112–114)*, which is paradoxical, in the sense that the CK fluxes into and out of the PCr pool must be equal when the CK reaction is under chemical equilibrium condition. One possible explanation is that the ATP metabolism involves other chemical exchange reactions besides the CK reaction, for example the ATP hydrolysis reaction *(115)*. Consequently, neglecting ATP hydrolysis may lead to an error in estimating the reverse CK flux. Therefore, it is necessary to consider *PCr→ATP↔Pi* as a three-spin chemical exchange system in order to accurately determine all kinetic parameters; in particular, the reverse rate constants and fluxes *(45, 46, 50, 115, 116)*.

We will discuss a newly introduced in vivo ^{31}P MT approach being able to noninvasively study the *PCr↔ATP↔Pi* exchange in the brain explicitly through the in vivo measurements of the following rate constants and fluxes associated with four coupled reactions in different subcellular compartments *(50)*:

(i) The forward flux ($F_f^{ATP_{ase}}$) of ATP_{ase} reaction (i.e., **R2** in **Figs. 15.1 and 15.13**) occurring inside the mitochondria;
(ii) The forward flux (F_f^{CK}) of CK reaction (i.e., **R4** in **Figs. 15.1 and 15.13**),
(iii) The reverse flux ($F_r^{ATP_{ase}}$) of ATP_{ase} reaction (i.e., **R5** in **Figs. 15.1 and 15.13**) occurring in the cytosol space;
(iv) The reverse flux (F_r^{CK}) of CK reaction (i.e., **R3** in **Figs. 15.1 and 15.13**).

4.1. In vivo ^{31}P MT Approach for Determining Entire PCr↔ATP↔Pi Exchange

A newly introduced in vivo ^{31}P MT approach for determining all kinetic parameters of the *PCr↔ATP↔Pi* exchange is called the Multiple Single-site Saturation (MSS) MT approach *(50)*. The MSS approach requires a total of four in vivo ^{31}P spectra: one control spectrum in the absence of RF saturation plus three single-site, RF-saturated spectra with the saturation frequency on PCr, γ-ATP and Pi, respectively *(50)*. The quantification of the ATP metabolic fluxes relies on solving three coupled Bloch equations based on the three-spin chemical exchange model. For simplifying the mathematical derivation, the symbols of a, b and c were used which stand for PCr, ATP and Pi, respectively. The Bloch equations describing the magnetizations of a, b and c and their changes as a function of time are given by *(43)*,

$$\mathrm{PCr} \underset{k_{-1}}{\overset{k_1}{\rightleftarrows}} \gamma\text{–}ATP \underset{k_{-2}}{\overset{k_2}{\rightleftarrows}} P_i$$

$$a \qquad\qquad b \qquad c$$

$$\frac{dM_a}{dt} = -\frac{(M_a - M_a^0)}{T_{1a}} - k_1 M_a + k_{-1} M_b \tag{15.12a}$$

$$\frac{dM_b}{dt} = -\frac{(M_b - M_b^0)}{T_{1b}} - k_{-1}M_b - k_2M_b + k_1M_a + k_{-2}M_c \tag{15.12b}$$

$$\frac{dM_c}{dt} = -\frac{(M_c - M_c^0)}{T_{1c}} - k_{-2}M_c + k_2M_b \tag{15.12c}$$

where M_a, M_b and M_c are the magnetizations of PCr, γ-ATP and Pi; M_a^0 , M_b^0 and M_c^0 are the magnetizations at Boltzmann thermal equilibrium; k_1 and k_{-2} are the pseudo first-order forward rate constants involving ATP production through the CK reaction and the ATP_{ase} reaction, respectively; k_{-1} and k_2 are the pseudo first-order reverse rate constants involving ATP utilization through the CK reaction and the ATP_{ase} reaction, respectively; T_{1a}, T_{1b} and T_{1c} are the intrinsic spin-lattice relaxation times of PCr, γ-ATP and Pi, which are B_0 dependent. Four rate constants involving the *PCr↔ATP↔Pi* exchange and their four associated ATP metabolic fluxes can be determined by the following three-step measurements with frequency-selective RF saturation on the γ-ATP (Step 1), Pi (Step 2) and PCr (Step 3) resonance peak, respectively.

Step 1: Progressive saturation of γ-ATP for determining intrinsic spin-lattice relaxation times of PCr and Pi, forward rate constants and fluxes

The first step of the ^{31}P MT measurements is to apply a frequency-selective RF pulse train for completely saturating the γ-ATP resonance peak with varied saturation time (t) (i.e., the progressive saturation experiment commonly used in the CST approach). For this case, the three-spin chemical exchange system of *PCr↔ATP↔Pi* can be treated as two independent two-spin chemical exchange systems (i.e., *PCr↔ATP* and *ATP↔Pi*). Solving **Eqs. (15.12a)** and **(15.12c)** with the boundary condition of $M_b = 0$ results in **Eqs. (15.13a)** and **(15.13b)** *(47)*,

$$M_a(t) = M_a^0\left[\left(\frac{k_1}{\alpha_a}\right)e^{-\alpha_a t} + \left(\frac{1}{\alpha_a T_{1a}}\right)\right] \tag{15.13a}$$

$$M_c(t) = M_c^0\left[\left(\frac{k_{-2}}{\alpha_c}\right)e^{-\alpha_c t} + \left(\frac{1}{\alpha_c T_{1c}}\right)\right] \tag{15.13b}$$

with

$$\alpha_a = \frac{1}{T_{1a}^{app}} = \left(k_1 + \frac{1}{T_{1a}}\right);\ \alpha_c = \frac{1}{T_{1c}^{app}} = \left(k_{-2} + \frac{1}{T_{1c}}\right).$$

Therefore, the parameters of apparent spin-lattice relaxation rates (α_a and α_c) and intrinsic spin-lattice relaxation times (T_{1a} and

T_{1c}) at a given B_0 can be determined via regressions as a function of t, where both k_1 and k_{-2} are replaced by their apparent spin-lattice relaxation rate and intrinsic spin-lattice relaxation time (see **Eqs. (15.13a)** and **(15.13b)**). Then, the forward rate constants (k_1 and k_{-2}) can be calculated by their relations with the apparent spin-lattice relaxation rate and intrinsic spin-lattice relaxation time as depicted above.

When a steady-state condition is approached with complete saturation of γ-ATP (i.e., $dM_a/\mathrm{dt} = 0$ and $dM_c/\mathrm{dt} = 0$), **Eqs. (15.13a)** and **(15.13b)** can be further simplified to the following formulae:

$$k_1 = \frac{M_a^0 - M_a^{*b}}{M_a^0}\alpha_a = \frac{M_a^0 - M_a^{*b}}{T_{1a}M_a^{*b}} \tag{15.13c}$$

$$k_{-2} = \frac{M_c^0 - M_c^{*b}}{M_c^0}\alpha_c = \frac{M_c^0 - M_c^{*b}}{T_{1c}M_c^{*b}} \tag{15.13d}$$

where M_a^{*b} and M_c^{*b} are the steady-state magnetizations of a and c when b is fully saturated. Therefore, the forward rate constants of k_1 and k_{-2} can be calculated by using **Eqs. (15.13c)** and **(15.13d)**. For this case, only two steady-state spectra acquired in the presence and absence of saturating b are needed to determine both k_1 and k_{-2}, if T_{1a} and T_{1c} have already been determined at a given B_0 through the progressive saturation measurement accordingly to **Eqs. (15.13a)** and **(15.13b)**. Finally, the forward fluxes for the CK and ATP_{ase} reactions can be determined by the following relations,

$$F_f^{CK} = k_1[PCr] \text{ and } F_f^{ATP_{ase}} = k_{-2}[Pi] \tag{15.13e}$$

Step 2: Steady-state saturation of Pi for determining CK reverse rate constant and flux

The second step of the in vivo ^{31}P MSS MT measurements is to apply a frequency-selective RF pulse train for completely saturating the Pi resonance peak with a sufficiently long saturation time resulting in steady-state magnetizations of a and b. For this case, $M_c = 0$, $dM_a/dt = 0$, $dM_b/dt = 0$, and **Eqs. (15.12a)** and **(15.12b)** yield

$$k_{-1} = \alpha_a\left(\frac{M_a^{*c}}{M_a^0} - \frac{1}{T_{1a}\alpha_a}\right)\frac{M_a^0}{M_b^{*c}} \tag{15.14a}$$

$$k_1 = \alpha_b\left(\frac{M_b^{*c}}{M_b^0} - \frac{1}{T_{1b}\alpha_b}\right)\frac{M_b^0}{M_a^{*c}} \tag{15.14b}$$

with

$$\alpha_b = \frac{1}{T_{1b}^{app}} = (k_2 + k_{-1} + \frac{1}{T_{1b}})$$

where M_a^{*c} and M_b^{*c} are the steady-state magnetizations of a and b when Pi is fully saturated. Only **Eq. (15.14a)** is needed in the MSS MT approach to determine the CK reverse rate constant (k_{-1}) by using the measured magnetizations (i.e., M_a^0, M_a^{*c} and M_b^{*c}) and other relaxation parameters (i.e., α_a and T_{1a}), which are determined by the measurement in Step 1. The reverse flux of CK reaction can be calculated by the following relation:

$$F_r^{CK} = k_{-1}[ATP]. \quad (15.14c)$$

Step 3: Steady-state saturation of PCr for determining ATP$_{ase}$ reverse rate constant and flux

The third step of the in vivo ^{31}P MSS MT measurements is to apply a frequency-selective RF pulse train for completely saturating the PCr resonance peak with a sufficiently long saturation time resulting in steady-state magnetizations of b and c. For this case, M_a=0, dM_b/dt=0, dM_c/dt=0, and **Eqs. (15.12b)** and **(15.12c)** yield

$$k_2 = \alpha_c \left(\frac{M_c^{*a}}{M_c^0} - \frac{1}{T_{1c}\alpha_c} \right) \frac{M_c^0}{M_b^{*a}} \quad (15.15a)$$

$$k_{-2} = \alpha_b \left(\frac{M_b^{*a}}{M_b^0} - \frac{1}{T_{1b}\alpha_b} \right) \frac{M_b^0}{M_c^{*a}} \quad (15.15b)$$

where M_b^{*a} and M_c^{*a} are the steady-state magnetizations of b and c when PCr is fully saturated. Only **Eq. (15.15a)** is needed in the MSS MT approach to determine the ATP$_{ase}$ forward rate constant (k_2) by using the measured magnetizations (i.e., M_c^0, M_c^{*a} and M_b^{*a}) and other relaxation parameters (i.e., $\alpha_c$and T_{1c}), which are determined by the measurement in Step 1. The reverse flux of the ATP$_{ase}$ reaction can be calculated via following relation:

$$F_r^{ATP_{ase}} = k_2[ATP]. \quad (15.15c)$$

Therefore, by combining these three measurement steps, one is able to measure all kinetic parameters which determine the entire exchange of *PCr↔ATP↔Pi* including all four ATP metabolic fluxes for both forward and reverse reaction directions as schematically summarized in **Figs. 15.1 and 15.13**. In total, a minimal number of three selective saturated in vivo ^{31}P spectra plus one control in vivo ^{31}P spectrum (for quantifying [ATP], [Pi], [PCr], i.e., M_a^0, M_b^0 and M_c^0) are required by the MSS MT approach for the measurements *(50)*.

4.2. Application of in vivo ^{31}P MSS MT Approach in Human Brain at 7T

The in vivo ^{31}P MT measurements can significantly benefit from the high-field advantages of improved MR detection sensitivity and spectral resolution. We have conducted extensive measurements at 7T for validating the in vivo ^{31}P MSS MT approach, and ultimately, determining the entire chemical exchange of $PCr \leftrightarrow ATP \leftrightarrow Pi$ in the human occipital lobe *(20, 50)*.

4.2.1. Progressive Saturation for Measuring Intrinsic T_1, Forward Rate Constants And Fluxes

Figure 15.10A illustrates the progressive saturation transfer measurements and related in vivo ^{31}P spectra when the γ-ATP resonance peak was completely saturated (i.e., Step 1 measurement used in the in vivo ^{31}P MSS MT approach). It shows a gradual decrease of Pi signal when the γ-ATP saturation time increases because of the chemical exchange between γ-ATP and Pi (i.e., $ATP \leftrightarrow Pi$) and the increased magnetization transfer effect *(20)*. This MT effect was observed on the chemical-exchangeable Pi resonance but not for those adjacent and non-chemical-exchangeable phosphate metabolites such as PDE and PME groups. Moreover, the MT effect on Pi disappeared when the RF saturation frequency was moved to the opposite side of the in vivo ^{31}P spectrum with the same chemical shift difference as shown in **Figure 15.10B. Figure 15.10C** plots the quantitative relation between the normalized Pi signal intensity (i.e., $M_a^{(t)}/M_a^0$ in **Eq. (15.13a)**) under the γ-ATP saturation with the saturation time of t. This plot can be used to calculate the intrinsic T_1 of Pi (i.e., T_{1a}) and the forward ATP_{ase} reaction rate constant (k_1) by the regression fitting according to **Eq. (15.13a)**. Then, the forward ATP_{ase} reaction flux can be determined according to **Eq. (15.13e)**. The same in vivo ^{31}P MT spectral data can be applied to quantify the MT effect on the PCr resonance as a function of the γ-ATP saturation time, ultimately, to determine the intrinsic T_1 of PCr (i.e., T_{1c}) and the forward CK reaction rate constant and flux according to **Eqs. (15.13b)** and **(15.13e)**.

4.2.2. Steady-State Saturation for Measuring Forward Rate Constants And Fluxes

One technical limitation of progressive saturation approach is the requirement of a number of saturated in vivo ^{31}P spectra with varied saturation time (e.g., **Fig. 15.10**). One alternative but more robust approach is the use of steady-state saturation if the intrinsic T_1 values have been determined at a given B_0 (see details in the discussion of Step 1 measurement above). **Figure 15.11** demonstrates such a measurement (i.e., Step 1) and the in vivo ^{31}P MT spectra acquired in the absence (control spectrum; **Fig. 15.11A**) and presence (saturated spectrum; **Fig. 15.11B**) of sufficiently long RF saturation for ensuring complete saturation of the γ-ATP resonance peak and approaching steady-state magnetizations for

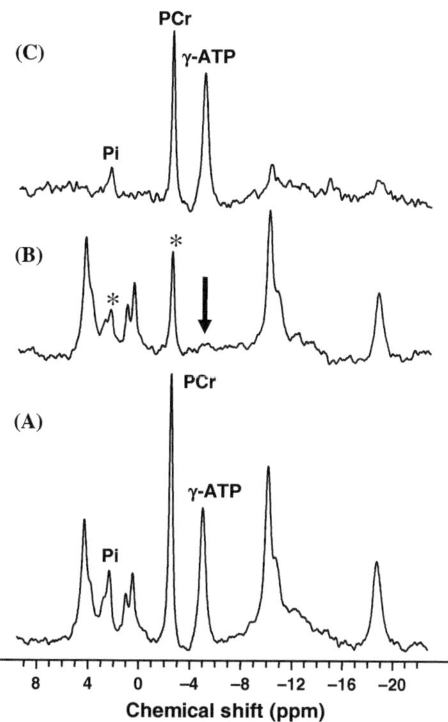

Fig. 15.11. In vivo ^{31}P spectra acquired from a health human occipital lobe in the absence (**A**) and presence (**B**) of complete γ-ATP saturation, and the difference spectrum (**C**) between the two. Only the Pi and PCr resonance peaks show the magnetization transfer effect due to the saturation. The intensity reduction of Pi can be used to determine the forward rate constant and flux for the ATPase reaction, and the intensity reduction of PCr can be used to determine the forward rate constant and flux for the CK reaction. Adapted from Lei et al. of Ref. *(20)*.

both Pi and PCr. **Figure 15.11C** shows the difference spectrum by subtracting the control and γ-ATP saturated spectra *(20, 50)*. The relative Pi signal reduction can be used to calculate the forward rate constant and flux for the ATP_{ase} reaction according to **Eqs. (15.13d)** and **(15.13e)**. Similarly, the relative PCr signal reduction can be used to calculate the forward rate constant and flux for the CK reaction according to **Eqs. (15.13c)** and **(15.13e)**.

4.2.3. In vivo ^{31}P MSS MT Measurements for Determining all ATP Metabolic Fluxes Involving PCr↔ATP↔Pi Exchange in Human Occipital Lobe

Figure 15.12 illustrates one example of in vivo ^{31}P MSS MT measurements in the human occipital lobe *(50)*. A total of four in vivo ^{31}P spectra were collected in the absence (**Fig. 15.12A**) and presence of complete RF saturation on the resonance peak of Pi (Step 2; **Fig. 15.12B**), PCr (Step 3; **Fig. 15.12C**) and γ-ATP (Step 1; **Fig. 15.12D**), respectively. It reveals that single-site saturation on one phosphate spin can lead to significant magnetization reductions in the other two coupled phosphate spins through the three-spin chemical exchange system

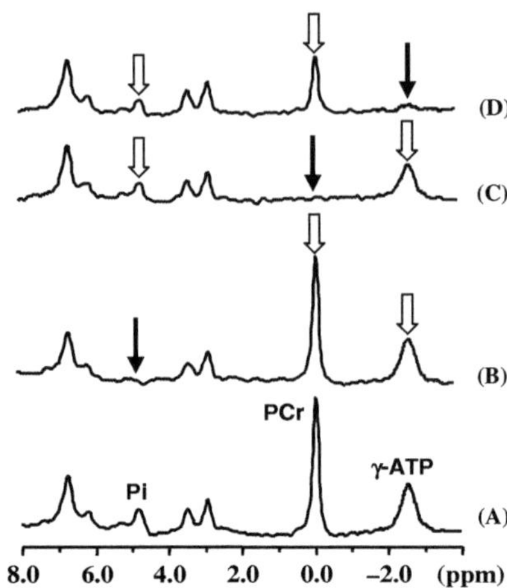

Fig. 15.12. In vivo ^{31}P MSS MT measurements and four spectra acquired from a representative human occipital lobe in the absence (**A**) and presence of complete RF saturation on the resonance peak of (**B**) Pi (Step 2), (C) PCr (Step 3) and (D) γ-ATP (Step 1), respectively. The narrow arrows indicate the saturation sites and the wide arrows indicate the signal reductions due to the magnetization transfer. Adapted from Du et al. of Ref. *(50)*.

of *PCr↔ATP↔Pi*. For instance, the PCr saturation results in a significant signal reduction for both γ-ATP and Pi as shown in **Fig. 15.12C**. All required steady-state saturated magnetizations can be determined by using the three saturated in vivo ^{31}P spectra as shown in **Fig. 15.12B–D**, and all equilibrium magnetizations can be determined by using the control in vivo ^{31}P spectrum as shown **Fig. 15.12A**. The values of these magnetizations were used to determine four rate constants (k_1, k_2, k_{-1} and k_{-2}) and four fluxes (F_f^{ATPase}, F_r^{ATPase}, F_f^{CK} and F_r^{CK}) associated with the *PCr↔ATP↔Pi* exchange in the human brain. These measured results and the ratios between the forward and reverse fluxes are summarized in **Fig. 15.13** indicating several important observations. First, there is no statistical difference between the forward and reverse exchange fluxes for both the CK reaction (i.e., F_f^{CK} versus F_r^{CK} ; $p = 0.38$) and the ATP_{ase} reaction (i.e., F_f^{ATPase} versus F_r^{ATPase} ; $p = 0.46$) *(50)*. Second, the ratios between the forward and reverse fluxes for the CK reaction ($F_f^{CK}/F_r^{CK} = 0.95 \pm 0.12$) and the ATP_{ase} reaction ($F_f^{ATPase}/F_r^{ATPase} = 1.08 \pm 0.21$) are close to unity. In general, the CK fluxes reflect the nonoxidative phosphorylation and the ATP_{ase} fluxes, in contrast, reflect oxidative phosphorylation. Both of these phosphorylation pathways can contribute to the total ATP production flux (i.e., $F_f^{total} = F_f^{CK} + F_f^{ATPase} = 73.9$ μmol/g/min) and the total ATP utilization flux (i.e., $F_r^{total} = F_r^{CK} + F_r^{ATPase} = 76.9$ μmol/g/min). The ratio between F_f^{total}

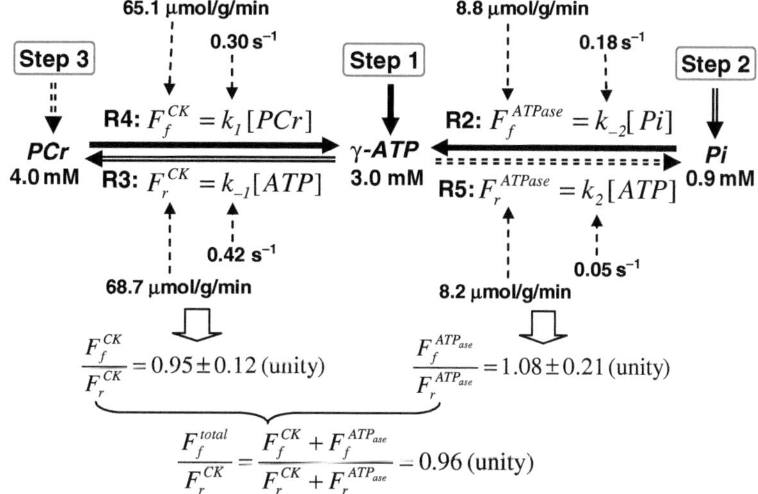

Fig. 15.13. Chart showing the three-step measurements of in vivo ^{31}P MSS MT approach for determining the entire ATP kinetic network and associated metabolic rate constants and fluxes, and the measurement results from the human occipital lobe. Step 1 measures two forward reactions (**R2** and **R4**) along the solid arrows. Step 2 measures the indirectly coupled reverse reaction (**R3**) along the double-line arrows. Step 3 measures another reverse reaction (**R5**) along the dotted-line arrows. All results point to the fact that flux ratios satisfy the chemical equilibrium condition.

and F_r^{total} is again close to unity (=0.96) indicating that the total ATP production flux equals the total ATP utilization flux. These results lead us to conclude that the fluxes measured by the in vivo ^{31}P MSS MT approach satisfy the chemical equilibrium conditions for the CK and ATP$_{ase}$ reactions in the human brain, and indicate that the MSS MT approach is able to explicitly determine four rate constants in the *PCr↔ATP↔Pi* kinetic process and their associated ATP metabolic fluxes in the brain noninvasively. Nevertheless, such equal forward and reverse flux relations are unable to be obtained for the saturation transfer measurement if the three-spin *PCr↔ATP↔Pi* exchange system was treated as two independent two-spin change systems (i.e., *PCr↔ATP* and *ATP↔Pi*) even when the same experimental MSS MT data were analyzed *(50)*. These findings clearly indicate that the chemical exchange system of *PCr↔ATP↔Pi* has to be treated as a three-spin exchange system in order to accurately determine the reverse rate constants and fluxes for both the CK and ATP$_{ase}$ reactions.

The *PCr↔ATP↔Pi* exchange system in a normal brain likely reaches chemical equilibrium (or near equilibrium) under most circumstances. As long as the equilibrium condition is satisfied, one can further simplify the in vivo ^{31}P MT measurement procedure by using the chemical equilibrium relation *(50)*. In this case, only two in vivo ^{31}P spectra (one control and another with full saturation of γ-ATP as illustrated in **Fig. 15.11**) are practically required to determine the entire *PCr↔ATP↔Pi* exchange process *(50)*. Thus, this simplification can make the in vivo ^{31}P MT measurement more robust for rapidly determining the ATP metabolic flux changes induced by either physiological or pathological perturbations.

5. Perspective and Discussion

Despite its low detection sensitivity compared to ^{1}H MRS, researchers in the past three decades have revealed the great merits of in vivo heteronuclear MRS for a broad range of biomedical applications. Two of them (i.e., ^{31}P and ^{17}O) have been discussed here in great detail and each method provides a unique tool for noninvasively studying specific metabolic pathways in the brain through the measurements of metabolite concentrations and biochemical reaction fluxes. The combination of different in vivo multinuclear MRS approaches may significantly enhance the MR capability to perform comprehensive studies concerning brain physiology, neurochemistry, bioenergetics and their relations to brain function. This point is addressed by answering the following questions:

5.1. Why is in vivo ^{31}P MT Approach Attractive and Promising for Studying Brain Bioenergetics?

A large amount of research efforts have been spent since early 1980s for exploring the possibility of determining the ATP metabolic fluxes using a variety of in vivo ^{31}P MRS MT approaches and great progress has been made *(20, 44–46, 49, 50, 103–110, 115, 117)*. Nevertheless, the course for advancing the in vivo ^{31}P MRS MT methodology and its clinical application has been significantly slowed down particularly in brain studies during the last decade. One of the major causes is probably in regard to two controversial observations in this research field. The first one is the inconsistency in the results of the measured reverse ATP fluxes in the literature (e.g., *(112–114)*). In our opinion, this inconsistency is likely caused by the lack of comprehensive spin-exchange models used in the early research work. And this controversy can be resolved by using more sophisticated three-spin exchange models for accurately determining the reverse ATP fluxes for both the CK and ATP_{ase} reactions *(45, 46, 50, 115, 116)*.

Another surprising observation has raised question regarding whether the measured forward ATP_{ase} reaction flux truly reflects the oxidative phosphorylation in the mitochondria in a living organ. The ATP synthesis reaction (i.e., **R2** in **Figs. 15.1 and 15.13**) catalyzed by the coupled activities of glycolytic enzymes glyceraldehyde-3-phosphate dehydrogenase (GAPDH) and phosphoglycerate kinase (PGK) have been shown to be a major contributor to the *Pi → ATP* reaction flux measured by the ^{31}P MT approaches in E. coli *(118)*, yeast *(119)*, liver *(120)*, and the myocardium *(121)*. For instant, in the perfused rat heart, GAPDH/PGK mediated exchange dominates the *Pi → ATP* flux measured by the ^{31}P MT approach in the myocardium *(121)*. Under these circumstances, ***only if*** the GAPDH/PGK effect was eliminated either directly using exogenous inhibitor iodoacetate or indirectly by eliminating all exogenous and endogenous sources of glucose, the myocardium *Pi → ATP* flux determined by the in vivo ^{31}P MRS MT measurement was found to be the same as the net rate of ***oxidative*** ATP synthesis calculated as the product of the cardiac oxygen consumption rate and the P:O ratio, and then the correlation between the measured flux and oxygen consumption rate became evident *(121)*. These observations and findings reveal the complexity of biological systems and potential limitation of in vivo ^{31}P MRS MT approaches for directly measuring the ATP metabolic flux related to oxidative phosphorylation in some organs such as heart and liver. Interestingly, in contrast to the observations in the myocardium, and analogous observations in yeast, E. coli and liver, the *Pi → ATP* flux measured by in vivo ^{31}P MRS MT in the human brain is similar to the net oxidative ATP synthesis rate which can be estimated by the available $CMRO_2$ data and the P:O ratio based on a tight link between the cerebral oxidative phosphorylation and oxygen utilization under normal physiological condition *(20, 50)*. The $CMRO_2$ value in the human occipital lobe has been measured to be 1.71 μmol/g/min by PET *(91)*. The ATP synthesis rate attributed by oxidative phosphorylation can be calculated by multiplying this $CMRO_2$ value by the P:O ratio of ~2.5 *(122)* and factor of 2, resulting in an estimated value of 8.6 μmol/g/min. This rate is almost identical to the cerebral ATP synthesis rate of F_f^{ATPase} = 8.8 μmol/g/min, which was directly measured by in vivo ^{31}P MRS MT *(50)*. A similar relation was also evident in the animal studies showing a tight correlation between the measured *Pi → ATP* flux and the estimated oxidative phosphor rate *(49, 123)*. These comparisons of results, thus, lead us to conclude that in the brain, the Pi→ATP flux measured by the in vivo ^{31}P MT approach equals the net oxidative ATP synthesis rate linking to the cerebral oxidative phosphorylation, and this flux should provide a vital index reflecting the major bioenergetics for supporting brain activity and function *(7, 12, 20, 123)*. This conclusion makes in vivo ^{31}P MT attractive for studying the central role of the cerebral bioenergetics associated with oxidative phosphorylation and brain activity, though the underlying mechanism for explaining the discrepancy in the ^{31}P MT measurements between the brain and other organs (e.g., heart and liver) remains to be explored *(20)*.

5.2. What Is the Possible Role that in vivo ^{31}P and ^{17}O MRS can Play for Studying Brain Activation and Function?

As mentioned earlier, both in vivo ^{31}P and ^{17}O MRS methods can provide noninvasive tools for studying brain metabolism and bioenergetics. Each of these methods has unique features for probing the metabolic functions in mitochondria. Specifically, the in vivo ^{17}O MRS imaging approach is useful for mapping the absolute $CMRO_2$ within a relatively short $^{17}O_2$ inhalation time. The $CMRO_2$ images should reflect the total metabolic rate of oxygen utilization (i.e., **R1** in **Fig. 15.1**) and its spatial distribution in the brain. In contrast, the in vivo ^{31}P MT approach is powerful for measuring all the cerebral ATP metabolic fluxes associated with four ATP reactions (i.e., **R2** to **R5** in **Fig. 15.1**) occurring in different subcellular compartments. All of these measured physiological parameters can be linked to brain ATP energy and ATP transportation. Therefore, the combination of in vivo ^{31}P and ^{17}O MRS methods may provide an essential in vivo MR tool being able to studying the crucial roles of both oxygen and ATP metabolisms in cerebral bioenergetics and brain functions.

One relevant question in the brain function and neuroimaging research fields is how much extra brain energy is needed during the brain stimulation and/or task performance. This topic is still in intense debate, especially regarding how much $CMRO_2$ is induced by brain stimulation *(7, 8, 91, 124–136)*. A more important question, perhaps, is how many more ATP molecules are consumed during brain activation as compared to the resting brain because of a close relation between the ATP utilization and energy consumption *(3, 7, 8)*. The answers to these questions are central for understanding the mechanisms underlying most modern neuroimaging techniques including two most popular methods of the functional MRI (fMRI) based on the blood oxygenation level dependent (BOLD) contrast *(137–142)* and PET *(7, 8, 143)*. One major hurdle for addressing the questions is the lack of sophisticated and robust neuroimaging modalities for directly assessing the absolute $CMRO_2$ and CMR_{ATP}, which is equivalent to $F_{f\text{ATPase}}$ in this article, and their changes elevated by brain activation. Therefore, the in vivo ^{31}P and ^{17}O MRS approaches as described here could fill this methodological gap and make significant contribution in understanding the relation between brain bioenergetics and function. This notion is convincingly supported by several lines of evidence provided by recent research progresses. One example of these progresses was to apply the in vivo ^{17}O MRS imaging approach for successfully mapping $CMRO_2$ images in the cat brain under resting and visual stimulation conditions *(144)*. A significant $CMRO_2$ increase was observed in the activated cat visual cortex. This initial observation indicates the possible central role of oxygen metabolism for supporting the elevated brain activity during activation.

Another significant progress is related to the in vivo ^{31}P MT application in the rat brain for studying the quantitative correlation between the CMR_{ATP} and the varied brain activity from mild anesthesia state to the isoelectric state *(123)*. In this study, we observed a strong correlation between the Pi→ATP flux measured by the in vivo ^{31}P MT approach and the brain activity quantified by EEG signal *(123)*. This result indicates a tight correlation between the ATP synthesis rate and brain bioenergetic demand under varied brain activity states. Therefore, the measurement of Pi→ATP flux should provide a sensitive energy index for quantifying the brain bioenergetics. In contrast, the steady-state HEP concentrations (e.g., [ATP], [PCr] and [Pi]) and pH are relatively stable in a wide range of brain activity level *(123)*. These results support the view that one of the major functions of the *PCr↔ATP↔Pi* exchange process is to sustain the balance between the ATP supply and demand in the brain resulting in a stable ATP concentration under the normal physiological condition. Such a balance is likely maintained by the rapid adjustments of both forward and reverse metabolic fluxes involved in the *PCr↔ATP↔Pi* exchange according to the change of brain energy. We conclude that the ATP metabolic flux measurements using the in vivo ^{31}P MT approach should be more useful for studying the brain bioenergetic change in response to brain activation as compared to the steady-state HEP concentration measurements using conventional in vivo ^{31}P MRS. This is evident from two high-field in vivo ^{31}P MRS studies showing a few percents of decrease in the PCr signal in contrast to a >30% increase in the CK forward flux in the human visual cortex during visual stimulation *(108, 145)*. It is conceivable that the in vivo ^{31}P MT approach, as a sole in vivo tool being able to directly and quantitatively measure the cerebral ATP metabolic fluxes, will play a vital role for studying the brain bioenergetics associated with brain function and activation. One particularly interesting question possibly to be addressed by this in vivo ^{31}P MT approach is whether more ATP utilization is required by the elevated neuronal activity, and if yes, how much more is needed? The answers should advance our understanding regarding the central roles of cerebral oxidative phosphorylation and mitochondria in supporting brain function.

5.3. What are the Potential Applications of in vivo ^{31}P and ^{17}O MRS for Clinical Research and Diagnosis?

In vivo MRS methodology can provide detailed metabolite fingerprints in the living organs, and it is perhaps one of the most classic molecular imaging modalities. Many metabolites and their abnormal changes, detectable by in vivo MRS, have been proved to be tightly linked to various brain diseases and disease progression.

Both of the in vivo ^{17}O MRS and ^{31}P MRS MT approaches as discussed here are especially useful for studying the brain oxidative metabolism and the mitochondrial metabolic function. They would provide opportunities for a variety of clinical brain research and potentially for clinical diagnosis because of the obvious role of oxidative metabolism in the pathology associated with many brain disorders and neurodegenerative diseases such as schizophrenia, Alzheimer's disease, Huntington's disease, Parkinson's disease, mitochondrial dysfunction and aging problems (e.g., *(146–150)*). One line of evidence is the histopathological findings indicating that the activity of cytochrome oxidase, the key mitochondrial enzyme that catalyzes the reduction of oxygen to form water, is significantly impaired in schizophrenic *(146)* and Alzheimer's patients *(147, 148)*. Another line of evidence is from the studies of diseases caused by mitochondrial DNA mutations suggesting that a variety of degenerative processes in Parkinson's disease and Alzheimer's disease may be associated with defects in mitochondrial oxidative phosphorylation *(151, 152)*. Thus, the in vivo ^{17}O MRS and ^{31}P MT approaches may play vital roles for investigating the neurodegenerative diseases associated with the mitochondrial abnormality and metabolic syndrome with great potential in clinical diagnosis and monitoring the brain functional recovery after medical treatment.

6. Conclusion

Although, in reality, in vivo heteronuclear MRS has not been widely applied compared to ^{1}H MRI/MRS due to its lower detection sensitivity, the currently available ultrahigh MRI/MRS scanners, technology and their associated high field merits have undoubtedly stimulated the in vivo heteronuclear MRS research and methodological development *(153)*. Significant progresses have been made for advancing in vivo MRS in brain research. One of the unique in vivo MRS utilities is to noninvasively measure the cerebral metabolic fluxes and their dynamic changes, which may lead to many important applications in medical research. Two interesting and exciting applications of the in vivo ^{17}O MRS approach and/or ^{31}P MRS MT approach are: (i) to study the dynamic changes in oxygen utilization and oxidative ATP metabolism during activation in normal brains; (ii) to explore the possibility for detecting abnormality in the oxidative metabolic fluxes during the progression of brain diseases prior to showing metabolite concentration abnormality. The outcomes from these applications will be essential for better understanding of brain function and dysfunction. Finally, the in vivo ^{17}O and ^{31}P MRS can be readily combined with many other MRI utilities for imaging brain anatomy, perfusion, diffusion and BOLD etc within the same scanning session. This combined MRS/MRI strategy can make it particularly powerful in neuroscience research.

Acknowledgments

The authors would like to thank Drs. Hellmut Merkle, Run-Xia Tian, Peter Andersen, Gregor Adriany, Pete Thelwall and Mr. John Strupp for their technical assistance, support and scientific discussion. Part of the reviewed work was supported by NIH grants of NS41262, EB02632, NS39043, EB00329, P30NS057091 and P41 RR08079, the W.M. Keck Foundation and the MIND Institute.

References

1. Boyer PD. What makes ATP synthase spin? Nature 1999;402(6759):247, 249.
2. Stock D, Leslie AG, Walker JE. Molecular architecture of the rotary motor in ATP synthase. Science 1999;286(5445): 1700–1705.
3. Siesjo BK. Brain energy metabolism. New York: Wiley; 1978. p 101–110.
4. Raichle ME. Circulatory and metabolic correlates of brain function in normal humans. In: Mountcastle VB, Plum F, Geiger SR, editors. Handbook of Physiology-The Nervous System. Bethesda: American Physiological Society; 1987. p 643–674.
5. Clarke DD, Sokoloff L. Circulation and energy metabolism of the brain. In: Siegel GJ, al. e, editors. Basic Neruochemistry: Molecular, Cellular and Medical Aspects. Philadephia: Lippincott-Raven Publishers; 1999. p 633–669.
6. Shulman RG, Rothman DL, Behar KL, Hyder F. Energetic basis of brain activity: Implications for neuroimaging. Trends Neurosci 2004;27(8):489–495.
7. Hyder F, Patel AB, Gjedde A, Rothman DL, Behar KL, Shulman RG. Neuronal-glial glucose oxidation and glutamatergic-GABAergic function. J Cereb Blood Flow Metab 2006;26(7):865–877.
8. Raichle ME, Mintun MA. Brain work and brain imaging. Annu Rev Neurosci 2006;29:449–476.
9. Hevner RF, Wong-Riley MT. Regulation of cytochrome oxidase protein levels by functional activity in the macaque monkey visual system. J Neurosci 1990;10(4):1331–1340.
10. Hevner RF, Duff RS, Wong-Riley MT. Coordination of ATP production and consumption in brain: Parallel regulation of cytochrome oxidase and Na^+, K^+-ATPase. Neurosci Lett 1992;138(1): 188–192.
11. Rolfe DF, Brown GC. Cellular energy utilization and molecular origin of standard metabolic rate in mammals. Physiol Rev 1997;77(3):731–758.
12. Attwell D, Laughlin SB. An energy budget for signaling in the grey matter of the brain. J Cereb Blood Flow Metab 2001;21(10):1133–1145.
13. Shulman RG, Rothman DL, Hyder F. Stimulated changes in localized cerebral energy consumption under anesthesia. Proc Natl Acad Sci USA 1999;96(6): 3245–3250.
14. Magistretti PJ, Pellerin L, Rothman DL, Shulman RG. Energy on demand. Science 1999;283(5401):496–497.
15. Wallimann T, Hemmer W. Creatine kinase in non-muscle tissues and cells. Mol Cell Biochem 1994;133–134:193–220.
16. Saks VA, Ventura-Clapier R, Aliev MK. Metabolic control and metabolic capacity: Two aspects of creatine kinase functioning in the cells. Biochim Biophys Acta 1996;1274(3):81–88.
17. Kemp GJ. Non-invasive methods for studying brain energy metabolism: what they show and what it means. Dev Neurosci 2000;22(5–6):418–428.
18. Shulman RG, Brown TR, Ugurbil K, Ogawa S, Cohen SM, den Hollander JA. Cellular applications of ^{31}P and ^{13}C nuclear magnetic resonance. Science 1979;205(4402): 160–166.
19. Ackerman JJH, Grove TH, Wong GG, Gadian DG, Radda GK. Mapping of metabolites in whole animals by ^{31}P NMR using surface coils. Nature 1980;283:167–170.
20. Lei H, Ugurbil K, Chen W. Measurement of unidirectional Pi to ATP flux in human visual cortex at 7 Tesla using in vivo ^{31}P magnetic resonance spectroscopy. Proc Natl Acad Sci U S A 2003;100: 14409–14414.
21. Gadian DG, Williams SR, Bates TE, Kauppinen RA. NMR spectroscopy: current status

and future possibilities. Acta Neurochir Suppl (Wien) 1993;57:1–8.
22. Weiner MW. NMR spectroscopy for clinical medicine. Animal models and clinical examples. Ann N Y Acad Sci 1987;508:287–299.
23. Hossmann KA. Studies of experimental cerebral ischemia with NMR spectroscopy. Arzneimittelforschung 1991;41(3A): 292–298.
24. Lei H, Zhu XH, Zhang XL, Ugurbil K, Chen W. In vivo ^{31}P magnetic resonance spectroscopy of human brain at 7 T: An initial experience. Magn Reson Med 2003;49: 199–205.
25. Petroff OA, Prichard JW, Behar KL, Alger JR, den Hollander JA, Shulman RG. Cerebral intracellular pH by 31P nuclear magnetic resonance spectroscopy. Neurology 1985;35(6):781–788.
26. Burt CT, Ribolow HJ. A hypothesis: Noncyclic phosphodiesters may play a role in membrane control. Biochem Med 1984;31(1):21–30.
27. Burt CT, Ribolow H. Glycerol phosphorylcholine (GPC) and serine ethanolamine phosphodiester (SEP): Evolutionary mirrored metabolites and their potential metabolic roles. Comp Biochem Physiol Biochem Mol Biol 1994;108(1):11–20.
28. Ross BM, Moszczynska A, Blusztajn JK, Sherwin A, Lozano A, Kish SJ. Phospholipid biosynthetic enzymes in human brain. Lipids 1997;32(4):351–358.
29. Podo F. Tumour phospholipid metabolism. NMR Biomed 1999;12(7):413–439.
30. Young RS, Chen B, Petroff OA, Gore JC, Cowan BE, Novotny EJ, Jr., Wong M, Zuckerman K. The effect of diazepam on neonatal seizure: In vivo ^{31}P and ^{1}H NMR study. Pediatr Res 1989;25(1):27–31.
31. Welch KM, Levine SR, Martin G, Ordidge R, Vande Linde AM, Helpern JA. Magnetic resonance spectroscopy in cerebral ischemia. Neurol Clin 1992;10(1):1–29.
32. Duncan JS. Imaging and epilepsy. Brain 1997;120(Pt 2):339–377.
33. Hetherington HP, Pan JW, Spencer DD. ^{1}H and ^{31}P spectroscopy and bioenergetics in the lateralization of seizures in temporal lobe epilepsy. J Magn Reson Imaging 2002;16(4):477–483.
34. Brown GG, Levine SR, Gorell JM, Pettegrew JW, Gdowski JW, Bueri JA, Helpern JA, Welch KM. In vivo ^{31}P NMR profiles of Alzheimer's disease and multiple subcortical infarct dementia. Neurology 1989;39(11):1423–1427.
35. Pettegrew JW, Klunk WE, Panchalingam K, McClure RJ, Stanley JA. Magnetic resonance spectroscopic changes in Alzheimer's disease. Ann N Y Acad Sci 1997;826:282–306.
36. Riehemann S, Volz HP, Smesny S, Hubner G, Wenda B, Rossger G, Sauer H. Phosphorus 31 magnetic resonance spectroscopy in schizophrenia research. Pathophysiology of cerebral metabolism of high-energy phosphate and membrane phospholipids. Nervenarzt 2000;71(5):354–363.
37. Jensen JE, Al-Semaan YM, Williamson PC, Neufeld RW, Menon RS, Schaeffer B, Densmore M, Drost DJ. Region-specific changes in phospholipid metabolism in chronic, medicated schizophrenia: ^{31}P-MRS study at 4.0 Tesla. Br J Psychiatry 2002;180:39–44.
38. Ramadan NM, Halvorson H, Vande-Linde A, Levine SR, Helpern JA, Welch KM. Low brain magnesium in migraine. Headache 1989;29(9):590–593.
39. Altura BM, Altura BT. Role of magnesium and calcium in alcohol-induced hypertension and strokes as probed by in vivo television microscopy, digital image microscopy, optical spectroscopy, 31P-NMR, spectroscopy and a unique magnesium ion-selective electrode. Alcohol Clin Exp Res 1994;18(5): 1057–1068.
40. Nioka S, Zaman A, Yoshioka H, Masumura M, Miyake H, Lockard S, Chance B. ^{31}P magnetic resonance spectroscopy study of cerebral metabolism in developing dog brain and its relationship to neuronal function. Dev Neurosci 1991;13(2):61–68.
41. Nioka S, Smith DS, Mayevsky A, Dobson GP, Veech RL, Subramanian H, Chance B. Age dependence of steady state mitochondrial oxidative metabolism in the in vivo hypoxic dog brain. Neurol Res 1991;13(1):25–32.
42. Chance B, Leigh JS, Jr., Nioka S, Sinwell T, Younkin D, Smith DS. An approach to the problem of metabolic heterogeneity in brain: Ischemia and reflow after ischemia. Ann N Y Acad Sci 1987;508:309–320.
43. Frosen S, Hoffman RA. Study of moderately rapid chemical exchange by means of nuclear magnetic double resonance. J Chem Phys 1963;39:2892–2901.
44. Alger JR, Shulman RG. NMR Studies of enzymatic rates in vitro and in vivo by magnetization transfer. Quart Rev Biophys 1984;17:83–124.
45. Ugurbil K. Magnetization transfer measurements of individual rate constants in the presence of multiple reactions. J Magn Reson 1985;64:207–219.
46. Ugurbil K. Magnetization transfer measurements of creatine kinase and ATPase rates in intact hearts. Circulation 1985;72(Supp. IV):IV94- IV96.

47. Degani H, Laughlin M, Campbell S, Shulman RG. Kinetics of creatine kinase in heart: a ^{31}P NMR saturation- and inversion-transfer study. Biochemistry 1985;24(20): 5510–5516.
48. Bottomley PA, Ouwerkerk R, Lee RF, Weiss RG. Four-angle saturation transfer (FAST) method for measuring creatine kinase reaction rates in vivo. Magn Reson Med 2002;47(5):850–863.
49. Shoubridge EA, Briggs RW, Radda GK. ^{31}P NMR saturation transfer measurements of the steady state rates of creatine kinase and ATP synthetase in the rat brain. FEBS Lett 1982;140(2):289–292.
50. Du F, Zhu XH, Qiao H, Zhang X, Chen W. Efficient in vivo ^{31}P magnetization transfer approach for noninvasively determining multiple kinetic parameters and metabolic fluxes of ATP metabolism in the human brain. Magn Reson Med 2006;57(1):103–114.
51. Mateescu GD, Yvars GM, Dular T. Water, Ions and O-17 Magnetic Resonance Imaging. In: Lauger P, Packer L, Vasilescu V, editors. Water and Ions in Biological Systems. Basel-Boston-Berlin: Birkhauser Verlag; 1988. p 239–250.
52. Pekar J, Ligeti L, Ruttner Z, Lyon RC, Sinnwell TM, van Gelderen P, Fiat D, Moonen CT, McLaughlin AC. In vivo measurement of cerebral oxygen consumption and blood flow using ^{17}O magnetic resonance imaging. Magn Reson Med 1991;21(2): 313–319.
53. Fiat D, Ligeti L, Lyon RC, Ruttner Z, Pekar J, Moonen CT, McLaughlin AC. In vivo ^{17}O NMR study of rat brain during $^{17}O_2$ inhalation. Magn Reson Med 1992;24(2): 370–374.
54. Zhu XH, Zhang Y, Tian RX, Lei H, Zhang N, Zhang X, Merkle H, Ugurbil K, Chen W. Development of ^{17}O NMR approach for fast imaging of cerebral metabolic rate of oxygen in rat brain at high field. Proc Natl Acad Sci U S A 2002;99(20):13194–13199.
55. Zhu XH, Zhang N, Zhang Y, Zhang X, Ugurbil K, Chen W. In vivo ^{17}O NMR approaches for brain study at high field. NMR Biomed 2005;18(2):83–103.
56. Ernst RR, Bodenhausen G, Wokaun A. Principles of Nuclear Magnetic Resonance in One and Two Dimensions. New York: Oxford University Press; 1991.
57. Hoult DI, Richards RE. The signal-to-noise ratio of nuclear magnetic resonance experiment. J Magn Reson 1976;24:71–85.
58. Wen H, Chesnick AS, Balaban RS. The design and test of a new volume coil for high field imaging. Magn Reson Med 1994;32:492–498.
59. Wang Z, Wang DJ, Noyszewski EA, Bogdan AR, Haselgrove JC, Reddy R, Zimmerman RA, Leigh JS. Sensitivity of in vivo MRS of the N-δ proton in proximal histidine of deoxymyoglobin. Magn Reson Med 1992;27(2):362–367.
60. Qiao H, Zhang X, Zhu XH, Du F, Chen W. In vivo ^{31}P MRS of human brain at high/ultrahigh fields: A quantitative comparison of NMR detection sensitivity and spectral resolution between 4 T and 7 T. Magn Reson Imaging 2006;24(10):1281–1286.
61. Boska MD, Hubesch B, Meyerhoff DJ, Twieg DB, Karczmar GS, Matson GB, Weiner MW. Comparison of ^{31}P MRS and ^{1}H MRI at 1.5 and 2.0 T. Magn Reson Med 1990;13(2):228–238.
62. Hardy CJ, Bottomley PA, Roemer PB, Redington RW. Rapid ^{31}P spectroscopy on a 4-T whole-body system. Magn Reson Med 1988;8(1):104–109.
63. Hetherington HP, Spencer DD, Vaughan JT, Pan JW. Quantitative ^{31}P spectroscopic imaging of human brain at 4 Tesla: assessment of gray and white matter differences of phosphocreatine and ATP. Magn Reson Med 2001;45:46–52.
64. Evelhoch JL, Ewy CS, Siegfried BA, Ackerman JJ, Rice DW, Briggs RW. ^{31}P spin-lattice relaxation times and resonance linewidths of rat tissue in vivo: Dependence upon the static magnetic field strength. Magn Reson Med 1985;2(4):410–417.
65. Mathur-De Vre R, Maerschalk C, Delporte C. Spin-lattice relaxation times and nuclear Overhauser enhancement effect for ^{31}P metabolites in model solutions at two frequencies: Implications for in vivo spectroscopy. Magn Reson Imaging 1990; 8(6):691–698.
66. Abragam A. The Principles of Nuclear Magnetism. Mott NF, Bullard EC, Wilkinson DH, editors. London: Oxford University Press; 1961.
67. Zhu XH, Merkle H, Kwag JH, Ugurbil K, Chen W. ^{17}O relaxation time and NMR sensitivity of cerebral water and their field dependence. Magn Reson Med 2001;45(4):543–549.
68. Thelwall PE, Blackband SJ, Chen W. Field dependence of ^{17}O T_1, T_2 and SNR - in vitro and in vivo studies at 4.7, 11 and 17.6 Tesla. In: Proc Intl Soc Mag Reson Med (ISMRM) 2003; Toronto. p 504.
69. Mateescu GD, Yvars G, Pazara DI, Alldridge NA, LaManna JC, Lust DW, Mattingly

M, Kuhn W. ^{17}O-^{1}H magnetic resonance imaging in plants, animals, and materials. In: Baillie TA, Jones JR, editors. Synthesis and Application of Isotopically Labeled Compounds. Amsterdam: Elsevier; 1989. p 499–508.

70. Hopkins AL, Barr RG. Oxygen-17 compounds as potential NMR T_2 contrast agents: Enrichment effects of $H_2{}^{17}O$ on protein solutions and living tissues. Magn Reson Med 1987;4(4):399–403.
71. Kwong KK, Hopkins AL, Belliveau JW, Chesler DA, Porkka LM, McKinstry RC, Finelli DA, Hunter GJ, Moore JB, Barr RG, Rosen BR. Proton NMR imaging of cerebral blood flow using $H_2{}^{17}O$. Magn Reson Med 1991;22(1):154–158.
72. Arai T, Nakao S, Mori K, Ishimori K, Morishima I, Miyazawa T, Fritz-Zieroth B. Cerebral oxygen utilization analyzed by the use of oxygen-17 and its nuclear magnetic resonance. Biochem Biophys Res Commun 1990;169(1):153–158.
73. Pekar J, Sinnwell T, Ligeti L, Chesnick AS, Frank JA, McLaughlin AC. Simultaneous measurement of cerebral oxygen consumption and blood flow using ^{17}O and ^{19}F magnetic resonance imaging. J Cereb Blood Flow Metab 1995;15(2):312–320.
74. Ronen I, Navon G. A new method for proton detection of $H_2{}^{17}O$ with potential applications for functional MRI. Magn Reson Med 1994;32(6):789–793.
75. Ronen I, Merkle H, Ugurbil K, Navon G. Imaging of $H_2{}^{17}O$ distribution in the brain of a live rat by using proton-detected ^{17}O MRI. Proc Natl Acad Sci U S A 1998;95(22):12934–12939.
76. Reddy R, Stolpen AH, Leigh JS. Detection of ^{17}O by proton T_1 rho dispersion imaging. J Magn Reson B 1995;108(3):276–279.
77. Chen W, Zhu XH, Ugurbil K. Imaging cerebral metabolic rate of oxygen consumption ($CMRO_2$) using ^{17}O NMR approach at ultra-high field. In: Shulman RG, Rothman DL, editors. Brain Energetics and Neuronal Activity. New York: John Wiley & Sons Ltd; 2004. p 125–146.
78. Arai T, Mori K, Nakao S, Watanabe K, Kito K, Aoki M, Mori H, Morikawa S, Inubushi T. In vivo oxygen-17 nuclear magnetic resonance for the estimation of cerebral blood flow and oxygen consumption. Biochem Biophys Res Commun 1991;179(2):954–961.
79. Fiat D, Kang S. Determination of the rate of cerebral oxygen consumption and regional cerebral blood flow by non-invasive ^{17}O in vivo NMR spectroscopy and magnetic resonance imaging: Part 1. Theory and data analysis methods. Neurol Res 1992;14(4): 303–311.
80. Fiat D, Dolinsek J, Hankiewicz J, Dujovny M, Ausman J. Determination of regional cerebral oxygen consumption in the human: ^{17}O natural abundance cerebral magnetic resonance imaging and spectroscopy in a whole body system. Neurol Res 1993;15(4):237–248.
81. Fiat D, Kang S. Determination of the rate of cerebral oxygen consumption and regional cerebral blood flow by non-invasive ^{17}O in vivo NMR spectroscopy and magnetic resonance imaging. Part 2. Determination of $CMRO_2$ for the rat by ^{17}O NMR, and $CMRO_2$, rCBF and the partition coefficient for the cat by ^{17}O MRI. Neurol Res 1993; 15(1):7–22.
82. Mateescu GD. Functional oxygen-17 magnetic resonance imaging and localized spectroscopy. Adv Exp Med Biol 2003;510: 213–218.
83. Stolpen AH, Reddy R, Leigh JS. ^{17}O-decoupled proton MR spectroscopy and imaging in a tissue model. J Magn Reson 1997;125(1):1–7.
84. Reddy R, Stolpen AH, Charagundla SR, Insko EK, Leigh JS. ^{17}O-decoupled 1H detection using a double-tuned coil. Magn Reson Imaging 1996;14(9):1073–1078.
85. de Crespigny AJ, D'Arceuil HE, Engelhorn T, Moseley ME. MRI of focal cerebral ischemia using ^{17}O-labeled water. Magn Reson Med 2000;43(6):876–883.
86. Zhu XH, Zhang Y, Zhang N, Ugurbil K, Chen W. Noninvasive and three-dimensional imaging of $CMRO_2$ in rats at 9.4 T: Reproducibility test and normothermia/hypothermia comparison study. J Cereb Blood Flow Metab 2007;27(6):1225–1234.
87. Zhang N, Zhu XH, Lei H, Ugurbil K, Chen W. Simplified methods for calculating cerebral metabolic rate of oxygen based on ^{17}O magnetic resonance spectroscopic imaging measurement during a short $^{17}O_2$ inhalation. J Cereb Blood Flow Metab 2004;24(8):840–848.
88. Ter-Pogossian MM, Eichling JO, Davis DO, Welch MJ. The measure in vivo of regional cerebral oxygen utilization by means of oxyhemoglobin labeled with radioactive oxygen-15. J Clin Invest 1970;49:381–391.
89. Lenzi GL, Jones T, Frackowiak RS. Positron emission tomography: state of the art in neurology. Prog Nucl Med 1981;7:118–137.
90. Mintun MA, Raichle ME, Martin WR, Herscovitch P. Brain oxygen utilization

measured with O-15 radiotracers and positron emission tomography. J Nucl Med 1984;25(2): 177–187.

91. Fox PT, Raichle ME, Mintun MA, Dence C. Nonoxidative glucose consumption during focal physiologic neural activity. Science 1988;241:462–464.
92. Kety SS, Schmidt CF. The determination of cerebral blood flow in man by the use of nitrous oxide in low concentrations. Am J Physiol 1945;143:53–66.
93. Kety SS, Schmidt CF. Nitrous oxide method for the quantitative determination of cerebral blood flow in man: theory, procedure and normal values. J Clin Invest 1948;27: 476–483.
94. Kety SS, Schmidt CF. Effects of altered arterial tensions of carbon dioxide and oxygen on cerebral blood flow and cerebral oxygen consumption of normal young men. J Clin Invest 1948;27:484–492.
95. Herscovitch A, Raichle ME. What is the correct value for the brain-blood partition coefficient for water? J Cereb Blood Flow Metab 1985;5:65–69.
96. Herscovitch P, Raichle ME, Kilbourn MR, Welch MJ. Positron emission tomographic measurement of cerebral blood flow and permeability-surface area product of water using [15O]water and [11C]butanol. J Cereb Blood Flow Metab 1987;7(5): 527–542.
97. Zhang X, Zhu XH, Tian R, Zhang Y, Merkle H, Chen W. Measurement of arterial input function of ^{17}O water tracer in rat carotid artery by using a region-defined (REDE) implanted vascular RF coil. Magma 2003;16(2):77–85.
98. Nakao Y, Itoh Y, Kuang TY, Cook M, Jehle J, Sokoloff L. Effects of anesthesia on functional activation of cerebral blood flow and metabolism. Proc Natl Acad Sci U S A 2001;98(13):7593–7598.
99. Hyder F, Kennan RP, Kida I, Mason GF, Behar KL, Rothman D. Dependence of oxygen delivery on blood flow in rat brain: A 7 tesla nuclear magnetic resonance study. J Cereb Blood Flow Metab 2000;20(3): 485–498.
100. Yee SH, Lee K, Jerabek PA, Fox PT. Quantitative measurement of oxygen metabolic rate in the rat brain using microPET imaging of briefly inhaled ^{15}O-labelled oxygen gas. Nucl Med Commun 2006;27(7): 573–581.
101. Zhang NY, Zhu XH, Chen W. Evaluation of the simplified method for mapping cerebral metabolic rate of oxygen based on ^{17}O NMR approach. In: Proc Intl Soc Mag Reson Med (ISMRM), 2006; Seattle. p 952.
102. Erecinska M, Thoresen M, Silver IA. Effects of hypothermia on energy metabolism in Mammalian central nervous system. J Cereb Blood Flow Metab 2003;23(5):513–530.
103. Weiss RG, Gerstenblith G, Bottomley PA. ATP flux through creatine kinase in the normal, stressed, and failing human heart. Proc Natl Acad Sci U S A 2005;102(3):808–813.
104. Balaban RS, Kantor HL, Ferretti JA. In vivo flux between phosphocreatine and adenosine triphosphate determined by two-dimensional phosphorous NMR. J Biol Chem 1983;258(21):12787–12789.
105. Degani H, Alger JR, Shulman RG, Petroff OA, Prichard JW. ^{31}P magnetization transfer studies of creatine kinase kinetics in living rabbit brain. Magn Reson Med 1987;5(1): 1–12.
106. Bottomley PA, Hardy CJ. Mapping creatine kinase reaction rates in human brain and heart with 4 Tesla saturation transfer ^{31}P NMR. J Magn Reson 1992;99:443–448.
107. Sauter A, Rudin M. Determination of creatine kinase parameters in rat brain by NMR magnetization transfer. J Biological Chem 1993;268:13166–13171.
108. Chen W, Zhu X-H, Adriany G, Ugurbil K. Increase of creatine kinase activity in the visual cortex of human brain during visual stimulation: A ^{31}P NMR magnetization transfer study. Magn Reson Med 1997;38:551–557.
109. Braunova Z, Kasparova S, Mlynarik V, Mierisova S, Liptaj T, Tkac I, Gvozdjakova A. Metabolic changes in rat brain after prolonged ethanol consumption measured by 1H and 31P MRS experiments. Cell Mol Neurobiol 2000;20(6):703–715.
110. Mlynarik v ZS, Brehm A, Bischof M, Roden M. An optimized protocol for measuring rate constant of creatine kinase reaction in human brain by ^{31}P NMR saturation transfer. 13th Proc Intl Soc Mag Reson Med 2005:2767.
111. Joubert F, Mateo P, Gillet B, Beloeil JC, Mazet JL, Hoerter JA. CK flux or direct ATP transfer: Versatility of energy transfer pathways evidenced by NMR in the perfused heart. Mol Cell Biochem 2004;256–257(1–2):43–58.
112. Matthews PM, Bland JL, Gadian DG, Radda GK. A ^{31}P-NMR saturation transfer study of the regulation of creatine kinase in the rat heart. Biochim Biophys Acta 1982;721(3):312–320.
113. Mora BN, Narasimhan PT, Ross BD. ^{31}P magnetization transfer studies in the monkey brain. Magn Reson Med 1992;26(1): 100–115.

114. Bittl JA, DeLayre J, Ingwall JS. Rate equation for creatine kinase predicts the in vivo reaction velocity: ^{31}P NMR surface coil studies in brain, heart, and skeletal muscle of the living rat. Biochemistry 1987;26: 6083–6090.
115. Ugurbil K, Petein M, Maiden R, Michurski S, From A. Measurement of an individual rate constant in the presence of multiple exchanges: Application to myocardial creatine kinase rates. Biochemistry 1986;25: 100–108.
116. Spencer RG, Balschi JA, Leigh JS, Jr., Ingwall JS. ATP synthesis and degradation rates in the perfused rat heart. 31P-nuclear magnetic resonance double saturation transfer measurements. Biophys J 1988;54(5): 921–929.
117. Joubert F, Mazet JL, Mateo P, Hoerter JA. ^{31}P NMR detection of subcellular creatine kinase fluxes in the perfused rat heart: Contractility modifies energy transfer pathways. J Biol Chem 2002;277(21):18469–18476.
118. Mitsumori F, Rees D, Brindle KM, Radda GK, Campbell ID. ^{31}P-NMR saturation transfer studies of aerobic Escherichia coli cells. Biochimica et Biophysica Acta 1988;969(2):185–193.
119. Campbell SL, Jones KA, Shulman RG. ^{31}P NMR saturation-transfer measurements in Saccharomyces cerevisiae: Characterization of phosphate exchange reactions by iodoacetate and antimycin A inhibition. Biochemistry 1987;26(23):7483–7492.
120. Thoma WJ, Ugurbil K. Saturation-transfer studies of ATP-Pi exchange in isolated perfused rat liver. Biochimica et Biophysica Acta 1987;893(2):225–231.
121. Kingsley-Hickman PB, Sako EY, Mohanakrishnan P, Robitaille PM, From AH, Foker JE, Ugurbil K. ^{31}P NMR studies of ATP synthesis and hydrolysis kinetics in the intact myocardium. Biochemistry 1987;26(23): 7501–7510.
122. Hinkle PC. P/O ratios of mitochondrial oxidative phosphorylation. Biochim Biophys Acta 2005;1706(1–2):1–11.
123. Du F, Zhang Y, Friedman M, Zhu XH, Ugurbil K, Chen W. Study of correlation between brain activity and ATP metabolic rates by means of ^{31}P magnetization transfer and electroencephalograph measurement. In: Proc Intl Soc Mag Reson Med (ISMRM), 2006; Seattle. p 2125.
124. Fox PT, Raichle ME. Focal physiological uncoupling of cerebral blood flow and oxidative metabolism during somatosensory stimulation in human subjects. Proc Natl Acad Sci USA 1986; 83:1140–1144.
125. Ribeiro L, Kuwabara H, Meyer E, Fujita H, Marrett S, Evans A, Gjedde A. Cerebral blood flow and metabolism during nonspecific bilateral visual stimulation in normal subjects. In: Uemura K, editor. Quantification of Brain Function in Tracer Kinetics and Image Analysis in Brain PET. New York: Elsevier Science; 1993. p 229–236.
126. Roland PE, Ericksson L, Stone-Elander S, Widen L. Does mental activity change the oxidative metabolism of the brain? J Neurosci 1987;7:2373–2389.
127. Marrett S, Fujita H, Meyer E, Ribeiro L, Evans A, Kuwabara H, Gjedde A. Stimulus specific increase of oxidative metabolism in human visual cortex. In: Uemura K, editor. Quantification of Brain Function in Tracer Kinetics and Image Analysis in Brain PET. New York: Elsevier Science; 1993. p 217–228.
128. Prichard J, Rothman D, Novotny E, Petroff O, Kuwabara T, Avison M, Howseman A, Hanstock C, Shulman RG. Lactate rise detected by ^{1}H NMR in human visual cortex during physiologic stimulation. Proc Natl Acad Sci (USA) 1992;88:5829–5831.
129. Shulman RG, Hyder F, Rothman DL. Cerebral energetics and the glycogen shunt: neurochemical basis of functional imaging. Proc Natl Acad Sci U S A 2001; 98(11):6417–6422.
130. Vafaee MS, Meyer E, Marrett S, Paus T, Evans AC, Gjedde A. Frequency-dependent changes in cerebral metabolic rate of oxygen during activation of human visual cortex. J Cereb Blood Flow Metab 1999;19(3):272–277.
131. Davis TL, Kwong KK, Weisskoff RM, Rosen BR. Calibrated functional MRI: Mapping the dynamic of oxidative metabolism. Proc Natl Acad Sci USA 1998;95:1834–1839.
132. Hoge RD, Atkinson J, Gill B, Crelier GR, Marrett S, Pike GB. Investigation of BOLD signal dependence on cerebral blood flow and oxygen consumption: The deoxyhemoglobin dilution model [In Process Citation]. Magn Reson Med 1999;42(5): 849–863.
133. Hyder F, Chase JR, Behar KL, Mason GF, Siddeek M, Rothman DL, Shulman RG. Increase tricarboxylic acid cycle flux in rat brain during forepaw stimulation detected with ^{1}H-{^{13}C} NMR. Proc Natl Acad Sci USA 1996;93:7612–7617.
134. Kim SG, Rostrup E, Larsson HB, Ogawa S, Paulson OB. Determination of relative CMRO2 from CBF and BOLD changes: Significant increase of oxygen consumption rate

during visual stimulation. Magn Reson Med 1999;41(6):1152–1161.
135. Chen W, Zhu XH, Gruetter R, Seaquist ER, Ugurbil K. Study of oxygen utilization changes of human visual cortex during hemifield stimulation using ^{1}H-{^{13}C} MRS and fMRI. Magn Reson Med 2001;45: 349–355.
136. Kasischke KA, Vishwasrao HD, Fisher PJ, Zipfel WR, Webb WW. Neural activity triggers neuronal oxidative metabolism followed by astrocytic glycolysis. Science 2004;305(5680):99–103.
137. Ogawa S, Lee T-M, Kay AR, Tank DW. Brain magnetic resonance imaging with contrast dependent on blood oxygenation. Proc Natl Acad Sci USA 1990;87:9868–9872.
138. Ogawa S, Tank DW, Menon R, Ellermann JM, Kim S-G, Merkle H, Ugurbil K. Intrinsic signal changes accompanying sensory stimulation: Functional brain mapping with magnetic resonance imaging. Proc Natl Acad Sci USA 1992; 89:5951–5955.
139. Kwong KK, Belliveau JW, Chesler DA, Goldberg IE, Weisskoff RM, Poncelet BP, Kennedy DN, Hoppel BE, Cohen MS, Turner R, Cheng HM, Brady TJ, Rosen BR. Dynamic magnetic resonance imaging of human brain activity during primary sensory stimulation. Proc Natl Acad Sci USA 1992;89:5675–5679.
140. Bandettini PA, Wong EC, Hinks RS, Tikofsky RS, Hyde JS. Time course EPI of human brain function during task activation. Magn Reson Med 1992;25:390–397.
141. Frahm J, Bruhn H, Merboldt KD, Hanicke W. Dynamic MR imaging of human brain oxygenation during rest and photic stimulation. J Magn Reson Imag 1992;2(5): 501–505.
142. Blamire AM, Ogawa S, Ugurbil K, Rothman D, McCarthy G, Ellermann JM, Hyder F, Rattner Z, Shulman RG. Dynamic mapping of the human visual cortex by high-speed magnetic resonance imaging. Proc Natl Acad Sci USA 1992;89:11069–11073.
143. Barinaga M. What makes brain neurons run. Science 1997;276:196–198.
144. Zhu XH, Zhang Y, Zhang NY, Zhang XL, Ugurbil K, Chen W. Direct imaging of $CMRO_2$ in cat visual cortex at rest and visual stimulation. In: Proc Intl Soc Mag Reson Med (ISMRM), 2006; Seattle. p 457.
145. Lei H, Zhu XH, Li YX, Ugurbil K, Chen W. Changes of cerebral metabolism in human primary visual cortex during functional activation observed by in vivo ^{31}P MRS at 7T. In: Proc Intl Soc Mag Reson Med (ISMRM), 2004; Kyoto, Japan. p 2361.
146. Maurer I, Zierz S, Moller H. Evidence for a mitochondrial oxidative phosphorylation defect in brains from patients with schizophrenia. Schizophr Res 2001; 48(1):125–136.
147. Wong-Riley M, Antuono P, Ho KC, Egan R, Hevner R, Liebl W, Huang Z, Rachel R, Jones J. Cytochrome oxidase in Alzheimer's disease: Biochemical, histochemical, and immunohistochemical analyses of the visual and other systems. Vision Res 1997;37(24):3593–3608.
148. Maurer I, Zierz S, Moller HJ. A selective defect of cytochrome c oxidase is present in brain of Alzheimer disease patients. Neurobiol Aging 2000;21(3):455–462.
149. Frackowiak RS, Herold S, Petty RK, Morgan-Hughes JA. The cerebral metabolism of glucose and oxygen measured with positron tomography in patients with mitochondrial diseases. Brain 1988;111 (Pt 5):1009–1024.
150. Beal MF. Does impairment of energy metabolism result in excitotoxic neuronal death in neurodegenerative illnesses? Ann Neurol 1992;31(2):119–130.
151. Wallace DC. Mitochondrial genetics: A paradigm for aging and degenerative diseases? Science 1992;256(5057):628–632.
152. Wallace DC. Mitochondrial diseases in man and mouse. Science 1999;283(5407): 1482–1488.
153. Ugurbil K, Adriany G, Akgün C, Andersen P, Chen W, Garwood M, Gruetter R, Henry P-G, Marjanska M, Moeller S, Van de Moortele P-F, Prüssmann K, Tkac I, Vaughan JT, Wiesinger F, Yacoub E, Zhu X-H. High Magnetic Fields for Imaging Cerebral Morphology, Function and Biochemistry. In: Robitaille PMLaB, L.J., editor. Biological Magnetic Resonance: Ultra High Field Magnetic Resonance Imaging. Volume 26. New York: Springer; 2006. p 285–342.

INDEX

D

E

F

M

N

O

P

R

S

V

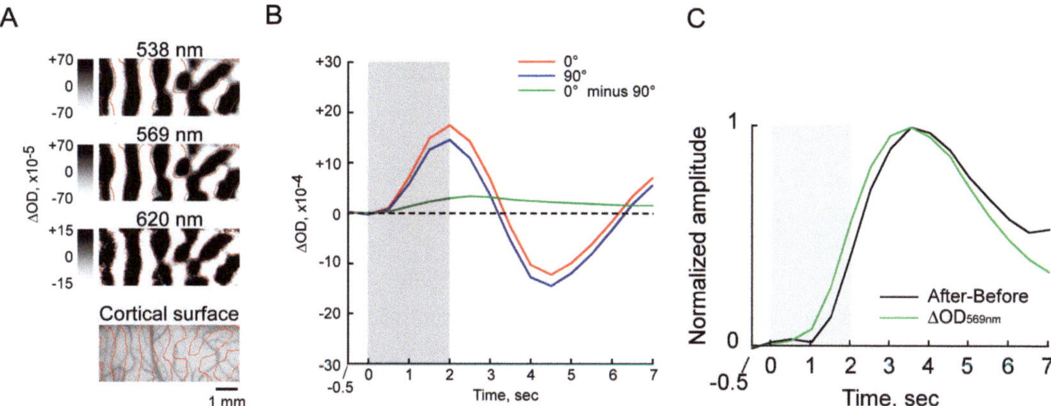

Fig. 6.1. Orientation columns visualized by OISI at different wavelengths. Typically in visual cortex, intrinsic signals consist of stimulus-specific and -nonspecific components. Orientation columns are reflected in the stimulus-specific component that is a local modulation of a stimulus-nonspecific component. In (A), we subtracted activation patterns obtained by one stimulus (90 deg. orientation) from the other (0 deg. orientation) to remove stimulus-nonspecific components. Please note that, depending on the size of the functional structures of interest, stimulus specificity of synaptic inputs, and spread of intrinsic signals, distinction between stimulus-specific and nonspecific components may not be required (for example, see **Figs. 6.3 and 6.4**). Columnar patterns obtained at different wavelengths are nearly the same as indicated by contours in red drawn for columnar patterns obtained at 620 nm. Although it is not shown here, similar patterns of functional structures were also observed using infrared light. Bottom image shows vessel patterns of the exposed cortical surface where the above recordings were made. (B) Time courses of stimulus-nonspecific (red and blue lines) and stimulus-specific (green line) components obtained at 620 nm. The visual stimulus was given from 0 to 2 s. (C) Time course of stimulus-nonspecific components obtained at 570 nm (Green line). The black line indicates time course estimated by subtraction between time courses obtained before and after the extrinsic absorption dye infusion (see Fukuda et al., 2005 for details). The visual stimulus was given from 0 to 2 s.

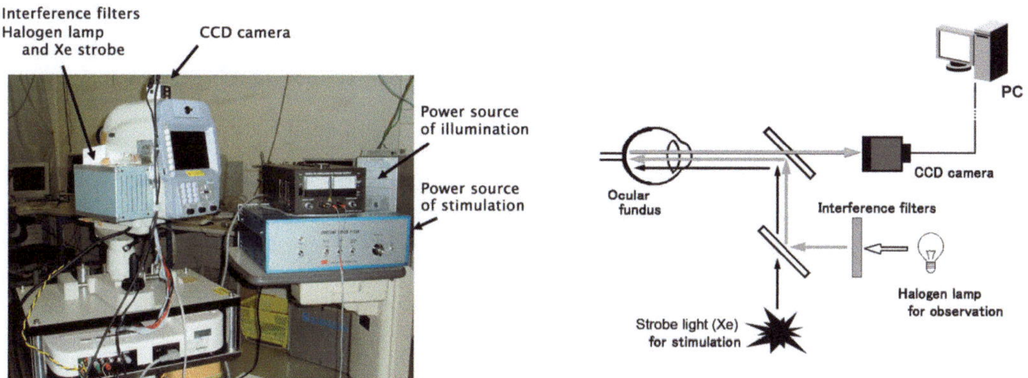

Fig. 6.2. Overall view of the intrinsic signals imaging system for retina (left) and schematic drawing of the experimental setup (right). Throughout the recording trial, the fundus was continuously illuminated with observation light through one of the bandpass filters.

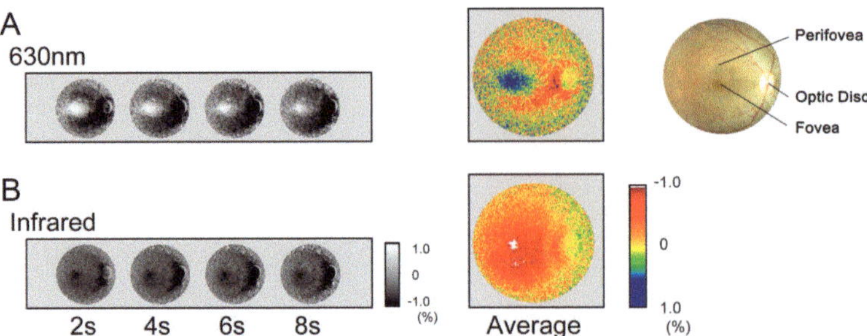

Fig. 6.3. Time courses of two-dimensional images of retina showing light reflectance changes following a flash stimulus observed with 630 nm (**A**) and infrared (**B**) light, measured in the posterior pole region of normal retina. The averaged reflectance changes during the first second after the flash are shown in pseudocolor maps on the right. Color indicates relative light reflectance changes from pre-stimulus level. Fundus photograph of a normal macaque retina is shown in the right.

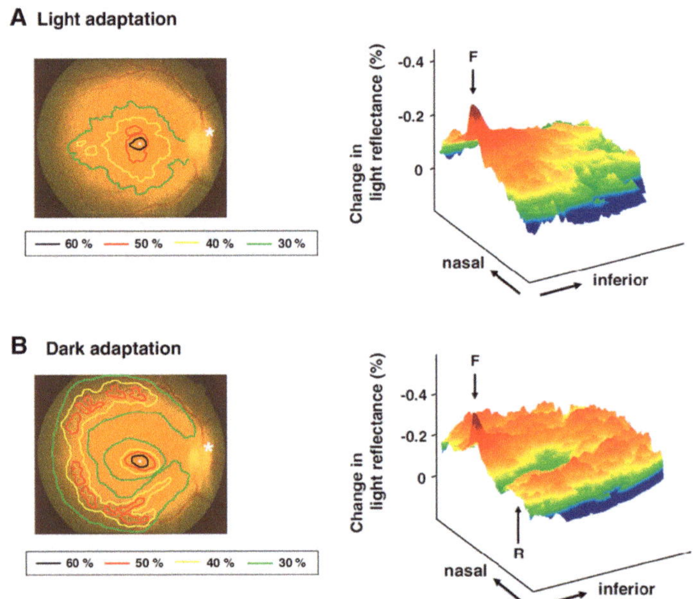

Fig. 6.4. Response topography of normal retina under light-adapted (**A**) and dark-adapted (**B**) conditions. The foveal center is indicated by a white dot and the optic disk is indicated by an asterisk. Regions with 60%, 50%, 40% and 30% of the peak signal intensity value at the fovea were outlined by different colors. Pseudocolor topographic maps of light reflectance changes in the inferior retina, profiled along the horizontal meridian, are shown on the right. The location of the fovea is indicated by F, and the crest of 'rod ring' is indicated by R.

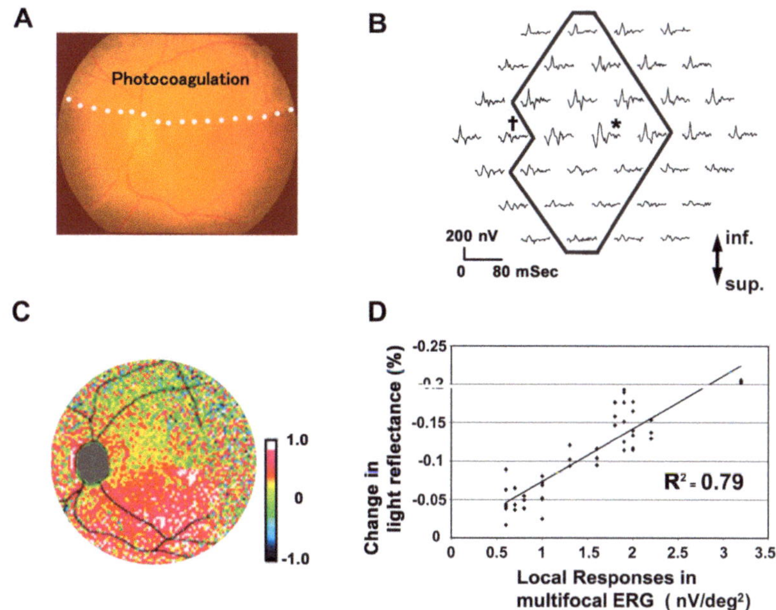

Fig. 6.5. (**A**) Photograph of the monkey's retina artificially damaged by Argon Laser Photocoagulation. The upper half of posterior retina was densely coagulated, sparing the macular area. (**B**) Array of 37 local responses of multifocal ERG, taken from the fundus in (**A**). ERGs with $*$ and † indicate the location of macula and optic disk, respectively. Note that the responses of the intact retina in the lower half region are inversely shown in the upper half of the arrays. (**C**) pseudocolor map of flash-evoked intrinsic signals measured with infrared light. Red color indicates light reflectance decrease (*darkening*) from pre-stimulus level. Flash-evoked darkening could not be observed in the damaged region (*upper half*). (**D**) Correlation between the change in light reflectance decrease and the focal responses in ERG at the corresponding retinal location under infrared light. Correlation coefficient: $r^2 = 0.79$ ($p < 0.001$, n=45).

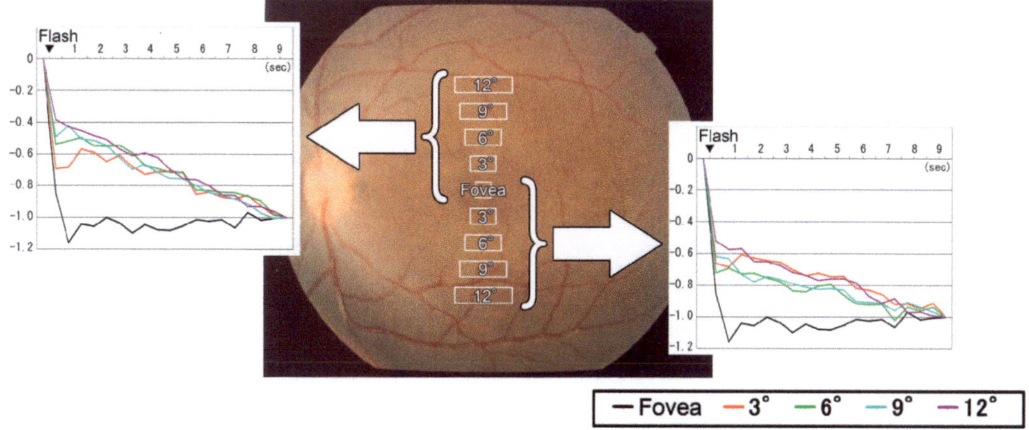

Fig. 6.6. Time courses of light reflectance changes in a single trial following a diffuse flash, measured at the fovea and different regions within twelve degrees superior or inferior to the fovea. Amplitudes are indicated as values relative to the light reflectance changes at the end of each trial (1.0). The four regions tested in each quadrant are indicated as distances from the fovea (3,6,9 and 12°).

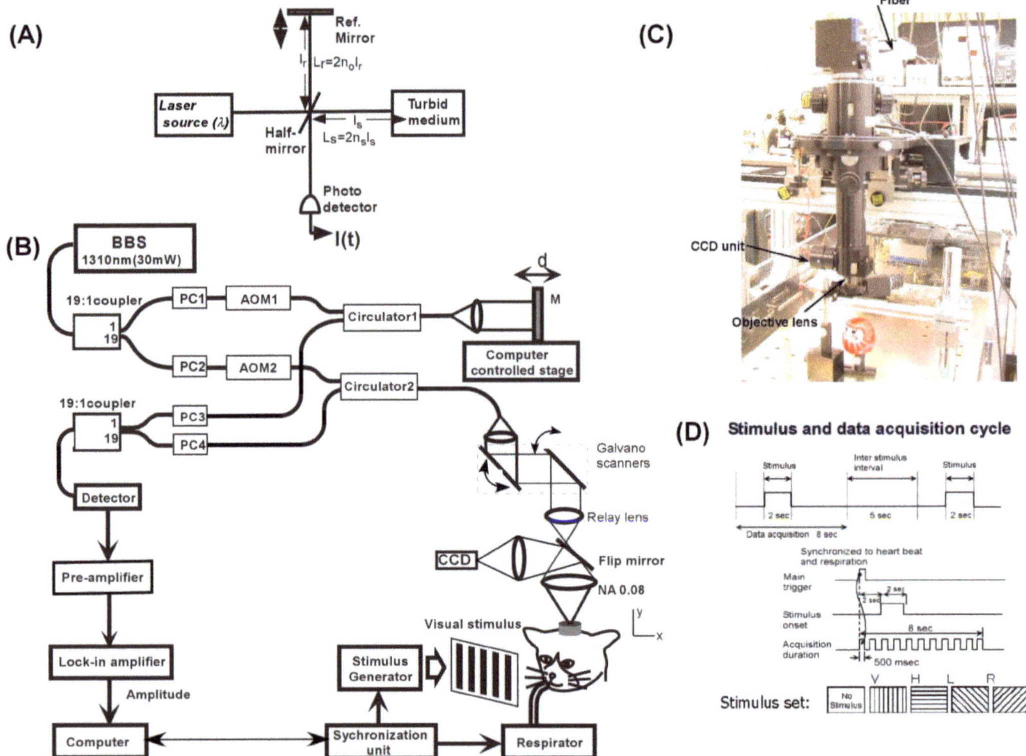

Fig. 6.7. A schematic of the basic principle of OCT (**A**) and the experimental system used (**B**) along with a picture of the probe unit (**C**) and a schematic of the scanning paradigm (**D**). In the figure (**B**), the abbreviations denote: BBS - Broad Band Source, AOM- Acousto-Optic Modulator, PC - Polarization Controller, M – Mirror and O – Objective lens.

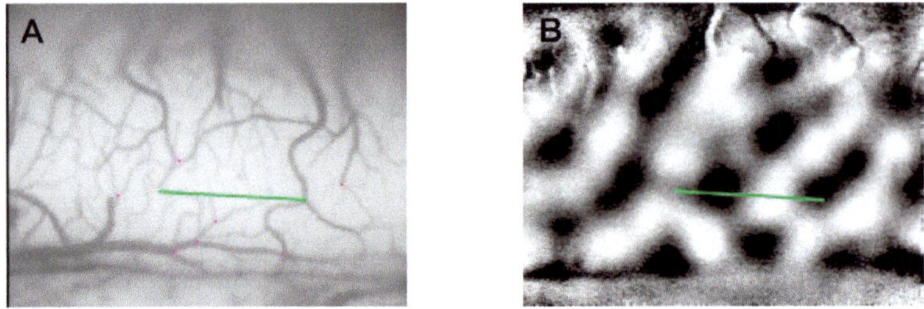

Fig. 6.8. Exposed cortical surface of cat visual cortex with (**B**) a thresholded activation map overlay . *Dark* and *light* patches represent the activated regions for horizontal and vertical gratings, respectively, and *green lines* indicate the region of OCT scans.

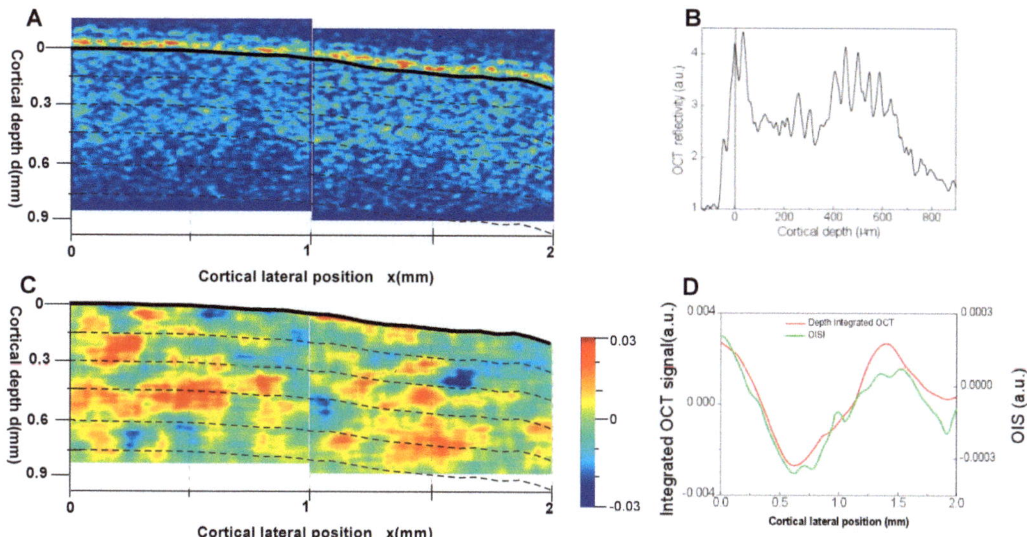

Fig. 6.9. (**A**) OCT scan, (**B**) Functional OCT map and (**C**) Consistency of OISI result with the integrated result of fOCT obtained across the line indicated in **Fig. 6.2**. In (**B**), *red* and *blue* patches represent the activated regions for horizontal and vertical gratings, respectively. Green line indicates the variation of OISI across the line indicated in **Fig. 6.2B**. In the graph (**C**), *red line* obtained by calculating the functional signal from integrating the OCT scans across the full scanned depth range of **Fig. 6.3B** while green line indicates the intensity variation across the green line of **Fig. 6.2B**.

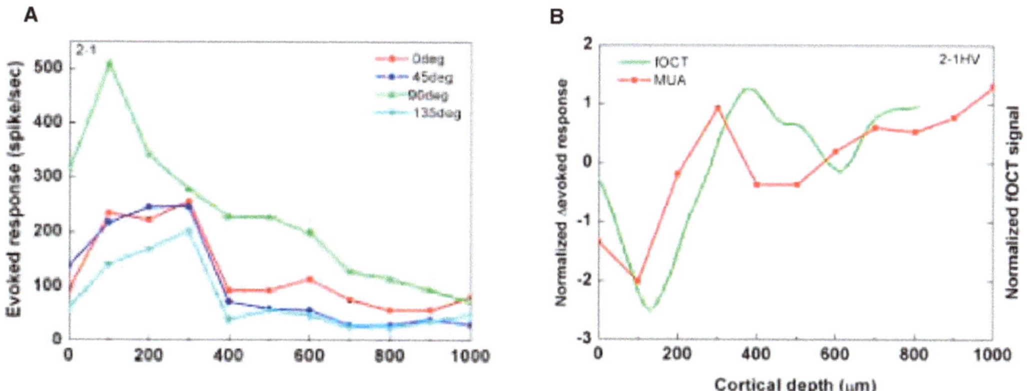

Fig. 6.10. An example of (**A**) MUA-evoked responses as a function of depth obtained from a single track for different orientation stimuli and (**B**) a comparison of difference of the evoked MUA response (shown in red) with the fOCT profile (shown in green) as a function of depth for the difference of 0° minus 90° orientation stimulus.

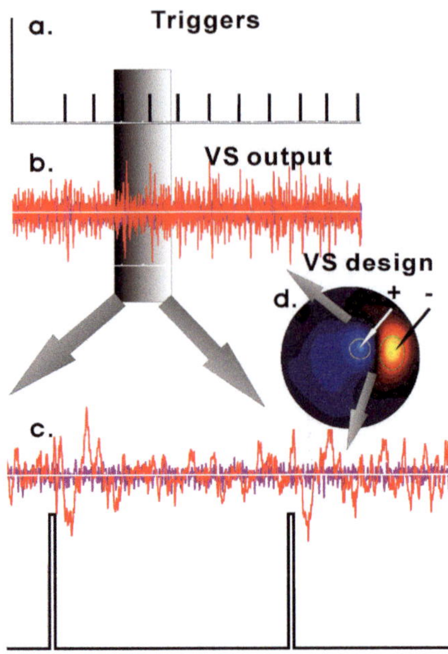

Fig. 8.1. Virtual sensor output for a strong signal. (**A**) The delivery of a median nerve stimulation is marked by a trigger and collected together with the MEG signal. (**B**) The MEG signal for the real measurement as marked by the triggers in (**A**) and for a recording of the noise level of the system (no subject in the MEG room). (**C**) A zoomed version of the VS output (*top*) and corresponding third and fourth triggers. (**D**) The VS is computed from the difference of means of the 7 most positive and 7 most negative MEG sensors.

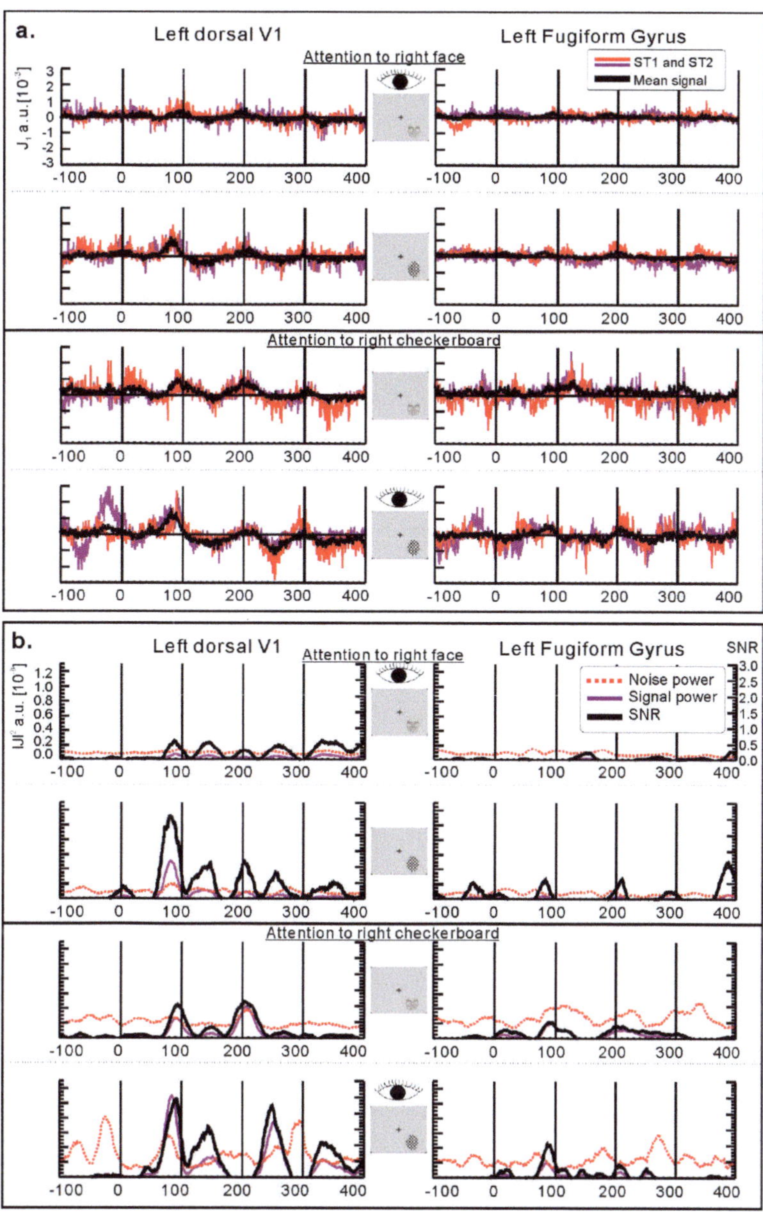

Fig. 8.2. (A) Two single trial activation curves extracted from tomographic MFT solutions with region of interest in the left dorsal V1 and the Left FG, for stimuli presented on the contralateral (*lower right*) part of the visual field. The stimuli in rows 1 and 3 were faces and in rows 2 and 4 checkerboard oval pattern. During a run, stimulus type and location varied randomly and the subject attended either to the faces on the right (*rows 1 and 2*) or to the checkerboards on the right (*rows 3 and 4*). For each case, two single trials and the average of the six trials used in the run are displayed. (**B**). The signal and noise power and their ratio (signal to noise ratio) are computed for the six trials and are displayed for each of the cases in (**A**).

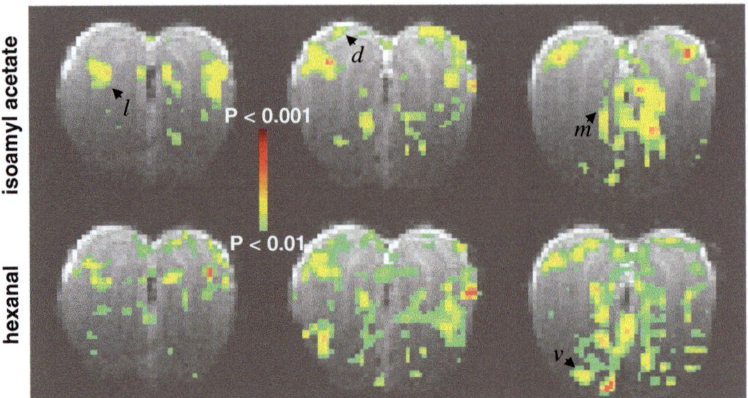

Fig. 10.2. BOLD responses from the olfactory bulb in Sprague-Dawley rats. The rat was exposed to isoamyl acetate (100%) and hexanal (100%) for a duration of 60 s each. The *t* maps were generated by comparing the mean signals in a 60-s baseline period before the odorant exposure. Odor-elicited activation patterns were imaged in 10 coronal slices with 500 μm thickness, of which three slices (4–6) are shown anterior-to-posterior (left to right). Both odors elicited distributed, yet non-identical responses throughout the bulb. Dorsal (*d*), lateral (*l*), and medial (*m*), and vental (*v*) foci are identified by arrows. All data shown are from single trial runs.

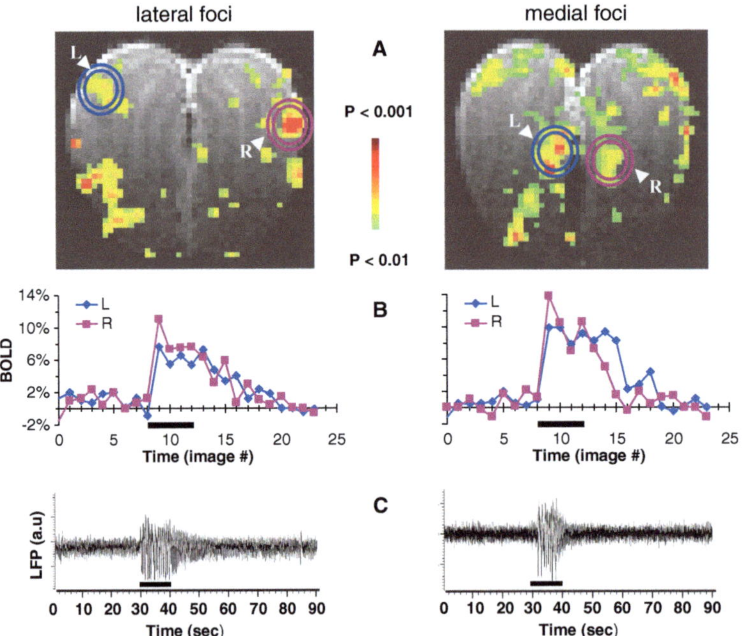

Fig. 10.3. Multi-modal responses from the olfactory bulb in Sprague-Dawley rats. Lateral (*left*) and medial (*right*) responses as measured by BOLD and LFP during hexanal (100%) exposure. (**A**) The BOLD activation maps from slices 8 (**A**) and 13 (**B**) situated from anterior-to-posterior in one subject where the entire olfactory bulb was imaged using 20 coronal slices with 250 μm thickness. The *t* maps were generated by comparing the mean signals in a 60-s baseline epoch before the odorant exposure. The circles shown in the BOLD maps identify the lateral (*left*) and medial (*right*) foci that were probed by electrophysiology (see below). (**B**) The BOLD responses from the circled (L,R) regions in each slice. Comparison of the two BOLD traces in the lateral (left) and medial (right) regions depicts the inter-bulbar differences (L,R). (**C**) The LFP responses measured from another animal but for the same odorant and the lateral/medial areas shown above. The lateral (*left*) and medial (*right*) LFP signals were made from the left and right olfactory bulbs, respectively. The exposure periods are indicated by the black bars in the middle and bottom rows. All data shown are from single trial runs.

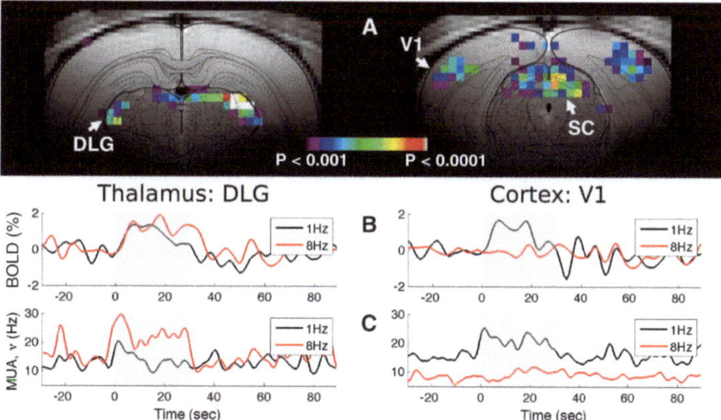

Fig. 10.4. Multi-modal responses from the visual cortex during bilateral light stimulation (white; 50 ms pulses; 30 lux) in Long Evans rats. (**A**) Cortical and subcortical BOLD responses with 1 Hz stimulation applied for 30-s duration. Data from five trials in one animal were averaged and the *t* maps were generated by comparing the mean signals in a 30-s baseline period before the 30-s stimulation block. The MRI data were overlaid on the structural outlines from the atlas of Paxinos and Watson *(53)*. Bilateral responses were observed in the dorsal lateral geniculate nuclei of the thalamus (DLG; left), the primary visual cortex (V1; *right*), and the dorsal layers of the superior colliculus (SC; *right*). At higher frequencies, there were weaker BOLD responses in the cortex. Results with green light were qualitatively similar to these results with white light. The time courses of (**B**) BOLD and (**C**) MUA responses to 1 Hz (*black line*) and 8 Hz (*red line*) stimuli in DLG (*left*) and V1 (*right*) from single trial runs. (**B**) The BOLD responses were calculated by averaging several activated voxels within a region-of-interest (in a single trial) from the maps shown above. (**C**) The MUA responses were obtained from another animal but under similar experimental conditions (*white*; 50 ms pulses; 23 lux). The coordinates of DLG and V1 were guided by prior fMRI results: DLG: [–4.8 mm posterior, –3.8 mm lateral to bregma, –4.0 mm ventral from dura]; V1: [–5.6 posterior, –3.9 lateral to bregma, –0.9 mm ventral from dura at an angle of 10° from the vertical axis]. The 30-s stimulation period is indicated by the shaded region.

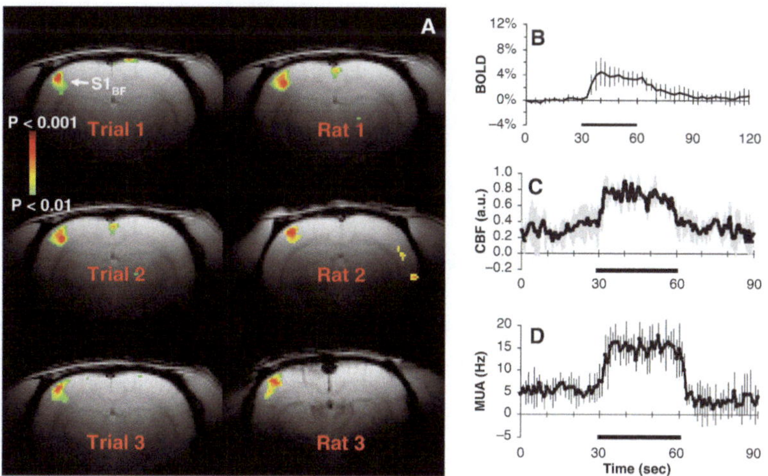

Fig. 10.5. Multi-modal responses from the contralateral whisker barrel field ($S1_{BF}$) during 8 Hz whisker stimulation in Sprague-Dawley rats. (**A**) Reproducibility of $S1_{BF}$ BOLD activation maps in the same subject (see Trial column; *left*) as well as other subjects (see Rat column; *right*). The *t* maps were generated by comparing the mean signals in a 30-s baseline epoch before the stimulation. All data shown are from single trial runs. Averaged time courses from the $S1_{BF}$ are shown for (**B**) BOLD ($n = 6$), (**C**) CBF ($n = 5$), and (**D**) MUA ($n = 5$). The 30-s stimulation period is indicated by the black bar. Vertical bars or gray shading represent standard deviations from the mean.

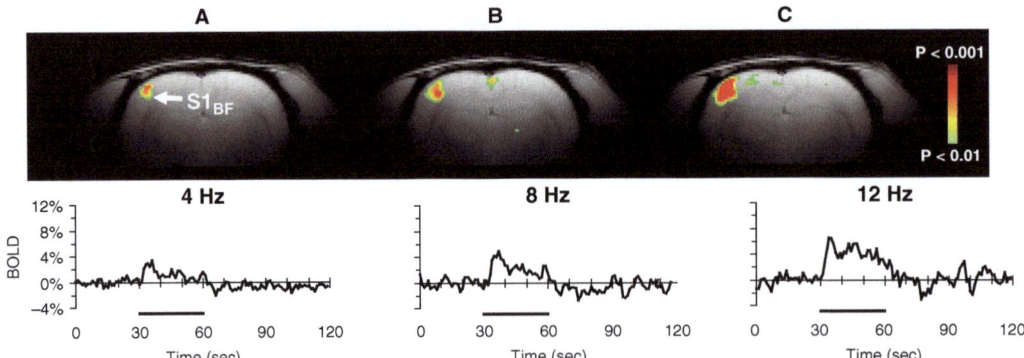

Fig. 10.6. Frequency-dependent BOLD responses from the contralateral whisker barrel field ($S1_{BF}$) in Sprague-Dawley rats. The *t* maps were generated by comparing the mean signals from 30-s epochs of baseline and stimulation. In the same subject, $S1_{BF}$ BOLD responses are shown for (**A**) 4 Hz, (**B**) 8 Hz, and (**C**) 12 Hz whisker stimulations. The top and bottom panels show the activation maps and time courses, respectively. All data shown are from single trial runs. The 30-s stimulation period is indicated by the black bar.

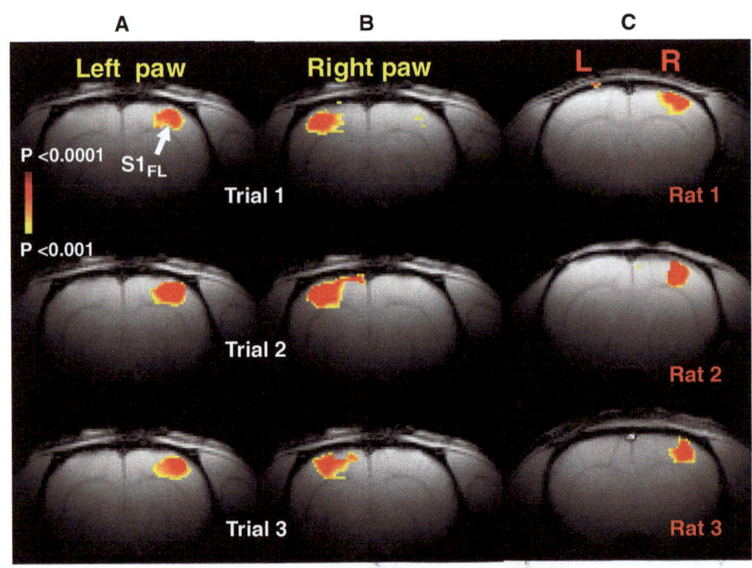

Fig. 10.7. BOLD responses from the contralateral forelimb area ($S1_{FL}$) during 3 Hz forepaw stimulation (0.3 ms pulses; 2 mA) in Sprague-Dawley rats. (**A**) Reproducibility of $S1_{FL}$ BOLD activation maps in the same subject during left paw and right paw stimulations (see Trial column; left and middle) as well as other subjects (see Rat column; *right*). The *t* maps were generated by comparison of the mean signals from 30-s baseline and stimulation periods. All data shown are from single trial runs.

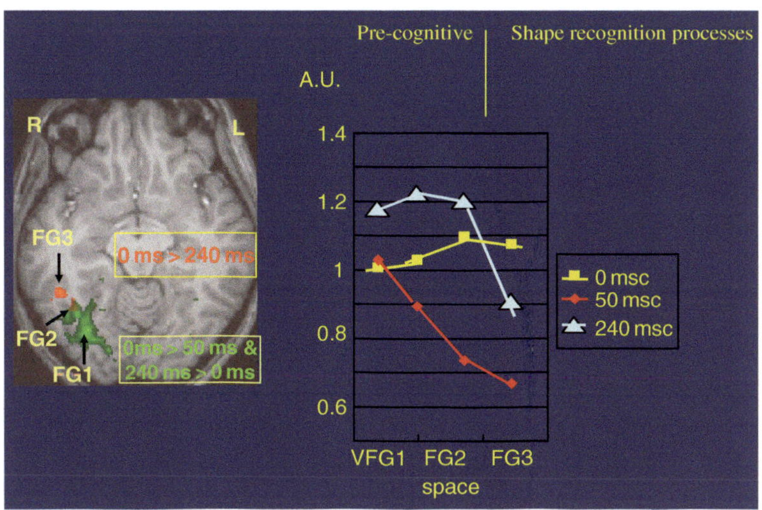

Fig. 11.8. Activation maps and trends of BOLD responses. Areas with positive responses for the difference of 0 and 50-ms (*orange*). Areas with positive response for the difference of 240 and 50-ms (*green*). FG1, FG2, and FG3 are areas in occipitotemporal region located at (30,-68,-9), (35,-57,-9), and (45,-46,-9). The responses were normalized by the response to the 0-ms ISI at V1.

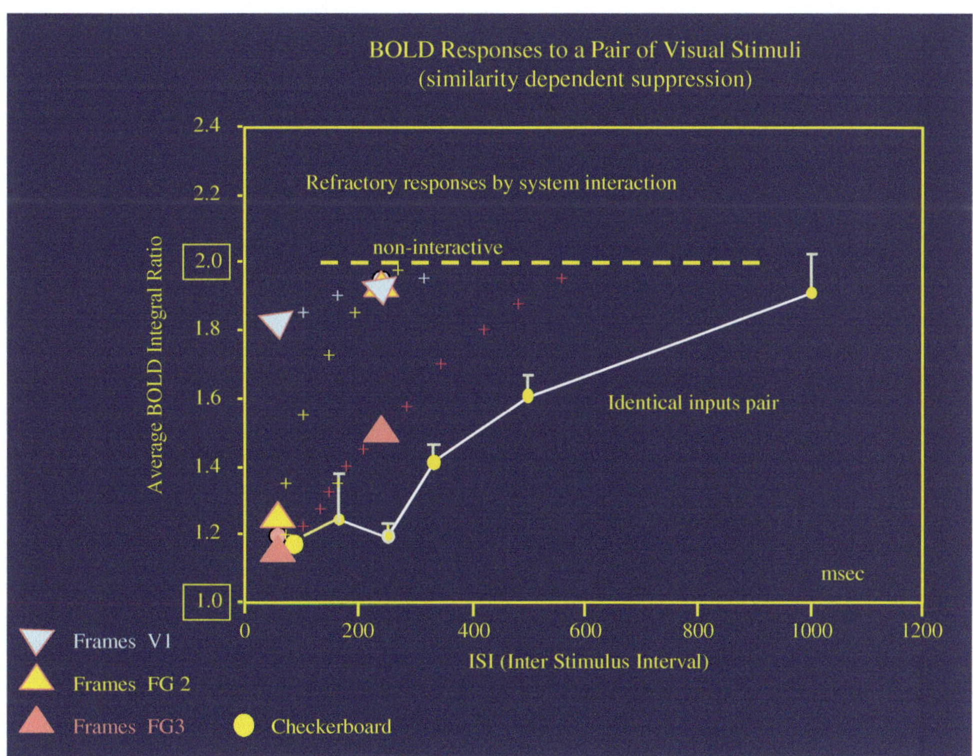

Fig. 11.9. Similarity dependent suppression of BOLD response.

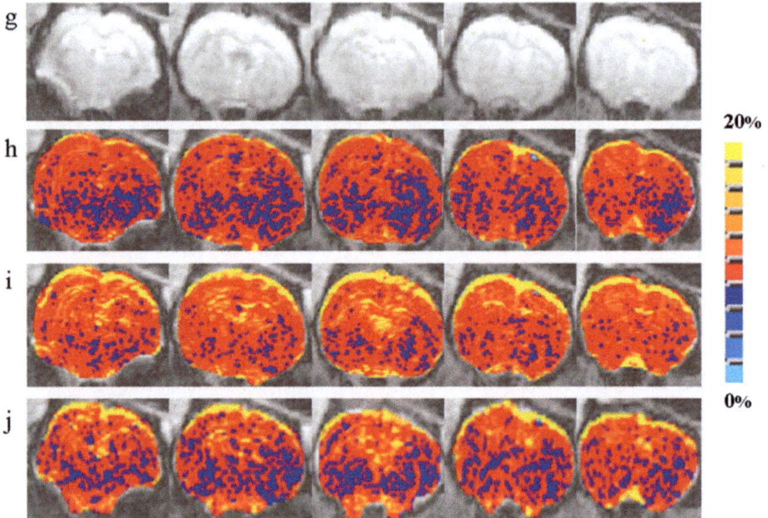

Fig. 12.1. Average power spectra of the BOLD signal, average filtered BOLD signal time series and standard deviation of the BOLD signal from the whole brain in a typical rat during normal, exsanguinated and blood replaced conditions. (**g**) anatomy (**h–j**) BOLD signal standard deviation maps during normal, exsanguinated and blood replaced conditions respectively.

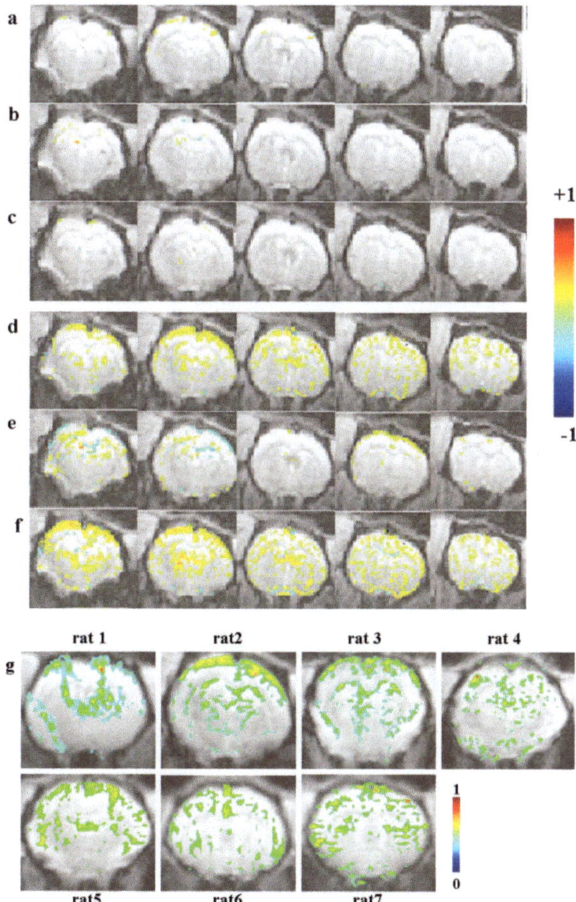

Fig. 12.3. Spatial correlation of low frequency BOLD signals from five contiguous slices from a typical rat. The anatomical underlay consists of a single EPI image of the brain and the functional overlay is the correlation coefficient in the absence of any stimulus during (**a–c**) normal and (**d–f**) exsanguinated conditions. Voxels from the whole brain were low-pass filtered (cut off 0.1 Hz) on a voxel wise basis and subsequently cross-correlated with the time course of a seed voxel obtained from (**a,d**) sensorimotor cortex (**b,e**) hippocampus and (**c,f**) thalamus. (**g**) Typical cross-correlation maps from a single coronal slice during exsanguinated conditions over all rats after cross-correlation with the time course of a seed voxel obtained from the sensorimotor cortex. A threshold ($\geq$0.3 for the correlation coefficient ($P < 10^{-6}$) was used to generate all correlation maps. The seed voxel location shows a high correlation coefficient value in the images.

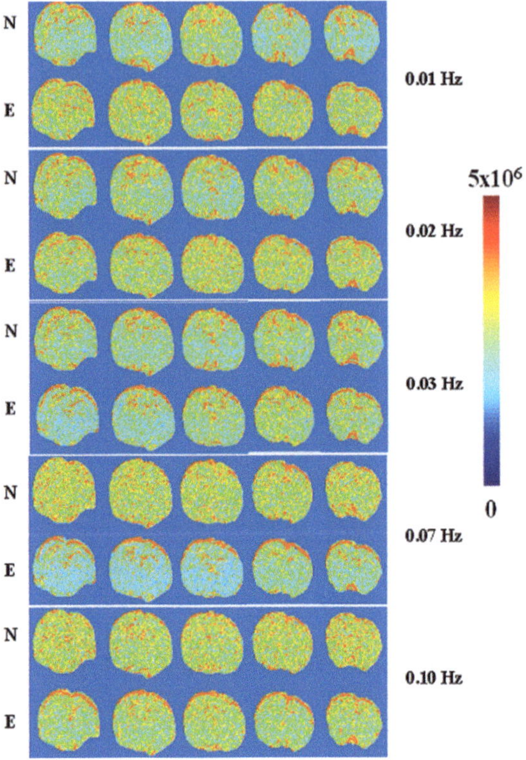

Fig. 12.4. Spatial maps of low frequency fluctuations in the BOLD signal at distinct frequencies in a typical rat during normal (N) and exsanguinated (E) conditions. Images were derived from specific frequencies namely 0.01, 0.02, 0.03, 0.07 and 0.10 Hz respectively from the power spectra of the BOLD signal time series on a voxel wise basis. Fluctuations in most of the low frequency bands increase spatially in the cerebral cortex during exsanguination.

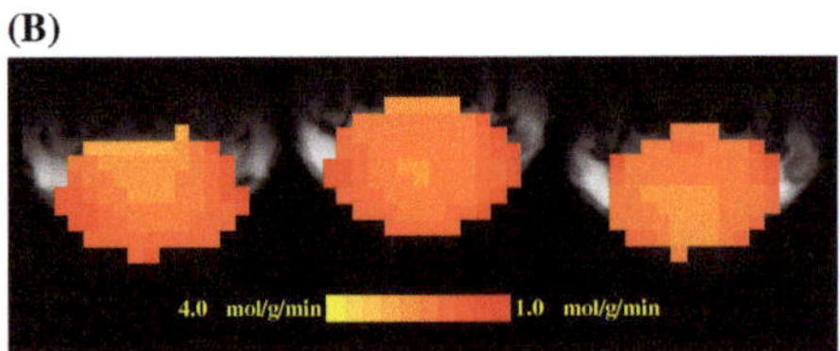

Fig. 15.5. (**B**) Three-dimensional coronal $CMRO_2$ images of rat brain measured by in vivo ^{17}O MRS approach during a two-minute $^{17}O_2$ inhalation. Adapted from Zhu et al of Ref. *(54)*.

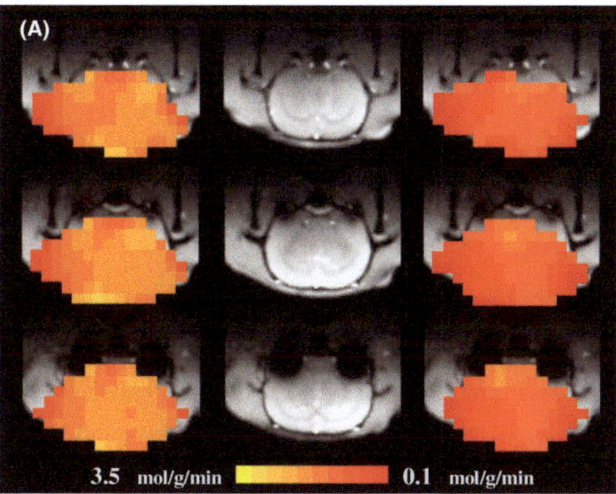

Fig. 15.8. (**A**) 3D $CMRO_2$ maps of a representative rat brain obtained at normothermia (*left column*) and hypothermia (*right column*), and their corresponding anatomic images (*middle column*).

MIX
Papier aus verantwortungsvollen Quellen
Paper from responsible sources
FSC® C105338

If you have any concerns about our products,
you can contact us on
ProductSafety@springernature.com

In case Publisher is established outside the EU,
the EU authorized representative is:
Springer Nature Customer Service Center GmbH
Europaplatz 3, 69115 Heidelberg, Germany

Printed by Libri Plureos GmbH
in Hamburg, Germany